30-

MODERN HYDROLOGY AND SUSTAINABLE WATER DEVELOPMENT

Author Biography

After graduating in geophysics from the Indian Institute of Technology (IIT), Kharagpur, Dr. S.K. Gupta did his PhD from IIT, Bombay in 1974. He is a recipient of the Vikaram Sarabhai National Award in Hydrology and Atmospheric Sciences. Dr. Gupta nucleated the Isotope Hydrology group at the Physical Research Laboratory, Ahmedabad and carried out research for more than past 3 decades. Presently, he is the Principal Coordinator of the National Programme for Isotopic Fingerprinting of Waters of India. Dr. Gupta has more than 150 publications in internationally refereed research journals and several books to his credit. Dr. Gupta has also been a Fulbright Fellow at the University of Hawaii at Manoa and an Alexander von Humboldt Fellow at the University of Heidelberg and a Visiting Fellow at the University of Canberra. He is also a Fellow of the National Academy of Sciences, India.

Modern Hydrology and Sustainable Water Development

S.K. Gupta

Physical Research Laboratory, Ahmedabad, Gujarat, India

A John Wiley & Sons, Ltd., Publication

Library of Congress Cataloguing-in-Publication Data

Gupta, Sushil K. (Sushil Kumar), 1946-
  Modern hydrology and sustainable water development / S.K. Gupta.
    p. cm.
  Includes bibliographical references and index.
  ISBN 978-1-4051-7124-3 (cloth)
1. Hydrology. 2.   Water resources development. I.   Title.
  GB661.2.G866 2010
  551.48–dc22
ISBN: 978-1-4051-7124-3
                                        2010008112

A catalogue record for this book is available from the British Library.

This book is published in the following electronic formats: eBook 9781444323979; Wiley Online Library 9781444323962

Set in 9.5/12 pt Garamond by Aptara® Inc., New Delhi, India
Printed and bound in Singapore by Markono Print Media Pte Ltd.

1   2011

DEDICATION

to

Prof. D. Lal, FRS
His life and work continue to inspire
my academic endeavours.

Contents

Foreword

Over the past 50 years the population of the world has increased from 3 billion to 6.5 billion and it is likely to rise by another 2 billion by 2025 and by another 3 billion by 2050. Following the current trends it is certain that the increasing population will mean a greater need for food. More people will dwell in cities and will strive for a higher standard of living. This will imply rapid urbanization, accelerating land-use change, depleting groundwater resources, increasing pollution of surface streams, rivers, and groundwater, and decaying infrastructure. To produce more food, there will be greater pressure on agriculture, which will call for more irrigation. There will be increasing demands for energy, which will also require more water. Thus, the demand for water in both rural and urban areas will rise and outpace the growth in population.

To make matters worse, there is the spectre of climate change. During the last one hundred years, the temperature has arisen by nearly $0.6°C$, and it is expected to rise by another $2°C$ during the next 100 years. This would translate into intensification of the hydrologic cycle, rising sea levels, more variable patterns of rainfall (more intense, more extreme), more changes in runoff (more frequently occurring floods and droughts), shorter snowfall seasons, earlier start of spring snowmelt seasons, melting of glaciers, increasing evaporation, deterioration in water quality, changes to ecosystems, migration of species, changes in plant growth, re-

action of trees to downpours, drying up of biomass during droughts, and quicker growing and subsequent wilting of crops. In other words, the entire ecosystem will undergo a significant change at local, regional, and global scales. One can only conjecture on the long-term consequences of such changes.

The impact of climate change on water resources management would entail serious ramifications. Larger floods would overwhelm existing control structures; reservoirs would not receive enough water to store for the use of people and agriculture during droughts; global warming would melt glaciers and cause snow to fall as rain; regimes of snow and ice, which are natural regulators that store water in winter and release it in summer, would undergo change; and there would be more swings between floods and droughts. It is likely that dams, after a lull of three decades, would witness a comeback.

Current patterns of use and abuse of water resources are resulting in the amount withdrawn being dangerously close to the limit and even beyond; an alarming number of rivers no longer reach the sea. The Indus, the Rio Grande, the Colorado, the Murray-Darling, and the Yellow River – are the arteries of some of the world's main grain-growing areas. Freshwater fish populations are in precipitous decline; fish stocks have fallen by 30% (WWF for Nature), larger than the fall in populations of animals in any ecosystem. Fifty percent of the world's wetlands were drained, damaged, or destroyed in the 20th century; in addition to the fall in the volume of fresh water in rivers, invasion of saltwater into deltas, and change in the balance between fresh water and salt water.

When compared to the global water resources situation, local water shortages are multiplying even faster. Australia has suffered a decade-long drought. Brazil and South America, who depend on hydroelectric power, have suffered repeated brownouts – not enough water to drive turbines. Excess pumping of water from feeding rivers led to the near-collapse of the Aral Sea in Central Asia in 1980; and global water crisis impinges on the supplies of food, energy, and other goods.

The water resources situation in the United States is facing the same trend, with decaying infrastructure built 50 to 100 years ago, such that 17% of treated water is lost due to leaky pipes. In Texas there is an ongoing drought, where ranchers have already lost nearly 1 billion dollars; worst hit are Central Texas and the Hill Country. December 2008–February 2009 has been the driest on record; 60% of the state's beef cows are in counties with severe to exceptional drought; in 2006, drought-related crop and livestock losses were the worst for any single year, totalling $4.1 billion. The effects of this drought are long-term.

Modern Hydrology and Sustainable Water Development, by Dr S.K. Gupta, is timely and addresses a number of key questions gravitating around the interactions between water, energy, environment, ecology, and socio-economic paradigms. The subject matter of the book will help promote the practice of hydrology focused on sustainable development, with due consideration to linkages between regional economic development, population growth, and terrestrial and lithological hydrologic systems. It states the challenges and opportunities for science, technology, and policy related to sustainable management of water resources development and in turn sustainable societal development.

Introducing the basic concepts and principles of hydrologic science in Chapter 1, the subject matter of the book is organized into 14 chapters, each corresponding to a specific theme and containing a wealth of information. Surface water in lakes, glaciers, streams, and rivers encompassing watershed concepts, stream flow components, hydrograph separation, landform and fluvial geomorphology, and the very wide range of time and space scales that hydrologic theories must span, is dealt with in Chapter 2. Subsurface flow is dealt with in the next two chapters, primarily encompassing groundwater hydrology and well hydraulics. The next chapter deals with methods of computer-aided modelling of surface and groundwater flow systems. Keeping in mind the impact of human activities on the hydro-environment, aqueous chemistry is the subject matter of Chapter 6. Tracer hydrology, developed during the last few decades and playing an important role in modern hydrology, constitutes the subject matter of Chapter 7. Chapter 8 deals with statistical analyses and techniques required for making hydrologic predictions and design. Fundamentals of remote sensing and GIS, another powerful field developed during the last few decades, are described in Chapter 9. Urban hydrologic processes are the theme of Chapter 10. Chapter 11 covers rainwater harvesting and groundwater recharge and is important given recurring water shortages around the world, especially in developing countries. This topic has been receiving a lot of emphasis today. Acknowledging the rightful place of the human dimension in water resource development and management, Chapter 12 goes on to discuss water ethics. A few case studies of field situations, linking many of the aspects discussed in the preceding chapters, are included in Chapter 13. A wrap-up of various chapters, concluding in a holistic manner, is presented in Chapter 14.

The book is well written and well organized. It reflects the vast experience of its author. It will help improve our understanding of the sensitivity of key water quantity and quality management targets to sustainable development. The book is timely and makes a strong case for sustainable development and management in relation to the science and practice of hydrology. It will be useful to students and faculty in engineering, and agricultural, environmental, Earth, and watershed sciences. Water resources planners, managers, and decision-makers will also find the book of value. Dr Gupta is to be applauded for preparing such a timely book on hydrology.

Professor Vijay P. Singh, Ph.D., D.Sc., Ph.D. (Hon.), P.E., P.H., Hon. D. WRE
Caroline and William N. Lehrer Distinguished Chair in Water Engineering
Professor of Civil and Environmental Engineering
Professor of Biological and Agricultural Engineering
Academician, GFA; President, FARA
President, G.B.S. Board
Editor-in-Chief, WSTL
Editor-in-Chief, ASCE Journal of Hydrologic Engineering
Department of Biological and Agricultural Engineering and
Department of Civil and Environmental Engineering
Texas A and M University
Scoates Hall, 2117 TAMU
College Station, Texas 77843-2117, USA

Preface

Water is essential not only to people but to all forms of life on Earth. The importance of water for the improvement of health, the production of food, and the support of industry is vital in the rapidly developing world of today. The wider political and social context of water resource development has always been important and will remain so in the future. Three very important examples of this wider context are: (i) the relationship between large-scale irrigation and society; (ii) the role of self-help in rural water supply schemes in developing countries; and (iii) large-scale long-distance transfers of water. Because of its importance to life and society, water is important to students and professionals in several fields. Chief amongst these are civil engineers with diverse specializations, geologists, agriculture and irrigation engineers, and personnel in charge of municipal and industrial water supplies. Environmentalists and planners often have vital interests in hydrology. Indirectly concerned persons can also be found in the fields of economics, mining and petroleum engineering, forestry, public health, and law amongst others.

Until comparatively recently the approach to hydrology was essentially a pragmatic one. However, during the past few decades, increased demands on water for various applications have stimulated many unsustainable practices of exploitation but, on the other hand, there has been development of new techniques for investigating the occurrence and movement of water in its various forms. Simultaneously, research has contributed to a better understanding of the subject of hydrology and new concepts of resource management have evolved.

Although it is impossible to present the subject of hydrology fitted to such a diversity of interests, the common need of all is an understanding of fundamental principles, methods, and problems encountered in the field as a whole. This book, therefore, represents an effort to make available a unified presentation of the various aspects of the science and practice of hydrology. It is intended to serve both as a reference book and as a textbook. The material derives its scientific underpinning from the basics of mathematics, physics, chemistry, geology, meteorology, engineering, soil science, and related disciplines. The aim has been to provide sufficient breadth and depth of understanding in each subsection of hydrology. The intention is that after going through the book, in the manner of undertaking a course in hydrology, a student should be able to make basic informed analyses of any hydrologic dataset and plan additional investigations, when needed, keeping in mind issues of water ethics and the larger issue of global change and the central position of water therein. Readers with a background in diverse disciples, using the book as a reference work in hydrology, should also be able to find the required information in the context of the practice and the basic science of hydrology. This book is subdivided into 14 chapters covering the most important conventional and modern techniques and concepts to enable a sustainable development of water resources in any region.

In this book, the utmost care has been taken to present the material free from errors of any kind; some errors and omissions may have escaped editorial scrutiny. If detected, the author requests that these be communicated to him by email: skg_skgupta@yahoo.co.in.

S.K. Gupta

Acknowledgements

This endeavour could not have succeeded without the institutional support of the Physical Research Laboratory (PRL), Ahmedabad – where I learned and researched in hydrology for more than three decades. This is the best opportunity of expressing my gratitude to this great institution. Dr Prabhakar Sharma has painstakingly corrected my manuscript. It is difficult to thank him enough for his encouragement and effort. Dr B.S. Sukhija also reviewed Chapter 7 on hydrologic tracing and provided important input. My patient wife Surekha has been the biggest supporter of this enterprise, always edging me on towards completion – I am indebted to her for all the care and support received. Having a colleague such as Dr R.D. Deshpande, always ready to provide help in every possible manner, has been a very fortunate situation for me.

While all those above have contributed to the successful completion of this project, any omission is solely my responsibility.

A note for students and teachers

The first 12 chapters of this book have been organized into three broad themes:

1) Water, its properties, its movement, modelling and quality (Chapters 1-6);
2) Studying the distribution of water in space and time (Chapters 7-9); and
3) Water resource sustainability (Chapters 10-12).

The first chapter introduces the fundamental concerns and concepts of hydrology. Starting with the basic physical and chemical properties of water, quality parameters, the physics of water flow, and measurement techniques, it also includes the hydrologic cycle. Two broad subdivisions of terrestrial water, namely, visible water sources at the surface and invisible water underground, are also introduced. The second chapter presents the subject of water in lakes, glaciers, and streams, watershed concepts, stream flow components, hydrograph separation, landform, and fluvial geomorphology. Also introduced is the fundamental equation in hydrology, namely, the concept contained in the equation of continuity, equations of free surface flow, saturated and unsaturated flow, and the very wide range of time and space scales that any hydrologic theory must span.

The invisible flow of water, namely, groundwater, is dealt with in the next two chapters. The composite nature of surface and groundwater, aquifer formations and their properties, hydraulic head, saturated flow equations, groundwater measurements, and pollution are described in Chapter 3. Well hydraulics, including steady and unsteady radial flow equations, for unconfined, confined, and leaky aquifers, including the methods of well testing, well losses, and hydraulics associated with partial penetration, form the subject matter of Chapter 4. Methods of surface and groundwater flow modelling, including finite difference, finite element, and analytical element methods, their comparative account, model calibration, parameter estimation, and sensitivity analysis, are discussed in Chapter 5.

Many human activities have adverse impacts on the hydro-environment in ways that decrease the usefulness of the resource substantially, either for human beings or other life forms. Aqueous chemistry is central to understanding: (i) sources of chemical constituents in water; (ii) important natural chemical processes in groundwater; (iii) variations in chemical composition of groundwater in space and time; and (iv) estimation of the fate of contaminants, both in surface and groundwaters, and for remediation of contamination. The focus of Chapter 6 is on principles of chemical thermodynamics, processes of dissolution and/or precipitation of minerals, and on undesirable impacts of human activities, including transport and attenuation of micro-organisms, non-aqueous phase liquids, geochemical modelling, and the relation between use and quality of water.

The second broad theme of studying distribution of water in space and time is addressed in the next three chapters. Fundamentals of the theory and practice of tracer hydrology developed during the last few decades and playing an important role in modern hydrology, particularly when the interest is in obtaining a direct insight into the dynamics of surface and subsurface water, are described in Chapter 7. The various tracers from dissolved gases and chemicals to radioactive and stable isotopes of dissolved constituents, including the water molecules themselves, have provided useful tools to understand transport processes, phase changes (evaporation, condensation, sublimation), and genesis of water masses and their quality.

Hydrologists are often required to make predictions on variability and long-term assurance of water availability and hydro-hazards, even as understanding of complex physical interactions and processes that govern natural hydrologic phenomena eludes. Recourse is then taken to understand the inherent statistical distribution of water in space and time. Statistical techniques and probability theory, as relevant to hydrology, form a simple description of statistics to time series analysis, and are the subject matter of Chapter 8.

Availability of many different earth-sensing satellites, with diverse calibrated sensors mounted on sophisticated platforms providing synoptic view and repetitive coverage of a given location, help to detect temporal changes and observations at different resolutions of several parameters important to water resource development, planning, and management. Together with computers, their vastly enhanced data handling capacity and powerful software to manipulate geographical data in the form of a geographical information system (GIS), a new field of analysing digital remote sensing data with other geo-referenced hydrologic/environmental/societal data has evolved. Diverse applications of this intertwined field include estimation of precipitation, snow hydrology, soil moisture monitoring, groundwater hydrology, urban issues, and monitoring of global change. These applications and fundamentals of the two techniques of remote sensing and GIS are described in Chapter 9.

The next three chapters address the broad theme of water resource sustainability. Hydrologic processes occurring within the urban environment where substantial areas consist of nearly impervious surfaces, and artificial land relief as a result of urban developments including formation, circulation, and distribution of water, and techniques

of waste treatment and disposal, are discussed in Chapter 10. Also included in this chapter are new approaches and technologies for sustainable urbanization.

Chapter 11 deals with the practice of rainwater harvesting from domestic scale for drinking water supply, to large catchments to support rain-fed agriculture and supplemental irrigation, and for groundwater recharge. Rainwater harvesting together with groundwater recharge using and/or conserving the captured rainwater at or near the place it occurs, offer great promise to improve or sustain water supplies in water-stressed regions.

Not to lose sight of the human dimension in water resource development, major concerns have been described in the form of water ethics in Chapter 12. This chapter also includes a section on global water tele-connections and virtual water that quantifies the amount of water embedded in production of goods and services. The concept of virtual water is very recent and may significantly influence regional and global commodity trade and water allocation for competing demands.

Four case studies from three continents, with field situations, covering regions of high water stress and linking many of the aspects discussed in previous chapters, are included in Chapter 13. Also covered are adaptation measures being employed in each region to mitigate the situation.

A wrap-up of the various chapters, concluding in a holistic manner, on how hydrologic investigations and analyses enable hydrologists to study local, regional, and global water cycle and manipulate and manage it with consideration for human welfare and sustainability, is presented in Chapter 14.

The various chapters are largely independent of each other. Cross references are given only to indicate the linkages between different lines of enquiry. Equations represent mathematical expressions of physical or chemical concepts and most of these have been derived using basic theorems of high school mathematics and calculus. Examples and Tutorials have been included in various chapters to enable students to gain a feel for numbers, units, and dimensions.

1

Fundamentals of hydrology

Hydrology deals with the scientific study of water, its occurrence, and movement through the Earth-atmosphere system (air, land, and ocean). It is a discipline in the realm of geosciences and derives upon knowledge and techniques from several other fields (i.e. civil engineering, hydraulics, meteorology, geology, forestry, soil science, etc.) to address water availability, in terms of both quantity and quality. In addition, mathematical modelling techniques, both analytical and numerical, are being extensively applied to describe problems of both surface- and ground-water hydrology.

We often perceive liquid water (H_2O) to be an ordinary substance and mistakenly take its availability for granted. However, it is an extraordinary substance and its unique properties make liquid water the most vital substance for sustenance of all life forms on the Earth. Although we drink it, wash and cook with it, and fish and other aquatic life forms live in it, we nearly always overlook the special relationship it has with our lives. The human body is about two-thirds water and without it we would die of dehydration within a few days. Where there is water there is life, and where water is scarce, life becomes a painful struggle.

Some important chemical and physical properties of water and their crucial relationship with living beings are described in this chapter.

1.1 Properties of water

1.1.1 Chemical properties

Many unique properties of water are largely a result of its chemical structure. Chemically, the water molecule (H_2O) is a tiny V-shaped molecule with one atom of oxygen bound to two atoms of hydrogen. Indeed, very few molecules in nature are smaller or lighter than water molecules. Two hydrogen atoms are 'attached' to one side of an oxygen atom, as shown in Fig. 1.1a, resulting in a water molecule exhibiting electrical polarity with a positive charge on the two hydrogen atoms and a negative charge on the oxygen atom. The polarity of a water molecule allows it to attach to other molecules easily, including another water molecule. Because of attraction between opposite electrical charges, water molecules tend to attract each. As shown in Fig. 1.1b, the side with the hydrogen atoms (positive charge) attracts the oxygen side (negative charge) of a different water molecule.

Water molecules attracting each other cause them to clump together. When a molecule of water is attached to another water molecule, it is called *cohesion*. When water is attached to other materials, it is called *adhesion*. If it wasn't for the Earth's

Modern Hydrology and Sustainable Water Development, 1st edition. © S.K. Gupta
Published 2011 by Blackwell Publishing Ltd.

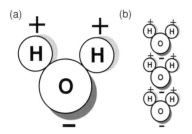

Fig. 1.1 (a) The H_2O molecule behaves like an electric dipole, which makes (b) water sticky due to attraction between the molecules. Redrawn from http://ga.water.usgs.gov/edu/waterproperties.html.

gravity, a drop of water would be ball shaped – a perfect sphere. The two properties of cohesion and adhesion give rise to the sticky nature of water.

Water is called the 'universal solvent' because it dissolves more substances than any other liquid. This feature also enables water to dissolve and carry minerals and nutrients in runoff, infiltration, groundwater flow, and also in the bodies of living organisms.

1.1.2 Physical properties

Water is the only substance that is found in nature in all three states – liquid, solid (ice), and gas (vapour) – at temperatures normally prevailing on Earth. Water changing from solid to liquid is called *melting*. When it changes from liquid to the vapour phase, the process is called *evaporation*. Water changing from vapour to liquid is called *condensation* (some examples are the 'moisture' that forms on the outside of a cold soda water bottle or when the moisture in the air condenses on grass and leaves during early mornings in winter). *Frost formation* occurs when water changes from vapour

directly to its solid form. When water changes directly from the solid to the vapour phase, the process is called *sublimation*. In fact, the Celsius scale of measuring temperature is derived based on the freezing point of water (at $0°C$) and its boiling point (at $100°C$ at sea level). The large range of temperature, $0-100°C$, between its melting and boiling points, enables water to exist in liquid form in most places on the Earth, including regions having extremes of temperature.

Water also has interesting thermal properties. When heated from $0°C$ (melting point of ice) to $4°C$, it contracts and becomes denser; most other substances expand and become less dense when heated. In the case of water this happens only beyond $4°C$. Conversely, when water is cooled in the temperature range $4-0°C$, it expands. It expands greatly as it freezes, adding about 9% by volume; as a consequence, ice is less dense than water and, therefore, floats on it.

Another unique property of water is its high *specific heat* (S), defined as the amount of heat required to raise the temperature by $1°C$ per unit mass of the matter. A related quantity is called the *heat capacity* (C) of a given mass of the substance. Specific heat and heat capacities of some common substances are given in Table 1.1, which shows that water has a very high specific heat. This means that water can absorb a lot of heat before it begins to get hot and its temperature changes significantly. This is why water is valuable to industries for cooling purposes and is also used in an automobile radiator as a coolant. The high specific heat of water, present in the air as moisture, also helps regulate the rate at which the temperature of the air changes, which is why the temperature changes between seasons are gradual rather than abrupt, especially in coastal areas.

Table 1.1 Specific heat and heat capacities of some common substances.

Substance →	Air	Aluminium	Copper	Gold	Iron	NaCl	Ice	Water
Specific heat $(Jg^{-1}°C^{-1})$	1.01	0.902	0.385	0.129	0.450	0.864	2.03	4.186
Heat capacity $(J°C^{-1})$ for 100 g	101	90.2	38.5	12.9	45.0	86.4	203	418.6

Note: The calorie, defined as the amount of heat required to raise the temperature of 1 *gram* of water by $1°C$, not in use today, has been replaced by the SI-unit, the *Joule*. The conversion is $4.186 J = 1 cal$.

Water is also unique in another thermal property - *latent heat*, the heat change associated with a change of state or phase. Latent heat, also called heat of transformation, is a measure of the heat given up or absorbed by a unit mass of a substance as it changes its form from solid to liquid, from liquid to gas, or vice versa. It is called latent because it is not associated with a change in temperature. Each substance has a characteristic heat of fusion, associated with the solid–liquid transition, and a characteristic heat of vaporization, associated with the liquid–gas transition. The *latent heat of fusion* for ice is $334\,J$ ($= 80\,cal$) *per gram*. This amount of heat is absorbed by each gram of ice in melting or is given up by each gram of water during freezing. The *latent heat of vaporization* of steam is $2260\,J$ ($= 540\,cal$) *per gram*, taken up by boiling water at $100°C$ to form steam or given up during condensation from vapour to liquid, that is, when steam condenses to form water. This is the reason that putting one's finger in a jet of steam causes more severe burning than dipping it in boiling water at $100°C$. For a substance passing directly from the solid to the gaseous state, or vice versa, the heat absorbed or released is known as the *latent heat of sublimation*.

Although pure water is a poor conductor of electricity, it is a much better conductor than most other pure liquids because of its self-ionization, that is, the ability of two water molecules to react to form a hydroxide ion, OH^-, and a hydronium ion, H_3O^+. Its polarity and ionization are both due to the high dielectric constant of water. In fact, water conducts heat more easily than any other liquid, except mercury. This fact causes large bodies of liquid water such as lakes and oceans to have a uniform vertical temperature profile over considerable depths.

Surface tension is the name given to the cohesion of water molecules at the surface of a body of water. Water has a high surface tension. For example, surface tension enables insects such as water striders to 'skate' across the surface of a pond. Surface tension is related to the cohesive properties of water. *Capillary action* is, however, related to the adhesive properties of water. Capillary action can be seen 'in action' by placing a very thin straw (or glass tube) into a glass of water. The water 'rises' up the straw. This happens because water

molecules are attracted to the molecules on the inner surface of the straw. When one water molecule moves closer to molecules of the straw, the other water molecules (which are cohesively attracted to that water molecule) also move up the inside surface of the straw. Capillary action is limited by gravity and the diameter of the straw. The thinner the straw, the higher will be the capillary rise. Plant roots suck soil water, making use of capillary action. From the roots, water (along with the dissolved nutrients) is drawn through the plant by another process, transpiration. *Transpiration* involves evaporation of water from leaves, branches, and stems, which results in a pump-like action sucking water from the soil by plant roots and giving it out into the atmosphere in the form of vapour.

The *triple point* of water refers to a unique combination of pressure / temperature at which pure liquid water, ice, and water vapour can coexist in a stable equilibrium. It is used to define the *Kelvin*, the SI unit of thermodynamic temperature. Water's triple point temperature is $273.16\,kelvin$ ($0.01°C$) and has a pressure of $611.73\,pascal$ ($0.0060373\,atm$). Because the average temperature on the surface of the Earth is close to the triple point of water, it exists in all three phases – gas, liquid, and solid. In the universe, no other planet is known to possess water in all its three phases in abundant quantities.

Fluid properties that control the movement of water are density, pressure, buoyancy, and viscosity. Of these, *viscosity* is the most significant property, being a measure of the resistance of a fluid to a change of shape or a resistance to flow, and it can be construed as internal friction of a moving fluid. Viscosity can be either dynamic (for fluids in motion) or kinematic (intrinsic). Dynamic viscosity is also a measure of the molecular 'stickiness' and is caused by attraction of the fluid molecules to each other. When viewed in the context of intermolecular hydrogen bonding as a cohesive force, it is perhaps surprising that water pours quite smoothly and freely, certainly more freely than, say, honey. Indeed, the viscosity of water is quite modest. So although gravity creates nearly the same stresses in honey and in water, the more viscous honey flows more slowly. Viscosity is temperature dependent but in most cases, due to narrow range of temperature variation, the effect on viscosity is minimal.

The other important factor governing dynamics of a body or hydrodynamics of a fluid is *inertia*, which is the resistance offered by a body or fluid to any change in its motion. Whether an object or fluid is stationary or moving, it will continue to maintain its state in accordance with Newton's First Law of motion. The effect of inertia is solely dependent on mass: if scale is increased then size, mass, and inertia increase. Thus, viscous forces dominate at small scales and inertial forces dominate at large scales. To a copepod, sea water is very sticky and it will stop moving immediately if it stops swimming. At this scale, viscosity is the dominant force. To a whale, sea water offers very little resistance and it will keep moving for quite a distance even if it stops swimming. At this scale, inertia is the dominant force. Since viscosity and inertia govern the motion of a fluid, the transition from viscosity-dominated flow to inertia-dominated flow also marks the transition from laminar to turbulent flow.

A summary of hydrologically relevant physical properties of water is given in Table 1.2.

1.2 Common water quality parameters

Water sustains plant and animal life. It plays a key role in governing the weather, and helps to shape the surface of the planet through erosion and other processes. The term water quality is used to describe physical, chemical, and biological characteristics of water, usually with reference to its suitability for a particular purpose. Quality of water is of concern, essentially in relation to its intended use. For example, water that is good for washing a car may not be good enough for drinking or may not even support aquatic life. Human beings want to know if the water is good enough for use at home, for serving in a restaurant, for swimming, or if the quality of the water occurring in the area is suitable for aquatic plants and animals.

Standards and guidelines have been established to classify water for designated uses such as drinking, recreation, agricultural irrigation, or protection and maintenance of aquatic life. Standards for drinking-water quality are prescribed to ensure

Table 1.2 Physical properties of water: summary table.

Property	Value
Molar mass	18.015 g
Molar volume	55.5 $mol\ l^{-1}$
Boiling point (BP)	100°C at 1 atm
Triple point	273.16 K at 4.6 $torr$
Surface tension	73 $dynes\ cm^{-1}$ at 20°C
Vapour pressure	0.0212 atm at 20°C
Latent heat of vaporization	40.63 $kJ\ mol^{-1}$, 22.6 × 10^5 $J\ kg^{-1}$
Latent heat of fusion	6.013 $kJ\ mol^{-1}$, 3.34 × 10^5 $J\ kg^{-1}$
Heat capacity (cp)	4.22 $kJ\ kg^{-1}\ K^{-1}$
Dielectric constant	78.54 at 25°C
Viscosity	1.002 $centipoises$ at 20°C
Density	1 $g\ cm^{-3}$
Density maximum	at 4°C
Specific heat	4180 $J\ kg^{-1}\ K^{-1}$ (T = 293–373 K)
Heat conductivity	0.60 $W\ m^{-1}\ K^{-1}$ (T = 293 K)
Critical temperature	647 K
Critical pressure	22.1 × 10^6 Pa
Speed of sound	1480 $m\ s^{-1}$ (T = 293 K)
Relative permittivity	80 (T = 298 °K)
Refractive index (relative to air)	
	1.31 (ice; 589 nm; T = 273 K; p = p_0)
	1.34 (water; 430–490 nm; T = 293 K; p = p_0)
	1.33 (water; 590–690 nm; T = 293 K; p = p_0)

that public drinking-water supplies are as safe as possible. The US Environmental Protection Agency (US-EPA) and similar agencies in other countries and/or international agencies such as the World Health Organization (WHO) are responsible for prescribing the standards for constituents in water that are known to pose a risk to human health. Other standards to protect aquatic life, including fish, and wildlife such as birds that prey on fish, have also been notified in most countries. In the following, some of the commonly used water quality parameters are described. In specific cases many other water quality parameters are used, depending on the problem or intended use.

1.2.1 Water quality – physical parameters

1.2.1.1 Transparency

The transparency of water is an important quality for drinking purposes, and even more important for aquatic plants and animals that thrive in water. Suspended particles in water behave similarly to dust in the atmosphere, reducing the depth to which light can penetrate. Sunlight provides the energy required for photosynthesis (the process by which plants grow by taking up atmospheric carbon dioxide and release oxygen). The depth to which light penetrates into a water body determines the depth to which aquatic plants can grow.

Transparency is the degree to which light penetrates in a water column. Two commonly used methods for measuring transparency of water are the Secchi disk and transparency tube. The Secchi disk was first used in 1865 by Father Pietro Angelo Secchi, Scientific Advisor to the Pope, to measure transparency of water. This simple method involves measurement of the depth at which a 20-*cm* black and white disk lowered into water just disappears from view, and reappears again when raised slightly. An alternative method of measuring transparency is by pouring water into a tube with a pattern similar to that of a Secchi disk on the bottom and noting the depth of water in the tube when the pattern just begins to disappear from view. The Secchi disk is used in deeper, still waters. The transparency tube can be used with either still or flowing waters and can also be used to measure shallow water sites or the surface layer of deep water sites.

1.2.1.2 Water temperature

Water temperature is largely determined by the amount of solar energy absorbed by water as well as the surrounding soil and air. More solar insolation leads to higher water temperatures. Effluent water discharged from chemical/manufacturing industries into a water body may also increase its temperature. Water evaporating from the surface of a water body can lower its temperature but only of the thin layer at the surface.

Water temperature can be indicative of its origin. Water temperature near the source will be similar to the temperature of that source (e.g. snowmelt will be cold, whereas most groundwaters are warm). Temperature of surface water farther from its source is influenced largely by atmospheric temperature.

Temperature can be easily measured by using a calibrated thermometer or some other type of calibrated probe and is read while the bulb of the thermometer or probe is still in water. Even though easy to measure, water temperature is a very important parameter because it enables better understanding of other hydrological measurements such as dissolved oxygen, pH, and conductivity.

Temperature influences the number and diversity of aquatic life. Warm water can be harmful to sensitive fish species, such as trout or salmon that require cold, oxygen-rich water. Warm water tends to have low levels of dissolved oxygen. Water temperature is also important for understanding local and global weather patterns. Water temperature changes differently compared to air temperature because water has a higher heat capacity than air. A water body also affects the air temperature of its surroundings through the processes of evaporation, transpiration (from the vegetation that may grow in water), and condensation.

1.2.2 Water quality – chemical parameters

1.2.2.1 Electrical conductivity

Pure water is a poor conductor of electricity. However, ionic (charged) impurities present in water, such as dissolved salts, that enable water to conduct electricity. As water comes in contact with rocks and soil, some of the minerals dissolve in it. Other impurities can enter water bodies through runoff (both surface and subsurface) and/or wastewater discharge.

The amount of dissolved mineral and salt impurities in water is denoted by its *total dissolved solids* (TDS) content. There are different types of solids dissolved in water, but the most common is sodium chloride (*NaCl*) – the common table salt. The TDS is specified as parts per million (*ppm*) or alternatively as mgl^{-1} (which is the same as *ppm*). This indicates how many units of soluble impurities are present in one million units, by mass, of water. For water that is used at home, a TDS of less than 500 *ppm* is desirable, although water with higher TDS can still be safe to drink. Water used for agriculture should have TDS below 1200 *ppm* so that salt-sensitive crops are not affected. Manufacturing, especially of electronics equipment, requires water with a very low TDS.

A quick and indirect measure of TDS is the amount of electricity that is conducted across a centimetre length/depth of water column of unit cross-sectional area. *Electrical conductivity* (EC) is measured as micro-siemens per cm ($\mu S\ cm^{-1}$) which is the same unit as a *micro-mho*. To convert electrical conductivity of a water sample ($\mu S\ cm^{-1}$) into an approximate concentration of the total dissolved solids (*ppm*) in the sample, one has to multiply the conductivity value (in $\mu S\ cm^{-1}$) by a suitable conversion factor. The conversion factor depends on the chemical composition of the dissolved solids and can vary between 0.54 and 0.96. For instance, sugars do not affect conductivity because they do not form ions when dissolved in water. The value 0.67 is commonly used as an approximation so that TDS (*ppm*) = conductivity (in $\mu S\ cm^{-1}$) $\times$ 0.67. It is desirable to use a conversion factor that has been determined for the particular water body instead of the approximation, since the type of impurities between water bodies can vary greatly.

1.2.2.2 Salinity

Another TDS related parameter is water *salinity*. This is a measure of its saltiness and is expressed as the amount of impurity present in parts per thousand of water (*ppt*) as against impurity expressed per million parts of water (*ppm*). The latter is used for relatively fresh waters and estimated by measuring the electrical conductivity. The salinity of the Earth's oceans averages 35 parts per thousand (*ppt*); fresh water measures 0.5 *ppt* or less. Coastal waters and surface waters of the ocean not far from shore can be less salty than 35 *ppt* due to freshwater input from land surface runoff or rain, or may be more salty due to high rates of evaporation in hot climates. Some seas and inland lakes are also saline. Examples include the Caspian Sea in central Asia, the Great Salt Lake in North America, and several lakes in the Great Rift Valley of East Africa. These water bodies are salty because water flowing into them has no outlet except evaporation, concentrating the dissolved salts. For freshwater bodies that have outlets, salts get flushed out instead of accumulating. Brackish water is saltier than fresh water, but not as salty as sea water. It is found in estuaries and bays where sea water and fresh water mix.

All animals and plants have salts inside the cells of their bodies. The concentration of these salts is about one-third that of sea water. Plants and animals in both fresh and salt water have special mechanisms to maintain a proper salt balance between their cells and the environment around them. Freshwater organisms are saltier than the water they live in. Animals, such as fish in salt water, are less salty than the sea water they live in. Organisms adapted to a given environment cannot be relocated to another without causing them serious injury or even death.

A quick and indirect measurement of salinity involves use of a hydrometer – an instrument that measures the specific gravity of a fluid. A hydrometer is a small float with a graduated scale engraved on its stem. The higher the dissolved salt content, the higher the hydrometer floats compared to its depth in pure water. As the water gets denser, more of the hydrometer is exposed. Marks along the calibrated hydrometer allow reading the specific gravity directly from the hydrometer. In addition to the amount of salt in water, hydrometers are also used to compare the densities of different liquids; for example, the amount of sugar present in fruit juice or the fat content of a milk sample.

Another related water quality parameter is *chlorinity*, which is a measure of the chloride concentration in water. Its usefulness arises from the fact that $NaCl$ is generally the most abundant dissolved salt in natural waters, particularly at the high end of the salinity. In sea water, the following six ions, that are well mixed and found in nearly constant proportions, account for over 99% of the dissolved material: chloride (Cl^-) – 55.0%; sodium (Na^+) – 30.6%; sulphate (SO_4^{-2}) – 7.7%; magnesium (Mg^{+2}) – 3.7%; calcium (Ca^{+2}) – 1.2%; and potassium (K^+) – 1.2%. Therefore, by measuring concentration of the most abundant constituent (Cl^-), which is also easily measurable by titration, it is possible to estimate total salinity using the equation:

$$\text{Salinity } (ppt) = \text{Chlorinity } (ppt) \times 1.80655 \quad (1.1)$$

Measurement of chlorinity by titration is a fairly simple procedure. First an indicator, potassium chromate, is added to a carefully measured volume of the sample. This reagent produces a yellow colour. Silver nitrate solution of known concentration is then added as titrant. Silver reacts with chloride present in the sample to form a white precipitate, silver chloride. When the entire chloride has been precipitated, any excess silver nitrate added forms red-coloured silver chromate, producing the pinkish-orange endpoint. Chloride concentration is calculated from the amount of sample taken and the concentration and amount of silver nitrate used to reach the endpoint.

1.2.2.3 Dissolved oxygen

Even though the water molecule (H_2O) is made up of two hydrogen atoms and one oxygen atom, molecules of oxygen gas (O_2) are naturally dissolved in water. The amount of *dissolved oxygen* (*DO*) that water can hold (under given conditions) is known as the solubility of dissolved oxygen. Factors affecting the solubility of dissolved oxygen include water temperature, atmospheric pressure, and salinity. Cold water can dissolve more oxygen than warm water. For example, at $25°C$, oxygen solubility is 8.3 $mg\ l^{-1}$, whereas at $4°C$ the solubility is 13.1 $mg\ l^{-1}$. As the temperature goes up, water releases some of its dissolved oxygen into the air.

Dissolved oxygen is naturally present in water, contributed by plants during photosynthesis, through diffusion from the atmosphere, or by aeration. The amount of dissolved oxygen is also affected by plants and life forms that exist in the water. Just as photosynthesis by terrestrial plants adds oxygen to the air, photosynthesis by aquatic plants contributes dissolved oxygen to water. Water may become supersaturated with oxygen, implying that the dissolved oxygen levels are higher than can be accounted for by its solubility. The excess dissolved oxygen eventually gets released back into the air or is removed through plant respiration. During respiration, biota (fish, bacteria, etc.) consume dissolved oxygen. Without adequate levels of dissolved oxygen in water, aquatic life is suffocated. Dissolved oxygen levels below 3 $mg\ l^{-1}$ are stressful to most aquatic organisms. Dissolved oxygen is also consumed during decay of organic matter in water and by some chemical reactions with anthropogenically added impurities.

Dissolved oxygen must be measured at the site (i.e. *in situ*). Samples cannot be taken back and analysed in the laboratory because the amount of dissolved oxygen in the water can change rapidly after the sample has been collected.

Dissolved oxygen test kits involve two parts – sample preservation (stabilization or fixing) and testing. Preservation involves addition of a chemical to the sample that precipitates in the presence of dissolved oxygen, followed by the addition of a chemical that produces colour in the solution. Testing involves adding titrant solution until the colour disappears. The dissolved oxygen value is calculated from the volume of titrant added.

1.2.2.4 Biochemical Oxygen Demand (BOD) and Chemical Oxygen Demand (COD)

A related measure of the amount of dissolved oxygen, in $mg\ l^{-1}$, necessary for decomposition of organic matter by micro-organisms (i.e. bacteria or carbonaceous organic pollution from sewage or industrial wastes), is the *Biochemical Oxygen Demand (BOD)*. Natural sources of organic matter include plant decay and litter fall. However, plant growth and decay may be accelerated by human activities when nutrients (due to pollution loading) and sunlight (due to removal of forest cover, etc.) are overly abundant. Oxygen consumed in the decomposition process robs other aquatic organisms of the oxygen they need for survival. Organisms that are tolerant of lower levels of dissolved oxygen may replace the diversity of more sensitive organisms. The total amount of oxygen consumed when a biochemical reaction is allowed to proceed to completion is called the *Ultimate BOD*. Determination of the ultimate *BOD* is quite time consuming, so the *5-day BOD* (referred to as *BOD*₅) has been universally adopted as a measure of relative pollution. BOD_5 measures uptake of oxygen by micro-organisms at $20°C$ over a period of 5 days, kept under darkness, and is the most common measure of *BOD*. Pristine rivers generally have a *BOD* of less than $1\ mg\ l^{-1}$. Moderately polluted rivers have a *BOD* in the range 2 to $8\ mg\ l^{-1}$. Adequately treated municipal sewage has a *BOD* of about $20\ mg\ l^{-1}$. In untreated sewage, the *BOD* is variable and ranges between 200 and $600\ mg\ l^{-1}$. Slurry from dairy farms has a *BOD* of around $8000\ mg\ l^{-1}$ and silage liquor of around $60,000\ mg\ l^{-1}$.

Another related parameter, the *Chemical Oxygen Demand (COD)*, is a measure of the chemically oxidizable material present in water and provides an estimate of the amount of organic and reducing material present in water. The basis of measurement of the *COD* is that nearly all organic compounds can be fully oxidized to carbon dioxide, with a strong oxidizing agent under acidic conditions. There are several to estimate *COD*. The most common is the *4-hour COD*.

It should be emphasized that there is no generalized correlation between the *BOD*₅ and the ultimate *BOD*. Likewise, there is no generalized correlation between the *BOD* and the *COD*, though the measured values may correlate for a specific waste contaminant in a particular wastewater stream.

1.2.2.5 pH

An important water quality indicator is its *pH*. It is a measure of the relative amount of free hydrogen and hydroxyl ions present in the water. As hydrogen ions are positively charged, their concentration alters the charge environment of other molecules in solution. For this reason, the *pH* of water is very important for living beings. Water that has more free hydrogen ions is acidic, whereas water that has more free hydroxyl ions is basic. The *pH* $(= log_{10}\ [H^+])$ is reported in 'logarithmic units', as -ve logarithm of free hydrogen ions concentration $[H^+]$. For pure water $[H^+] = 10^{-7}\ mol\ l^{-1}$, giving it a *pH* of 7. Each number represents a 10-fold change in the level of acidity/basicity of the water. The *pH* ranges from 0 to 14; values less than 7 indicate acidity, 7 is neutral, whereas a *pH* greater than 7 indicates a base. Water with a *pH* of 5 is ten times more acidic than water having a *pH* of 6. The *pH* values of some common liquids are shown in Fig. 1.2. Rain has an acidic *pH* of about 5.6 because the atmosphere, through which rain drops fall, contains natural carbon dioxide and sulphur dioxide.

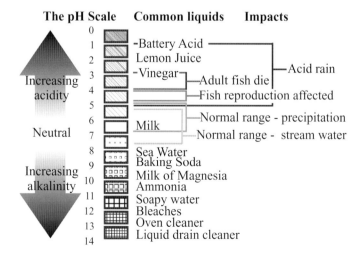

Fig. 1.2 The *pH* scale with *pH* values of some of the common liquids and environmental impacts these substances may cause. Redrawn from http://ga.water.usgs.gov/edu/phdiagram.html. © U.S. Geological Survey.

Since pH can be affected by chemicals dissolved in water, *pH* is an indicator of water that is changing chemically. Pollution can change the *pH* of water, which can be harmful to animals and plants living in that water. For instance, water from an abandoned coal mine can have a *pH* of 2, which is highly acidic and may seriously affect fish and other aquatic life forms living in the receiving water body. However, some substances help to maintain *pH* in a solution when an acid or base is added. Such substances are called, which play an important role in helping organisms to maintain a relatively constant *pH*.

1.2.2.6 Alkalinity

Alkalinity is a measure of resilience of water to the lowering of *pH* when an acid is added. Acidic conditions generally result from rain or snow, though soil sources and waste water are also important contributors in some areas. Alkalinity results from dissolution of rocks containing calcium carbonate, such as calcite and limestone. Alkalinity is expressed as the amount of calcium carbonate ($CaCO_3$) present in water, although other substances can also contribute to alkalinity. The unit of alkalinity is either parts per million (*ppm*) or its equivalent, $mg\ l^{-1}$.

When a lake or stream has low alkalinity, typically below 100 $mg\ l^{-1}$ as $CaCO_3$, a large influx of acids from a large rainfall or rapid snowmelt event could (at least temporarily) reduce the *pH* of the water to levels that are harmful for amphibians, fish, or zooplankton.

Alkalinity test kits are based on the technique of adding a *pH*-sensitive colour indicator to a sample and then adding an acid titrant solution drop by drop until a colour change is observed.

1.2.2.7 Nitrate

Plants in both fresh and saline waters require three major nutrients for growth: carbon, nitrogen, and phosphorus. In fact, most plants tend to use these three nutrients in the same proportion, and cannot grow if there is deficiency of any one of them. Carbon is relatively abundant in the air as carbon dioxide. Carbon dioxide dissolves in water and so a lack of either nitrogen or phosphorus generally limits the growth of aquatic plants. In some cases, trace nutrients such as iron can also be a limiting factor. Availability of sunlight can also limit plant growth. Nitrogen exists in water bodies in various forms as: dissolved molecular nitrogen (N_2); organic compounds; ammonium (NH_4^+); nitrite (NO_2^-); and nitrate (NO_3^-). Of these, nitrate is usually the most important nutrient for plant growth.

Nitrate in natural waters originates in the atmosphere through rain, snow, fog, or dry deposition by wind, from groundwater inputs, and from surface- and sub-surface runoff generated from surrounding land cover and soils, as well as from the decay of plant or animal matter in soils or sediments that contain nitrates. Human activities can greatly affect the amount of nitrate entering water bodies.

Many plants that use nitrogen are the microscopic algae or phytoplankton. Addition of excessive amounts of a limiting nutrient such as nitrogen to a water body, such as a lake or stream, leads to growth and proliferation of algae and aquatic plants. Excessive growth results in a condition called *eutrophication*. This can create taste and odour problems in lakes used for drinking water, as well as causing nuisance to persons living around the water body.

Although plants and algae add oxygen to the water, useful to aquatic plants and animals, their overgrowth can potentially lead to reduced amount of sunlight penetrating the water body. As plants and algae die and decay, bacteria multiply and consume the dissolved oxygen in the water. The amount of available dissolved oxygen in the water may become very low and adversely affect fish and other aquatic life forms.

Nitrate (NO_3^-) is very difficult to measure directly, whereas nitrite (NO_2^-) is easier to measure. In order to measure nitrate, kits have been developed that convert the nitrate present in a water sample to nitrite. Therefore, nitrate measurement involves adding a chemical (such as cadmium) to the water sample that converts nitrate to nitrite. A second chemical that reacts with the nitrite is then added to cause a change in colour of the water sample which is proportional to the amount of nitrite in the sample. Measurement by a nitrate kit, therefore, gives the combined concentration of nitrite (if present) and nitrate. Most natural waters have nitrate levels under 1 $mg\ l^{-1}$ nitrate-nitrogen, but concentrations over 10 $mg\ l^{-1}$ nitrate-nitrogen are found in some areas.

High nitrate levels in water can cause *methemoglobinemia*, a condition found especially in infants younger than six months. The stomach acid of an infant is not as strong as in older children and adults. This causes an increase in bacteria that can readily convert nitrate to nitrite that reacts with haemoglobin in the blood of humans and other warm-blooded animals to produce methemoglobin. Methemoglobin destroys the ability of red blood cells to transport oxygen. Methemoglobinemia has not been reported where water contains less than 10 $mg\ l^{-1}$ of NO_3-N. This level has therefore been adopted by the US Environmental Protection Agency as a standard for the Primary Drinking Water Regulation, mainly to protect young infants. The Maximum Contaminant Level (MCL) in drinking water as nitrate is 45 $mg\ l^{-1}$, whereas the MCL as NO_3-N is 10 $mg\ l^{-1}$.

1.2.3 Water quality – pathogens

Pathogenic organisms, namely bacteria, viruses, and parasites, cause diseases such as cholera, typhoid, jaundice, and dysentery.

Pathogens are difficult to detect in water because of their small numbers and also they cannot survive for very long outside a human or animal body. However, if intestinal pathogens are present in an individual, they will be found in faeces along with faecal bacteria. Coliform bacteria are a group of relatively harmless micro-organisms that occur in large numbers in the intestines of humans and warm- as well as cold-blooded animals. A specific subgroup of coliform bacteria is the faecal coliform bacteria, the most common being *Escherichia coli* (often written as *E. coli*). These organisms are distinguished from the total coliform group by their ability to grow at elevated temperatures and are associated only with the faecal material of humans and warm-blooded animals. Presence of faecal coliform bacteria in aquatic environments indicates that the water has been contaminated with human or animal faecal material. This suggests that the source water may have been contaminated by pathogens or disease-producing bacteria or viruses which also exist in faecal material. Presence of faecal contamination is a potential health risk to individuals in contact with this water. Faecal coliform bacteria may occur in water as a result of mixing of domestic sewage or by nonpoint sources of human and animal waste.

1.2.4 Freshwater macro-invertebrates

Millions of small organisms inhabit the fresh waters of lakes, streams, and wetlands. Macro-invertebrates, consisting of a variety of insects and insect larvae, crustaceans, molluscs, worms, and other small, spineless organisms live in the mud, sand, or gravel of the substrate or on submerged plants and logs. They play a crucial role in the ecosystem, providing an essential link in the food chain and the source of food for many higher animals. Macro-invertebrates, such as freshwater mussels, help to purify water. Scavengers feed on decaying matter in water, while certain macro-invertebrates prey on smaller organisms.

Macro-invertebrates can give a clue to the biochemical conditions prevailing in a water body. Many macro-invertebrates are sensitive to changes in *pH*, dissolved oxygen, temperature, salinity, transparency, and other changes in their habitat. Habitat refers to the environment that includes everything that an animal needs to survive and proliferate. Macro-invertebrates record the history of a water body because many are sessile or stay within a small area and live for a couple of years while the water flows by. Changes in the habitat (including water chemistry) are likely to cause changes in the macro-invertebrate assemblage. Chemical measurements in a water body at a given instant reflect on the ambient water quality at that time. Biological quality, on the other hand, governs the biodiversity that exists and will flourish in a water body over a period of time.

For some additional discussion on water quality in relation to its use, please refer to Section on 'Relation between use and quality of water', in Chapter 6.

1.3 Hydrologic cycle and global water distribution

The Earth, often referred to as the 'water' planet, is covered by one of our most precious resources – water. On the Earth, water is moving around

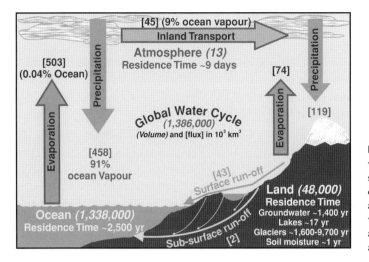

Fig. 1.3 The global water cycle, involving water in all three of its phases – solid (as snow/ice), liquid, and gas – operates on a continuum of temporal and spatial scales, and exchanges large amounts of energy as water undergoes phase changes and moves around from one part of the Earth system to another. See also Plate 1.3.

continually, changing its form: in the atmosphere as water vapour; in oceans, lakes, and rivers as liquid water; and in polar ice caps and mountain glaciers as ice. In fact, the Earth is very much a 'closed system', meaning that, as a whole, it neither gains nor loses significant amounts of matter, including water. This means that the same water that existed on the Earth millions of years ago still exists there and is being recycled around the globe continually as part of the global hydrologic cycle (Fig. 1.3). This cycle circulates water on Earth through what is known as the *hydrosphere*, which is the region containing the entire water in the atmosphere as well as on the surface of the Earth.

The hydrologic cycle begins with evaporation from the oceans. The evaporated water that forms clouds condenses to produce rain or snow (collectively called precipitation), over the oceans as well as the continents. In fact, about 87% of the water that evaporates and returns to the atmosphere is from the oceans. The remaining 13% of water that returns to the atmosphere is from the continents. This includes evaporation from lakes, rivers, and soil and rock surfaces, and *transpiration* from vegetation. On an average, the residence time of water vapour in the atmosphere is only about nine *days*. Evaporation from open water such as lakes and surface reservoirs does not vary much, but transpiration by plants varies considerably for different species; for example, the amount of water tran-

spired by desert plants with small canopy cover is far less than that transpired by dense forests.

The hydrologic cycle is usually depicted on a global scale, starting its journey from the oceans, moving through the atmosphere and surface/subsurface reservoirs and returning to the oceans through river flows as well as groundwater discharge from coastal regions. However, the hydrologic cycle operates at several different scales, from the hydrologic cycle of the entire Earth to that of the backyard of an individual household. An estimate of global water distribution is given in Table 1.3.

On a global scale, nearly 97% of all water occurs in the oceans; only 3% is fresh water. A significant fraction of this fresh water, about 69%, is locked up in mountain glaciers and polar ice caps, mainly in Greenland and Antarctica. It is rather surprising that of the remaining 31% fresh water, almost all occurs below our feet, ubiquitously as groundwater. No matter where we are located on Earth, the chances are that, the ground below, at some depth, is saturated with water. Of all the fresh water, only about 0.3% is contained in rivers and lakes. Despite this, rivers and lakes are the only water sources we are familiar with, and it is also where most of the water that we use in our daily lives comes from. In general, modern hydrology is concerned with the land component of the global hydrologic cycle and involves study of the movement of water both above and below the Earth's land surface, the

Table 1.3 An estimate of volume of water and its distribution on the Earth.

Water in	Volume (1000 km^3)	% Total Water	% Fresh Water	Residence Time
Oceans, seas, and bays	1,338,000	96.5	–	~4000 yr
Ice caps, glaciers, and permanent snow	24,064	1.74	68.7	10–100,000 yr
Groundwater	23,400	1.7	–	Weeks–100,000 yr
Fresh	(10,530)	(0.76)	30.1	
Saline	(12,870)	(0.94)	–	
Soil moisture	16.5	0.001	0.05	2 weeks–several yr
Ground ice and permafrost	300	0.022	0.86	
Lakes	176.4	0.013	–	~10 yr
Fresh	(91.0)	(0.007)	0.26	
Saline	(85.4)	(0.006)	–	
Atmosphere	12.9	0.001	0.04	~10 days
Swamp water (in wetlands)	11.47	0.0008	0.03	1–10 yr
Rivers	2.12	0.0002	0.006	~2 weeks
Biological water	1.12	0.0001	0.003	~1 week
Total	1,385,984	100.0	100.0	

Source: Gleick (1996) and Chilton and Seiler (2006) for residence time.

physical and chemical interactions with earth materials, and the biological processes that affect its quality.

Due to a variety of geographic, geomorphic, and meteorological factors, the distribution of water on Earth is uneven (Table 1.4). To a large extent, water shapes the landforms through erosion and deposition of sediments. Water is fundamental to existence and sustenance of life on Earth. Water makes up a substantial part of the bodies of all living organisms, as they need water for survival as well as growth. Therefore, the uneven distribution of water at any given place on the land manifests its unique imprint on almost every aspect, including landscape, ecology, and environment, and has governed the evolution of human societies and cultures from time immemorial.

The uneven distribution of water across the populated regions of the world is one of the main reasons for the emerging global water crisis, though rising population is undoubtedly the primary cause. It is rather intriguing that two-thirds of the world's population lives in areas that receive only one-quarter of the annual global rainfall. In contrast, the most water-rich areas of the world, such as the Amazon and Congo River basins, are sparsely populated. Some of the most densely populated regions of the world, such as the Mediterranean, the

Middle East, India, and China will face severe water shortages in the coming decades, largely due to rapidly growing populations. Even areas of the United States (particularly the southwest and parts of the Midwest) are vulnerable to water shortages.

1.3.1 Global precipitation distribution

Precipitation involves transfer of water from the atmosphere to the Earth's surface in various forms, such as rain, snow, sleet, hail, or dew. Since precipitation is the primary and annually renewable source of all fresh water on the Earth, let us understand its basic features. Precipitated water may be intercepted and taken up by plants; it may infiltrate the soil; it may flow over the land surface or percolate through the subsurface to reach streams, lakes, wetlands, and ultimately rejoin the ocean. Some of the discharged water evaporates from exposed surfaces and is transpired by plants to re-enter the atmosphere, thus completing the hydrologic cycle which continues year after year. Precipitation falls everywhere on Earth, but its distribution is highly variable, both in space and time. Largely due to greater surface area, the oceans receive three times more precipitation than the continents (Table 1.4). On land, in parts of vast deserts such as the Sahara, it may not rain for sev-

Table 1.4 Estimated annual balance of water over continents and oceans. P = Precipitation; E = Evapotranspiration; $P - E$ = Runoff; R_0 = Runoff from continents to oceans; $(P - E)/P$ = Runoff ratio. Source: Piexoto and Oort (1992) and Sellers (1965) for figures in parenthesis.

Region	Surface area (10^6 km^2)	P (mm a^{-1})	E (mm a^{-1})	$P - E$ (mm a^{-1})	R_0 (mm a^{-1})	$(P-E)/P$
Europe	10.0	657 (600)	375 (360)	282 (240)		0.43
Asia	44.1	696 (610)	420 (390)	276 (220)		0.40
Africa	29.8	696 (670)	582 (510)	114 (160)		0.16
Australia	8.9	803	534	269		0.33
North America	24.1	645 (670)	403 (400)	242 (270)		0.38
South America	17.9	1564 (1350)	946 (860)	618 (490)		0.40
Antarctica	14.1	169 (30)	28 (0)	141 (30)		0.83
All Land Areas	**148.9**	**746 (720)**	**480 (410)**	**266 (310)**		**0.36**
Arctic Ocean	8.5	97 (240)	53 (120)	44 (120)	307	0.45
Atlantic Ocean	98.0	761 (780)	1133 (1040)	−372 (−260)	197	−0.49
Indian Ocean	77.7	1043 (1010)	1294 (1380)	−251 (−370)	72	−0.24
Pacific Ocean	176.9	1292 (1210)	1202 (1140)	90 (70)	69	0.07
All Oceans	**361.1**	**1066 (1120)**	**1176 (1250)**	**−110 (−130)**	**110**	**−0.10**
Global	**510.0**	**973 (1004)**	**973 (1004)**	**0 (0)**		**0**

Note: Over the land, the runoff ratio indicates the fraction of precipitation that contributes to runoff. For various ocean basins, a balance between $(P - E)$ and R_0 is achieved via ocean currents.

eral years. At the other extreme, over tropical rain forests, precipitation might exceed 800 *cm a$^-$*. Some extreme precipitation events recorded around the world are given in Table 1.5.

On the Earth's surface, subtle changes in atmospheric pressure and temperature drive atmospheric circulation, including associated water vapour, from regions of high pressure to low pressure. Precipitation often occurs due to the rising motion of a moist air parcel to higher altitudes, resulting in condensation of moisture at the lower temperatures prevailing there (see tempera-

ture profile in troposphere; Fig. 1.6a). Areas with low pressure are generally found in regions where rising air motion predominates in the atmosphere, while areas with high pressure are generally found in regions where sinking motion predominates. Because of the Earth's spherical shape, the intensity of incoming solar radiation per unit area of the Earth's surface is maximum at the equator, and minimum at the poles. This results in a belt of maximum heating near the equator, with resultant rising motion giving rise to a belt of low pressure; and cooling at the poles, with resultant sinking motion and areas

Table 1.5 Extremes of precipitation recorded around the world.

Record	Location	Amount (*mm*)	Date
1-*year* Rainfall	Cherrapunji, India	26,470	1861
1-*month* Rainfall	Cherrapunji, India	9300	1861 (July)
Average Annual Rainfall	Mt Waialeale, Hawaii, USA	11,680	
24 *hr* Rainfall	Belouve, La Reunion Island	1350	Feb 28, 1964
Lowest *Annual* Average Rainfall	Arica, Chile	0.8	
Greatest 1 *Month* Snowfall	Tamarack, California, USA	9910	1911 (Jan)
Greatest Snowfall Single Storm	Mt Shasta, California, USA	4800	Feb 13–18, 1959

Source: http://www.physicalgeography.net/fundamentals/8g.html

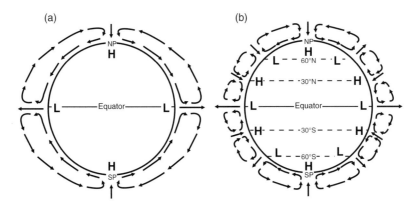

Fig. 1.4 (a) For a non-rotating spherical Earth, three pressure belts with low pressure at the equator and high pressure at the two poles are generated due to higher solar radiation incident at the equator compared to the poles. Consequently there is a rising motion at the equator and a sinking motion at the poles. (b) The Earth's rotation induces two other zones of sinking air at about 30° latitude on either side of the equator, and two additional belts of rising air at about 60° latitude north and south of the equator.

of high pressure. A non-rotating Earth would have just these three pressure belts (Fig. 1.4a).

The Earth's rotation induces two other zones of sinking air at about 30° latitude on either side of the equator, and two additional belts of rising air at about 60° latitude north and south of the equator (Fig. 1.4b). These are associated, respectively, with semi-permanent belts of high pressure, and belts of migratory low-pressure centres that bring precipitation to the middle latitudes. Broadly, the part of

the globe between the belts of high pressure and migratory low-pressure centres, is the temperate zone. The part north of the migratory low-pressure central belt is the polar zone, and the part between the two belts of high pressure is the tropical zone.

The relationship between these pressure belts, the average vertical motion in the atmosphere (sinking or rising), and production of precipitation, is brought into sharp focus by inspection of average daily zonal precipitation (Fig. 1.5). Most

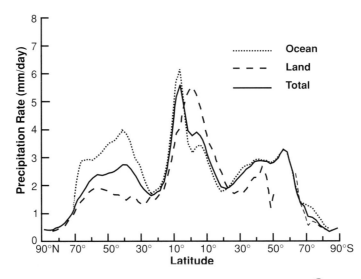

Fig. 1.5 Zonal mean precipitation, land versus ocean, 1979–2003. After Adler *et al.* (2003). © American Meteorological Society.

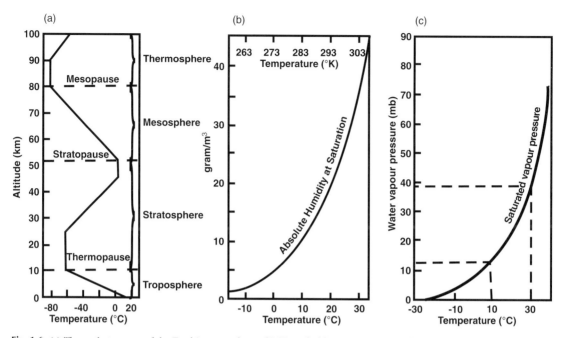

Fig. 1.6 (a) Thermal structure of the Earth's atmosphere. (b) Water holding capacity, given by absolute humidity at saturation decreases with temperature as (c) the amount of saturated vapour decreases.

of the world's major deserts are found in the semi-permanent belts of high pressure, where the sinking motion gives rise to dry weather. This includes the high-latitude polar regions of the Earth – the cold deserts. Precipitation is high, in general, in the equatorial belt characterized by low pressure, where rising motion and high moisture content give rise to cloud formation and consequently rainfall occurs almost year-round. One major exception to this is the region of the southern slopes of the Himalayan Massif. This huge mountain range serves as an elevated heat source during the Northern Hemisphere summer. An area of low pressure forms there, and induces a persistent southerly circulation towards its centre. Moisture-laden air is forced to rise as it meets the mountain range, which results in heavy rainfall. A second zone of relatively high precipitation is found in the temperate belt of migratory lows. Here, as the average air temperature is lower, the average amount of water vapour in the atmosphere is small, and so total precipitation is also generally low, except where persistent forced ascending motion of the air over mountain ranges results in regionally higher rainfall.

1.3.1.1 How does precipitation actually occur?

In the troposphere (layer of atmosphere 10-12 *km* above the Earth's surface), air temperature exhibits a sharp gradient, as shown in Fig. 1.6a. As a result, air temperature decreases significantly as one goes up in altitude, at a rate of approximately $10°C$ km^{-1}. This rate of decrease of air temperature is called the environmental or ambient *lapse rate*. A rising mass of air, therefore, encounters colder atmospheric conditions and accordingly its capacity to hold water vapour decreases (Fig. 1.6b) as saturation condition is approached (Fig. 1.6c), which favours condensation.

Three major types of lifting mechanism for air parcels operate in the atmosphere. When two air masses of different temperatures meet along a front, as shown in Fig. 1.7, (i) the warmer air rises up leading to frontal precipitation; (ii) heating of air in contact with the land surface also provides upward buoyancy and leads to convective precipitation; and (iii) the rising motion occurs when air is forced up along a mountain barrier, causing orographic precipitation (Fig. 1.7).

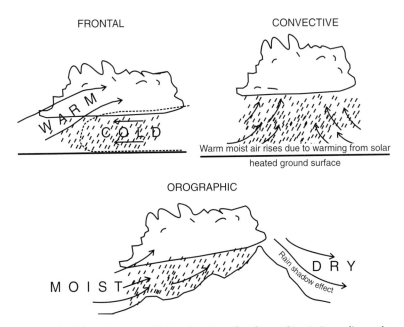

Fig. 1.7 Major mechanisms that lift an air mass to higher elevations, thereby resulting in its cooling and consequent loss of vapour holding capacity and condensation, resulting in precipitation.

1.3.2 Water vapour

Water in the gaseous phase – the water vapour – is the most significant atmospheric trace constituent relevant to climate, weather, hydrology, and atmospheric chemistry. Water vapour is produced by evaporation of liquid water or by sublimation of ice. Under normal atmospheric conditions, water continuously evaporates as well as condensing. Water vapour is lighter (or less dense) than dry air. Therefore, at a given temperature it is buoyant with respect to dry air. A volume of moist air will rise due to buoyancy if placed in a region of dry air.

A commonly used indicator of the moisture content of air (or humidity) is the *dew point*, which is actually the temperature at which an air mass becomes saturated. It is not the ambient air temperature but the temperature to which the air must be cooled in order for it to become saturated, assuming that no drastic changes in pressure or moisture content occur. In this sense, the dew point is an indicator of absolute moisture content that can be calculated using Fig. 1.6b. It is obvious that the higher the humidity, the higher will be the dew point. The dew point is associated with relative humidity, being the ratio (in %) of the moisture content at the ambient temperature to the saturation moisture content at the same temperature. A high value of relative humidity indicates that the dew point is close to ambient air temperature. If relative humidity is 100%, the dew point is equal to the prevailing air temperature.

Most of the water vapour in the atmosphere is contained in the troposphere. It is the most abundant atmospheric greenhouse gas, and without it the Earth's surface temperature would be well below the freezing point of water. Water vapour condenses to form clouds, which may act to either heat or cool, depending on circumstances. Phase changes involving water vapour – the condensation and evaporation processes – involve exchange of latent heat energy that affects the vertical stability of the atmosphere, the structure and evolution of meteorological storm systems, and the energy balance of the global climate system (Chahine 1992). In general, atmospheric water vapour strongly influences, and is in turn influenced by, weather. Water vapour is also the prime source of the atmospheric hydroxyl radical, an oxidizing agent that cleanses the atmosphere of many air pollutants.

The total amount of water vapour present in the atmosphere is $\sim 13 \times 10^{15}$ *kg*, which is a negligible fraction, $\sim 10^{-5}$ times the water contained in the oceans (Table 1.3). Nevertheless, changes in world amount and distribution of atmospheric water vapour, which can happen relatively quickly, cause as well as provide evidence of other global environmental changes. For example, temperature and water vapour changes often go hand in hand, so that any increase in water vapour can provide an indirect evidence of warming. As changes in climate are accompanied by concomitant changes in atmospheric circulation and in the transport of water vapour, global and regional precipitation patterns will also change. If all the water vapour in the atmosphere were to condense and fall as rain, it would cover the entire Earth to an average depth of about 25 *mm*. This is called *precipitable water*. Since water vapour is not evenly distributed over the globe, the layer would be about 50 *mm* deep near the equator and less than 5 *mm* near the poles (Randel *et al.* 1996). The average annual precipitation over the globe is about 1 *m*. This means there must be a swift transfer of atmospheric water vapour to the surface via precipitation and back again to the atmosphere via evaporation; a water molecule spends on an average only 9–10 days in the atmosphere before returning to the surface as precipitation. This rapid turnover, combined with the variation of temperature with height (Fig. 1.6a) and geographic location, causes water vapour to be distributed unevenly in the atmosphere, both horizontally and vertically. With global warming, an increase in the rate of evaporation must be counterbalanced by an increase in the rate of precipitation, because the atmosphere can hold only a limited amounts of water. This modification of the global hydrologic cycle is likely to lead to an increase in the frequency of extreme precipitation events, and correspondingly an increase in the proportion of precipitation falling in extreme events (Trenberth 1988, 1999), which has indeed been observed in some regions (IPCC 2001). Such changes would have important consequences for agriculture, flood control, and other aspects of human life.

1.3.3 Surface water and groundwater

Precipitation falling on the continents either runs off the land surface and joins streams, lakes, and wetlands, or soaks into the ground. Water that re-

mains on the Earth's surface, such as in streams, lakes, and wetlands, is called surface water. Water that soaks into the ground is either stored as soil moisture or recharges the groundwater.

Streams are channels that carry surface water from the continents back to the oceans as part of the hydrologic cycle. The surface water ponded in lakes and wetlands is also part of the hydrologic cycle because lakes and wetlands contribute water to the atmosphere through evaporation. They also receive water from and lose water to the adjoining aquifer system or drain into other surface water bodies.

Some of the water that seeps into the ground becomes part of soil moisture and the excess water (after satisfying the field capacity of the top soil) infiltrates downward under the influence of gravity and becomes part of groundwater. Water held in soils usually does not move very far down as it is transpired back to the atmosphere by vegetation.

Unlike soil water, which does not move down very far, groundwater moves in flow systems (aquifers) that can range in size from a few metres in horizontal extent to several hundred kilometres. Aquifers eventually discharge groundwater to surface water bodies such as streams, lakes, and wetlands as well as directly to bays, estuaries, and oceans, but the latter amount is much less than the amount that discharges into streams, lakes, and wetlands.

Some other aspects of water on the Earth, including the factors that control its distribution both globally and locally, will be considered in the subsequent chapters. The most important characteristic of liquid water that we often see is its constant motion. Therefore in the next chapter we shall be concerned with the study of water and its motion on the surface of the Earth. But, before dwelling on this and other aspects of hydrology, a small digression concerning the units and dimensions important for any quantitative analysis will be in order.

1.4 Units and dimensions

In the foregoing, some of the physical properties of water are described that make it unique for sustaining life on Earth. Furthermore, each of these physical properties has a quantitative value attached to it, which comprises two parts: a numerical value and its associated unit. The value of the physical property is intrinsic and not related to the choice of unit used, but this value makes no sense if it is given without specifying the relevant unit.

For example, the height of a person is not changed by how we choose to describe it – $5'10''$ or 177.8 *cm*, or 3.9 *cubits*! One may equate these measurements, i.e. $5'10'' = 177.8$ *cm* = 3.9 *cubits*. Clearly it would make no sense to equate just these numbers unless one specifies their associated unit. It is, therefore, clear that *units* indicate the measure that has been used to specify a physical quantity, and are probably best explained by examples. Examples include the *metre*, *yard*, or *inch* (for measuring distance), the *hour* or *second* (for measuring time), the *kilogram* or *ounce* (for mass), the *kelvin* or degree *centigrade* (for temperature), and the *joule*, *kilocalorie*, or *British Thermal Unit* (or *B.T.U.*) for measuring energy.

An important term used in physical sciences is, dimension, which refers to a set of equivalent units. For example, length is a dimension that could be measured in *metres*, *feet*, or even *furlongs*. Alternatively, one may say enthalpy has the dimension of energy, and may therefore be expressed in units of *joules* or *calories*. It follows that while writing equations such as Eqn 1.1, these must be dimensionally consistent. In other words, the type of units on the left-hand side of an equation must match the type of units on the right-hand side. The units do not have to be the same, but they must have the same dimension. Thus one can equate 177.8 *cm* with 3.9 *cubits* because both numbers together with their units measure length, but it would make no sense to try to equate 3 *hours* with 1.85 *m*, since one expresses time and the other measures

length. It follows that an equation should be valid, regardless of the units used.

There are a number of systems of units for measuring physical dimensions (see Table 1.6). The SI or MKS system is based on *metre*, *kilogram*, and *second* as the basic units of length, mass, and time, respectively. The CGS system, is based on *centimetre*, *gram*, and *second* as basic units. The English or FPS system, no longer in common use, is based on *foot*, *pound*, and *second*.

A physical equation has two aspects: (i) the dimensionality which is independent of the system of units used; and (ii) a consistent system of units.

For example, the basic equation of force from Newton's Second Law of motion, $F = m \times a$, has the dimension mass [M] × length [L]/time squared $[T^2]$, which yields MLT^{-2} as the dimension of force. The centripetal force $F = mv^2/r$ has the dimension $M \times [L\ T^{-1}] \times [L\ T^{-1}]/L$. Combining the terms yields MLT^{-2}. This agrees with the dimensions expected for force.

In order to be able to add or subtract two physical quantities, they must have the same dimension. The resulting physical quantity has the same dimensions. Physical quantities with the same dimensions in different systems of units can be added or subtracted by multiplying one of the quantities by a suitable conversion factor to maintain compatible units.

Multiplication of two physical quantities results in a new physical quantity that has the sum of the exponents of the dimensions of the two original quantities. Division of one physical quantity by another results in a new physical quantity that has the dimension of the exponent of the first quantity minus the exponent of the second quantity. Taking the square root of a physical quantity results in

Table 1.6 Different systems of units for measuring physical quantities.

Physical Quantity	Symbol	Dimension	SI or MKS	CGS	English/ FPS
			Unit System		
length	s or l	L	metre	centimetre	feet
mass	m	M	kilogram	gram	pound
time	t	T	second	second	second
force	F	MLT^{-2}	newton	dyne	poundal
energy	E	$ML^2\ T^{-2}$	joule	erg	B.T.U.

a new physical quantity having a dimension with an exponent half its original dimension. Raising a physical quantity to a power results in a new physical quantity having a dimension with the exponent multiplied by the power to which it is raised.

Volume is a basic physical property of water, giving a measure of the space occupied by it in certain units. Although the dimension of volume is L^3, very often in hydrology, volume is also expressed in units of depth [L] alone, particularly in relation to rainfall, snowfall, and irrigation. In such cases, one has to multiply the depth of water by the surface area $[L^2]$ involved to get the absolute value of its volume. Another useful indicator of water availability is flow rate that defines the volumetric flow per unit time $[L^3 \ T^{-1}]$ across a real or hypothetical surface.

A given dataset is accurate when the sample average (average of several observations) is equal to its true average. A given dataset is precise when all the observations fall within a narrow range. Results may be accurate, though imprecise, when measurements show a wide scatter. Results may be precise, though inaccurate, when measurements fall within a narrow range, but when their mean does not equal the true mean (Chapter 8, Figs 8.8–8.12).

1.5 Significant figures and digits

In a rigorous sense, if the catchment area is given as 2000 km^2, it is meaningful to trust only the leftmost digit (2 in this case). However, if one is also certain about the next digit (the leftmost 0), it is better to express the area as $2.0 \times 10^3 \ km^2$. In general, it is advisable to write numbers with the argument in decimal notation with appropriate number of digits to indicate reliability of figures, multiplied by the relevant power of 10. The uncertainty determines the number of digits that may be quoted. For instance, it is in order to quote 12.5 $\pm$ 0.1 cm, but it is not consistent to write 12.57 $\pm$ 0.1 cm. However, while making the calculations. all digits are to be retained; rounding off is done only for the final result. Nevertheless, results written down in the course of a mathematical calculation expressed with the number of digits that can be justified from physical considerations. The entire calculation itself, however, is to be carried out without any intermediate rounding off.

2

Surface water hydrology

Precipitation falling on land areas and moving on the land surface as sheet flows or as stream flow, or accumulating in low lying areas such as swamps and lakes, is called surface water. Surface water is naturally replenished by precipitation and lost through discharge to evaporation and sub-surface flow to groundwater. Surface water is the largest source of fresh water on the Earth. This is the most visible part of water that hydrology is concerned with. A swamp or a lake is formed when water collects on the surface in low lying areas with little or no natural drainage and slope of the land. When precipitation, in the form of net snow, exceeds the amount that melts and the process of net snow accumulation continues year after year over long periods, formation of glaciers occurs. A stream is formed when water collects on the surface in low lying areas with good natural drainage and sloping land.

2.1 Lakes

A *lake* (from Latin *lacus*) is an inland body of water larger and deeper than a pond, and is generally fed by one or more streams, glacial meltwater, and/or groundwater discharge. The majority of lakes on the Earth have fresh water, and most lie in the Northern Hemisphere at higher latitudes. Natural lakes are generally found in mountainous areas, rift zones, and areas with ongoing or recent glaciations. Other lakes are found in endorheic basins or along the courses of mature rivers. In some parts of the world, there are many lakes because of irregular drainage patterns, which are remnants of the last Ice Age, e.g. in Canada which has 60% of the world's lakes. In addition to providing recreation for humans, lakes and their shores also provide important wildlife habitats for both aquatic and terrestrial animals. They help protect water quality; eroded sediments, debris, and other pollutants washed from watersheds are deposited by inflowing streams, as a result of which out-flowing streams often carry less of these pollutants. Eventually lakes fill in with the material washed in by streams. Even without human influence, a natural, deep, clear lake will become shallow, weed filled, and greenish in colour from growth of algae. Over time, it will become a pond, then a marsh, and finally a forest. This natural aging process of lakes – based on increased growth and productivity – is called *eutrophication*. In terms of relative age and productive state, a young lake with low productivity is termed *oligotrophic*, a middle-aged lake is *mesotrophic*, and an older lake that is highly enriched is called *entrophic*. Normally, this aging process takes hundreds to thousands of years. In lakes affected by human actions, changes occur more quickly – sometimes change that would normally take centuries may occur over a person's lifetime.

No two lakes are exactly alike. They may differ in size, depth, number, and size of inflowing and out-flowing streams, and also in their shoreline configuration. Each of these physical factors in turn

Modern Hydrology and Sustainable Water Development, 1st edition. © S.K. Gupta
Published 2011 by Blackwell Publishing Ltd.

(a)

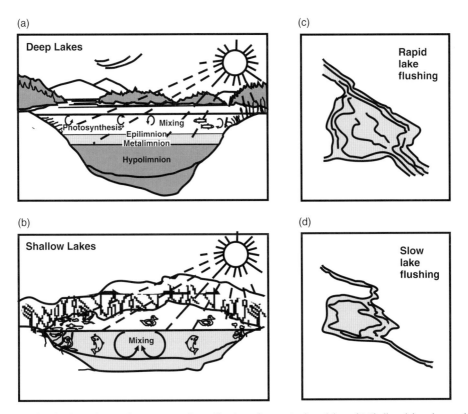

(b)

(c)

(d)

Fig. 2.1 (a) Wind and solar radiation driven seasonal stratification of water in deep lakes. (b) Shallow lakes do not show such depth stratification. The ratio of lake volume to inflowing volume determines if the lake is (c) rapidly flushing or (d) slowly flushing. Redrawn after Michaud (1991). © Washington State Department of Ecology. See also Plate 2.1.

influences the lake's characteristics. Some of these include survival of certain fish species in the lake, growth of weeds on the shoreline or lake water turning green in the summer, and whether the lake water is warm enough for swimming or suitable as a drinking-water source.

In a deep lake (Fig. 2.1a), water near the surface may be very different physically, chemically, and biologically from water near the bottom. The top portion of the lake is mixed by the wind and warmed by the sun. This upper more or less uniformly mixed warm layer is called the *epilimnion*. Because of the available sunlight and warmer temperatures, many organisms live here. The more organisms there are photosynthesizing, breathing, eating, and growing, the higher the growth rate and productivity of the lake. The bottom portion of a deep lake receives little or no sunlight. The water is colder; it is not mixed by wind; and decay of dead

organic matter, called *decomposition*, is the main physical, biological, and chemical activity. This uniformly cold and relatively undisturbed water layer is called the *hypolimnion*. The middle layer of water in a lake, which marks the transition between the top epilimnion and the bottom hypolimnion, is called the *metalimnion* (also called *thermocline*). In this layer the temperature changes rapidly with depth. A shallow lake (Fig. 2.1b) is more likely to be homogeneous – the same from top to bottom. The water is well mixed by wind, and physical characteristics such as temperature and oxygen vary little with depth. Because sunlight reaches all the way to the lake bottom, photosynthesis and growth occur throughout the water column. As in a deep lake, decomposition in a shallow lake is higher near the bottom than at the top, because when plants and animals die they sink. A larger portion of the water in a shallow lake is likely to be influenced by sunlight, therefore photosynthesis and growth are proportionately higher, compared to a deep lake.

Lakes range in size from little bigger than ponds to reservoirs over 100 *km* long. Although there are no precise parameters governing lake size comparisons, the size does affect a number of important relationships. Some parameters are: ratio of lake surface area to length of shoreline, percentage of total water volume that is influenced by sunlight, and ratio of size of the watershed to size of lake. These relationships affect how lakes bahave. A small lake with a higher ratio of shoreline to water volume may be more susceptible to damage from human activities on the shoreline or in the watershed.

The size and number of inflowing and outflowing streams in a lake determine how long it takes for a drop of water entering a lake to leave it – a process called *flushing*. Some lakes flush in days, while others take years. In some lakes, the inflow is not even visible; all of it comes from groundwater seeps and precipitation. In rapidly flushing lakes (Fig. 2.1c), quality of incoming water is the single most important factor influencing lake water quality. On the other hand, in slow flushing lakes (Fig. 2.1d), internal lake processes and groundwater determine water quality. In terms of pollution, the faster a lake flushes the better its water quality, because pollutants are flushed from the lake before they can cause much damage. A more rapidly flushing lake may also respond faster to pollution control activities in its watershed.

Another important lake characteristic is the shape of the shoreline. Shallow bays and inlets tend to be warmer and more productive than other parts of a lake. A lake with many of these features will be different than, say, a bowl-shaped lake with a smooth, round shoreline.

Commonly tested water quality parameters of lake water are temperature, dissolved oxygen, *pH*, Secchi disk depth, nutrients, total suspended solids and turbidity, chlorophyll and faecal coliform bacteria.

Temperature exerts a major influence on biological activity and growth; up to a certain level, the higher the water temperature, the greater the biological activity. Temperature also governs the types of organisms that can grow in a lake. Fish,

insects, zooplankton, phytoplankton, and other aquatic species have an ideal range for survival and growth. When temperatures deviate significantly from their preferred range, the number of individuals of the different species falls until they are completely wiped out. Temperature is also important because of its influence on water chemistry. The rate of chemical reactions generally increases at higher temperatures, which in turn affects biological activity. An important example of the effects of temperature on water chemistry is its impact on dissolved oxygen. Warm water holds less oxygen than cool water, so it may be saturated with oxygen but still not contain enough for survival of aquatic life forms. Some compounds are also more toxic to aquatic life at higher temperatures.

The most obvious reason for temperature change in lakes is the change in seasonal air temperature. Daily variation also may occur, especially in the surface layers, which are warmed during the day and cooled at night. In deeper lakes during summer, the water separates into layers of distinctly different temperature, a process called *thermal stratification*. The surface water is warmed by the sun, but the bottom of the lake remains unaffected. Once the stratification develops, it tends to persist until the air temperature cools again in the fall season. Because the layers do not mix, they develop different physical and chemical characteristics. For example, dissolved oxygen concentration, *pH*, nutrient concentrations, and species of aquatic life in the upper layer can be quite different from those in the lower layer. When the surface water cools again in the fall season, to about the same temperature as the lower water, the stratification is lost and the layers mix vertically. This process is called *fall turnover*. A similar process may occur during spring as colder surface waters warm to the temperature of bottom waters and the lake mixes. This is called *spring turnover*.

Because the sun can heat a greater proportion of the water in a shallow lake than in a deep lake, a shallow lake may warm up faster and to a higher temperature. Lake temperature is also affected by the size and temperature of inflows (e.g. a glacial-fed stream, springs, or a lowland creek) and by how quickly water flushes through the lake. Even a shallow lake may remain cool if fed by a comparatively large, cold stream. Fortunately, lake water is complex; it is full of chemical buffers that prevent major changes its *pH*. Small or localized changes in *pH* are quickly neutralized by various chemical reactions, so little or no measurable change in pH may occur. This ability to resist change in *pH* is called the *buffering capacity*. This capacity of the lake controls not only future localized changes in *pH*, but also the overall range of *pH* change under natural conditions – the *pH* of natural waters ranges hovers between 6.5 and 8.5.

Nutrients in lakes are essential for biological growth. Whereas in a garden or field, growth and productivity are thought beneficial, this is not necessarily so in a lake. Proliferating algae and other plants caused by the nutrients is fine up to a certain level, but may become a nuisance (see below).

Chlorophyll is the green pigment in plants that enables them to create energy from light for photosynthesis. As with other parameters, the amount of algae present in a lake greatly affects the lake's physical, chemical, and biological environment. Algae produce oxygen during daylight hours but use up oxygen at night and also when they die and decay. Decomposition of algae also causes release of nutrients to the lake, which may result in growth of more algae. Their photosynthesis and respiration cause changes in *pH* of the lake, and presence of algae in the water column is the main factor affecting Secchi disk readings indicating transparency and depth of light penetration. Algae, of course, also can cause aesthetic problems in a lake; a green 'scum', swimmers' itch, and a rotting smell are common problems.

Faecal coliform bacteria are microscopic life forms that live in the intestines of warm-blooded animals such as humans and cattle. They survive in waste or faecal material excreted from the intestinal tract. When faecal coliform bacteria are present in high numbers in a water sample, it means that the water may have received faecal matter from some source. Although not necessarily agents of disease, faecal coliform bacteria indicate the potential presence of disease-carrying organisms, which thrive live in the same environment as faecal coliform bacteria.

Many lakes are artificial, constructed for hydroelectric power generation, recreational, industrial or agricultural use, or domestic water supply.

2.2 Glaciers

A *glacier* is a huge mass of ice deposited on land (either mountain or polar glacier) or floating in the sea next to land. Moving extremely slowly, a glacier essentially behaves as a huge river of ice, often merging with other glaciers, similarly to streams merging. Glaciers are formed over long periods from accumulation of snow in areas where the amount of snowfall exceeds the amount that melts. Glacial ice is the largest reservoir of fresh water on Earth, second only to oceans as the largest reservoir of total water. Glaciers cover vast areas of the polar regions and are found in mountain ranges of every continent except Australia. In the tropics, glaciers are restricted to the highest mountains. There are two main types: *alpine glaciers* and *continental glacier*.

Alpine glaciers – Most glaciers that form on a mountain are known as alpine glaciers. Outside the polar regions, glaciers cannot form unless the land surface elevation is above the snowline, the lowest elevation at which snow can remain year round. Most such glaciers are located in the high mountain regions of the Himalayas of Southern Asia or the Alps of Western Europe, South America (the Andes), California (the Sierra Nevada), and Mount Kilimanjaro in Tanzania. The Baltoro Glacier in the Karakoram Mountains, Pakistan, $62\,km$ in length, is one of the longest alpine glaciers on Earth.

Continental glaciers – An expansive, continuous mass of ice considerably bigger than an alpine glacier is known as a continental glacier. There are three primary subtypes: (a) *ice sheets* are the largest of any glacier type, extending over $50\,000$ km^2. The only places on Earth that have such large ice masses are Antarctica and Greenland. These regions contain vast quantities of fresh water. The volume of ice is so large that if the Greenland ice sheet melted, it would cause sea levels to rise globally by about 6 m. If the Antarctic ice sheet melted, sea levels would rise by about 65 m. Antarctica is home to 92% of all glacial ice worldwide. The Lambert Glacier in Antarctica is about 96 km wide, over 400 km long, and about 2.5 km thick. (b) An *ice cap* is similar to an ice sheet, though smaller and forming a roughly circular, dome-like structure that completely blankets the landscape. (c) An *ice field* is a smaller version of an ice cap that fails to cover the land surface and is elongated relative to the underlying topography.

Glaciers have a tendency to move, or 'flow', downhill. While the bulk of a glacier flows in the direction of lower elevation, every point of the glacier can move at a different rate, and in a different direction. The general motion is in response to the force of gravity, and the rate of flow at each point of the glacier is affected by several factors.

There are two types of glacial movement: sliders and creepers. Sliders travel along a thin film of water located at the bottom of a glacier. Creepers, on the other hand, form internal layers of ice crystals that move past one another based on the surrounding conditions (e.g. weight, pressure, temperature). The top and middle layers of a glacier tend to move faster than the rest. Most glaciers are both creepers and sliders, plodding along in both fashions. The speed of glacial displacement is partly determined by friction, which makes the ice at the bottom of the glacier move more slowly than the upper portion. In an alpine glacier, friction is also generated at the side walls of the valley, which slows movement of the edges relative to the centre. This was confirmed by experiments in the 19th century, in which stakes were planted in a straight line across an alpine glacier, and as time passed, those in the centre moved farther. Glacier speed can vary from being virtually at rest to a kilometre or more per year. On average, though, glaciers move at the slow pace of a couple of metres per year. In general, a heavier glacier moves faster than a lighter one, a steep glacier faster than a less steep one, and a warmer glacier faster than a cooler one.

A large mass, such as an ice sheet/glacier, depresses the crust of the Earth and displaces the mantle below – a phenomenon known as *isostatic depression*. The depression is about a third the thickness of the ice sheet. After the glacier melts, the mantle begins to return to its original position, pushing back the crust to its original position. This post-glacial rebound from the last ice age, which lags behind melting of the ice sheet/glacier, is still under way in measurable amounts in Scandinavia and the Great Lakes region of North America. This rise of part of the crust is known as *isostatic adjustment*.

Because glaciers are massive bodies, the landscape they dominate is carved and shaped in significant and long-lasting ways through glacial erosion. As a glacier moves it grinds, crushes, and envelopes rocks of all shapes and sizes, wielding the ability to alter any landform in its path, a process known as *abrasion*.

A simple analogy when thinking about how glaciers shape the land is to envisage the large rocks it carries as chisels, gashing and scraping out new formations in the ground below. Typical formations that result from the passing of a glacier include U-shaped valleys (sometimes forming fjords when seawater fills them), long oval hills called drumlins, narrow ridges of sand and gravel called eskers, and hanging waterfalls, among others. The most common landform left by movement of glacier is known as a *moraine*. There are a variety of these depositional hills, but all are characterized by unstratified material including boulders, gravel, sand, and clay. Glaciers have shaped much of the Earth's surface through the processes described above and are just as closely connected with the Earth's geomorphologic evolution.

In many parts of the world, large populations inhabiting various river basins depend critically on the glacial meltwater supply to streams, particularly during the dry season. In this context, glaciers in the Himalayan-Karakoram region are the largest storehouse of fresh water in the lower latitudes, and the important role of their snow and ice mass in maintaining the flows of the Indus, Ganges, and Brahmaputra rivers during the dry season has been well perceived and understood since ancient times.

An understanding of glacier hydrology is central to understanding glacier behaviour. Glacier hydrology controls many of the major dynamic and geologic processes operating on glaciers. The behaviour of water in glaciers also reveals the structure of the ice and of the glacier on different spatial of scales, and indicates how this structure changes through time in response to seasonal and long-term changes in the glacier and its environment. Major issues in modern glaciology, including surges, ice streams, and deforming beds, centre on the role of water at the glacier bed. Our understanding of traditional glaciological problems such as sliding is based on improved knowledge of the effect of melt water within the glacier mass. However, substantial portions of our understanding of sub-glacial drainage are based on theoretical modelling in the absence of direct observation. Observations of englacial and sub-glacial hydrology rely on techniques such as dye tracing, water-pressure monitoring, and chemical analysis of meltwater which, in turn, provide a reliable theoretical basis for modelling and interpretation. Understanding of surface meltwater routing through the glacier to produce stream outflow continues to be a stimulating area of research.

A common apprehension is that with rising global temperatures, glaciers will begin to melt, releasing some or all of huge amounts of water held inside. As a result, the oceanic processes and structures we have adapted to will abruptly change, with unknown consequences.

2.3 Streams

Stream is a general term for any natural channel through which water flows. A stream can flow all year around (perennial stream) or only when it rains (ephemeral stream). The term includes the smallest gully flow to the largest river. With the exception of groundwater piping, overland flow (sheet flow) is necessary for channel formation. The rate at which a channel forms depends on vegetation, storm characteristics, slope, permeability, and erodibility of the surface. Sheet flow is dynamically unstable and small irregularities in the surface will cause local concentrations of flow. Eddy formation and flow diversion around obstructions or into hollows initiate scouring of surface materials. Once developed, a scoured trough, or rill, is perpetuated by the continued concentration of flow and migrates upstream by headward erosion of the *nick point* – the inflection where scouring is highest. Eventually a rill grows into a gully, and if the gully deepens to intersect the water table becomes a permanent stream. A gully may take only a few days or years to form. However, a large integrated stream network can take thousands to several millions of years to develop.

Because sheet flow is easily deflected, it can concentrate in linear depressions, joints, mud cracks, etc. Any irregularity on a slope can potentially influence channel development. Large slope depressions created by landslides are also favourable sites for water flow to concentrate. Over a period of time, streams may establish paths in structural troughs, faults, rift valleys, or linear depressions on weak lithologic formations.

2.4 Watershed concept

As already mentioned, the *hydrologic cycle* operates at many scales, from the hydrologic cycle of the entire Earth to the hydrologic cycle of the backyard of an individual household. Beyond household or small community water requirements, the smallest unit that water managers are concerned with is the hydrologic cycle on a watershed scale. In simple terms, *watershed* is the area of land surface that receives rain and/or snow and drains or seeps into a marsh, stream, river, lake, or the ground (Fig. 2.2). Watersheds are typically defined based on topographic data and surface stream flow. Houses, farms, ranches, forests, small towns, and big cities can make up watersheds. Some watersheds extend beyond administrative/geographical limits of a county, state, and even across international borders. Watersheds have a variety of shapes and sizes – some are millions of km^2 in area; others are just a few *hectares*. Just as creeks drain into rivers, watersheds generally form part of a larger watershed.

Because surface water and fresh groundwater are the only components of the hydrologic cycle that are extensively used by humans, the primary interest in the hydrologic cycle by water managers is focused on these two resources. The most common approach is to use watersheds or basins to conduct water balance estimates for the region. Although it is important to know how much water is stored in groundwater, lakes, and wetlands, understanding the movement of water within and from individual watersheds is far more important, and poses a real challenge. Most research in the hydrologic sciences is devoted to understanding the movement of water and the transport of chemicals and sediment by water flowing through watersheds, as both surface- and groundwater.

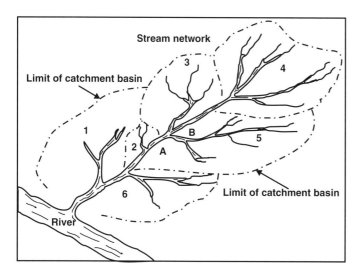

Fig. 2.2 A network formed by streams draining six watersheds, '1' to '6'. At point 'A' water is received from drainage of the rain or snowmelt in all three watersheds '3', '4' and '5'. At point 'B', which is only a short distance upstream of point 'A', the water is delivered only by catchment '4'. Therefore the amount of water available will be larger at 'A'. Redrawn from http://edna.usgs.gov (accessed on 30 March 2007).

To ensure availability of adequate water for human use, water managers need to be able to estimate the amounts of water that enter, pass through, and leave individual watersheds. This is a challenge because the relative magnitudes of the transfers of individual components in the hydrologic cycle can vary greatly. For example, in mountainous areas, precipitation is more difficult to measure compared to in plains or valleys. Mountain snowpacks and the amount of meltwater that these can deliver may vary widely, thereby affecting estimation of natural water budgets at lower elevations. Similarly, evaporation rates from an agricultural field compared to nearby woodland or a wetland may differ widely. Furthermore, the discharge of groundwater into surface water bodies may vary in different parts of watersheds, because different rock and sediment types may be present. Some other important reasons for making hydrologic measurements are to enable: (i) prediction of impacts of certain activities from prior knowledge or experience; (ii) infer impacts from available field evidence; and (iii) experimentally investigate the impacts of anthropogenic activities.

Understanding of the hydrologic cycle is a prerequisite to formulating any water management strategy. When the flow of water is manipulated to meet various human needs, it is necessary to understand how these actions will affect the local and regional hydrologic cycle and, ultimately, the availability and quality of water to downstream

users. Proper understanding of the hydrologic cycle is therefore essential, if optimal use of the water resources is to be achieved, while avoiding detrimental impacts on the environment.

The hydrologic cycle is also an important and practical concept for maintaining a healthy and sustainable Earth. To a large extent, water shapes the Earth's surface through erosion and deposition of sediments and minerals. The survival of humankind and other terrestrial life forms critically depends on availability of fresh water. Therefore, managing water resources by thoroughly understanding the hydrologic cycle at various scales, ranging from the entire Earth to the smallest of watersheds, should be our prime concern.

2.5 Instrumentation and monitoring

2.5.1 Precipitation

Of all the components of the hydrologic cycle, the most commonly measured element is *precipitation*, both as rain and snow. The objective is to quantify precipitation over the entire catchment area, as well as its spatial and temporal distribution. Several methods exist for measuring the magnitude and intensity of precipitation that include: (i) ground-based measurements at individual point sites; (ii) ground-based remote sensing measurements over large areas (weather radar); and (iii) aircraft and satellite based sensors over

(a)   (b)

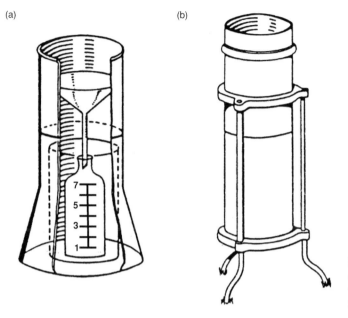

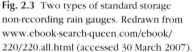

Fig. 2.3 Two types of standard storage
non-recording rain gauges. Redrawn from
www.ebook-search-queen.com/ebook/
220/220.all.html (accessed 30 March 2007).

larger areas. Each measurement is different, based
on temporal and spatial scale of measurements and
also the technique employed.

Point measurements of precipitation serve as the
primary source of data for estimation of areal averages. However, even the best measurement of
precipitation at a point is only representative of
a limited area, the size of which is a function
of the length of accumulation period, the physiographic homogeneity of the region, local topography, and the precipitation-producing process.
Radar and, more recently, satellites are used to
define and quantify the spatial distribution of precipitation. In principle, a suitable synthesis of all
three sources of data (automatic gauges, radar, and
satellite) can be expected to provide sufficiently
accurate areal precipitation estimates on an operational basis for a wide range of precipitation data
users.

2.5.1.1 Rain guauge (non-recording type)

Rain gauges and rainfall recorders are the com-
monest instruments used to measure rainfall at individual points. A rain gauge measures the depth
of accumulated water on a flat surface, which is
assumed to be the same in the area around the rain
gauge. The unit of precipitation is depth, usually

expressed in millimetres (volume/area) collected
over a horizontal surface in a given time interval.
Therefore, for precipitation:

$$1 \; mm = 1 \; l/m^2 = 10 \; m^3/ha \qquad (2.1)$$

Rain gauges have different capacities, depending
on the desired time interval of readings (daily
or monthly) and the expected rainfall. A rain
gauge shows the total quantity of rainfall or snow
collected between two intervals (usually a day, or a
month, if the rain gauge is in a relatively inaccessible place). The components (Fig. 2.3) of a standard
non-recording rain gauge are: (i) a sampling orifice
made of brass; (ii) a slowdown funnel that forms
the top part of the gauge and an inner part that
is made of copper; and (iii) a graduated glass that
gives a direct reading of the depth of rain that falls
on the area. The requirements of a good rain gauge
are: (a) the sampling orifice should have a sharp
edge; (b) the rim should be exactly vertical; and (c)
entry of water splashed from the outside area into
the collection chamber should be prevented. The
narrow neck prevents evaporation between two
measuring intervals. For long reading intervals,
evaporation loss can be significant and may be
reduced by using an oil film, keeping the exposed
area minimal, reducing internal ventilation, and

maintaining low temperature. Choosing a site where the collector should be placed is very important, and must be representative for the area and should have no obstacles in the neighbourhood that may affect collection of rainwater into the gauge. The receiving surface of a rain gauge must be horizontal and placed at approximately 1.5 m above ground level to minimize splash collection.

2.5.1.2 Recording rain gauges

Recording gauges are equipped with paper charts and/or data logger equipment. The need for continuous recording of precipitation arose from the need to know, not just how much rain has fallen, but also when it fell and over what interval, to enable determination of its intensity.

2.5.1.2.1 Siphon rain recorder (float recording gauges)

In this system, rain is stored in a cylindrical collecting chamber containing a float. Movements of the float are recorded on a chart recorder by a pen trace. When there is no rain, the pen draws a continuous horizontal line on the chart. During rainfall the float rises and the pen traces upward slopes on the chart, depending upon the intensity of rainfall. When the chamber is full, the pen arm lifts off the top of the chart and the rising float releases a trigger, interrupting the balance of the chamber, which tips over and activates the siphon. A counter-weight brings the empty chamber back into an upright position and the pen returns to the bottom of the chart.

2.5.1.2.2 Tipping-bucket rain gauge

Rain caught in the collector flows down a funnel into a two-compartment bucket of fixed capacity. When the bucket fills, it tips to empty out and a twin adjoining bucket begins to fill. As the bucket is tipped, it activates an electrical circuit, and the ensuing pulse is recorded by an electronic counter. In many stations, particularly in remote locations, measurements are recorded on magnetic tape or solid-state event recorders and the cassette is usually changed at monthly intervals.

2.5.1.2.3 Optical precipitation gauge

The optical precipitation gauge can measure both rain and snow. It is mounted on a small pole and sends a beam of infrared light from one end to a detector fixed at the other end. When rainfall/snowfall occurs, the light beam is interrupted. The precipitation rate is measured by recording how often the beam is interrupted. The precipitation rate is used to calculate the total amount of rainfall/snowfall in a given period. These gauges can record directly onto a data storage device that can be uploaded onto a computer.

Errors in precipitation measurement arise as gauges tend to give lower estimates of rainfall due to: (i) evaporation between readings ($\sim -1\%$); loss due to adhesion of rainwater on the material of the gauge ($\sim -0.5\%$); inclination of the gauge ($\sim -0.5\%$); and drifting away of raindrops/snowflakes due to wind (-5 to -8%). Splash contributes about $+1\%$ to the measurement error.

2.5.1.3 Measurement of snowfall and snow cover

Snowfall is the depth of freshly fallen snow deposited over a specified period (generally 24 *hours*). This, however, does not include deposition of drifting or windblown snow. For the purpose of depth measurement, the term snowfall should also include ice pellets, glaze, hail, and sheet ice formed directly or indirectly from precipitation. Snow depth usually includes the total depth of snow on the ground at the time of observation. The water equivalent of a snow cover is the depth of the water column that would be obtained by melting of the entire collected snow.

Direct measurements of the depth of fresh snow cover or snow accumulated on the ground are made with a snow ruler or similar graduated rod, which is pushed through the snow and reaches the ground surface. The standard method of measuring the water equivalent of a snow cover is by gravimetric measurement using a snow tube to obtain a sample core. Cylindrical samples of fresh snow may be taken with a suitable snow sampler and then either weighed or melted. Details of the available instruments and sampling techniques are described in WMO (1994).

Snow gauges measure snowfall water equivalent directly. Essentially, any of the non-recording precipitation gauges can be used to measure the water equivalent of solid precipitation. Snow collected in these types of gauges should be either weighed or melted immediately after each observation. The recording-weighing gauge will collect solid forms of precipitation as well as liquid rain, and measure the water equivalent.

2.5.1.3.1 Radioisotopic snow gauges

Nuclear gauges measure the total water equivalent of the snow cover and/or provide a density profile. They offer a non-destructive method of sampling and are adaptable to *in situ* recording and/or telemetry systems. Nearly all such systems operate on the principle that water, snow, or ice attenuates radiation. Gauges that are employed to measure total water content consist of a radiation detector and a source, either natural or artificial. One part (e.g. detector/source) of the system is located at the base of the snow pack and the other at a height greater than the maximum expected snow depth. As snow accumulates, the count rate decreases in proportion to the water equivalent of the snow pack. A portable gauge (Young 1976), which measures the density of the snow cover by backscattering rather than transmission of the gamma rays, offers a practical alternative to digging deep snow pits, while instrument portability enables assessment of areal variations of density and water equivalent.

2.5.1.3.2 Natural gamma radiation

The method of snow surveying by gamma-radiation is based on the attenuation by snow of gamma radiation emanating from natural radioactive elements present in the top layer of the soil. The greater the water equivalent of the snow, the more will be the attenuation of the radiation. Terrestrial gamma surveys involve a point measurement at a remote location, a series of point measurements, or a traverse over a selected region. The method can also be used on board an aircraft. The equipment includes a portable gamma-ray spectrometer that utilizes a small scintillation crystal to measure the rays in a wide spectrum encompassing three spectral windows (corresponding to potassium, uranium, and

thorium emissions). With this method, measurements of gamma levels are required at the selected point, or along the traverse, prior to the deposition of snow cover. In order to obtain absolute estimates of the snow water equivalent, it is necessary to correct the readings for soil moisture changes in the top 10 to 20 *cm* of the soil for variations in background radiation due to cosmic rays, instrumental drift, and washout of radon gas (a source of gamma radiation) in precipitation with subsequent build-up in the soil or snow. Also, in order to determine the relationship between spectrometer count rates and water equivalent, supplemental snow water equivalent measurements are initially required. Snow tube measurements are taken as the common reference standard.

2.5.1.4 Estimation of areal precipitation

The relationship between hydrologic processes and scale is one of the most complex issues in hydrology. Hydrologic variables are often measured at a point or on a plot scale, while land-use managers and hydrologic modellers are often interested in rainfall/runoff/evaporation/infiltration processes operating at the watershed/regional scales.

Estimating temporal and spatial variation of precipitation over a geographical area is essential when calculating water balance for indirect estimation of flow and the net recharge over the watershed area. It is also important for modelling environmental phenomena, for example, variation in amount of rainwater falling over a river basin. Extrapolating the point measurements to determine the total amount of water delivered to a particular catchment involves certain assumptions and a scheme of areally weighting rainfall through graphical procedures (e.g. *Thiessen Polygon Method*) or by using contours of equal rainfall (*isohyets*). The Thiessen method assigns an area, called a *Thiessen polygon*, to each gauging point in order to weigh the effects of non-uniform rainfall distribution. In this region, if we choose any point at random within the polygon, the point is closer to this particular gauge than to any other gauge. In effect, the precipitation is assumed to be constant, equal to the gauge value, throughout the area of a given polygon. To create

(a)

(b)

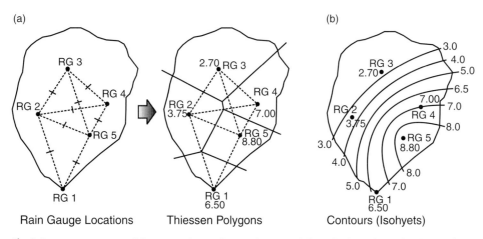

| Rain Gauge Locations | Thiessen Polygons | Contours (Isohyets) |

Fig. 2.4 (a) Construction of Thiessen polygons for weighting rainfall areally through graphical procedures. (b) Drawing of contours of equal rainfall (isohyets) is another method for weighting rainfall areally.

Thiessen polygons, adjoining lines are created between each of the rain gauges. Perpendicular bisectors are then drawn to form polygons around each gauge (Fig. 2.4a). The area bounded by the sides of the polygon/basin/catchment boundary is the assigned Thiessen polygon for the respective gauge within it, and its area ratio to the total basin gives the weight to be assigned to the gauge measurement while computing mean areal precipitation for the catchment. This method is used for non-orographic regions where rain gauges are not spaced uniformly.

The isohyetal method uses observed precipitation data as a basis for drawing contours of equal precipitation (isohyets), and the average precipitation of adjacent isohyets is weighted by the area between the two isohyets. The method is suitable for large areas, especially those where orographic effects in precipitation may be present.

Although somewhat cumbersome, Thiessen and isohyetal methods are basically subjective; the final result depends on the skill of the analyst in terms of draftsmanship, knowledge of the terrain, as well as the storm characteristics. Computer programs for these methods are also available from Diskin (1970) for the Thiessen method and Kwan *et al.* (1968) for the isohyetal method.

Weather radars, even though far from giving an accurate measurement, have become increasingly important tools for estimating the spatial distribution of rainfall. These provide detailed information

on temporal and spatial distribution of rainfall and can be particularly valuable for spells of heavy rainfall. RADAR, an acronym for Radio Detection and Ranging, transmits a focused microwave signal designed to detect precipitation-sized particles in the atmosphere (rain, snow, hail, etc). Electromagnetic waves transmitted by a radar travel in a narrow beam, through the atmosphere, at the speed of light. The radar antenna directs the beam around the horizon as well as up and down at various angles until most of the sky (within a given radius around the radar) has been scanned. After radar sends out a signal, it receives a return signal, or echo, which occurs when the transmitted signal is reflected by objects (raindrops, ice, snow, trees, buildings, etc.) along its path. Part of the reflected signal is received by the radar antenna. The amplitude of the reflected signal is a measure of reflectivity that can be correlated with intensity of precipitation. The direction in which the antenna is pointed determines direction of the target; time interval between transmission of a signal and reception of an echo determines distance to a target. Modern weather radars can also evaluate the returned signal to detect target motion towards or away from the radar using the Doppler effect.

Satellite platforms also enable estimation of rainfall intensity and its spatial and temporal distribution over large areas. Various satellites have on board sensors for indirect measurement of precipitation or snow. A Geostationary Operational

Environmental Satellite (GOES) network has sensors to detect radiation in the infrared and visible spectral ranges. GOES satellite imagery is also used to estimate rainfall during thunderstorms and hurricanes for flash flood warnings, as well as to estimate snowfall accumulation and areal extent of snow cover. INSAT and METSAT series of satellites have provided services over India similar to GOES. The SSM/I (Special Sensor Microwave Imager) has an onboard passive microwave imager on the defence satellite network. TRMM (Tropical Rainfall Measurement Mission) has both an active radar and passive microwave imager.

2.5.2 Measurement of evaporation and evapotranspiration

Evaporation (E) involves the change of water from its liquid (or solid) phase to its vapour phase. It is exclusively a surface phenomenon and should not be confused with boiling. For a liquid to boil, its vapour pressure must equal the ambient atmospheric pressure, whereas for evaporation to occur, this is not the case. The two main factors influencing evaporation from the Earth's surface area are: (i) supply of energy (from solar radiation); and (ii) transport of vapour away from the surface, for example by surface winds.

Transpiration (T) involves transfer of water to the atmosphere in the vapour phase from plant leaves (through stomatal openings). It is affected by plant physiology and environmental factors, such as: (i) type of vegetation; (ii) stage and growth of plants; (iii) soil type and its moisture content; and (iv) climate (warm/cold) and atmospheric conditions (relative humidity, temperature, wind).

Evapotranspiration (ET) involves transfer of water in the vapour phase from the land surface – a combination of evaporation from open water surfaces (e.g. lakes, rivers, puddles) and transpiration by plants. In practice, the terms E and ET are often used synonymously – being the total evaporation flux from the land surface. Therefore, one must use the term evaporation appropriate to the context, to determine what it means in the specific case (i.e. whether it is just from an open water surface or from the entire land surface).

Potential evaporation (PE) is the maximum amount of water that can evaporate from an open water surface under a given climatic regime, with an unlimited supply of water available for evaporation. *Potential evapotranspiration* (PET) is the ET that would occur from a well vegetated surface when moisture supply is not a limiting factor (often calculated as the PE). *Actual evapotranspiration* (AET) is lower than its potential value, as the soil progressively dries up.

The change from the liquid to the vapour phase requires considerable energy. This energy is provided by solar insolation, which supplies the latent heat of evaporation required for the phase change and energizes water molecules to leave the water surface and escape into air. Thus, phase change is a diffusive process governed by Fick's Law of Diffusion:

$$F_w = -D \frac{\Delta W}{\Delta x}. \tag{2.2}$$

Accordingly, a the one-dimensional case, the flux density F_w [$ML^{-2}T^{-1}$] of molecules responds to a concentration gradient, indicating change in the concentration ΔW [ML^{-3}] of a given species with distance Δx [L]. D, in this case, is the mass transport diffusion coefficient for water.

Three major factors governing evapotranspiration at a given location are: (i) input solar energy (which is dependent on latitude and season); (ii) water availability (moisture content of soils, presence of open water surface); and (iii) turbulent transport (dependent upon wind speed, degree of sheltering).

Evaporation from an open water surface, involving a direct phase change from liquid to vapour with abundant availability of water, is the simplest system to understand and study as it is easier to measure the change in water volume. This is applicable to oceans and seas, lakes, reservoirs, ponds, rivers, streams, detention storage (on roads, puddles, roof tops, gutters), and water intercepted on vegetation leaves. Evaporation from an open water surface is affected by the exposed surface area, depth, and temperature of the water body. Therefore, deeper, larger, and colder water bodies will have lower evaporation rates as compared to shallow, small, warm waters. Well protected bodies with low fetch (upwind area) and wind speeds also have lower evaporation rates. Higher concentration of dissolved constituents (salts) also reduce evaporation. Evaporation from an open water surface can be measured by employing an evaporation pan with adjustments appropriate to a given site.

Evaporation from bare soils and other Earth surface materials is complex because: (i) soil water content in a soil profile varies with depth; (ii) soil properties modify vapour transport through the soil matrix; (iii) interaction with the water table could be significant and must be accounted for; and (iv) surface conditions (roughness, portion under shade, albedo) are important. This is applicable to evaporation from natural soils as well as agricultural soils and porous geological materials. Evaporation from soils is considered to occur in two stages. In Stage I, for soil surface at or near saturation, evaporation is controlled by the heat input and turbulent transport (winds) at the surface and occurs around its maximum rate. In Stage II, with the drying out of the upper soil surface, limitation on the availability of water occurs and transport of water vapour through soil becomes crucial. This stage is also known as the soil-controlled or falling evaporation stage.

Transpiration is a complex process to visualize and measure and is influenced by (i) atmospheric conditions such as relative humidity, temperature, CO_2 concentration in the air and wind speed; and (ii) type of vegetation cover (broad leaved/coniferous trees, grasses, shrubs, etc.). In the process of transpiration, water is extracted from the entire root zone, which can extend vertically and laterally in the vicinity of plants and is limited by energy (solar radiation) for photosynthesis, water availability (soil moisture for uptake by plants), and also by turbulent transport (wind speed near leaf surfaces). Depending on the parameter of interest, various methods exist for measuring evapotranspiration rates. The methods vary from direct to indirect measurement and involve mass and/or energy balance; theory of turbulence is employed for interpretation.

2.5.2.1 Pan evaporation

Perhaps the simplest way of measuring evaporation is with an evaporation pan. Different sections (square, circular, etc.) can be used to construct an evaporation pan and depending on the situation these can be used in different positions (on ground surfaces, above ground, partially buried, floating, etc.). But in many countries the Class A Evaporation Pan has been adopted for many years as a standard to measure evaporation. Pan evaporation is a method that integrates the effects of several climatic elements: temperature, humidity, insolation and wind. Evaporation is highest on hot, windy, and dry days, and is greatly reduced when the air is cool, calm, and humid. The Class A evaporation pan is cylindrical, with a diameter of 120.65 *cm* and a depth of 25.4 *cm*. The pan should rest securely on a properly levelled wooden base. Evaporation is measured daily as the depth of water column that evaporates from the pan. A bird guard constructed of wire mesh is used to prevent access to the water by birds or animals. Measurements are taken with a fixed point gauge and measuring tube. The fixed point gauge consists of a pointed rod placed vertically at the centre of a cylindrical stilling well. The measuring tube has a cross-sectional area of one hundredth the area of the evaporation pan, and is subdivided into 20 equal divisions, each of which is equivalent to 0.2 *mm* of water in the pan. At the start of the measurement cycle, the water level in the pan is at the reference point. At the end of 24 *hours*, normally at 9 a.m. every day, enough water is added using the measuring tube, to fill the pan again exactly to the apex of the fixed point gauge. The deficit is the evaporation that may have occurred since the last measurement after accounting for any precipitation (*P*) input. The recording process can be automated by adding a data logger and ultrasonic depth sensor. Historical records of daily pan evaporation are available from the National Climatic Data Centre (NCDC) for US Weather Bureau Class A land pans.

However, hydrologists and engineers are not really interested in what evaporates from a pan; instead, what is of practical use is the regional evaporation from the land surface or the evaporation from a nearby lake. Unfortunately, pan evaporation is often a poor indicator of these quantities, due in part to pan boundary effects and limited heat storage. Evaporation from an *open water surface*

(E) is usually estimated from the pan evaporation (E_p) as:

$$E = K E_p \qquad (2.3)$$

where K is the pan coefficient. Similar expressions are also used in practice to estimate potential evapotranspiration (*PET*) from pan data.

2.5.2.2 Percolation gauges

Percolation gauges measure both evaporation and transpiration, generally referred to as evapotranspiration (*ET*) from vegetated surfaces, in contrast to tanks and pans used for measuring E. There are several designs and, in general, these are regarded as research tools. A cylindrical or rectangular tank, about 1 *m* deep, is set into the ground and filled with a representative soil sample supporting the vegetated surface (Shaw 1988). A pipe from the bottom of the tank leads surplus percolating water (Drainage, D) into a collecting container. Evapotranspiration is given by:

$$ET = P - D \qquad (2.4)$$

Percolation gauges do not take into account changes in soil moisture storage. The measurements should be taken over a period of time when the difference in the soil moisture storage would be small, so are neglected in Eqn 2.4.

2.5.2.3 Lysimeter

A lysimeter is a device that isolates a volume of soil or earth between the soil surface and a given depth and includes a sampling system for percolating water at its bottom and a device for weighing the lysimeter underneath. This improves the accuracy of *ET* measurements of percolation gauges by accounting for changes in soil moisture storage. Although the purpose of a lysimeter, as originally conceived, is to determine transport and loss of solutes by leaching – a name derived from the Greek words 'lusis' = solution and 'metron' = measure – it is also used for determining actual evapotranspiration and groundwater recharge and, therefore, for estimating water balance. Weighable lysimeters

provide a good estimate of evapotranspiration and are employed for this reason:

$$ET = P - D \pm \Delta(Weight) \qquad (2.5)$$

All units of measurements are with reference to the area of the lysimeter orifice at ground level. Lysimeters have been used extensively in the past to provide baseline information for development, calibration, and validation of various methods for estimation of evapotranspiration.

2.5.2.4 Other methods

Soil water depletion is used to obtain a direct estimate of actual evapotranspiration (*AET*) from a soil. In this method, the soil column is sub-divided into several layers and the soil moisture content of each layer is independently estimated at each measurement time, giving the soil water content $(\theta_1 - \theta_2)_i$ in each layer i from Time 1 to Time 2, that is, during the interval Δt. The *AET* can be estimated by summing up the change in soil moisture over time in all the layers of the soil column and accounting for infiltration (I) and drainage (D), as below:

$$AET = \frac{\Delta SM}{\Delta t}$$
$$= \frac{\sum_{i=1}^{n}(\theta_1 - \theta_2)_i \, \Delta S_i + I - D}{\Delta t} \qquad (2.6)$$

For large regions (e.g. a watershed) and for long time periods, the *water balance method* is usually applied. This method quantifies *ET* by measuring changes in water volumes. Precipitation and discharge measurements over a watershed can be used to estimate *AET*:

$$AET = P - Q \pm \Delta G \pm \Delta \theta \qquad (2.7)$$

where P is the precipitation over the watershed; Q the discharge of the stream draining the watershed; and ΔG and $\Delta \theta$ refer to changes in groundwater and soil water content, respectively. For long-term estimates, assuming no changes in storage volume of either G or θ:

$$AET = P - Q \qquad (2.8)$$

Several other methods exist for estimating the rates of *PE* and *AET*, without taking into account changes in water storage volumes. Some methods are empirical (developed from experiments at a particular site) and applied to other locations. Some other methods are physically-based and attempt to quantify the limits on the estimated evaporation. There are some methods that are purely theoretical in nature.

One of the widely used empirical methods is the *Thornthwaite method* that predicts the monthly potential evapotranspiration (*PE*; in *mm*), based on mean monthly air temperature (*T*) in °*C*:

$$PE = 16 \left[\frac{10\,T}{I} \right]^a \qquad (2.9)$$

where the Heat Index (*I*) is computed from the temperature (*T*) for all 12 months in a year (*j* = 1 to 12), using:

$$I = \sum_{1}^{12} \left[\frac{T_j}{5} \right]^{1.514} \qquad (2.10)$$

The location-specific parameter, *a*, is computed from heat index (*I*), using Eqn 2.11:

$$a = 6.75 \times 10^{-7}\, I^{3*} - 7.71 \times 10^{-5}\, I^2$$
$$+ 1.792 \times 10^{-2} + 0.49239 \qquad (2.11)$$

More complex evapotranspiration equations/methods exist and are also in use. Some of these are: the Jensen-Haise Radiation method; the Hargreaves equation; the Penman equation; the Penman-Monteith equation (Combination method); the Shuttleworth-Wallace equation; and the Complementary relationship between potential evaporation at local scale and actual evaporation at regional scale. Most of these equations require data on temperature, radiation, and soil conditions. For details on these, reference is made to Ward and Trimble (2003).

2.5.3 Infiltration and its measurement

As precipitation occurs, some of it is intercepted by vegetation and other surfaces before reaching the ground. This process is called interception and the precipitation fraction thus withheld from reaching the ground surface is called interception storage. Once the precipitation reaches the ground surface, starts infiltrating into the soil. Infiltration can be defined as the downward flux of water from the soil surface into the soil profile via pore spaces. The infiltration rate is a measure of how rapidly water enters the soil. Water entering too slowly may lead to ponding on level fields or may cause erosion by surface runoff on sloping fields. Therefore, measuring infiltration of water into the soil is important for estimating and ensuring the efficiency of irrigation and drainage, optimizing the availability of water for plants, improving the yield of crops, and minimizing soil erosion.

The infiltration rate is expressed as amount of water penetrating the soil per unit surface area per unit time [LT^{-1}]. This rate changes with time as the soil pore spaces are filled with water. If the precipitation rate exceeds the infiltration rate, runoff usually occurs unless there is some physical barrier. It is related to the saturated hydraulic conductivity of the near-surface soil.

The infiltration process becomes more complex, both with increasing non-uniformity of soil and increasing rainfall rate. Infiltration through a soil profile is governed by two major opposing forces, namely, capillary force and gravity. The *capillary force* in pore spaces is due to adhesion whereas cohesion tends to hold water tightly in the pore spaces and is also referred to as suction or matric tension force or pressure. Surface tension can be strong within pore spaces due to their small size. A balance of forces exists in pores between tension and gravity (Fig. 2.5a). Pore radius (*r*) controls the capillary rise of the water in a pore. Large pores, with less tension, empty out first during drying whereas small pores, with higher tension, fill first during soil wetting. The difference in response between wetting and drying is called *soil hysteresis*. The variation of infiltration rate depending on initial soil wetness condition, as shown in (Fig. 2.5b), is due to dependence of hydraulic conductivity on tension or suction force. *Hydraulic conductivity* is the ability of a soil to transmit water under the unit hydraulic gradient. It is also called *soil permeability*. Soils with small pores having high tension or suction forces tend to have low conductivity. Soils with large pores having low tension or suction forces tend to have high conductivity. A relationship also exists between water content and suction pressure for each soil type (Fig. 2.6a). Below the field capacity, the plant available water (*PAW*)

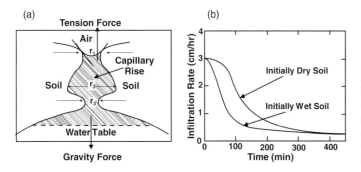

Fig. 2.5 (a) A force balance exists in pores between tension and gravity. Pore radius (r) controls the height of the water in the pore. (b) Infiltration varies with time (reckoned since the commencement of rainfall) and with antecedent soil water content conditions. Redrawn from Vivoni (2005a).

decreases rapidly and becomes almost negligible as wilting point is approached due to increasing tension/suction force (Fig. 2.6b).

Infiltration varies both in space and time. It is affected by several factors that include: (i) water input (from rainfall, snowmelt, irrigation, and ponding); (ii) soil profile properties (porosity, bulk density, conductivity); (iii) antecedent soil water content and its depth profile; (iv) soil surface topography and roughness; and (v) soil freeze and thaw conditions. Infiltration process is characterized by a general decrease in infiltration rate as a function of

time and a progressive downward movement and diffusion of an infiltration front into a soil profile.

2.5.3.1 Ring infiltrometer

A *ring infiltrometer* is commonly used for *in-situ* measurement of infiltration characteristics. In a ring infiltrometer test, a constant depth of water should be maintained in the rings and the volume of water infiltrated with time must be recorded during the test. The infiltrometer employs one or two concentric rings, and accordingly it is called a single

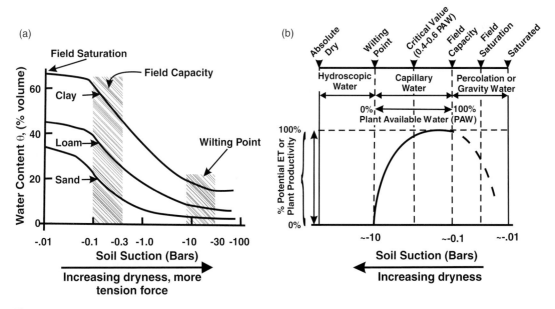

Fig. 2.6 (a) Decrease in water content increases the suction pressure or tension and a different relation exists for each soil type. (b) Below the field capacity of the soil the plant available water (*PAW*) decreases rapidly and becomes almost zero as wilting point is approached. Redrawn from Vivoni (2005a).

or double ring infiltrometer. Water poured into a single ring flows downwards and also laterally in the soil. The water flowing between the two rings can also flow both downwards and laterally in the soil. But across the inner ring, water flows both inwards and outwards. If the height of the water column in the two rings is the same, the lateral flow of water from the outer ring inwards will balance any lateral flow of water from the inner ring outwards. Because of this balancing of lateral flow, the water level drop in the inner ring is due solely to the downward flow of the water. Several researchers (Bouwer 1963; Burgy and Luthin 1956; Tricker 1978) have studied the effects of ring diameter and depth of water ponded in the rings on the reliability of infiltration measurement.

2.5.3.2 Tension infiltrometer

A *tension infiltrometer* measures hydraulic properties of unsaturated soils. Water held under tension infiltrates into a dry soil through a highly permeable nylon membrane. The time dependent infiltration rate is used to calculate unsaturated hydraulic conductivity and related hydraulic properties.

2.5.3.3 Empirical and physical equations for estimating infiltration

In watershed studies for hydrologic modelling, empirical as well as physical equations enable estimation of infiltration as a function of time for a given set of soil properties. Empirical models rely on regression equations based on large datasets. In general, empirical equations need to be applied with care, as these are not valid for conditions that have not been envisaged or taken into account while formulating these models. These are simple to apply and can provide reasonable estimates and hence are popular. Use of these equations enables determination of infiltration with limited measurements. Some of the commonly used equations are the Green-Ampt equation, the Richards equation, and the Horton equation.

2.5.3.3.1 Green-Ampt equation

The Green and Ampt (1911) method of infiltration estimation is a physical infiltration model based on

soil properties. It predicts infiltration under ponded condition of a soil. Theoretical conceptualization of the *Green-Ampt* process involves a sharp wetting front dividing saturated soil above from the prevailing conditions below. The infiltration rate, $f(t)$, proceeds at the rainfall rate, $f(0)$, when the surface is not ponded, and at the limiting potential rate, $f(t)$. Ponding occurs when the rainfall rate is higher than the hydraulic conductivity, K [LT^{-1}], of the wetted part of the soil profile, and rainfall depth and cumulative infiltration, $F(t)$, exceed available moisture storage. The hydraulic conductivity, K, is given by *Darcy's Law* according to which water flux, q [LT^{-1}], through a soil matrix is proportional to the change in potentiometric or piezometric head, h [L], per unit distance z [L], i.e.:

$$\text{Darcy's Law } q = -K \frac{\partial h}{\partial z} \quad \text{or}$$

$$K = -q \cdot \left(\frac{\partial h}{\partial z}\right)^{-1} \tag{2.12}$$

The *Darcy flux*, q, is positive upwards, while *infiltration flux*, f, is positive downwards, making $f = -q$. Typical conceptualization of soil moisture variation and the definitions of ponded depth and the wetted depth are shown in Fig. 2.7. Soil water content [L^3L^{-3}] is θ; L_p is ponding depth [L]; potentiometric or piezometric head, h, is the sum of the pressure head p and the elevation head z. At atmospheric pressure, the piezometric head $h = 0$, at a higher pressure than atmospheric pressure, h is positive, and below atmospheric pressure, h is negative, and is known as the matric suction head, $-\psi$.

The method involves an iterative process between the infiltration rate and the cumulative infiltration, using the equation:

$$f(t) = \frac{dF(t)}{dt} = K \cdot \left[\frac{\Psi_{wf} \Delta\theta}{F(t)} + 1\right] \tag{2.13}$$

where Fig. 2.7 defines the terms in a typical Green-Ampt profile as depth of ponding, L_p; depth to wetting front, $-L_s$; and negative matric suction at wetting front, $-\psi_{wf}$. The cumulative infiltration, $F(t)$, is obtained by integrating Eqn 2.13 and is given by Eqn 2.14. Wetting front depth, L_s, is equivalent to the cumulative infiltration length, $F(t)$, divided by the change in soil moisture, $\Delta\theta$,

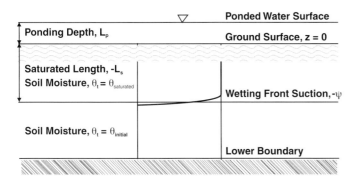

Fig. 2.7 Green-Ampt model of a soil profile with layers identified on the right-hand side, and soil moisture variables identified on the left-hand side of the diagram. The middle of the diagram shows the wetting front. Redrawn from Endreny (2007). © Tempus Publications.

computed by taking the difference between θ_{sat} and θ_{init}, which typically corresponds to moisture content at the wilting point or higher:

$$F(t) = Kt + \Psi_{wf}\Delta\theta \ln\left(1 + \frac{F(t)}{\Psi_{wf}\Delta\theta}\right) \qquad (2.14)$$

Green-Ampt infiltration parameters, such as K, ψ_{wf}, and θ, are available for different soil types (Chin 2000). These parameters for a subset of common soil types are presented in Table 2.1.

2.5.3.3.2 Richards equation

The *Richards equation* (Richards 1931) for vertical flow is:

$$\frac{\partial\theta}{\partial t} = \frac{\partial}{\partial z}\left(K(h)\frac{\partial h}{\partial z} + 1\right) \qquad (2.15)$$

where θ is the soil water content; h is the initial matric potential of the soil (in *cm*); and $K(h)$ is the hydraulic conductivity at the matric potential h. The Richards equation has been widely used in infiltration modelling. While solving the various infiltration equations, there are three types of upper boundary conditions that most real situations are simplified to: constant water content (or matric potential); constant influx; or a constant change of influx. These conditions are named as the first, second, and third boundary conditions. Numerical methods can solve the above soil water flow equation with the first, second, or third type of upper boundary conditions. Note that the Green-Ampt model assumes a first upper boundary condition with constant water content.

2.5.3.3.3 Philip equation

Philip (1957) suggested another solution to the Richards equation for ponded infiltration into a deep soil profile with uniform initial water content:

$$f(t) = \frac{1}{2}St^{-1/2} + K \qquad (2.16)$$

Table 2.1 Green-Ampt hydraulic conductivity, wetting front suction, and volumetric soil moisture parameter values for common soil texture types (Chin 2000). The change in volumetric soil moisture, $\Delta\theta$, is computed by taking the difference between θ_{sat} and θ_{init}, which is typically at wilting point or higher.

Soil Texture Class	Hydraulic Conductivity K_{sat} (*m h^{-1}*)	Wetting Front Suction Head ψ_{wf} (*m*)	Volumetric Soil Moisture at Saturation θ_{sat} (*m^3 m^{-3}*)	Volumetric Soil Moisture at Wilting Point θ_{wp} (*m^3 m^{-3}*)
Sandy Loam	0.011	0.110	0.453	0.085
Silt Loam	0.007	0.170	0.501	0.135
Loam	0.003	0.089	0.463	0.116
Clay Loam	0.001	0.210	0.464	0.187
Sandy Clay	0.001	0.240	0.479	0.251

where S is the soil sorptivity and K (h) is the hydraulic conductivity. The cumulative infiltration $F(t)$ over a period of time is obtained by integrating Eqn 2.16 as:

$$F(t) = S\,t^{1/2} + K\,t \qquad (2.17)$$

2.5.3.3.4 Horton equation

The Horton (1940) model is based on the observation that infiltration capacity declines rapidly during the early part of a storm and then tends towards an approximately constant value after a couple of hours for the remainder of the storm. Water that has already infiltrated fills the available soil storage spaces and reduces the capillary forces drawing water into the pores. The model gives an empirical formula involving three parameters to predict infiltration rate, $f(t)$, at time t after infiltration begins. Accordingly, infiltration starts at a constant rate, $f(0)$, and levels off to the rate, $f(c)$, when the soil reaches saturation level. The infiltration rate, $f(t)$, is then given by Eqn 2.18:

$$f(t) = f(c) + \big(f(0) - f(c)\big)\,e^{-kt} \qquad (2.18)$$

where 'k' is the decay constant specific to the soil. The other method of using Horton's equation is to find the total volume of infiltration 'F' after time 't', obtained by integration of Eqn 2.18 between the time interval 0 to t:

$$F(t) = f(c).t + \frac{(f(0) - f(c))}{k}\left(1 - e^{-kt}\right) \qquad (2.19)$$

There are several other infiltration models. These include the SCS Runoff Curve model developed by the Soil Conservation Service (Mockus 1972; SCS 1986). In this method, the effects of land use and treatment are taken into account. This method was empirically developed from studies of small agricultural watersheds (see Section 2.7.1.2).

2.6 Runoff processes and flow measurement

Runoff occurs when rain is not fully absorbed by the ground on which it falls, the excess flowing down the gradient. During this process, runoff carries topsoil, nutrients, and any other water-transportable material (e.g. petroleum, pesticides, or fertilizers) that it encounters. Eventually runoff water drains into a stream or river. All of the land surface from where runoff water flows into a particular river or stream is known as the watershed of the stream. Watershed runoff is the entire water transported out of a watershed. It includes: (i) overland flow on the land surface; (ii) interflow through the soil profile; (iii) groundwater effluent into rivers and springs; and (iv) direct precipitation onto channels. Runoff is generally beneficial, but under high rainfall conditions it can lead to flooding and may have significant deleterious impacts on urban, agricultural, and industrial areas.

Runoff occurs after filling up of any interception on vegetation and other surfaces and depression storages on the ground. The rate of rainfall exceeding infiltration rate into the ground is called the *infiltration excess overland flow*, or unsaturated overland flow (Hortonian mechanism). This occurs more commonly in arid/semi-arid regions or in paved areas, where rainfall intensities are high and the soil infiltration capacity is reduced because of surface sealing by particles of fine material, such as clay, which swell on wetting. The amount of runoff may be reduced in a number of ways during its flow: a small portion of it may evaporate; water may become temporarily stored in microtopographic depressions; and a portion of it may become run-on, which is the infiltration of runoff as it flows overland. The level of antecedent soil moisture is one of the factors that affects runoff until the time top soil becomes saturated. When the soil gets saturated and the depression storage is filled, and if rainfall continues further, the runoff generated is called *saturation excess overland flow* or saturated overland flow (Dunne mechanism). This occurs more commonly in regions near river valleys with high water tables and permeable soils. After water infiltrates the soil on an up-slope portion of a hill, it may flow laterally through the soil or aquifer, and exfiltrate (flow out of the soil) closer to a stream channel. This is called *subsurface return flow or interflow*. Low permeability layers between ground surface and the main water table often form a perched water table that contributes to the interflow. A schematic of important runoff generation mechanisms is shown in Fig. 2.8.

Infiltration-Excess Runoff

Subsurface Storm Runoff

Saturation-Excess Runoff

Perched Interflow Runoff

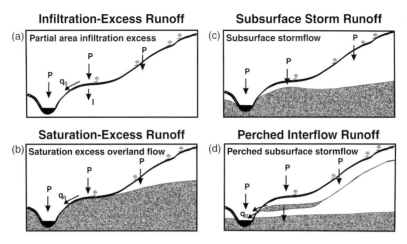

Fig. 2.8 Various stream runoff generation processes and their variation with local conditions. (a) Infiltration-excess runoff over largely unsaturated soil due to limited soil hydraulic conductivity. (b) Saturation-excess runoff when the soil becomes saturated. (c) Subsurface runoff due to formation of groundwater mound. (d) Interflow or throughflow due to lateral movement through unsaturated matrix or macro pores or perched water table. After http://hydrology.neng.usu.edu/RRP/usedata/4/87/ch2.pdf (accessed 7 March 2010).

The stream hydrograph is a primary record in a river network draining a basin. Hydrologists have long observed stream discharge and attempted to relate it to rainfall characteristics. A time-series of precipitation values in a watershed is called a *Rainfall Hyetograph* and a similar time-series of stream discharge from the watershed is called a *Stream Flow Hydrograph* (Fig. 2.9). In a hyetograph, the storm duration ($t_{we} - t_{w0}$) represents the total du-

ration of the rain storm from the time of the beginning (t_{w0}) to the time of the end (t_{we}) of the rainfall input. The centroid (centre of mass) of the hyetograph (t_{wc}) corresponds to the time of 50% of the rainfall input. In the hydrograph, the discharge begins at time t_{q0} and ends at time t_{qe}. The total duration of discharge or time base is $T_b = t_{qe} - t_{q0}$. The peak of the hydrograph occurs at time t_{pk} and the centroid of the hydrograph occurs at time t_{qc}.

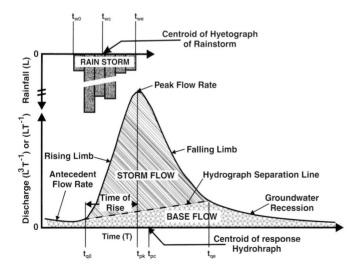

Fig. 2.9 Definition sketch for describing hyetograph of a rainstorm and response hydrograph of a watershed. Redrawn from Vivoni (2005b).

Each of the above time intervals provides an indication of the watershed characteristics, rainfall input duration, and rainfall distribution. *Time of Concentration (T_C)* is defined as the time it takes for water to travel from the hydraulically farthest part of the contributing area in the basin to the outlet, after the soil has become fully saturated and minor depression storages have been filled. T_c is computed by summing all the travel times for consecutive components of the drainage conveyance system. The hypothetical time at which the input and output to the watershed become equal and all portions of the basin contribute to runoff, is defined as *Time to Equilibrium (T_{eq})*. Equilibrium runoff can only occur in small basins with a short time of concentration and so occurs rarely in nature.

Stream flow is perhaps the most useful component of the global/local water cycle from a practical point of view. Therefore, to properly understand the runoff generation process in a given watershed and also to ensure an equitable distribution of available water between the various competing demands such as power generation, irrigation, municipal, industrial, recreation, aesthetic, and for fish and wildlife, one needs realistic estimates of river flow. Therefore, basic concepts related to water flow and measuring instruments/methods are now described.

2.6.1 Flow measurements – basic concepts

There are a number of methods to measure the amount of water flowing in a stream or a canal. The objective is to estimate Q [L^3T^{-1}], the volume of water flowing per unit time through any given cross-section of a river channel. This is usually calculated using the area-velocity method:

$$Q = w.b.v = A.v \qquad (2.20)$$

where w is the overall width, [L]; b is the mean depth, [L], of the cross-section; A is the average cross-sectional area; and v is the flow velocity [LT^{-1}] at the measuring location. The choice of method of measurement depends on several factors, namely: (i) desired accuracy of the result; (ii) magnitude of flow; and (iii) available equipment.

A simple method to measure approximate flow velocity in small streams involves measurement of the distance a floating object – a leaf, a log, or a bottle – travels in a given time interval. Assuming that this measurement gives an average velocity of the flowing water, and the width and depth of the flow can be independently estimated or measured, the flow rate is estimated by multiplying velocity with cross-sectional area of the stream through which the flow occurs. However, when high accuracy is desired, flumes and weirs are commonly used. Most devices measure flow indirectly. Flow measuring devices are commonly classified into ones that sense or measure velocity and those that measure pressure or head. The head (or velocity) is measured from which discharge is obtained by using equations or is read directly from previously calibrated charts or tables.

2.6.1.1 Flumes and weirs

The word flume acquires meaning, depending upon the context. The wooden chutes, or troughs, for carrying water to power the wheel of a mill and/or to transport logs down steep mountain sides are called artificial channel flumes. When used for measuring open-channel flow, flumes are designed sections that cause water flow to accelerate either by the converging sidewalls or by the raised bottom, or a combination of both. Flumes range in size from very small – a few *cm* wide – to large structures over 15 *m* wide that are installed in ditches, laterals, and in large canals to measure flow. These flumes cover a discharge range of <1 *litre s^{-1}* to over 100 *m^3 s^{-1}*, although no particular upper limit exists. A weir is also an overflow structure built perpendicular to an open channel flow direction in such a way as to make all the water to pass through a specially shaped opening or notch (Fig. 2.10). The weir results in an increase in the water level, or head, which is measured on the upstream side of the structure. In a nutshell, measuring flumes and weirs are robust, stable structures with well defined cross-sectional area. The flow rate over a flume or a weir is a function of the head at specific points along it.

Three types of weir notches, or shapes of overflow section, are commonly used: (i) V-notch or triangular; (ii) rectangular contracted; and (iii) Cipolletti (Fig. 2.10). If the notch plate is mounted

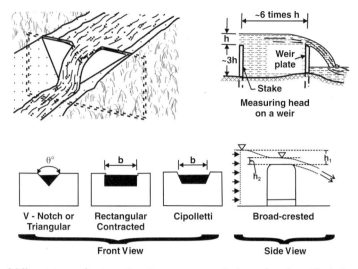

Fig. 2.10 Schematic of different types of weir sections in common use. Redrawn from http://hydrology.neng.usu.edu/RRP/userdata/4/87/ch2.pdf (accessed 7 March 2010).

on the supporting bulkhead, such that the water does not come into contact with the downstream side of the weir plate or supporting bulkhead, but springs clear, the weir is a sharp-crested or thin-plate weir. Thin-plate weirs are often used in water supply, wastewater, and sewage systems. A weir, in the form of a relatively long raised channel control crest section, is a broad-crested structure. Broad-crested weirs can be observed in dam spillways where the broad edge is below the water surface across the entire width of the stream. A properly built and operated weir of a given shape has a unique depth of water for a given discharge at the measuring station in the upstream pool. The reliability of measurements is governed by construction and installation of the weir. If properly constructed and installed, weirs are one of the simplest and most accurate methods of measuring the channel flow.

A discharge measurement consists of measuring depth or head relative to the crest at a properly selected upstream location in the weir pool, and using a table or equation for the specific kind and size of weir to compute the discharge. Commonly, a staff gauge having a graduated scale, with the zero marked at the same elevation as the weir crest, is employed to measures the head. Putting staff gauges in stilling wells dampens wave disturbances when reading the head. Using Vernier

hook point gauges in stilling wells produces much greater accuracy than staff gauges. Some modern systems also use ultrasonic level transmitters and/or pressure transmitters.

The flow rate measurement in a rectangular weir is based on the Bernoulli Equation and can be expressed as:

$$Q = 2/3\,C_d[2\,g]^{1/2}\,b^{3/2} \qquad (2.21)$$

where Q = flow rate; h = head at the weir; b = width of the weir; g = gravity; and C_d = discharge constant for the weir that must be determined by calibration and analysis of the weir for known flow rates. For standard weirs, C_d is well defined or is a constant for measuring within a specified range of head.

For a triangular or V-notch weir, the flow rate can be expressed as:

$$Q = 8/15\,C_d[2\,g]^{1/2}\,\tan(\theta/2)h^{5/2} \qquad (2.22)$$

where θ = V-notch angle.

For a broad-crested weir, the flow rate can be expressed as:

$$Q = C_d h_2\,b[2\,g(h_1 - h_2)]^{1/2} \qquad (2.23)$$

2.6.1.2 Stage method

Another commonly used method to estimate surface water flow is to use a rating curve between water depth and flow rate. A rating curve correlates depth of flow to flow rate, based on measurement of both water depth and flow over a range of flows at the location of interest. Flow is estimated using a current velocity meter and measurement of the cross-sectional area of the stream. Appropriate locations for flow measurement sites need to be selected in a fairly straight reach of the stream, where the geometry is relatively stable and there is no backwater effect arising from a downstream dam/weir or any other conditions that can introduce errors. This is often done under a bridge or at a stream crossing. Another commonly used method of measuring the stage of a river is through the use of a stilling well located on its bank or on a bridge pier. The well is connected to the stream by several intake pipes, such that any change of water level in the stream is instantaneously communicated to the well.

Once a rating curve is constructed, only water depth measurement is required for flow estimation. Water depths are commonly measured manually using a staff gauge. Alternatively, float, pressure, or electronic depth measuring devices with an automatic recorder can be used for continuous automatic measurement of depth.

2.6.1.3 Index velocity method

The other approach, based on index velocity determination, uses a velocity meter, either magnetic or acoustic, to measure the index velocity of the flow at a given gauging station. This index velocity is used to calculate average velocity of the flow in the stream. A rating curve, similar to that used for stage-discharge relationship, is constructed to relate the indicated index velocity with stream discharge. These data can be used analogously to those obtained from the traditional stream gauging method.

2.6.1.4 Dilution stream flow gauging

Dilution gauging methods measure stream flow on the basis of the rate of dispersal of an introduced

tracer, monitoring the concentration of the tracer downstream. Tracers are substances that label a mass of water and move with it as part of the flow. The tracer dilution method is used in streams with highly *turbulent flow* conditions where conventional stream velocity measurements are not applicable. Chemical tracers, such as table salt ($NaCl$), that readily dissolve in water, can be monitored using an electrical conductivity (EC) meter or a specific ion-selective electrode. A fluorimeter can be used to monitor mixing of fluorescent dyes, such as Rhodamine WT. Alternatively, water samples can be collected at different time intervals for laboratory analysis.

The slug-injection method involves introduction of a tracer of known volume and concentration into a stream, as a one-shot pulse (or slug) at a selected location. The concentration of the tracer is monitored as a function of time at a downstream location, before and after the slug is introduced. The resulting concentration hydrograph represents the passage of the dispersed tracer slug at the measurement point, with stream discharge (Q) represented as the area under the curve and calculated by using the equation:

$$Q = \frac{V \cdot C}{\int_{t_1}^{t_2} (C_t - C_0)\, dt} \qquad (2.24)$$

where V is the volume of the tracer added and C is the tracer concentration in the introduced solution; C_0 is the background concentration of the tracer in the stream; C_t is the changing tracer concentration measured downstream as a function of time; and t_1 and t_2 are the initial and final time of measurement.

The constant-injection method, as the name suggests, relies on the tracer solution being continuously introduced into the stream at a constant rate. The tracer concentration at the downstream measurement point rises and eventually stabilizes to a constant level. The stream flow (Q) can be calculated from:

$$Q = \frac{(C - C_1)}{(C_1 - C_0)} \, Q_t \qquad (2.25)$$

where C_1 is the stabilized tracer concentration at the downstream measurement point, and Q_t is the tracer injection rate at time t.

For some additional discussion on injected chemical tracers, see Section 6.6.2.

2.6.1.5 The Manning formula

This is an empirical formula to estimate average cross-sectional velocity for open channel flow under the influence of gravity. It was developed by an Irish engineer, Robert Manning, almost a hundred years ago. The *Manning formula* is expressed as:

$$v = \frac{1}{n} R_b^{2/3} \cdot S^{1/2} \tag{2.26}$$

where v is the cross-sectional average velocity (ms^{-1}); n is the Manning coefficient of roughness; S is slope of pipe/channel (m/m); and R_b is the hydraulic radius (m) of the stream given by:

$$R_b = \frac{A}{P} \tag{2.27}$$

with A being the cross-sectional area of flow (m^2) and P being the wetted perimeter (m). It is to be noted that the hydraulic radius is not just half of the hydraulic diameter, despite a similarity in the names. It is a function of the shape of the pipe, channel, or river that carries the water. In wide rectangular channels, the hydraulic radius is approximated by the flow depth. The Manning's n can vary significantly, that is, for a concrete, finished channel, $n = 0.014$; for a clean, straight, uniform earth channel, $n = 0.02$; for an unmaintained channel with dense weeds, $n = 0.07$ (normal); and for a floodplain with dense willows, $n = 0.15$.

2.7 Rainfall-runoff analysis and modelling

This is concerned with the physical processes operating in a watershed, leading to the transformation of rainfall into stream runoff. This transformation is complex, being non-linear, scale dependent, and site specific, depending on the characteristics of the land surface, vegetation, and stream channel and man-made structures (dams, roads, etc.). Various factors affect the amount and type of runoff generated in a watershed. These are: (i) watershed size; (ii) channel network shape and orientation; (iii) topography; (iv) geological properties; and (v) soil and land-cover properties, including co-variation of these properties. A watershed can be treated as a system that acts as a transfer function between the rainfall input (the Hyetograph) and the corresponding runoff output (the Hydrograph), transforming the excess or effective rainfall input, P_{eff}, into the event storm runoff or stream flow output, Q_{eff}, as shown schematically in Fig. 2.11.

The runoff hydrograph provides a good indication of the processes operating in a basin. Perhaps the runoff hydrograph is more characteristic of a watershed than any other measurement. For example, the runoff hydrograph of a rainstorm from an urban watershed, with a large proportion

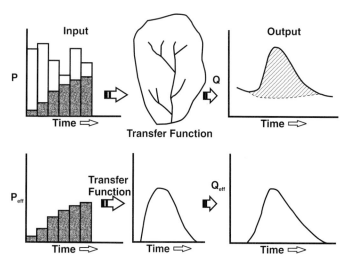

Fig. 2.11 The rainfall-runoff transformation can be thought of as a complex transfer function determined by basin characteristics converting excess or effective rainfall input into effective event or storm runoff output. Redrawn from Vivoni (2005b).

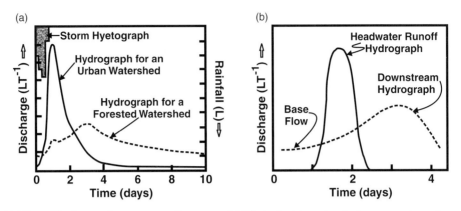

Fig. 2.12 (a) Effect of land-cover change on runoff response. (b) Effect of catchment scale (temporary storage) on runoff response. Redrawn from Vivoni (2005b).

of impermeable constructed surfaces inhibiting soil infiltration, reaches its peak quickly and also falls off rapidly for the same reason. In contrast, a forested watershed, due to significant infiltration of rain during the storm and its subsequent release through interflow, gives rise to gradually rising and falling hydrograph limbs (Fig. 2.12a). The response of small watersheds is determined primarily by hill slope runoff processes, while runoff in large catchments is determined by temporary storage in the surface depressions (ponds, lakes, etc.) and the stream network (determined by length, cross-sectional area, flow resistance, and surface–groundwater interactions). All these processes retard the response of the catchment, as its size increases (Fig. 2.12b).

2.7.1 Constructing a hydrograph

There are a number of ways to estimate the of water that runs off from a given surface. In addition to data collected in the field, one can also use computer models and simulations to estimate runoff. Depending on the objective of investigation, different models can be used. Typically, the variables of interest are: (i) total volume of runoff; (ii) peak runoff rate; and (iii) time to attain peak runoff.

2.7.1.1 Rational method

The peak discharge is the primary variable for designing storm water runoff structures, such as pipe systems, storm inlets and culverts, and small open channels. It is also used for some hydrologic planning, such as small surface water detention facilities in urban areas.

Ideally, one would like to have a 50-year flood record available at every site where a peak discharge estimate is needed. If such data is available, a frequency analysis of the flood record can be used to characterize the flood hazard at a given site. More often than not, flood discharge data of streams is rarely available. Therefore, it is necessary to use either a prediction method developed from flood frequency analyses of gauged data in the region, or an uncalibrated prediction equation designed for use at ungauged sites.

The most widely used uncalibrated equation is the Rational Method. Mathematically, the rational method relates the design peak discharge (Q, $m^3.s^{-1}$) to the basin drainage area (A, m^2), the rainfall intensity (i, $mm\ h^{-1}$), and the runoff coefficient (C):

$$Q = 0.0028\,C\,i\,A \qquad (2.28)$$

for the design return period of the flood and for duration equal to the 'time of concentration' of the watershed. In English Units with Q (ft^3/s), i (in/hr), and A (*acres*), the constant $= 1$, giving $Q = C\,i\,A$.

The commonly used equations are described in (SCS 1986), the June 1986 *Technical Release 55 – Urban Hydrology for Small Watersheds* (TR-55), the April 2002 *WinTR-55 – Small Watershed Hydrology* computer program, and *Technical Release 20 – Computer Program for Project Formulation: Hydrology* (TR-20) published by the NRCS, USA.

Use of the Rational Method is based on a set of assumptions: (i) rainfall intensity and duration is uniform over the basin area under study; (ii) storm duration is equal to or greater than the time of concentration of the watershed; and (iii) the runoff coefficient, C, is dependent upon physical characteristics of the watershed, e.g. soil type. *The Modified Rational Method* is an adaptation of the Rational Method. It uses the same input data and coefficients as the Rational Method along with a further assumption that, for the selected storm frequency, the duration of peak-producing rainfall is also the entire storm duration.

Variations of the general formula Eqn 2.29, to estimate time of concentration (T_c), have been developed for different land uses/geometries, (see Table 2.2):

$$T_c = k L^a n^b S^y i^{-z} \qquad (2.29)$$

where T_c = time of concentration; k = constant;

a, b, y, and z = exponents; L = length of channel/ditch from headwater to outlet, in ft.

T_c is generally associated with weather and geological parameters such as rainfall intensity, slope, and flow length. Most formulae have been developed in the united States and, therefore, use Imperial/English Units. Conversion to metric units requires adjustment of the constant term.

For urban areas, the time of concentration, T_c, is typically calculated by breaking the flow path into reaches of overland flow, T_0, and travel time, T_t, in the storm system, paved gutter, or drainage channel, so that $T_c = T_0 + T_t$. For non-urban areas, T_0 is typically much longer and T_t is the time of travel in natural swales and waterways. A minimum T_c of 10 and 5 min should be used for undeveloped and developed areas, respectively. The rational method uses runoff coefficients in the same way as the SCS curve number method uses CN (see Section 2.7.1.2) for estimating the runoff volume. These runoff coefficients have been

Table 2.2 Some commonly used formulae for calculation of time of concentration (T_c) for use with Rational and other methods of estimating peak discharge.

Method	Formula for T_c (*min*)	Remarks
Kirpich (1940)	$T_c = 0.00778 \, L^{0.77} \, S^{-0.385}$	Steep slope: 3–10%. Reduction factor applied for impervious area (0.4 for overland flow on concrete or asphalt surface)
Kerby (1959)	$T_c = 0.83 \left(N L \, S^{-0.5} \right)^{0.47}$ N = flow retardation factor	For small watersheds where overland flow is important. If the watershed length exceeds 1200 ft, a combination of Kerby's and the Kirpich equations may be appropriate.
Izzard (1946)	$T_c = \dfrac{41.025 \, (0.0007i + C) \, L^{0.33}}{S^{0.333} i^{0.667}}$ C = retardation coefficient i = rainfall intensity, *in./h*	Roadway and turf surfaces $i \times L$ <500 ft. Hydraulically derived formula; values of C range from 0.007 for very smooth pavement, to 0.012 for concrete pavement, to 0.06 for dense turf.
FAA (1970)	$T_c = \dfrac{1.8 \, (1.1 - C) \, L^{0.50}}{S^{0.333}}$	Developed from airfield drainage data assembled by US Corps of Engineers. Overland flow in urban basins
Aron and Egborge (1973)	$T_c = \dfrac{0.94 L^{0.6} n^{0.6}}{i^{0.4} S^{0.3}}$ n = Manning roughness coefficient	From kinematic wave analysis (L <300 ft)
SCS lag time (SCS 1986)	$T_c = \dfrac{1.67 L^{0.8} \left[(1000/CN) - 9 \right]^{0.7}}{1900 S^{0.5}}$ CN = Curve Number	Small urban basins <2000 *acres*
SCS Avg. vel. charts (SCS 1986)	$T_c = \dfrac{1}{60} \sum \dfrac{L}{V}$ V = average velocity in $ft \, s^{-1}$ for various surfaces	Developed as a sum of individual travel times. V can be calculated using Manning's equation.

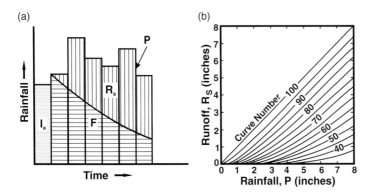

Fig. 2.13 (a) Conceptual diagram of the SCS-CN method showing partitioning of precipitation into Initial Abstraction, I_a; Retention or Infiltration, F; and Storm Runoff R_s. (b) Runoff predictions for different P and CN. Redrawn from Vivoni (2005c).

determined over the years, with primary focus on urban watershed applications.

2.7.1.2 SCS-CN model

One of the most widely used models is the US Soil Conservation Service Curve Number model – also known as *SCS-CN model* (Mockus 1972; SCS 1986). It predicts storm flow based on total amount of rainfall rather than its intensity, but takes into account the effects of soil properties, land cover, and its antecedent moisture content. It is based on the simple approach of partitioning the total precipitation into three components (Fig. 2.13):

(i) *Initial abstraction, I_a*, the amount of storage that must be satisfied before any flow can take place. This is poorly defined in terms of process, but is roughly equivalent to interception and the infiltration that occur before runoff can commence. Thus, $(P - I_a)$ is the 'excess precipitation' (after the initial abstraction) or the 'potential runoff';

(ii) *Retention, F*, the amount of rainfall after the initial abstraction is satisfied, but which does not contribute to the storm flow. This is roughly equivalent to the water that is infiltrated;

(iii) *Storm runoff, R_s*;

so that:

$$R_s = P - I_a - F \quad or \quad F = P - I_a - R_s \quad (2.30)$$

It is further assumed that the maximum amount of precipitation that a watershed can hold is S_{max}, the sum of initial abstraction, I_a, and the cumulative amount of infiltration over a heavy storm of long duration. During the storm, and particularly at the end of the storm, as more of the potential storage is exhausted (i.e. cumulative infiltration, F, converges to S_{max}), more of the 'excess rainfall' or 'potential runoff', $P - I_a$, will be converted to storm runoff. Thus:

$$\frac{R_s}{[P - I_a]} = \frac{F}{S_{max}} \quad (2.31)$$

Eliminating F from Eqn. 2.30 and Eqn 2.31:

$$R_s = \frac{[P - I_a]^2}{P - I_a + S_{max}} \quad (2.32)$$

Another generalized approximation made on the basis of measuring storm runoff in small, agricultural watersheds under 'normal conditions of antecedent wetness, gives $I_a = 0.2 \times S_{max}$. With this, Eqn 2.32 becomes:

$$R_s = \frac{[P - 0.2 \times S_{max}]^2}{[P - 0.8 \times S_{max}]} \quad (2.33)$$

for all values of $P > I_a$, and for all $P \leq I_a$; $R_s = 0$. This reduces the problem of predicting storm runoff depth to estimating a single maximum retention capacity of the watershed, S_{max}. An index of 'storm-runoff generation capacity', varying from 0 to 100 (i.e. in terms of percentage of rainfall), has been developed using Eqn 2.34. This index is called Curve Number (*CN*):

$$CN = \frac{1000}{S_{max} + 10} \quad (2.34)$$

The curve number index (CN) is related to back-calculated values of S_{max} (in *inches*) from measured storm hydrographs and Eqn 2.31 to relate to field measurements. *CN*s are evaluated for many watersheds and related to: (i) soil type (SCS soil types classified into soil hydrologic groups on the basis of their measured or estimated infiltration behaviour); (ii) vegetation cover and/or land-use practices; and (iii) antecedent soil-moisture content (SCS 1986). A spatially weighted average CN is computed for a watershed using:

$$Weighted \; CN = \frac{\sum Area_i \times CN_i}{\sum Area_i} \qquad (2.35)$$

Thus, the entire rainfall-runoff response for various soil-plant cover types is represented by a single index, the Curve Number (*CN*). This is used to estimate S_{max}, from the relation:

$$S_{max} \; (in \; inches) = \frac{1000}{CN} - 10, \quad or$$

$$S_{max} \; (in \; cm) = \frac{2540}{CN} - 25.4 \qquad (2.36)$$

Estimation of R_s for a given value of P is made using either Eqn 2.33 or Fig. 2.13b. A higher value of curve number indicates response from a watershed with a fairly uniform soil cover with low infiltration capacity. A lower curve number indicates response expected from a watershed with a permeable soil, with a relatively high spatial variability of infiltration capacity. Some of the limitations of the SCS-CN method are: (i) no consideration is given to rainstorm characteristics, i.e. – its intensity and/or duration; and (ii) no specification is given on the watershed for which the method is applicable, except that empirical relations are established for 'small' watersheds. With the use of computers, it is relatively easy to separate a watershed into sub-watersheds and to perform the runoff calculation for each unit separately, and then combine the individual runoff hydrographs.

When tested against measured storm runoff volumes, the method is highly inaccurate. But it has been widely adopted for runoff prediction and is acceptable to regulatory agencies and professional bodies, possibly because: (i) it is simple to use; (ii) the required data is available in SCS county soil maps in printed as well as digital forms, at least

in the United States; (iii) the method is included in handbooks and computer programs are readily available; (iv) it appears to give 'reasonable' results in the sense that big storms yield a lot of runoff, (fine-grained, wet soils, with thin vegetation covers yield more storm runoff in small watersheds compared to sandy soils under forest cover, etc); and (v) non-availability of any other better method. The method is already 'in-built' in various larger 'computer models', such as HEC-HMS. The task for a watershed analyst or regulator is only to decide how to sensibly interpret and use the results.

2.7.1.2.1 Computation of travel time and time of concentration

Any of the methods from Table 2.2 with coefficients from literature, appropriate for weather and surface/channel parameters such as rainfall intensity, slope, and flow length (SCS 1986), are then used to estimate T_c, as the sum of the overland flow time (T_0) and the travel time (T_t) through the hydrological conveyance system of a watershed. For example, the Kerby (1959) method is often used for overland flow (T_0) and the Kirpich (1940) method for calculating T_t and then using the relation $T_c = T_0 + T_t$.

2.7.1.3 Unit hydrograph method

This approach disregards the physical processes that transport the water that is generated in the watershed and through the network of channels to its outlet. Rather, the watershed is mathematically treated as an arbitrary function that transforms an input (e.g. rainfall time series) into an output (e.g. discharge of a river) like a black box. The transfer function representing the watershed and converting an input signal, and the rate of runoff generation (also called rainfall excess) into an outflow hydrograph at the outlet, is represented by a unit response function – implicitly expressing the sum total of the responses of the channel network – called the *unit hydrograph (UH)*. The use of transfer function for hydrograph generation was first conceived by Sherman (1932). This is a pulse response function that is derived using observed hydrological data.

The unit hydrograph for a given catchment is defined as the hydrograph due to a unit depth of rainfall excess or runoff generation (e.g. 1 *mm* or 1 *cm*) caused by a storm of specified duration. The concept behind the approach is that, based on observed data on both input rainfall and outflow runoff for many storms, an average representative unit hydrograph can be constructed. Having done this empirically, the future events can be predicted by combining the estimated inputs of rainfall excess and the unit hydrograph.

Unit hydrograph approach is based on two key assumptions:

1. *Linearity*: the unit hydrograph is independent of the rainfall intensity and rainfall excess is uniformly distributed through the catchment and has constant intensity during the period under consideration.
2. *Time invariance*: the unit hydrograph does not change with time, either during an event, or between different events, i.e. catchment characteristics remain invariant with time.

These two assumptions lead to the *Principle of Superposition*. Given that the base time (or duration) of the unit hydrograph (due to the unit depth of rainfall, say 1 *mm*) is kept constant, the stream flow hydrograph arising from a composite event can be estimated by simply adding the contributions of different parts of the composite storm, as illustrated in Fig. 2.14. Therefore, if $u(t)$ is the unit

hydrograph, for a given catchment and rainfall duration, Δt, and the rainfall excess, P, occurs over duration of $n\Delta t$, then the catchment response is a linear summation over n storms. For example, if one has a 10-*min* unit hydrograph, to generate 30-min storm intensity, it is possible to combine three 10-*min* unit hydrographs to get the desired composite response.

2.7.1.3.1 SCS unit hydrograph method

The *SCS Unit Hydrograph* method is appropriate for drainage areas larger than 100 acres and complex watersheds. Determination of the storm runoff hydrograph for a specified total rainfall, using the SCS method, involves three primary steps:

1. Develop the unit hydrograph for the watershed, based on the watershed characteristics and time of concentration.
2. Determine the excess precipitation values using the CN and rainfall values.
3. Calculate the storm runoff hydrograph by applying the excess precipitation values in Step 2 to the unit hydrograph values in Step 1.

The time of concentration, T_c, influences the shape and peak of the runoff hydrograph. The T_c for a watershed is determined as described in the section on 'Rainfall-runoff analysis and modelling' above. To determine the unit hydrograph for a specific

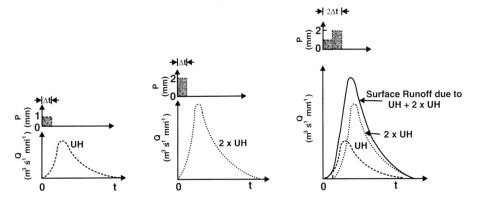

Fig. 2.14 The unit hydrograph approach indicating linearity and invariance of watershed runoff response leading to the Principle of Superposition for estimating response of a watershed to a composite event.

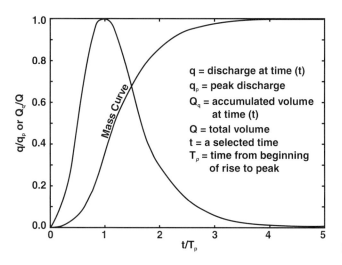

Fig. 2.15 SCS Dimensionless Unit Hydrograph.

watershed, the following relationships are used:

$$T_p = (\Delta D/2) + 0.6 \times T_c \quad and$$

$$q_p = 484 \, (A/T_p) \tag{2.37}$$

where q_p = peak discharge (*cfs*); A = watershed area (*square miles*); T_p = time to peak (*hours*); T_c = time of concentration (*hours*); and ΔD = duration of excess rainfall (*hours*). The constant 484 represents the peaking factor related to the ratio of the unit hydrograph time base to the time to peak and unit conversions to provide q_p in units of *cfs*. The unit hydrograph values are obtained by reading q/q_p and t/T_p values from the dimensionless SCS unit hydrograph (Fig. 2.15) and multiplying by the computed q_p for selected duration of the excess rainfall, ΔD. The duration of excess rainfall should be selected so that ΔD is approximately equal to $0.133 T_c$.

2.8 Stream processes

As mentioned in the section on 'Runoff processes and flow measurement' above, passage of runoff through a channel system is accompanied by erosional processes, and transport of sediments and nutrients. In fact, transport of runoff water (together with dissolved, suspended, or otherwise transported materials) is a major agent that changes a landscape. The biogeochemical interactions and transformations that occur during this process

are vital to support aquatic and riparian ecology. Floods, on the other hand, are the most well-known natural hazards associated with channel runoff. In the following, the focus is on headwater streams and rivers. The former, having a small area, results in short duration headwater floods and has strong linkages to climate and physiography. Headwater streams form tributaries of larger rivers where stream processes also occur. Streams erode the landscape in the headwater region Δ deposit the material downstream in the plains. Much of our understanding of stream flow processes is derived from fluvial geomorphology, which is concerned with the study of erosion and deposition along hill slopes and channels in the context of landscape changes. The spatial scales in hydrology, with focus on the river basin and its internal streams and reaches, have a considerable overlap with geomorphology. However, the temporal scales of interest in hydrology, focused on individual floods and return periods of floods and droughts, are shorter than those of concern in geomorphology, which span geological time scales (Fig. 2.16).

Some of the commonly used terms for stream sizes (in order of increasing size) are Headwater Creek → Stream → River. And for the stream flow (in order of increasing water flow during the whole year): Ephemeral → Intermittent → Perennial. In order of increasing drainage basin size, the terms used are Sub-watershed → Watershed/ Catchment → River Basin.

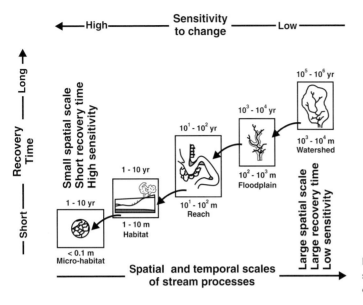

Fig. 2.16 Spatial and temporal scales of stream processes and their response to change. Redrawn from Vivoni (2005d).

Hill slopes, streams, and drainage basins together constitute the fluvial network that transports water, nutrients, and sediment through landscape that itself changes in time due to erosion and deposition in its different parts. The fluvial network continually concentrates flow from upland to lowland areas from Sheet flow → Rill → Gully → Stream → River. In lowland areas, streams interact with floodplains and meander through valleys. Erosion by streams exposes alluvial fill material deposited over geological time scales, resulting in the formation of terraces.

2.8.1 Entrainment, transportation, and deposition of fluvial sediments

If a stream is dissipated by turbulence, a small part is used in the important task of eroding and transporting sediment. The movement of water and materials in a stream depends on the energy of the flowing water and the friction between the layers of water and the surface over which the water flows. Depending on the balance of forces, a given flow may be laminar, turbulent, or transitional.

2.8.1.1 Laminar and turbulent flow

When a fluid flows slowly through a stream or a pipe, as shown in the upper pipe in Fig. 2.17, the flow may be considered to consist of various layers that move at different velocities relative to each other. Fluid at the centre of a pipe moves with the maximum velocity, while the fluid at the edge of the pipe is almost stationary. Such a movement of a fluid is called a *Laminar Flow*. Since each layer moves with a different velocity relative to its neighbour, a frictional force F exists between the various layers. This force depends on the area A of the liquid surface and also on the rate of shear strain, i.e.:

$$F \propto A \text{ . } Rate\,of\,Shear\,Strain \tag{2.38}$$

with rate of shear strain given by the velocity gradient in the pipe:

$$F = \eta A \cdot \frac{dv}{dr} \tag{2.39}$$

where η is the coefficient of viscosity, defined as:

$$\eta = \frac{Shear\,stress}{rate\,of\,shear\,strain} \tag{2.40}$$

The units of η are Nsm^{-2} or '*dekapoise*' $[ML^{-1}T^{-1}]$.

Consider a cylinder of fluid of radius r centred on the axis of a pipe of radius a and length l. The surface area of the fluid cylinder is $2\pi rl$. Hence the

---

Water flow is either laminar or turbulent

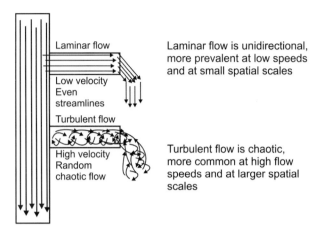

Laminar flow

Low velocity
Even
streamlines

Turbulent flow

High velocity
Random
chaotic flow

Laminar flow is unidirectional, more prevalent at low speeds and at small spatial scales

Turbulent flow is chaotic, more common at high flow speeds and at larger spatial scales

Fig. 2.17 Depending on viscosity and speed, the flow of water can be laminar or turbulent.

force exerted by fluid outside the cylinder on the fluid inside the cylinder is:

$$F = -\eta \cdot 2\pi r l \cdot \frac{dv}{dr} \qquad (2.41)$$

The frictional force F opposes fluid motion inside the cylinder. The negative sign indicates that v decreases as r increases. For steady flow, a driving force, due to difference in the pressures (P_1 and P_2) at the ends of the fluid element, must exist to counteract the resisting viscous force. The net force is given by:

$$F = (P_1 - P_2) \cdot \pi r^2$$

$$= -\eta \cdot 2\pi r l \cdot \frac{dv}{dr} \qquad (2.42)$$

The velocity of a cylindrical shell (radius r), in a cylinder of radius a, is obtained by integrating dv/dr:

$$v(r) = \frac{P_1 - P_2}{4\eta l}(a^2 - r^2) \qquad (2.43)$$

The total volume per unit time (i.e. the flow rate or volume flux) through the pipe is obtained by adding up the flow due to all such shells of radius r and thickness dr, i.e.:

$$\frac{dv}{dt} = \int_0^a 2\pi v r \, dr$$

$$= \frac{\pi a^4 (P_1 - P_2)}{8\eta l} \qquad (2.44)$$

This is the *Poiseuille's Law*, which holds for laminar flow but not for turbulent motion.

Turbulent flow is characterized by randomness and is generally thought of as being chaotic, as shown in the lower pipe in Fig. 2.17. When the velocity of fluid is increased, inertial forces also increase, the sticky effect of viscosity is dampened, and the laminar nature of flow becomes chaotic and turbulent. The turbulent fluid flow can be described in terms of a dimensionless quantity known as the *Reynolds number*, *Re*, defined as:

$$\mathrm{Re} = \frac{dv\rho}{\eta} \left[Dimension = \frac{L \cdot LT^{-1} \cdot ML^{-3}}{ML^{-1}T^{-1}} = 1 \right]$$
$$(2.45)$$

where ρ is the density of the fluid; d is the diameter of the tube; and $v (= Q/A)$ is the average velocity of water in the tube. At a certain velocity, the critical velocity, V_c, there is transition from laminar to turbulent flow. For fluid flow along a cylindrical pipe, V_c is given by:

$$V_C = \frac{R_c \eta}{2a\rho} \qquad (2.46)$$

where R_c is the critical Reynolds number, which for most fluids is approximately 2000.

As turbulently flowing water comes in contact with a solid surface, viscous forces give rise to friction between the water flow and the surface, causing a decrease in the flow velocity so that it becomes more laminar. The layer of water which is in

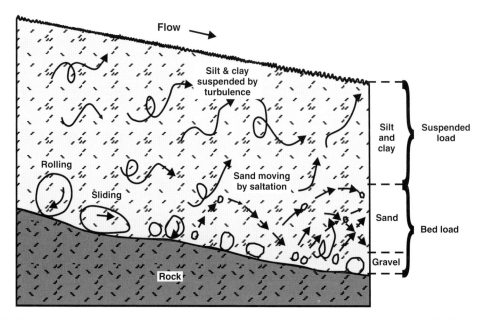

Fig. 2.18 Modes of sediment load transportation in a stream. After http://www.csus.edu/indiv/c/cornwell/surface/sp-2007/river%20processes%202.pdf (accessed 7 July 2009).

immediate contact with the solid surface is stationary; it does not move and exhibits a 'no-slip' condition. The region above the surface, where the flow changes its nature, is called the boundary layer.

2.8.1.2 Entrainment and erosion

Turbulent flow with eddies and vortices is important in the sediment scouring process and pothole formation. The process that initiates motion of a particle into a fluid in motion is called entrainment. Impact or momentum of water mass or the already entrained particles in suspension, or moving along the bed by saltation (skipping) or traction (sliding or rolling), contribute to entrainment of new particles from surfaces exposed to water as part of the operation of erosion. The process of entrainment determines the type and magnitude of erosion that occurs along the channel banks or floor in two principal ways: (i) *corrasion* – mechanical breakdown of rock due to wearing and grinding caused by material carried in transport across the rock surface and (ii) *cantilevers* – differential corrasion produces overhangs that collapse.

2.8.1.3 Transportation

Sediments entrained in a flowing stream can be transported along the bed as bed load, in suspension as suspended load, or along the top (air-water) surface of the flow as wash load. The total amount of sediment actually carried by a stream is known as its load. Suspended load comprises particles transported mainly or entirely in suspension through the supporting action of turbulence. The bed load, on the other hand, comprises sediment which moves by skipping, sliding, and rolling along the channel bed (Fig. 2.18). Particles transported as bed load essentially remain within a few grain diameters of the channel bed. The size of the largest particle that a stream can entrain under any given set of hydraulic conditions is known as competence.

For flows having a free surface or such an interface that gravity forces play an important role in causing the flow, another dimensionless number, called the Froude number, is defined to describe the type of turbulent flow:

$$F\# = v^2/(gd) \qquad (2.47)$$

where v = velocity [L]; d = average depth of flow [L]; and g = acceleration due to gravity [LT$_{-2}$]. The

Froude number is often viewed as the ratio of the inertial force (the force acting to continue motion) and the gravitational force (the force acting to stop the motion); as such it reflects the strength of the flow. *F#* is also viewed as the ratio of kinetic ($\sim v^2$) and potential (*gd*) energy. When *F#* >1, the flow is described as the shooting type in which waves cannot propagate upstream. When *F#* <1, the flow is tranquil wherein waves can propagate both upstream and downstream. At the transition, when *F#* = 1, all upstream propagating waves are 'stuck' here.

2.8.1.4 Deposition

If entrainment of sediments represents a threshold for erosion, a similar threshold must exist when sediments in transport are deposited. The relationship between grain size, entrainment, transportation, and deposition is shown in the Hjulström diagram (Fig. 2.19), which shows the minimum critical velocity necessary for erosion, transportation, and deposition of clasts of various sizes and cohesion.

Once a particle is entrained in a fluid it begins to sink again under the pull of gravity. The distance it travels depends on the drag force of the fluid and the settling velocity of the particle. The settling velocity of a particle is calculated using *Stoke's*

Law, which can be considered as the sum of the gravitational pull downwards versus the drag force of the fluid pushing upwards. It can be shown that for free-falling spherical grains in a liquid, the terminal settling velocity is given by:

$$v_{stoke} = C\,D^2 \tag{2.48}$$

where v_{stoke} is the terminal Stoke's velocity of the particle of diameter *D* (assuming it to be perfectly spherical); density ρ_s settling in fluid of density ρ_f; and dynamic viscosity μ. C is a constant given by $(\rho_s - \rho_f)g/18\mu$. At $20°C$ in water, for a sphere of density 2.65 $g\ ml^{-1}$, C = 3.59×10^4.

Just as a long episode in which more sediments leave the bed than are returned results in a distinct period of *degradation*, a similar long episode in which lower amounts of sediments leave the bed than are returned results in a distinct period of *aggradation*. Fluvial deposition is important for several reasons:

1. On a long-term basis, continued deposition results in landforms that reflect distinct periods of geomorphic history.
2. On a short-term basis, deposition creates bottom features such as dunes, bars, and riffle-pool sequences that are closely interrelated with channel pattern and the character and distribution of flow within the channel.

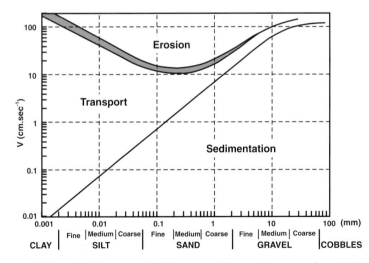

Fig. 2.19 Hjulström curve. Higher velocities are required to move and transport coarser sediments. Note that still higher velocities are required to entrain cohesive clay and fine silt particles than coarse sand, which is cohesionless. However, once fine sediment is entrained, a much lower velocity is required to keep it in suspension. Redrawn from http://www. geology.iupui.edu/academics/classes/G415/powerpoints/Fluvial_Mechanics.ppt (accessed on 20 July 2009).

3. Finally, short-term and long-term mechanics of deposition have implications beyond the boundaries of geomorphology, for example, gold mining and contaminant plume migration.

2.9 Stream characteristics

In its downward journey, runoff produced at individual points in a catchment routed through a stream network may reach a lake and continue from there as an out-flowing stream or terminate on reaching an internal closed basin or sea. A stream or river network comprises and is characterized by: (i) a set of nodes; (ii) a set of links or reaches; and (iii) a numbering system to order the stream links and nodes. The nodes are classified into: (a) exterior nodes; (b) interior nodes; and (c) an outlet node (Fig. 2.20). Exterior nodes occur in headwater basins. Interior nodes include all inner basins with multiple tributaries. The outlet node is the root of the network through which water flows out of the basin. The links or reaches are classified into: (a) interior links connecting interior nodes; or (b) exterior links connecting an exterior node to an interior node. The numbering system used to order the stream links and nodes allows classification of individual reaches and to study river network structure and find common laws that are useful for hydrologic and geomorphologic studies. Two ordering systems (Fig. 2.20) are popular, *Horton-Strahler* and *Shreve* ordering (Shreve 1966; 1967). The former was developed by Horton (1945) and improved upon later by Strahler (1957).

In both systems, channels originating at a source (headwater) are first-order (1st) streams (Fig. 2.20). But in the Horton-Strahler scheme, when streams of similar order join, 1 is added to the stream of order and when streams of different order merge, the highest order is used. In this framework, a sub-basin is of the same order as its main stream. On the other hand, in the Shreve scheme, when streams of similar order join, their orders are simply added to obtain the downstream order. It is important to appreciate that stream order is dependent upon the scale of the map. With decrease in the map scale, as more details are added to the river network (i.e. new tributaries), a river may increase its stream order.

Several network laws relating to stream order, their number, length, and drainage basin area have been identified through the study of river ordering (Fig. 2.21a). These laws are briefly stated in Table 2.3 and are found to be quite robust over many different basins. The near universality of these empirical stream networking 'laws' seem to indicate that some properties of fluvial geomorphology and the processes that shape the landscape may be invariant over several orders of spatial and temporal scales and that these empirical equations, in a way, represent the applicable scaling laws. Studies of other natural branching networks have revealed patterns similar to the stream order model. For example, the bifurcation ratio of three has also been discovered in the rooting systems of plants, the branching structure of woody plants, the veins in plant leaves, and the human circulatory system.

Components of Stream Network and Ordering Systems

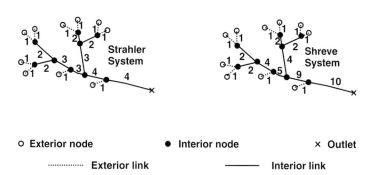

Fig. 2.20 Nomenclature of stream network components and two popular systems of stream ordering with assigned order numbers.

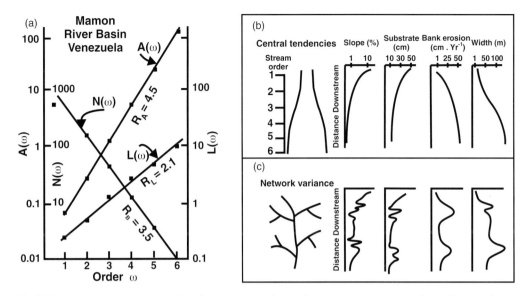

Fig. 2.21 (a) Horton's stream networking laws relating order number, average segment lengths, and average basin areas. Redrawn from www.ees.nmt.edu/vivoni/ surface/ lectures/Lecture18.pdf. (b) Linear approaches, such as the river continuum concept (RCC), predict gradual and continuous downstream change, with central tendencies, in physical and biological processes. (c) In the nonlinear, network-variance model, the branching character of river networks, coupled with stochastic watershed disturbance, interrupts the downstream continuum of physical and biological processes to generate hypothetical deviations from downstream central tendencies in the geomorphic properties along the main stem of the network. Redrawn from Benda *et al.* (2004). © American Institute of Biological Sciences.

Table 2.3 Stream network laws (after Vivoni, 2005d).

Law of stream	Relation*	Usual Range	Author
Numbers	$R_B = \dfrac{N_\omega}{N_{\omega+1}}$ or $N_\omega = a_1 e^{-b_1 \omega}$	$3 < R_B < 5$	(Horton 1945)
Segments Lengths	$R_L = \dfrac{L_{\omega+1}}{L_\omega}$ or $L_\omega = a_2 e^{b_2 \omega}$	$1.5 < R_L < 3.5$	(Horton 1945)
Areas	$R_A = \dfrac{A_{\omega+1}}{A_\omega}$ or $A_\omega = a_3 e^{b_3 \omega}$	$3 < R_A < 6$	(Schumm 1956)
Main Stream Length	$l \, \alpha \, A^h$	$h \approx 0.6$	(Hack 1957)

Relations between bank-full (effective) stream discharge and channel pattern

Slope	$s = a Q^b$		
Width	$w = c Q^f$		(Leopold and Wolman 1957)
Depth	$d = g Q^j$		
Velocity	$v = k Q^m$		

Since $Q = w.d.v$; $c.g.k = 1$ and $f + j + m = 1$

$\omega = 1, 2, 3 \ldots$ is the stream order number following the Horton-Strahler scheme.
R_B = Bifurcation ratio. R_L = Length ratio. R_A = Drainage area ratio.
N_ω = Number of streams of order ω. L_ω = Average length of streams of order ω.
A_ω = Average drainage area of streams of order ω. l = Length of the main stream.

It is worth noting that several principles of fluvial geomorphology have guided the development of fluvial ecology. A prominent example is the well-known river continuum concept (RCC) propounded by Vannote *et al.* (1980) based on early principles of fluvial geomorphology (Leopold *et al.* 1964). The RCC emphasizes spatially and temporally averaged downstream changes in channel morphology over many orders of magnitude. It predicts gradual adjustments of biota and ecosystem processes in rivers in accordance with the geomorphic perspective of gradual downstream changes in hydrologic and geomorphic properties (Fig. 2.21b), although downstream interruptions in channel and valley morphology, caused by alternating canyons and floodplains, tributary confluences, and landslides, have long been observed. Some have viewed these interruptions simply as adjustments to the original RCC (Bruns *et al.* 1984; Minshall *et al.* 1985), whereas others have opined that they serve as the basis for a new view of a river as a 'discontinuum' (Montgomery 1999; Perry *et al.* 1987; Rice *et al.* 2001). In the nonlinear, network-variance model (Benda *et al.* 2004), the branching character of river networks, coupled with stochastic watershed disturbance, interrupts the downstream continuum of physical and biological processes to generate hypothetical deviations from downstream central tendencies in the geomorphic properties along the main stem of the network. Tributary junctions represent locations in a network where channel and valley morphology can change and any local heterogeneity can be amplified relative to the central tendency expected under the river continuum concept (Fig. 2.21c). Spatial and temporal heterogeneity in resources and habitat may, amongst other things, contribute to increased local species richness (Huston 1994). Tributary junctions may, therefore, represent biological hotspots within a given river network.

2.10 River and reservoir routing

Hydrologists are often concerned with making predictions of the river runoff. This could be the runoff during an actual storm event, or due to a hypothetical 'design storm', or predicting the runoff yield in a river over long periods of time (long-term water balance), including effects of land-use or climate changes. Besides the obvious need for flood forecasting that can save countless lives, some riverside industries are required to adjust their operations, such as water intake, based on forecasts of river levels. Utilities, such as water and sewage treatment plants, can also have their operations affected by river and stream levels. Many types of aquatic recreation, for example, boating, fishing, etc., can be affected by river and stream levels. Dam operators need to know how much water reaches the reservoir at various times, to properly operate the reservoir and maintain its water level.

Channel routing is a term applied to methods of accounting for the effects of channel storage on the runoff hydrograph, as a flood wave is routed through a given reach of the channel (Fig. 2.22a). The effects of channel storage and flow resistance of the reach are reflected in changes in the shape and timing of the hydrograph as it passes from upstream to downstream. As it proceeds downstream, the hydrograph peak is attenuated and also its shape broadens. The basic premise used in any flow-routing procedure is that the total inflow is equal to the total outflow plus the change in storage (Eqn 2.49) – as governed by the continuity (or the mass balance) equation. Therefore, routing procedure has two components – the routing method and the physical channel characteristics of the stream reach.

Flow routing methods can be grouped into two different types, lumped- and distributed-flow routing. Lumped, or hydrologic routing, discretizes the system in the time domain (i.e. flow is calculated as a function of time at one location); while distributed, or hydraulic routing, discretizes the system in both the time and space domains (i.e. flow is calculated as a function of space and time throughout the system). Since detailed information on the physical properties of the river channels, including width, depth, slope, roughness, curvature, etc. along the entire length of all the channels that constitute the channel network is difficult to obtain and discretize, use of hydrodynamic models requires considerable expertise. In many applications, hydrodynamic models are unnecessary. Nevertheless, there are certain applications, such as

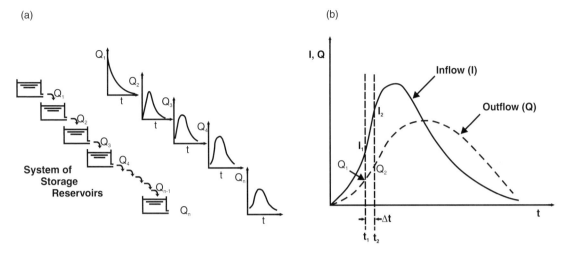

Fig. 2.22 (a) Conceptual model of runoff routing. (b) Schematic of upstream and downstream flood hydrograph for the Muskingum method. After http://www.ctre.iastate.edu/pubs/stormwater/documents/2C-10ChannelandStorage(Reservoir) Routing.pdf (accessed 17 May 2007).

flood forecasting, where such methods are indispensable, and are widely used.

For a hydrologic system, inflow, outflow, and storage are related by the continuity equation:

$$\frac{dS}{dt} = I(t) - Q(t) \approx \frac{\Delta S}{\Delta t} \qquad (2.49)$$

where I = inflow [L^3]; Q = outflow [L^{3}T^{-1}]; S = storage [L^3]; and t = time. This approach is the widely used Muskingum method and the related modified Muskingum-Cunge methods for channel routing.

2.10.1 Muskingum method

In this method, runoff routing is accomplished by routing the generated runoff through a system of linear or nonlinear reservoirs (Fig. 2.22a). Each channel reach essentially represents a dynamic storage which receives runoff from the adjacent hill slopes and also from a reservoir upstream. The outflow from the channel storage feeds into the storage downstream. The volume of water S in the reservoir is assumed to be some function of the outflow Q from each reservoir; when this function is linear, that is, $S = kQ$, it is called a linear reservoir, while if $S = kQ^m$, it is called a nonlinear reservoir. Thus, the routing problem through the river channel network is transformed into a routing problem

through a network of linear or nonlinear reservoirs. In this respect, storage in the reservoirs is representative of the volume of water held in transit within the channels, and the tributaries, rills, and drainage lines which contribute water to the main stream channels. The governing routing equation now becomes a simple continuity (or mass balance) equation (Eqn 2.49) for each storage reservoir. For the hydrographs shown in Fig. 2.22b, the continuity equation can be expressed in terms of the inflow (upstream) and outflow (downstream) at two times: t_1 and t_2 and $\Delta t = t_2 - t_1$. The numerical form of the routing equation (Eqn 2.49) can, therefore, be written as:

$$(S - S_1)/\Delta t = {}^1\!/_2 (I_1 - I_2) - {}^1\!/_2 (Q_1 - Q_2)$$

$$(2.50)$$

assuming that the inflow hydrograph is known for all times, t, and the initial outflow and storage, Q_1 and S_1, are also known at time, t_1. Furthermore, if $S_I = kI^m$ and $S_O = kQ^m$, the inflow storage is related to the inflow rate and the outflow storage to the outflow rate, and a weighting factor x is assigned to account for the relative effect of the inflow and outflow on storage, so that:

$$S = x \cdot S_I + (1 - x) \cdot S_O \quad and$$

$$for\ m = 1; \quad S = k \cdot [x \cdot . I + (1 - x) \cdot Q] \quad (2.51)$$

Substituting Eqn 2.51 into Eqn 2.50, and rearranging to solve for Q_2, gives:

$$Q_2 = C_0 . I_2 + C_1 . I_1 + C_2 . Q_1 \qquad (2.52)$$

with

$$C_0 = (0.5 \Delta t - kx)/k(1 - x) + 0.5 \Delta t \qquad (2.53)$$

$$C_1 = (0.5 \Delta t + kx)/k(1 - x) + 0.5 \Delta t \qquad (2.54)$$

$$C_2 = [k(1 - x) - 0.5 \Delta t]/k(1 - x) + 0.5 \Delta t$$

$$\qquad (2.55)$$

Eqn 2.52 is the *Muskingum routing equation* and C_0, C_1, and C_2 are the routing weighting factors. Note that $C_0 + C_1 + C_2 = 1$.

Given an inflow hydrograph and an initial flow condition, a chosen time interval ($t_{pk}/\Delta t \geq 5$), and routing parameters k and x, the routing coefficients can be calculated from Eqn 2.53 through Eqn 2.55 and the outflow hydrograph from Eqn 2.52. The routing parameters k and x are related to the flow and channel characteristics, k being interpreted as the travel time of the flood wave from the upstream to downstream end of the channel reach; k is therefore a function of channel length and flood wave speed. The parameter x accounts for the storage portion of the routing – for a given flood event, there is a value of x for which the storage in the calculated outflow hydrograph matches the measured outflow hydrograph. In the Muskingum method, x is used as a weighting factor and restricted to a range of values from 0.0 to 0.5. For $x > 0.5$, the outflow hydrograph becomes greater than the inflow hydrograph (hydrograph amplification). For $k = \Delta t$ and $x = 0.5$, the outflow hydrograph retains the same shape as that of the inflow and is just translated downstream at a time equal to k. For $x = 0$, the Muskingum routing reduces to a linear reservoir routing. The k and x parameters in the Muskingum method are determined by calibration using stream flow records.

Where measured hydrograph data exists for a given channel reach, coefficients for the routing function can be determined. However, for most design situations, the measured hydrograph data is not available. In this case, the upstream hydrograph is synthesized using methods discussed in Section 2.7.1, and the resulting downstream runoff hydrograph is computed.

2.10.2 Muskingum-Cunge method

An alternative method, related to the Muskingum procedure, is the Muskingum-Cunge method, which uses a kinematic wave (conservation of mass/momentum) approach. The main advantage of the *Muskingum-Cunge method* is that the routing coefficients are evaluated from physical characteristics of a channel and can be determined without the flood hydrograph data. In this method, for a channel section, it is assumed that:

$$Q = e\,A^m \qquad (2.56)$$

where Q = discharge [$L^3 T^{-1}$]; A = flow area [L^2]; and e and m are constants related to geometry and roughness. The relationship in Eqn 2.56 can be obtained from a stream channel rating curve. A rating curve can be constructed using the Manning formula. Both e and m change with discharge, and average values need to be used. To apply the Muskingum-Cunge model, first a reference flow condition is chosen, which may be the base flow value, being the peak flow of the inflow hydrograph, or the average inflow rate. From these reference conditions and the channel characteristics, the Muskingum constants k and x are estimated from:

$$k = (L/m) \cdot V_0 \qquad (2.57)$$

$$x = 0.5 \left[1 - (Q_0/T_0)/(S_0\, m\, V_0\, L) \right] \qquad (2.58)$$

where S_0 = longitudinal slope of channel [LL^{-1}]; L = length of channel reach [L]; Q_0 = reference discharge [$L^3 T^{-1}$]; T_0 = top width [L] of the flow at Q_0; A_0 = flow area [L^2] at Q_0; $V_0 = Q_0/A_0$ = cross-sectional average velocity [LT^{-1}] corresponding to the discharge value Q_0. Using these values of k and x, the coefficients C_0, C_1, and C_2 can be calculated, and the Muskingum Eqn 2.52 is used to route the hydrograph. If the reference discharge is updated after every time step, the accuracy of the method can be improved by using variable routing coefficients.

2.10.3 Storage (reservoir) routing

Storage routing is similar in concept to channel routing. The inflow hydrograph of a detention basin corresponds to the hydrograph at the upstream location, and the hydrograph for the outflow from the basin corresponds to the downstream hydrograph. The variables associated with storage routing are: (i) input (upstream) hydrograph; (ii) outflow (downstream) hydrograph; (iii) stage–storage volume relationship for a given storage; (iv) physical characteristics of the outlet structure (i.e. weir length, riser pipe diameter, orifice diameter, number of outlet stages, length of the discharge pipe, etc.); (v) coefficients of energy loss (at weir and orifice) coefficients; (vi) storage volume versus time relationship; (vii) depth (stage) – discharge relationship; (viii) target peak discharge from the reservoir; and (ix) volume and time for extended detention.

2.11 Scales and scaling

The science of hydrology so far described has largely been developed from 'experimental knowledge', for example, infiltration (Horton 1933), evaporation and transpiration (Monteith 1965; Penman 1948), water flow through soils (Richards 1931), and open channel flow (Manning 1891). All these processes were observed at small spatial scales ranging from single points to only a few square metres or square kilometres in extent, and from rainfall recorded typically with a single rain gauge.

There is a growing realization that anthropogenically induced atmospheric forcing could change global patterns of weather and climate over a decadal time span. The resulting change in global, water resources, ecological and agricultural production patterns has led to hydrologists being confronted with a 'scale jump' in conceptualization and modelling from small space-time scales to global and decadal time scales. The scale problem, however, is not unique to hydrology. In fact, it transcends all facets of organizational endeavour in the natural (and other) sciences, as well as in day-to-day governance.

Scale may be defined as a characteristic dimension (or size), in either space or time or both, of: (i) an observation; (ii) a process; or (iii) a model of the process. Intuitively, scale is an indication of

order of magnitude change rather than a specific value. Three dominant types of scales relevant in hydrology are:

1. *Process scales* are defined as the scales at which natural phenomena occur. These scales are not fixed, but vary with the process.
2. *Observation scale* is that scale at which we choose to collect samples of observations and to study the phenomenon concerned. Observation scales are determined by: (i) logistics (e.g. access to places of observation); (ii) technology (e.g. cost of state-of-the-art instrumentation); and (iii) individuals' perception (i.e. what is perceived to be important for a study at a given point in time).
3. *The operational scale* is the working scale at which management actions and operations focus. This is the scale at which information is available. Operational scales seldom coincide with process scales, because they are determined by administrative rather than by purely scientific considerations of convenience and functionality.

Scaling, on the other hand, represents the link between processes at different levels in time and space. Here, a distinction also needs to be made between *up-scaling*, i.e., the process of extrapolation from the site-specific scale at which observations are made or at which theoretical relationships apply, to a coarser scale of study and *downscaling*, i.e., taking output from a larger-scale observation or model and deducing the changes that would occur at a finer resolution (e.g. in a timespan of 1 hr, or on a small catchment). Scaling, therefore, entails changes in processes, upwards or downwards, from a given scale of observation and thus includes the constraints and feedbacks that may be associated with such changes. Included in the concept of scaling are changes in spatial and temporal variability, in patterns of distribution, and in sensitivity. Scaling thus goes beyond simple aggregation (up) or disaggregation (down) of values at one level to achieve values at a more convenient level of consideration (DeCoursey 1996).

Six causes of scale problems with regard to hydrological responses have been identified as (Bugmann 1997; Harvey 1997):

- *Spatial heterogeneity in surface processes*: Natural landscapes display considerable heterogeneity (or patchiness) in topography, soils, rainfall, evaporation, and land use. These factors influence the types of processes that dominate and the rates at which they occur.
- *Non-linearity in response*: Some processes occur episodically (e.g. rainfall), other cyclically (e.g. evaporation), still others ephemerally (e.g. lateral flows), or continually (e.g. groundwater movement). Another perspective is that certain responses are rapid (e.g. surface runoff), others at the time scales of days (e.g. lateral flows), or months (e.g. groundwater movement). These different rates of process responses introduce a high degree of non-linearity to the system, which is exacerbated when the natural system is dominated by anthropogenic systems such as land-use changes or construction of reservoirs.
- *Processes require threshold scales to occur:* The threshold for the generated runoff to flow down and reach a natural channel is subject to hydraulic laws (e.g. rainfall intensity exceeding the infiltration capacity of the soil), which are determined, *inter alia*, by channel length, shape, roughness, and slope. However, these thresholds would be modified markedly by human interventions through dam construction, canalization, or water transfers into or out of the system.
- *Dominant processes change with scale:* On small catchments, for example, hill slope processes are influenced by slope, soil, and/or land-use properties which, together with the occurrences and characteristics of localized small-scale storm events, would largely determine the shape of the hydrograph. On large catchments, on the other hand, the hydrograph shape is often determined largely by hydraulic characteristics of channels and reservoirs, as well as by occurrences and properties of large-scale regional rains of frontal and cyclonic origin.
- *Evolution of properties:* Hydrological and other natural properties evolve/change at all space-time scales through mutual interactions, for example, through catchment response to unusual floods/droughts and other landscape evolution processes.

- *Disturbance regimes:* Scaling problems immediately arise as a consequence of disturbance regimes being superimposed on a natural system, for example, by the construction of dams, draining of fields, or changes in land use such as intensification of agricultural activities or increased urbanization.

According to Blöschl and Sivapalan (1995), process scales possess three characteristics of both time and space. Temporally, one needs to distinguish *intermittent processes*, which have a certain lifetime (e.g. a thunderstorm), from *periodic processes*, which have a cycle or period (e.g. the rainy season, or evaporation being dominantly a daytime process), and *stochastic processes*, that is, probabilistic processes that have a correlation length that is often expressed by the recurrence interval (e.g. a 1 in 10 year flood or drought occurrence). Spatially, on the other hand, process scales exhibit: (i) *spatial extent* (e.g. the area over which thunderstorm-related rainfall occurs); (ii) *space period* (e.g. the area over which a certain seasonality of rainfall occurs); and (iii) *correlation space* (e.g. the area over which the 1 in 10 year drought has left its signature).

The intrinsic relationships between time and spacial scales are well illustrated by 'scope diagrams', of which Fig. 2.23 is an example, taken for processes relevant in hydrology (Blöschl and Sivapalan 1995). It is seen that:

1. Processes and their scales overlap; process scales are, therefore, relative rather than absolute in both time and space.
2. Events of small spatial scale are associated with short temporal scale (e.g. thunderstorms).
3. Small-scale events display more variability than large-scale events (e.g. thunderstorms are usually highly localized and very rapid changes in rainfall intensity occur within them).

2.12 The invisible resource: groundwater

The part of precipitation that seeps down through the soil profile until it reaches rock material that is fully saturated with water becomes ground-water. This water, in most cases, is invisible from the surface. Water under the ground surface is stored in

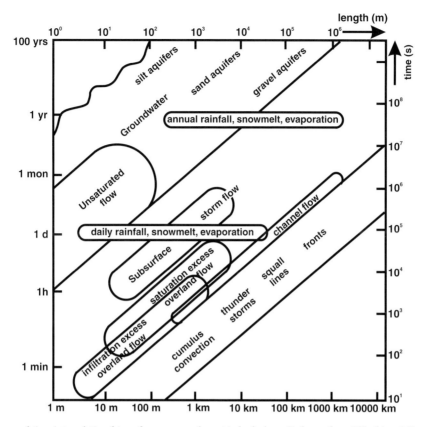

Fig. 2.23 Space and time interrelationships of processes relevant in hydrology. Redrawn from Blöschl and Sivapalan (1995). With permission from John Wiley and Sons Inc.

the void spaces between rock particles in clastic sediments with primary porosity and in the fractures, fissures, joints, and solution cavities in hard rock aquifers characterized by secondary porosity. Contrary to common perception, there are no underground rivers or lakes, except in some karstic regions that can have large channels formed by dissolution of limestone/dolomite rock types. Groundwater slowly moves underground, generally down gradient (due to gravity), and may seep into streams and lakes, and eventually discharge into ocean. Groundwater is an important part of the hydrological cycle. Some water underlies the Earth's surface almost everywhere, beneath hills, mountains, plains, and deserts. It is not always accessible, or may not be of good enough quality for use without prior treatment, and it is sometimes difficult to locate or to estimate and characterize it. This water may occur close to the land surface, as in a marsh, or

it may be several hundreds of metres below the surface, in arid areas such as in North Gujarat in India.

The time that has elapsed since the water first reaches the saturated rock layer (i.e. water table) is called the age of groundwater. Groundwater at very shallow depths might be just a few hours old; at moderate depths, it may be 100 years old; and at great depths or after having traversed long distances from places of its entry in the aquifer, it may be thousands of years old.

Groundwater can occur only at shallow depths below the Earth's surface, because there must be space between the rock particles for groundwater to occur. As the subsurface material becomes more compact with depth, the weight of the overlying rocks compresses the rocks underneath, which reduces the open pore spaces deeper in the Earth. That is why groundwater can only be found within a few kilometres depth below the Earth's surface.

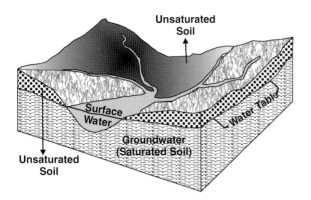

Fig. 2.24 Simplified sketch showing how the ground is saturated below the water table. In this zone, groundwater fills the tiny pore spaces between rock particles and cracks (fractures) in the rocks. The ground above the water table may be wet to a certain degree, but it does not stay saturated. The soil and rock in this unsaturated zone contain air and some water and support the vegetation on the ground.

A simplified, schematic depiction of how the ground is saturated below the water table is shown in Fig. 2.24. The layer above the water table may be saturated temporarily after a rain spell (or below fields when irrigation water is applied), but it does not remain saturated all the time. The soil and rock in this unsaturated zone contain partly air and partly water and support vegetation. The saturated zone below the water table has water that completely fills the void spaces (pores) between rock particles and the cracks (fractures) within the rocks.

The rock below the Earth's surface is called bedrock, which consists of many types of formations, such as sandstone, granite, limestone, etc. Bedrocks have varying amounts of void spaces in them where groundwater can accumulate. Bedrock can also be broken down and fractured, creating void spaces that can be filled with water. And some rocks, such as limestone and dolomite, are dissolved by rain water – which over time results in the formation of large cavities that may be filled with water and also convey it like a pipeline flow.

Sometimes during excavation for road construction, the layered structure of subsurface is seen in vertical cross-section, especially in areas of sedimentary rocks. Some layers have rocks that are more porous than others, and water moves relatively freely (in a horizontal manner) through them. At some road cuttings, groundwater can also be seen oozing out from the exposed layers.

Groundwater also forms part of the general hydrological cycle and the total amount of groundwater on Earth is about hundred times more than the entire world's surface water contained in rivers and lakes. The next chapter is, therefore, entirely devoted to the occurrence and study of groundwater.

2.13 Tutorial

Ex. 2.1 The total amount of water in the atmosphere of the Earth is estimated as $1.29 \times 10^4 \ km^3$. The total surface area of the Earth is $5.1 \times 10^8 \ km^2$, comprising $1.5 \times 10^8 \ km^2$ of land area and $3.6 \times 10^8 \ km^2$ of oceanic area. Estimate the depth of a water column in cm if the water in the atmosphere is completely rained out on (a) land area only; (b) oceanic area; (c) equally on land and oceans simultaneously. The amount of water cycled through the atmosphere every year is estimated as $4.95 \ km^3$. Estimate the column depth in cm if this water were to uniformly cover the Earth's surface. Also estimate the residence time of atmospheric moisture in *days*.

[Ans. 8.6 *cm*, 3.6 *cm*, 2.5 *cm*, 97 *cm*, 9.4 *days*]

Ex. 2.2 Estimated cross-sectional area of a stream channel is $0.045 \ m^2$. Average flow velocity measured at a gauging station is $30 \ m \ min^{-1}$. Calculate the volume of water flowing past the gauging station in m^3 in a week.

[Ans. $1.36 \times 10^4 \ m^3$]

Ex. 2.3 A pipe and the fluid flowing through it have the following properties: water density $\rho = 1000 \ kg \ m^{-3}$; pipe diameter $d = 0.5 \ m$; and (dynamic) viscosity $\eta = 0.55 \times 10^3 \ N \ m^{-2}$. Estimate the critical velocity when transition from laminar to turbulent flow can be expected to occur, that is, Re is ≥ 2000. What would be the limiting velocity for laminar flow if this were a pipe in a domestic central heating system, with a typical pipe diameter of 0.015 *m*?

[Hint: Use Eqn 2.45 and Eqn 2.46. Ans. $2.2 \times 10^{-3} \ m \ s^{-1}$, $7.33 \times 10^{-2} \ m \ s^{-1}$]

Table 2.4 Measurements data for Ex 2.5.

Width of the Vertical Section (m) (A)	Distance of Vertical Centre from a Bank (m) (B)	Avg. Depth of Vertical Section (m) (C)	Average Velocity (m s^{-1}) (D)	Area of Vertical (m^2)	Section Discharge (m^3 s^{-1})
0.76	2.18	0.38	0.27		
0.76	2.94	1.14	0.48		
0.76	3.70	1.75	0.68		
0.76	4.46	2.13	1.01		
0.76	5.22	2.20	1.16		
0.76	5.98	3.28	1.20		
0.76	6.74	2.86	1.05		
0.76	7.50	1.70	0.38		
0.76	8.26	0.64	0.27		
0.76	9.02	0.52	0.10		

Ex. 2.4 Four measurements of depth and flow velocity (in m and m s^{-1}, respectively) have been recorded in a stream 1 m deep using a current meter as: (0.1, 0.42), (0.4, 0.35), (0.6, 0.29), and (0.8, 0.17). Identify the relation between depth of water and flow velocity by plotting a depth profile. Estimate flow velocity at depths of 0.2, 0.3, 0.5, 0.7, and 0.9 and use these and the given data to determine the average velocity of the stream. Determine the depth at which the average velocity is seen in the profile.

[Ans. Avg. 0.29 m s^{-1}, Depth 0.6 m]

Ex. 2.5 In an experiment to estimate stream discharge, measurements as given in Table 2.4 were made. Complete the table and estimate total river discharge. Hint: Area = Width × Depth, Discharge = Area × Average Velocity. Total discharge = $\sum$ Discharge through the section.

[Ans. 10.908]

Ex. 2.6 For the following soil properties, determine the amount of water when ponding occurs, the time to ponding; plot the cumulative infiltration function and plot infiltration rate (f) versus infiltration volume (F): $K_{sat} = 1.97$ cm hr^{-1}; $\theta_i = 0.318$; $\theta_s = 0.518$; $i = 7.88$ cm hr^{-1}; and $\psi_f = 9.37$ cm.

Solution: If rainfall intensity is constant, eventually exceeding the infiltration rate, then at some moment the surface will become saturated and pond-

ing will occur when the infiltration rate equals the precipitation rate (i). The depth infiltrated at that moment (F_s) is obtained by setting $f = i$ in 22.13 and solving:

$$F_s = \frac{[(\theta_s - \theta_i)\psi_f]}{[(i/K_s - 1)]}$$

$$= \frac{[(0.518 - 0.318) 9.37]}{[(7.88/1.97 - 1)]} = 0.625 \ cm \quad (2.59)$$

Time of ponding is obtained by dividing the total amount of water infiltrated until ponding (F_s) by the precipitation rate (i), i.e. $t_s = F_s/i$.

Thus $t_s = (0.625/7.88) = 0.079$ $hrs = 4.74$ $minutes = 284$ $seconds$.

Until 0.625 cm has infiltrated, the rate of infiltration is equal to the rainfall rate. Then substitute F into Eqn 2.14 and 2.13 and solve for time t and infiltration rate f. First plot F as a function of t and then combine the results of Eqn 2.14 and 2.13 to plot f as a function of t.

Ex. 2.7 Construct a rating curve from the data given in Table 2.5, using double log paper.

Use the rating curve (i.e. the best fit line, which is the relationship between depth and discharge) to estimate the discharge at the gauge heights of 0.610 m, 1.981 m, and 2.438 m.

Ex. 2.8 Construct a hydrograph (plot of discharge and time) for a flood event of August 14–22, 1990 at the Millersville gauging site using the data given

Table 2.5 Measurements data for Ex 2.7 and Ex 2.8.

Gauge Height (m)	Discharge ($m^3\ s^{-1}$)	Gauge Height (m)	Discharge ($m^3\ s^{-1}$)
0.576	1.702	0.792	3.908
0.539	1.376	1.100	10.506
0.695	2.945	0.920	6.994
2.283	64.279	0.741	3.596
0.707	2.806	0.668	2.699

in Table 2-6. To convert gauge height to discharge, use the rating curve constructed from Table 2.5.

Plot the precipitation data in Column 5 on the same graph using a new y-axis on the right-hand side. Use the graph to determine the lag time for this flood.

Ex. 2.9 Find the 100-year peak flow rate for a 40-acre drainage area consisting of single-family residential land use with an average plot size of $^1/_4$ acre with good grass cover and maximum ground slope of $\sim$2%. The upper 200 feet of the watershed has a slope of 2%, and the lower 1100 feet is a grass-covered waterway with a slope of 1%.

Table 2.6 Measurements data for Ex 2.8 for construction of hydrograph.

Date	Time	Gauge Height (m)	Discharge ($m^3\ s^{-1}$)	Rain (cm)
14 August	08:00	2.121		0
	16:00	2.003		0
15 August	08:00	1.798		0.39
	16:00	1.935		0.63
16 August	08:00	2.667		0.72
	16:00	2.935		0.57
17 August	08:00	2.950		0.08
	16:00	2.801		0
18 August	08:00	2.512		0
	16:00	2.073		0
19 August	08:00	1.771		0
	16:00	1.585		0
20 August	08:00	1.481		0
	16:00	1.405		0
21 August	08:00	1.338		0
	16:00	1.292		0
22 August	08:00	1.228		0
	16:00	1.170		0

Solution: Determination of the Runoff Coefficient: from Table of Runoff Coefficients Based on Surface Type for Rational Equation, for single-family residential land use with $^1/_4$-acre plots, slope of 2%, and 2–10 year rainfall; the runoff coefficient is 0.45. For a 100-year rainfall, this value must be multiplied by 1.25, resulting in a runoff coefficient of 0.56.

Determination of T_c: calculate the overland flow segment, T_0, with average grass cover ($N = 0.40$) using the Kerby Equation (Table 2.2):

$$T_0 = 0.83 \left(\frac{0.40 \times 200}{0.02^{0.5}} \right) = 16.3\ \text{min}$$

Calculate the travel time for drainage segment, using the Kirpich Formula (Table 2.2):

$$T_t = 0.00778 \times 1100^{0.77} \times 0.01^{-0.385} = 10.1\ \text{min}$$

$$T_c = 16.3 + 10.1 = 26.4\ \text{min}$$

Determine Rainfall Intensity for $T_c = 26\ min$ from Rainfall Intensity-Duration-Frequency Relationships for different return periods (from the *Rainfall Frequency Atlas*).

Using linear interpolation with bounding values $i_{15min} = 8.84\ in/hr$; $i_{30min} = 6.06\ in/hr$. $i_{26min} = 6.8\ in/hr$.

Determine Q_{100} from the Rational equation Eqn 2.28 in English units, i.e.:

$$Q = C \times i \times A = 0.56 \times 6.8 \times 40 = 152\ cfs$$

$$\simeq 4.30\ \text{cubic metres per second}$$

Ex. 2.10 Find the five-year peak flow rate at a point with a 20-acre drainage area. The lower portion of the watershed is a 15-acre parking area. The remainder of the watershed is a flat (slope <2%) grass-covered area. Using the methods described Section 2.7.1.1, the time of concentration determined for the entire watershed is 22 min and the time of concentration is 16 min for the parking area only.

Solution: Determination of Runoff Coefficients: from Table of runoff coefficients, the appropriate runoff coefficient for a new asphalt parking area is 1.0, and that for a flat, grass-covered area is 0.15. A runoff coefficient for the entire watershed can be calculated by area-weighting of these runoff coefficients:

$$C_{watershed} = \frac{C_{asphalt} \times A_{asphalt} + C_{grass} \times A_{grass}}{A_{asphalt} + A_{grass}}$$

$$= \frac{1.0 \times 15 + 0.15 \times 5}{15 + 5} = 0.79$$

From the Table of Rainfall Intensity-Duration-Frequency Relationships for different return periods; for a five-*year* return interval, rainfall intensities corresponding to $T_c = 15$ and 30 *min* are 5.16 and 3.54 *in/hr*, respectively. Using linear interpolation with bounding values $i_{20\ min} = 4.62\ in/hr$, calculate Peak Flow Rates:

For the parking area only:

$$Q_5 = 1.0 \times 5.16 \times 15 = 77.4\ cfs$$

For the entire watershed:

$$Q_5 = 0.79 \times 4.62 \times 20 = 73.0\ cfs$$

The appropriate Q_5 to use for the design point is 77.4 *cfs*. Because of the shorter time of concentration (higher design rainfall intensity) and the higher runoff coefficient, the parking area alone will produce a higher rate of peak runoff rate than the drainage area considered as a whole.

Ex. 2.11 Determine the 100-yr, 2-*hr* peak flow rate for a watershed with the following characteristics: Area = 1.2 sq miles (767 acres); Hydrologic Soil Group (HSG) = C; Land uses: Residential ($^1/_4$-acre plots) = 267 acres; Residential ($^1/_2$-acre plots) = 300 acres; Commercial = 100 acres; Park/open space (good) = 100 acres.

Solution: Using the methods described Section 2.7.1.2.1 and Ex. 2.9, T_c is estimated as 45 *min*. To determine the Unit Hydrograph using SCS method (section 2.7.1.3.1), select the time interval such that: $\Delta D = 0.133\ T_c = 0.133\ (45) = 5.99$ minutes $\approx 6\ min$.

From Eqn 2.37: $T_p = 6/2 + 0.6(45) = 30\ min = 0.5$ and:

$$q_p = 484(1.2/0.5) = 1162\ cfs$$

Read q/q_p and t/T_p values from Fig. 2.15 and multiply by the computed q_p value to develop the unit hydrograph shown in Table 2.7.

Calculation for estimation of excess precipitation: Calculate composite CN for watershed using Runoff Curve Number (*CN*) Values for Fully Developed and Developing Urban Areas (SCS 1986) and HSG C: (i) Residential ($^1/_4$ acre) = 267 acres − $CN = 83$; (ii) Residential ($^1/_4$ acre) = 300 acres − $CN = 80$; (iii) Commercial = 100 acres − $CN = 94$; (iv) Park/open space (good) = 100 acres − $CN = 74$. The composite *CN* can be calculated based on area weighting as:

$CN_{composite}$

$$= \frac{(267 \times 83) + (300 \times 80) + (100 \times 94) + (100 \times 74)}{267 + 300 + 100 + 100}$$

$$= 82$$

Read total rainfall for the given region (say Midwest USA) for a 100-*year*, 2-*hour* event from Rainfall Depth-Duration-Frequency Relationships using the *Rainfall Frequency Atlas* = 4.74 *in*. To

Table 2.7 Table for use with Ex 2.11.

t (min)	t/T_p	q/q_p	q (cfs)	t (min)	t/T_p	q/q_p	q (cfs)
0				66			
6				72			
12				78			
18				84			
24				90			
30				96			
36				102			
42				108			
48				114			
54				120			
60							

Table 2.8 Table for developing precipitation distribution over a storm duration for use with Ex 2.11.

Cumulative Storm Time (%)	Cumulative Precipitation (%)	Storm Time (min)	Cumulative Precipitation (in)	Cumulative Storm Time (%)	Cumulative Precipitation (%)	Storm Time (min)	Cumulative Precipitation (in)
(1)	(2)	(3)	(4)	(1)	(2)	(3)	(4)
0	0	0		55	84	66	
5	16	6		60	86	72	
10	33	12		65	88	78	
15	43	18		70	90	84	
20	52	24		75	92	90	
25	60	30		80	94	96	
30	66	36		85	96	102	
35	71	42		90	97	108	
40	75	48		95	98	114	
49	79	54		100	100	120	
50	82	60					

Col (1) and Col 2 from tables. Col (3) = Col (1) $\times$ 120/100. Col (4) = Col (2) $\times$ 4.74/100.

develop the precipitation distribution of this event over the storm duration, appropriate tabulated values of cumulative storm time versus cumulative storm rainfall for the given region are used as in Table 2.8:

Calculation of S_{max} based on the CN value (using Eqn 2.36): = 1000/82 − 10 = 2.19 *in.*

Using precipitation values in the above Table 2.8, determine excess precipitation for each time period using Eqn 2.33 and tabulate this in Table 2.9.

Calculation for Runoff Hydrograph:

The runoff hydrograph is calculated by multiplying the ordinates of the unit hydrograph by the

Table 2.9 Table for tabulation of excess precipitation for use with Ex 2.11.

Storm Time (min)	Cumulative Precipitation (in)	Accumulated Runoff (in)	Incremental Runoff (in)	Storm Time (min)	Cumulative Precipitation (in)	Accumulated Runoff (in)	Incremental Runoff (in)
(1)	(2)	(3)	(4)	(1)	(2)	(3)	(4)
0				66			
6				72			
12				78			
18				84			
24				90			
30				96			
36				102			
42				108			
48				114			
54				120			
60							

Col (1) is Col (3) of Table 2.8. Col (2) is Col (4) of Table 2.8.
Col (3) = [(Col (2) − 0.2 $\times$ S$_{max}$)^2]/(Col (2) + 0.8 $\times$ S$_{max}$) (S$_{max}$ = 2.19, calculated using Eqn 2.33). Note if value in Col (2) is less than 0.2 $\times$ 2.19, Col (3) = 0
Col (4) = Col (3) Row (i) − Col (3) Row (i-1)

Table 2.10 Table for deriving runoff hydrograph for use with Ex 2.11.

Time (min)	Unit hydrograph (cfs)	Excess Precipitation (in) /Time (min)									Storm Hydrograph (cfs)
		6	12	18	24	30	36	42	48	...	
(1)	(2)	(3)	(4)	(5)	(6)	(7)	(8)	(9)	(10)	(11)	(12)
0	0	0									
6			0								
12				0							
18					0						
24						0					
30							0				
36								0			
42									0		
48										0	
54										...	
60										...	
66										...	
72										...	
78										...	
84										...	
90										...	
96										...	
102										...	
108										...	
114										...	
120										...	

Col (1) = storm time; $\Delta D = 6$ min per time step; Col (2) = calculated q (Col 4 from Table 2.7)
Col (3) through (11) = Excess precipitation (top value), from Col 4 of Table 2.9, for each time step (ΔD).
Col (12) = sum of Cols (3 to 11).

excess precipitation for each time increment. This applies to all time increments in which there is excess precipitation to derive a hydrograph. These incremental hydrographs are lagged appropriately, as a result of which, starting of the incremental hydrograph corresponds to the time of the excess precipitation it represents. Incremental hydrographs are superimposed to derive the runoff hydrograph (Table 2.10).

Ex. 2.12 Given two storm events: Excess rainfall of 1.5 $mm\ hr^{-1}$ for the first hour and 0.7 $mm\ hr^{-1}$ in the second hour, and the stream flow as given in the following Table 2.11.

Find the 1 hr unit hydrograph,

Solution: 1-hr unit hydrograph means the runoff pattern resulting from 1 $mm\ hr^{-1}$ of excess rain for 1 hr) at 1 hour, 2 hours, 3 hours, 4 hours, etc.

Let's assign the ordinates of a unit (1 $mm\ hr^{-1}$ of direct runoff) hydrograph:

U1 is the 1 hour unit hydrograph value at time = 1 hr

U2 is the 1 hour unit hydrograph value at time = 2 hr

U3 is the 1 hour unit hydrograph value at time = 3 hr

Table 2.11 Table for use with Ex 2.12

T (hr)	1	2	3	4
Q ($m^3 s^{-1}$)	2.12×10^{-2}	6.09×10^{-2}	4.08×10^{-2}	7.93×10^{-3}

Table 2.12 Table for use with Ex 2.12.

Time Interval	Rainfall	U1	U2	U3	U4	Direct Runoff
1	P1	P1U1				P1U1
2	P2	P2U1	P1U2			P2U1 + P1U2
3			P2U2	P1U3		P2U2 + P1U3
4				P2U3	P1U4	P2U3 + P1U4

U4 is the 1 hour unit hydrograph value at time $= 4$
hr

To start, we know that the discharge of $2.12 \times 10^{-2} \ m^3 s^{-1}$, which occurred at time 1 hr, is the result of runoff from the first hour; it is the hydrograph ordinate value at time 1 hr of a 1.5 $mm \ hr^{-1}$ hydrograph. Thus
$$1.5 \times U1 = 2.12 \times 10^{-2} \ m^3 s^{-1} \text{ at } 1 \ hr$$
The discharge of $6.09 \times 10^{-2} \ m^3 s^{-1}$ occurring at time 2 hr is the addition of the hydrograph value at time 1 hr of a 0.7 $mm \ hr^{-1}$ hydrograph plus the hydrograph value at time *2 hrs* of the 1.5 $mm \ hr^{-1}$ hydrograph; so
$$0.7 \times U1 + 1.5 \times U2 = 6.09 \times 10^{-2} \ m^3 s^{-1} \text{ at } 2 \ hr$$
Similarly the discharge of $4.08 \times 10^{-2} \ m^3 s^{-1}$ at time $= 3$ hours...
$$0.7 \times U2 + 1.5 \times U3 = 4.08 \times 10^{-2} \ m^3 s^{-1}$$
and for the discharge of $7.93 \times 10^{-3} \ m^3 s^{-1}$ at time $= 4$ hours
$$0.7 \times U3 + 1.5 \times U4 = 7.93 \times 10^{-3} \ m^3 s^{-1}$$
The first of the above equations can be solved quickly, and each of the other equations becomes an equation with one unknown and can therefore be solved quickly by hand:

$U1 = 1.41 \ m^3 s^{-1}$
$U2 = 3.4 \ m^3 s^{-1}$
$U3 = 1.13 \ m^3 s^{-1}$
$U4 = 0$

The 1.5 $mm \ hr^{-1}$ hydrograph is the unit hydrograph $\times$ 1.5, 'beginning' at time $= 0$; likewise, the hydrograph of the 0.7 $mm \ hr^{-1}$ event is $0.7 \times$ the 1 hour unit hydrograph, beginning at time $= 1 \ hr$.

We can derive a new design hydrograph from the unit hydrograph by setting up a table to help solve the four linear equations, as in Table 2.12.

Derive the design hydrograph for a storm event with estimated excess rainfall of 2 $mm \ hr^{-1}$ in the first hour and 2 $mm \ hr^{-1}$ in the second hour.

Ex. 2.13 In a dilution gauging experiment, a slug of 10 *litres* fluorescent dye solution of Rhodamine WT with concentration of 237 $\mu g \ l^{-1}$ of the dye was used. Table 2.13 gives measured data of concentration versus time elapsed since injection from a location 25 m downstream.

Estimate stream flow at the measurement station.

Solution: In terms of Eqn. 2.24, $V = 10 \ l$ and $C = 237 \ \mu g \ l^{-1}$. To compute the denominator, from the data in Table 2.13, $\sum Conc = 28.65 \ \mu g \ l^{-1}$; $\Delta T =$ Time interval between first detection at 5:00 *min* and last detection at 7:40 *min* $= 2:40 \ min = 2.67 \ min$.

Number of observations when plume was $>0.0 = 9$

$C_a =$ Average plume conc. during time detected $= (\sum Conc / $Number of observations when plume was $>0.0.) = 28.65/9 = 3.2 \ \mu g \ l^{-1}$

Using Eqn. 2.24, $Q = (V \times C)/(\Delta T \times C_a)$
$Q = (10 \times 237)/(2.67 \times 3.2) = 277.38 \ l \ min^{-1} = 4.62 \times 10^{-3} \ m^3 s^{-1}$.

Ex. 2.14 Find the 100-year peak flow rate for a 40-*acre* drainage area consisting of single-family residential land use with an average plot size of $^1/_4$ acre with good grass cover and maximum ground slope of $\sim$2%. The upper 200 feet of the watershed has a slope of 2%, and the lower 1100 feet is a grassed waterway with a slope of 1%.

Solution: Determination of Runoff Coefficient: From Table of Runoff Coefficients Based on Surface Type for Rational Equation, for a single-family residential land use with $^1/_4$-acre plots, slope of 2% and 2–10 year rainfall, the runoff coefficient is 0.45.

Table 2.13 Concentration vs. lapsed time data for use with Ex 2.13.

Elapsed Time (*min sec*)	Measured Concentration ($\mu g \, l^{-1}$)	Lapsed Time (*min sec*)	Measured Concentration ($\mu g \, l^{-1}$)	Lapsed Time (*min sec*)	Measured Concentration ($\mu g \, l^{-1}$)
3:00	0	5:00	5.3	6:40	3.0
3:20	0	5:20	4.6	7:00	1.8
3:40	0	5:40	4.6	7:20	0.5
4:00	0	5:00	5.3	7:40	0.15
4:20	0	6:00	4.6	8:00	0
4:40	0	6:20	4.1	8:20	0

For a 100-year rainfall, this value must be multiplied by 1.25, resulting in a runoff coefficient of 0.56.

Determination of T_c: Calculate the overland flow segment, T_0, with average grass cover ($N = 0.40$) using the Kerby Equation (Table 2.2):

$$T_0 = 0.83 \left(\frac{0.40 \times 200}{0.02^{0.5}} \right) = 16.3 \, \text{min}$$

Calculate the travel time segment, using the Kirpich Formula (Table 2.2):

$$T_t = 0.00778 \times 1100^{0.77} \times 0.01^{-0.385} = 10.1 \, \text{min}$$

$$T_c = 16.3 + 10.1 = 26.4 \, \text{min}$$

Determine Rainfall Intensity for $T_c = 26 \, min$ from Rainfall Intensity-Duration-Frequency Relationships for different return periods (from the *Rainfall Frequency Atlas*).

Using linear interpolation with bounding values $i_{15min} = 8.84 \, in/hr$. $i_{30min} = 6.06 \, in/hr$. $i_{26min} = 6.8 \, in/hr$.

Determine Q_{100} from Rational Eqn 2.28 in English units, i.e.:

$$Q = C \times i \times A = 0.56 \times 6.8 \times 40 = 152 \, cfs$$

$$= 4.30 \, \text{cubic metres per second}$$

3

Groundwater hydrology

It is generally not realized how widely and heavily groundwater is used throughout the world, almost to the point of over exploitation. A possible reason for this lack of awareness is that groundwater is usually an invisible resource. Within the hydrologic cycle, groundwater is the slowest moving component, with residence times ranging from a few days to thousands of years (Fig. 3.1). Due to this lack of awareness, the main features of groundwater systems are poorly known and often misunderstood. It is seen from (Table 1.3) that in terms of volume, groundwater is the most important component of the terrestrial hydrologic cycle. Excluding the water permanently locked up in polar ice caps and glaciers, groundwater accounts for nearly all usable freshwater. Globally, nearly 95% of the rural population relies on groundwater for their drinking water supply. About half of irrigated cropland uses groundwater. About one-third of industrial water needs are met from groundwater. About 40% of river flow (on average) depends on groundwater. Thus, in terms of its use to human life and economic activity, groundwater has a dominant role. Therefore, its proper management and conservation are of vital importance.

This chapter deals with the geological environment of groundwater occurrence, recharge, and movement, its exploitation, and methods of groundwater investigation and modelling for local and regional resource estimation, understanding systematics, and planning for sustainable groundwater development. Providing an adequate

technical framework necessitates defining some of the terms related to groundwater.

3.1 Occurrence of groundwater

The definition of groundwater as the water contained beneath the surface in rocks and soil is conceptually simple and convenient, although the actual situation is somewhat complex. Water beneath the ground surface includes that contained in the soil, in the intermediate unsaturated zone below the soil, in the capillary fringe, and below the water table (Fig. 3.2). The soil is composed of the broken down and weathered rock and decayed plant debris at the ground surface. The region between the soil and the water table is referred to as the unsaturated or the *vadose* zone. The unsaturated zone contains both air and water, while in the saturated zone all of the voids are filled with water. The *water table*, the boundary between the unsaturated and saturated zones, is often misused as a synonym for groundwater. To be precise, it represents the upper surface of groundwater, where hydraulic pressure is equal to atmospheric pressure. The water level encountered in an idle well, or in a well a long time after any water was pumped from it, is often the same level as the water table (Fig. 3.2). In groundwater parlance, some authors prefer to use the term borehole for drilled wells, reserving the term well for a large diameter dug well. Because the term tubewell is also commonly

Modern Hydrology and Sustainable Water Development, 1st edition. © S.K. Gupta
Published 2011 by Blackwell Publishing Ltd.

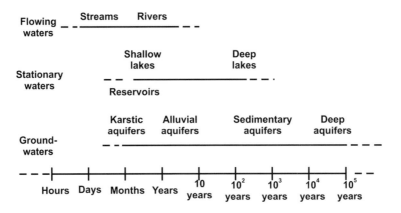

Fig. 3.1 Variation in residence time of water in various components of the hydrologic cycle. Note the longer residence time of most groundwaters compared to surface water bodies. See also Table 3.1. Redrawn from Meybeck *et al.* (1989). © Blackwell Publishing.

used for drilled water wells, the general term well is preferred in the present text to refer collectively to borehole, borewell, tubewell, handpump, and dugwell.

Strictly speaking, therefore, groundwater refers only to water in the saturated zone below the water table, and the total water column beneath the Earth's surface is usually called subsurface water (Fig. 3.2). The saturated and unsaturated zones are hydraulically connected, and the po-

sition of the water table fluctuates seasonally in response to recharge from rainfall and also as a result of groundwater abstraction. Geological formations having pore spaces that are saturated and permit easy movement of the groundwater are called *aquifers*. Materials through which water can pass easily are said to be *permeable*, and those that scarcely allow water to pass through, or only with difficulty, are termed *impermeable* or *semi-permeable*, respectively.

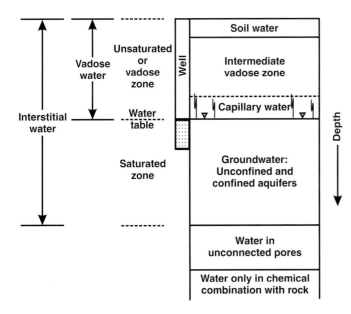

Fig. 3.2 Schematic sketch for defining various zones of subsurface water.

3.1.1 Aquifer formations

Some amounts of groundwater occur in most geological formations because nearly all rocks in the uppermost part of the Earth's crust, of whatever type, origin, or age, possess openings called pores or voids. Geologically, rock formations can be subdivided into three main types according to their origin and method of formation:

Sedimentary rocks are formed by deposition of weathered rock material, usually under water in lakes, rivers, and in the sea, and occasionally by the wind as aeolian deposits. In unconsolidated granular materials, such as sands and gravels, voids constitute the pore spaces between the grains (Fig. 3.3a, b and c). These may become consolidated physically by compaction due to the overburden of earth materials and chemically by cementation (Fig. 3.3d), to form typical sedimentary rocks such as sandstone, limestone, and shale, that have considerably reduced void spaces between the grains.

Igneous rocks are formed from molten magma rising from great depths and subsequent cooling to form crystalline rocks either below the ground or on the land surface. The former group includes igneous rocks such as granites and volcanic types of rocks such as basalts. The latter group are as-

sociated with various types of volcanic eruptions and also include hot ashes. Most igneous rocks are highly consolidated and, being crystalline, usually have few void spaces between the grains.

Metamorphic rocks are formed by deep burial, compaction, melting, and alteration/recrystallization of other rocks during periods of intense geological activity in the past. Metamorphic rocks, such as gneisses and slates, are normally well consolidated with few void spaces in the matrix between the grains. In the highly consolidated rocks, such as lavas, gneisses, and granites, the only void spaces may be fractures resulting from cooling or stresses generated by movement of the Earth's crust in the form of folding and faulting. These fractures may be completely closed or may have limited interconnected openings of relatively narrow aperture (Fig. 3.3f). Weathering and associated decomposition of igneous and metamorphic rocks may significantly increase the void spaces in the rock matrix as well as in the fractures. Fractures may enlarge to become open fissures as a result of dissolution by the chemical action of flowing groundwater or infiltrating rainwater (Fig. 3.3e). Limestone, largely made up of calcium carbonate, and evaporites, composed of gypsum and other salts, are particularly susceptible to active

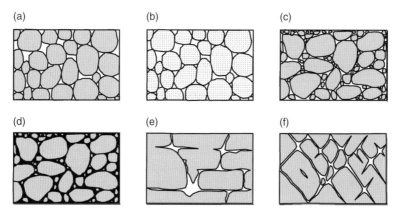

Fig. 3.3 Texture and porosity of typical aquifer forming rocks/sedimentary deposits. Primary porosity: (a) Well-sorted unconsolidated sedimentary deposits with high porosity. (b) Well-sorted sedimentary deposit comprising pebbles that may themselves be porous, making the deposit as a whole highly porous. (c) Poorly-sorted sedimentary deposits having low porosity. (d) Sedimentary deposits with primary porosity reduced by secondary deposition of mineral matter (cementing material) between the grains. Secondary porosity: (e) Rock with porosity increased by dissolution due to chemical action of infiltrating (or flowing) water. (f) Rock with porosity increased by fracturing. Redrawn from Todd (2006). © John Wiley & Sons Inc.

dissolution, which can produce caverns, swallow holes, and other features characteristic of karstic aquifers.

In contrast to the main geological subdivision of rock types according to their origin, earth materials in hydrology are usually classified on the basis of their potential to bear and transmit water, into four broad groups: (i) unconsolidated materials, generally referred to as clastic sediments; (ii) porous sedimentary rocks; (iii) porous volcanic rocks; and (iv) fractured rocks. In unconsolidated materials, water is transported through the primary openings in the rock/soil matrix. Consolidation is a process where loose materials become compacted and coherent. Sandstones and conglomerates are common consolidated sedimentary rocks formed by compaction and cementation. Carbonate rocks (i.e. limestone and dolomite) are sedimentary rocks that are formed by chemical precipitation. Water is usually transported through secondary openings in carbonate rocks that are progressively enlarged over time by dissolution of rock by the action of carbonic acid contained in water which arises from dissolution of atmospheric carbon dioxide. Movement of water through volcanics and fractured rock is dependent upon the interconnection and density of flow pathways.

A method of distinguishing an aquifer and the way in which groundwater occurs in it, when considering both its development and protection of groundwater quality, is shown in Fig. 3.4. An *unconfined aquifer* contains a phreatic surface (water table) as the upper boundary that fluctuates in response to recharge (e.g. from rainwater infiltration or seepage of irrigation water) and discharge (e.g. pumped from a well). Unconfined aquifers are generally close to the land surface, with continuous layers of materials of high intrinsic permeability extending from the land surface to the base of the aquifer, usually an impermeable rock or clay formation. An unconfined aquifer, therefore, continually interacts with the overlying unsaturated zone and also with nearby streams and channels. In contrast, at greater depths, the effective thickness of an aquifer often lies between two impermeable layers (Fig. 3.4), so is known as a *confined aquifer*. If the overlying layer has low permeability that retards the movement of water, it is known as an *aquitard*. While aquitards do not yield water freely to wells, these may transmit appreciable amount of water to and from adjacent aquifers, particularly under conditions of hydraulic stress. When sufficiently thick, aquitards may constitute important groundwater storage; sandy clay is an example. This situation causes the confined aquifer beneath an aquitard to be partially confined or semi-confined. When groundwater is pumped from a semi-confined (leaky) aquifer, water moves both horizontally within the aquifer and vertically through the semi-permeable layer. If the overlying

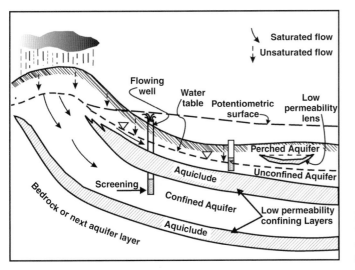

Fig. 3.4 Schematic cross-section showing unconfined and confined aquifers and the confining beds called either aquicludes or aquitards, depending on their permeability.

layer is relatively impermeable and also does not yield appreciable quantities of water to wells, it is called an *aquiclude*; clay is an example. A relatively impermeable formation that neither bears nor transmits water is known as an *aquifuge*; unweathered granite belongs to this category. In the latter two cases, the aquifer sandwiched between two impermeable layers becomes fully confined; at any point in the confined aquifer, the water pressure is higher than atmospheric, because of the higher elevation of the outcrop from where the aquifer is recharged through lateral subsurface flow. If a borehole is drilled into the confined aquifer, water rises up to a level that balances the hydraulic pressure in the aquifer.

An imaginary surface joining the water level in boreholes tapping the same confined aquifer is called the *potentiometric or piezometric surface*, which can be above or below the groundwater surface (water table) in the overlying unconfined aquifer (Fig. 3.4). If the pressure in a confined aquifer is such that the potentiometric surface is above ground level, then a borehole drilled into the aquifer will overflow (Fig. 3.4). Recharge to confined aquifers is predominantly from areas where the confining bed is breached either by an erosional unconformity, fracturing, depositional absence, or from the outcrop region known as the 'recharge area'. For a phreatic aquifer, which is the first unconfined aquifer formed below the surface, the recharge is from all over the ground surface and so potentiometric surface and water table coincide.

A special type of unconfined aquifer, where a groundwater body is separated above the water table by a layer of unsaturated material, is called a *perched aquifer*. A perched aquifer occurs when water moving down through the unsaturated zone encounters an impermeable formation (Fig. 3.4). Clay lenses in sedimentary deposits often have shallow perched water bodies overlying them. Wells tapping perched aquifers generally yield small quantities of water temporarily, which may be used for domestic water supply for individual households or small communities. For groundwater development, unconfined aquifers are often favoured because their much higher storage coefficient makes them more efficient for exploitation than confined aquifers. Unconfined

aquifers, being shallower, are cheaper to drill and require less energy to pump out water.

Thus, occurrence and movement of groundwater are controlled by the geological environment in which they occur, the geometric arrangement of different formations, and by the hydraulic conditions prevailing in the subsurface.

3.2 Movement of groundwater

Groundwater is usually not static but moves slowly through aquifers, both laterally and vertically. External forces acting on water in the subsurface include gravity, pressure exerted by the atmosphere and by the overlying water column, and molecular attraction between aquifer solids and water. In the subsurface, water can occur as: (i) water vapour, which moves from regions of higher pressure to lower pressure; (ii) liquid water, which wets dry soil particles by absorption; (iii) water retained on particles under the molecular force of adhesion; and (iv) water that is not subjected to attractive forces towards the surface of solid particles and is under the influence of the force of gravity.

3.3 Hydraulic head

In the saturated zone, groundwater flows through interconnected voids in response to the difference in fluid pressure and elevation. The driving force is measured in terms of the *hydraulic head* (or potentiometric head), which is defined by *Bernoulli's equation*:

$$h = z + \frac{p}{\rho g} = \frac{\vartheta^2}{2g} \tag{3.1}$$

where h = hydraulic head [L]; z = elevation above datum [L]; p = fluid pressure [$ML^{-1}T^{-2}$] with constant density ρ [ML^{-3}]; g = acceleration due to gravity [LT^{-2}]; and ϑ = fluid velocity [LT^{-1}]. Pressure head (or fluid pressure) h_p is defined as:

$$h_p = \frac{p}{\rho g} \tag{3.2}$$

By convention, the pressure head is expressed in terms of atmospheric pressure. At the water

(a)   (b)

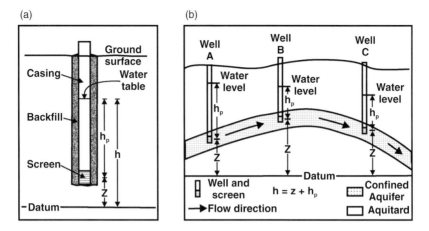

Fig. 3.5 Relationship between hydraulic head, pressure head, and elevation head within a well tapping (a) unconfined aquifer; (b) confined aquifer.

table the pressure is equal to 1 *atm*; therefore, the head is reckoned as 0. In the unsaturated zone, water is held in tension and the pressure head is less than the atmospheric pressure (h_p < 0; negative matric suction head, $-\psi$; see Section 2.5.3.3.1). Below the water table, in the saturated zone, the pressure head is greater than the atmospheric pressure (h_p > 0). As groundwater velocities are usually very low, the velocity component of the hydraulic head can be neglected. Thus, the hydraulic head can usually be expressed as:

$$h = z + h_p \qquad (3.3)$$

Fig. 3.5 depicts the relationship between hydraulic head, pressure head, and the topographic elevation given by Eqn 3.3 in a well tapping the unconfined and confined aquifers. Hydraulic heads are normally measured with respect to an arbitrary datum, which is often taken as mean sea level. The elevation head is the height above the reference datum of the midpoint of the section of borehole or well that is open to the aquifer and the pressure head is the height of the water column above this midpoint.

Groundwater moves through the sub-surface from areas of higher hydraulic head to areas of lower hydraulic head (Eqn 3.3). Rate of groundwater movement depends upon the slope of the hydraulic head (hydraulic gradient) and intrinsic aquifer and fluid properties.

3.3.1 Darcy's law, hydraulic conductivity and permeability

Flow of groundwater through an aquifer is governed by *Darcy's Law*, which states that the rate of flow is directly proportional to the hydraulic gradient:

$$Q/A = q = -K\,(h_1 - h_2)/l = K\,\Delta h/\Delta l \qquad (3.4)$$

where Q is the rate of flow [L^3T^{-1}] through area A [L^2] under a *hydraulic gradient* $\Delta h/\Delta l$ [LL^{-1}], which is the difference in hydraulic heads ($h_1 - h_2$) between two measuring points; and q is the volumetric flow rate per unit surface area [LT^{-1}]. The direction of groundwater flow in an aquifer is at right angles to lines of equal head. A simple experimental set-up to demonstrate Darcy's Law is shown in Fig. 3.6, also indicating the elevation and pressure components of the hydraulic head referred to above. Note that Eqn 3.4 is the same as (Eqn. 2.12), the latter describing infiltration in a vertical direction. The minus sign in Darcy's Law indicates that the flow is in the direction of decreasing hydraulic head. The constant of proportionality in the equation, K, known as *hydraulic conductivity*, has dimensions of LT^{-1}, as hydraulic gradient ($\Delta h/\Delta l$) is dimensionless. The parameter K is a measure of the ease with which water flows through the sand contained in the cylinder in the laboratory experiment of Fig. 3.6 or through the various materials that form aquifers and aquitards. The similarity

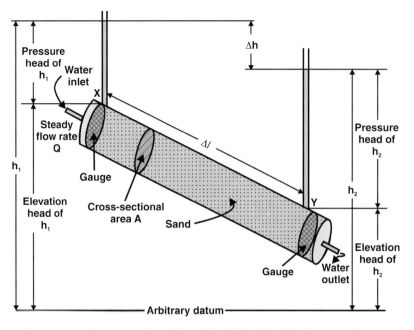

Fig. 3.6 Experimental apparatus to demonstrate Darcy's Law. Also indicated are the elevation and pressure components of the hydraulic head referred to in the text and in Fig. 3.5. Modified from Todd (2006). © John Wiley & Sons Inc.

between Darcy's Law and other important laws of physics governing the flow of both electricity and heat through conductors may be also noted.

The hydraulic conductivity of a given medium is a function of the properties of the medium as well as of the fluid. Using empirically derived proportionality relationships and dimensional analysis, the hydraulic conductivity of a given medium transmitting a given fluid is:

$$K = \frac{k\rho g}{\mu} \qquad (3.5)$$

where k = intrinsic permeability of porous medium [L^2]; ρ = fluid density [ML^{-3}]; μ = dynamic viscosity of fluid [$ML^{-1}T^{-1}$]; and g = acceleration due to gravity [LT^{-2}]. A commonly used unit for *permeability* is the *darcy* (D); more commonly *millidarcy* (mD) is used. Other units are cm^2 and m^2 (1 *darcy* $\approx 10^{-12}m^2 = 10^{-8}cm^2$).

The intrinsic permeability of a medium is a function of the shape and diameter of the pore spaces. Several empirical relationships describing intrinsic permeability have been presented in the literature. Fair and Hatch (1933) used a packing factor, shape factor, and the geometric mean of the grain size to

estimate intrinsic permeability. Krumbein (1943) used the square of the average grain diameter to approximate the intrinsic permeability of a porous medium. Values of fluid density and dynamic viscosity are dependent upon water temperature. Fluid density is additionally dependent upon total dissolved solids (TDS).

Darcy's Law adequately describes laminar groundwater flow under most naturally occurring hydrogeological conditions, i.e. for fractured rocks as well as granular materials. The overall permeability of a rock mass depends on a combination of the size of the pores and the degree to which the pores are interconnected. For clean, granular materials, hydraulic conductivity increases with grain size. Typical ranges of hydraulic conductivity and permeability for the main types of geological materials are shown in Fig. 3.7.

3.3.2 Porosity and specific yield

The *specific discharge* per unit area, or *Darcy flux* ($q = Q/A$), in Eqn 3.4 gives the apparent volumetric flow velocity through a given cross-section of the aquifer that includes both solids and voids. But,

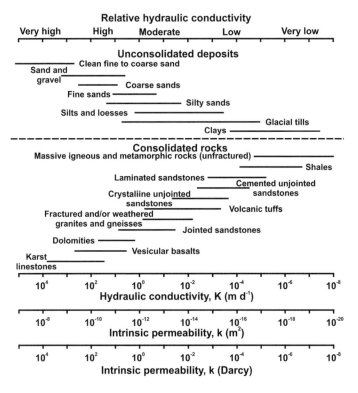

Fig. 3.7 Range of hydraulic conductivity (K) and intrinsic permeability (k) values for different geological materials. Modified from Chilton and Seiler (2006). © IWA Publishing.

since the flow can only occur through the voids or pore spaces, to obtain an estimate of the flow velocity through the pores (the *pore water velocity*), one needs to know the volume ratio of the connected pores to the total volume of the medium. This ratio is defined as the *effective* or *dynamic porosity*, n_e, of the medium and may be significantly lower than the *total porosity*, n, of the medium defined simply as the ratio of void space (V_V) to the total volume (V_T). Thus $n = V_V/V_T$ and average *pore water velocity* is obtained by using:

$$v_x = q/n_e = -Ki/n_e \qquad (3.6)$$

where i is conventionally used to represent the hydraulic gradient $\Delta h/\Delta l$. For unconfined aquifers, n_e is close to *specific yield*, and S_y is defined as the ratio of the water that will drain from a saturated rock under the force of gravity to the total volume of the aquifer material. Another similar term, *specific retention, S_r*, is defined as the ratio of the volume of water that a unit volume of the material can retain against the draining force of gravity to the total volume of the material. The porosity of a

rock is, therefore, equal to the sum of the specific yield and specific retention of a medium. A typical relationship between specific yield and specific retention with the total porosity for different soil types is illustrated in Fig. 3.8. Primary porosity is an inherent property of the soil or rock matrix, while secondary porosity is developed in a material after its emplacement through processes such as dissolution and fracturing. Representative ranges of porosity for sedimentary materials are given in Table 3.1.

Materials in which inter-granular flow occurs, such as clastic sediments, have effective porosities of 0.15 to 0.25, so in these aquifers the actual groundwater flow velocity is 4 to 6 times the specific discharge. Average pore water velocity obtained using Eqn 3.6 is always higher than specific discharge, and increases with decreasing effective porosity. The average pore water velocity in the direction of groundwater flow does not really represent the true velocity of flow through the pore spaces. These microscopic velocities are generally higher because the intergranular flow pathways are irregular, tortuous and longer than

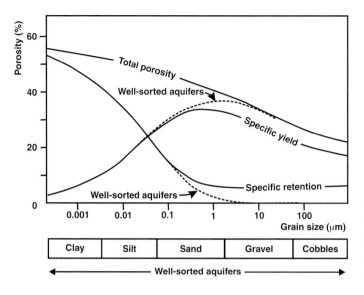

Fig. 3.8 A typical relationship between total porosity, specific yield, and specific retention for aquifers with unconsolidated sediments having primary porosity. Redrawn from Davis and DeWiest (1966). © John Wiley & Sons Inc.

the average linear dimension of the macroscopic pathways Average linear velocity, v, is a key parameter in groundwater protection, as it defines the travel times for water and solutes within aquifers. For unconsolidated granular aquifers, typical natural groundwater flow velocities range from a few $mm\ d^{-1}$ for silts and fine sands to 5-10 $m\ d^{-1}$ for clean and coarse gravels.

Fractured rocks are characterized by low porosity and localized high hydraulic conductivity. Very high flow velocities of up to several km/d may result (Orth *et al.* 1997; US-EPA 1997), especially where some fractures are enlarged by dissolution (Fig. 3.3e). Extensive development of solution cavities in limestone areas can result in karstic ter-

rain, which is typified by channels, sinkholes, depressions, and caves, into which all traces of surface flow may disappear. Such conditions can be quite favourable for groundwater supplies from springs and boreholes, but aquifers of this type are often highly vulnerable to various types of pollution (Malard *et al.* 1994). Groundwater may flow through the pore spaces between aquifer grains or through fractures (Fig. 3.3), or a combination of both, such as, in a jointed sandstone or limestone. Hydrogeologists commonly refer to these as dual-porosity aquifers that have primary porosity and permeability from the presence of inter-granular pores and additional secondary porosity and permeability due to fracture systems. Serious note should be taken of the presence of any highly fractured rocks located close to water supplies in view of the risk of rapid transport of pollutants from potential contaminant sources. Such an environment is considered prone to a very high risk of groundwater pollution.

3.3.3 Groundwater flow and transmissivity

Transmissivity, T, is a measure of the amount of water that can be transmitted horizontally through a unit width by the fully saturated thickness of

Table 3.1 Range of porosity values for unconsolidated sedimentary materials. Source: Heath (1983).

Material	Porosity	Material	Porosity
Clay	0.45–0.55	Fine to medium mixed sand	0.30–0.35
Silt	0.40–0.50	Gravel	0.30–0.40
Medium to coarse mixed sand	0.35–0.40	Gravel and sand	0.20–0.35
Uniform sand	0.30–0.40		

an aquifer under a unit hydraulic gradient. Transmissivity is equal to the hydraulic conductivity multiplied by the saturated thickness of the aquifer, and is given by:

$$T = Kb \qquad (3.7)$$

where K = hydraulic conductivity [LT^{-1}]; and b = saturated thickness of the aquifer [L]. Since transmissivity depends on hydraulic conductivity and saturated thickness, its value differs at different locations within aquifers composed of heterogeneous materials and bounded by sloping confining beds, or under unconfined conditions where the saturated thickness varies in response to fluctuations in the water table.

3.3.4 Aquifer storage

The *storage coefficient* or *storativity* S (dimensionless) is the volume of water that an aquifer will take in or give out from storage per unit surface area per unit change in head. At the water table, water is released from storage by gravity drainage. Below the water table, water is released from storage due to reduction of hydrostatic pressure within the pore spaces which results from withdrawal of water from the aquifer. The total load above an aquifer is supported by a combination of the solid matrix of the aquifer and by the hydraulic pressure exerted by the water in the aquifer. Withdrawal of water from the aquifer results in a decline in pore water pressure, as a result, more of the overburden must be supported by the solid matrix. This causes compression of rock particles in the aquifer matrix, which brings them closer to each other, leading to a reduction in effective porosity of the aquifer. Further, the decrease in hydraulic pressure causes the pore water to expand. Both these conditions, namely compression of the mineral skeleton and expansion of the pore water, cause water to be expelled from the aquifer.

The *specific storage*, S_s, is the amount of water per unit volume of a saturated formation that is taken in/released from storage owing to compression of the mineral skeleton and expansion of the pore water per unit change in hydraulic head. The specific storage [L^{-1}] is given by:

$$S_s = -\frac{dV_w}{V_t}\frac{1}{db} = \rho_w g \left(\alpha + n\beta \right) \qquad (3.8)$$

where dV_w is the amount of water stored in or expelled from the entire volume, V_t, of the aquifer for a head change by db; ρ_w = density of water [ML^{-3}]; g = acceleration due to gravity [LT^{-2}]; α = compressibility of the aquifer skeleton [$1/(ML^{-1}T^{-2})$]; n = porosity [L^3L^{-3}]; and β = compressibility of water [$1/(ML^{-1}T^{-2})$]. The minus sign arises because S_s is positive when head decline, db, is negative and dV_w is positive.

Within a confined aquifer, the entire thickness of the aquifer remains fully saturated whether water is released or taken in. Therefore, water is released due to the compaction of the mineral skeleton and expansion of the pore water. The *storage coefficient*, S, is given as:

$$\text{(Confined aquifer)} \quad S = bS_s \qquad (3.9)$$

where b = vertical thickness of the aquifer [L].

Let us consider a vertical column in a confined aquifer that has a unit cross-sectional area and spans the whole thickness, b, of the aquifer. If the head in the aquifer declines by one unit, the volume of water expelled from the column, by the very definition of storativity, is equal to S. In other words, S is the decrease in the volume of water stored in the aquifer per unit surface area per unit decline of head. It follows from Eqn 3.8 and Eqn 3.9 that the volume of water removed from storage with cross-section A and subject to head change db is:

$$dV_w = -S\,A\,db \qquad (3.10)$$

The minus sign on the right-hand side of the above equation arises because a decline in head results in positive dV_w. Values of storage coefficient in confined aquifers are generally less than 0.005 (Todd 2006). Values between 0.005 and 0.10 generally indicate a leaky confined aquifer.

In an unconfined aquifer, the degree of saturation varies as water is added to or removed from it. As the water table falls, water is released by gravity drainage coupled with compaction of the aquifer matrix and, to a lesser degree, by expansion of the pore water. The volume of water released by gravity drainage is given by the *specific yield* of the aquifer. The storage coefficient of an unconfined aquifer is, therefore, given by the sum of the specific yield and the volume of water released due to the specific storage as:

(Unconfined aquifer) $S = S_y + bS_s \approx S_y$ (3.11)

The magnitude of specific storage is quite small, generally less than 10^{-5} m^{-1}. As the value of specific yield is usually several orders of magnitude higher than specific storage, the storage coefficient of an unconfined aquifer is approximately the same as its specific yield. The storage coefficient of unconfined aquifers typically ranges from 0.10 to 0.30.

3.3.5 Flow in stratified media

In the above discussion it was assumed that hydraulic conductivity of porous media is homogenous and isotropic, implying that the value of K does not vary with location within the aquifer and is the same in all directions. However, geological materials are rarely homogeneous in all directions. In the field it is not uncommon to find discontinuous heterogeneity resulting from different geological structures, such as bedrock outcrop contacts, clay lenses, and buried oxbow stream cut-offs. Sedimentary formations of deltaic, alluvial, and glacial origin often exhibit heterogeneity that steadily varies in some particular direction.

As geological strata are formed, individual particles usually rest with their flat surfaces down in a process called imbrication, giving rise to anisotropy. Consequently, flow is generally less restricted in the horizontal direction than in the vertical and K_x is greater than K_z for most situations. Therefore, properties such as hydraulic conductivity can often be approximated to a constant value in one direction. Layered heterogeneity occurs when strata of homogeneous, isotropic materials are interbedded. Such layered conditions commonly occur in alluvial, lacustrine, and marine deposits.

3.3.5.1 Equivalent hydraulic conductivity for stratified systems

Let us consider an aquifer consisting of two horizontal layers, each being individually isotropic, with different values of thickness and hydraulic conductivity (Fig. 3.9). For horizontal flow parallel to the layers, the flow q_1 in the upper layer per unit width is:

$$q_1 = K_1 i b_1 \qquad (3.12)$$

Since i must be the same for each layer for horizontal flow, it follows that the total horizontal flow q_x is:

$$q_x = q_1 + q_2 = i(K_1 b_1 + K_2 b_2) = i K_x(b_1 + b_2)$$

$$(3.13)$$

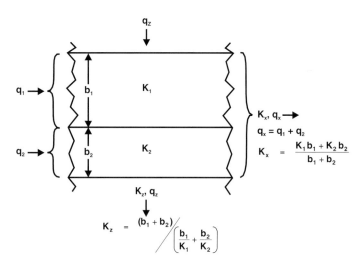

Fig. 3.9 Sketch showing two horizontal strata, each with different thicknesses and hydraulic conductivity. Equivalent hydraulic conductivities of the system in horizontal and vertical directions are also indicated.

where K_x is the effective or equivalent hydraulic conductivity for the entire system. Solving for K_x yields:

$$K_x = \frac{K_1 b_1 + K_2 b_2}{b_1 + b_2} \qquad (3.14)$$

which can be generalized for n layers as:

$$K_x = \frac{K_1 b_1 + K_2 b_2 + \cdots + K_n b_n}{b_1 + b_2 + \cdots b_n} = \frac{\sum K_i b_i}{\sum b_i}$$

$$(3.15)$$

This equation defines the equivalent horizontal conductivity for a stratified system.

For vertical flow through the two layers in Fig. 3.9, the flow per unit horizontal area in the upper layer is:

$$q_z = K_1 \frac{db_1}{b_1} \quad so\ that \quad db_1 = \frac{b_1}{K_1} q_z \qquad (3.16)$$

where db_1 is the head loss within the first layer. From consideration of continuity of flow, q_z must be the same for other layers, so that the total head loss is given by:

$$db_1 + db_2 = \left[\frac{b_1}{K_1} + \frac{b_2}{K_2} \right] q_z \qquad (3.17)$$

For an equivalent homogenous system:

$$q_z = K_z \left[\frac{db_1 + db_2}{b_1 + b_2} \right] \quad so\ that$$

$$db_1 + db_2 = \left[\frac{b_1 + b_2}{K_z} \right] q_z \qquad (3.18)$$

and equating Eqn 3.18 with Eqn 3.17:

$$K_z = (b_1 + b_2) \Big/ \left(\frac{b_1}{K_1} + \frac{b_2}{K_2} \right) \qquad (3.19)$$

which can be generalized for n layers as:

$$K_z = (b_1 + b_2 \cdots + b_n) \Big/ \left(\frac{b_1}{K_1} + \frac{b_2}{K_2} + \cdots + \frac{b_n}{K_n} \right)$$

$$= \frac{\sum b_i}{\sum \frac{b_i}{K_i}} \qquad (3.20)$$

This equation defines the equivalent vertical hydraulic conductivity for a stratified system.

As mentioned above, in an alluvial aquifer, horizontal hydraulic conductivity is higher than that in the vertical direction. This also follows from

the above derivation of Eqn 3.14 and Eqn 3.19 for $K_x > K_z$, i.e.:

$$\frac{K_1 b_1 + K_2 b_2}{b_1 + b_2} > (b_1 + b_2) \Big/ \left(\frac{b_1}{K_1} + \frac{b_2}{K_2} \right) \qquad (3.21)$$

which reduces to:

$$\frac{b_1 b_2 (K_1 - K_2)^2}{(b_1 + b_2)(K_1 b_1 + K_2 b_2)} > 0 \qquad (3.22)$$

As the left-hand side of this expression is always positive, it must be >0, thereby establishing that $K_x > K_z$. The ratio K_x/K_z, usually falls in the range of 2–10 for alluvium, but values up to 100 or more occur where clay layers are present (Morris and Johnson 1967). For consolidated geological formations, anisotropy is governed by orientation of strata, presence of fractures, solution openings, or other structural formations, which may not necessarily have horizontal alignment.

In applying Darcy's law to two-dimensional flow in anisotropic media, an appropriate value of K must be selected in accordance with the direction of flow. For directions other than horizontal (K_x) and vertical (K_z), the K value can be obtained from:

$$\frac{1}{K_\beta} = \frac{\cos^2 \beta}{K_x} + \frac{\sin^2 \beta}{K_z} \qquad (3.23)$$

where K_β is the hydraulic conductivity in the direction making an angle β with the horizontal.

3.3.6 General groundwater flow equations

The first step in developing a mathematical model of almost any system is to formulate what are known as general equations. These are differential equations that are derived from the physical principles governing the process to be modelled. In the case of subsurface flow, the governing physical principles are Darcy's law and the mass conservation principle. In a general form, Darcy's law (Eqn 3.4 above) may be written as:

$$q = -K \frac{\partial b}{\partial s} \qquad (3.24)$$

where q, K, and b are as defined above (Section 3.3.1), and s is the distance along the average direction of flow. For a representative elemental volume (REV), in the Cartesian coordinate system as

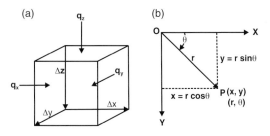

Fig. 3.10 (a) A hypothetical representative elemental volume (REV) within the saturated zone with dimensions $\Delta x \times \Delta y \times \Delta z$. (b) A point $P(x, y)$ in x-y plane in the Cartesian co-ordinate system is represented by co-ordinates r and θ in the radial coordinate system as shown.

shown in Fig. 3.10a, the law of mass conservation (continuity principle) is given by:

Rate of mass accumulation/release

= Rate of mass inflow

    − Rate of mass outflow   (3.25)

Let us now assume that the macroscopic flow in the vicinity of this REV is one-dimensional in the x direction: $q_x \neq 0$; $q_y = q_z = 0$. The mass flux (mass/time) of water entering through the left-hand side face of the REV is:

Rate of mass inflow $= \rho_w(x)\, q_x(x) \Delta y \Delta z$ (3.26)

where $\rho_w(x)$ is water density at coordinate x and $q_x(x)$ is the specific discharge at the point x. The corresponding outflux from the right-hand side face of the REV is:

Rate of mass outflow

    $= \rho_w(x + \Delta x)\, q_x(x + \Delta x)\, \Delta y \Delta z$ (3.27)

When these two fluxes are identical, the flow is in a steady state. When the two are different, the flow is transient and there must be change in the mass of water stored in the REV. According to the definition of specific storage, S_s (Eqn 3.8), change in the volume of water stored in an element of volume, V_t, when the head changes by an amount db, is:

$$dV_w = -S_s\, db\, V_t \tag{3.28}$$

For the time interval, dt, this becomes:

$$\frac{\partial V_w}{\partial t} = -S_s \frac{\partial b}{\partial t} \Delta x \Delta y \Delta z \tag{3.29}$$

The rate of change in the mass of water stored in the REV is, therefore:

$$\frac{\partial m}{\partial t} = \rho_w \frac{\partial V_w}{\partial t} = -\rho_w S_s \frac{\partial b}{\partial t} \Delta x \Delta y \Delta z \tag{3.30}$$

The rate $\partial b/\partial t$ is expressed as a partial derivative because, in this case, b is a function of two variables, x and t. Substituting Eqn 3.26, Eqn 3.27 and Eqn 3.30 into Eqn 3.25, one obtains for the rate of change of mass in the REV:

$$\frac{\partial m}{\partial t} = -\rho_w S_s \frac{\partial b}{\partial t} \Delta x \Delta y \Delta z = \rho_w(x)\, q_x(x) \Delta y \Delta z$$

$$- \rho_w(x + \Delta x)\, q_x(x + \Delta x) \Delta y \Delta z \tag{3.31}$$

Dividing by $\Delta x\, \Delta y\, \Delta z$ and rearranging:

$$\rho_w S_s \frac{\partial b}{\partial t}$$

$$= \left[\frac{\rho_w(x + \Delta x) q_x(x + \Delta x) - \rho_w(x) q_x(x)}{\Delta x} \right]$$

$$\tag{3.32}$$

The right-hand side is a derivative in the limit Δx tending to zero:

$$\rho_w S_s \frac{\partial b}{\partial t} = \frac{\partial(\rho_w q_x)}{\partial x} = \rho_w \frac{\partial q_x}{\partial x}$$

$$+ q_x \frac{\partial \rho_w}{\partial x} \approx \rho_w \frac{\partial q_x}{\partial x} \tag{3.33}$$

because the term $\rho_w \frac{\partial q_x}{\partial x}$ is generally orders of magnitude greater than $q_x \frac{\partial \rho_w}{\partial x}$.

In most situations, Eqn 3.33 is valid and governs the flow of groundwater. Derivations of more rigorous general flow equations are given by Freeze and Cherry (1979), Verruijt (1969), and Gambolati (1973, 1974).

Substituting the definition of q_x, given by Darcy's law, (Eqn 3.24) into Eqn 3.33 gives the one-dimensional general equation for saturated groundwater flow:

$$S_s \frac{\partial b}{\partial t} = \frac{\partial}{\partial x}\left(-K_x \frac{\partial b}{\partial x} \right) \tag{3.34}$$

For three-dimensional flow, the general equation can similarly be derived as:

$$S_s \frac{\partial h}{\partial t} = \frac{\partial}{\partial x}\left(K_x \frac{\partial h}{\partial x}\right) + \frac{\partial}{\partial y}\left(K_y \frac{\partial h}{\partial y}\right) + \frac{\partial}{\partial z}\left(K_z \frac{\partial h}{\partial z}\right)$$

(3.35)

If there is any volumetric recharge or discharge from the REV (e.g. due to a pumping well), it can be represented by W, and the volumetric flux per unit volume $[L^3/T/L^3 = 1/T]$, and Eqn 3.35 becomes:

$$S_s \frac{\partial h}{\partial t} + W = \frac{\partial}{\partial x}\left(K_x \frac{\partial h}{\partial x}\right)$$
$$+ \frac{\partial}{\partial y}\left(K_y \frac{\partial h}{\partial y}\right) + \frac{\partial}{\partial z}\left(K_z \frac{\partial h}{\partial z}\right)$$

(3.36)

A mathematical representation of head ($h(x$, y, z, $t) = \ldots$) must follow the partial differential equation (Eqn 3.35 or Eqn 3.36), if it is to be consistent with Darcy's law and mass balance equation. This is the general form of the saturated flow equation, describing flow in all three dimensions (x, y, z), transient flow ($\partial h/\partial t \neq 0$), heterogeneous conductivities (e.g. $K_x = f(x)$), and anisotropic hydraulic conductivities ($K_x \neq K_y \neq K_z$). If hydraulic conductivities are assumed to be both homogenous (independent of location, i.e. x, y, and z) and isotropic (i.e. $K_x = K_y = K_z = K$), Eqn 3.35 simplifies to:

$$K\left(\frac{\partial^2 h}{\partial x^2} + \frac{\partial^2 h}{\partial y^2} + \frac{\partial^2 h}{\partial z^2}\right) = \nabla^2 h = S_s \frac{\partial h}{\partial t}$$

(3.37)

The symbol ∇^2 is called the *Laplacian operator*, and represents the sum of second derivatives:

$$\nabla^2 O = \frac{\partial^2 O}{\partial x^2} + \frac{\partial^2 O}{\partial y^2} + \frac{\partial^2 O}{\partial z^2}$$

(3.38)

The above derivation includes storage changes associated with transient flow. If, however, the flow is in steady state, i.e. $\partial h/\partial t = 0$, the general equation (Eqn 3.37) in homogenous media, with isotropic K, reduces to:

$$\nabla^2 (h) = 0$$

(3.39)

This is the famous partial differential equation named after French mathematician Pierre de Laplace (1749–1827), having numerous applications in a wide variety of fields such as fluid flow, heat conduction, electrostatics, and electricity.

There exist hundreds of known solutions to the *Laplace equation*, depending on the initial and boundary conditions; many of them apply directly to groundwater flow conditions. Similarly, for steady flow in homogenous media, with isotropic K, Eqn 3.36 with recharge or discharge, W, reduces to:

$$\nabla^2 (h) = W$$

(3.40)

Eqn 3.40 is known in physics and engineering as the *Poisson equation*.

In a homogenous and isotropic aquifer for two-dimensional horizontal flow (i.e. $q_z = 0$ so that $\partial h/\partial z = 0$) with fixed saturated thickness equal to b (e.g. in confined aquifers), Eqn 3.37 reduces to:

$$K\left(\frac{\partial^2 h}{\partial x^2} + \frac{\partial^2 h}{\partial y^2}\right) = S_s \frac{\partial h}{\partial t} \quad \text{or}$$

$$T\left(\frac{\partial^2 h}{\partial x^2} + \frac{\partial^2 h}{\partial y^2}\right) = S \frac{\partial h}{\partial t}$$

(3.41)

from the respective definitions of T and S in Eqns 3.7 and 3.9.

3.3.6.1 Radial coordinates

For axis-symmetric groundwater flow to wells, use of radial coordinates (Fig. 3.10b) is advantageous. It can be shown that in the radial coordinate system, Eqn 3.41 becomes:

$$\frac{\partial^2 h}{\partial r^2} + \frac{1}{r}\frac{\partial h}{\partial r} = \frac{S}{T}\frac{\partial h}{\partial t}$$

(3.42)

where r is the coordinate representing radial distance from the centre of the well. For steady state flow this reduces to:

$$\frac{\partial^2 h}{\partial r^2} + \frac{1}{r}\frac{\partial h}{\partial r} = 0$$

(3.43)

In an unconfined aquifer, saturated thickness varies with time as the hydraulic head changes. Therefore, ability of the aquifer to transmit water (i.e. transmissivity) is not constant during the course of pumping. In this case, Eqn 3.35 can be rewritten as:

$$\frac{\partial}{\partial x}\left(K_x h \frac{\partial h}{\partial x}\right) + \frac{\partial}{\partial y}\left(K_y h \frac{\partial h}{\partial y}\right)$$
$$+ \frac{\partial}{\partial z}\left(K_z h \frac{\partial h}{\partial z}\right) = \frac{S_y}{K}\frac{\partial h}{\partial t}$$

(3.44)

where S_y = specific yield (which is dimensionless).

If change in elevation of the water table is small in comparison to the saturated thickness of the aquifer, the variable thickness h can be replaced with an average thickness b that is assumed to be constant over the aquifer and Eqn 3.44 can be linearized in the form:

$$\frac{\partial^2 h}{\partial x^2} + \frac{\partial^2 h}{\partial y^2} + \frac{\partial^2 h}{\partial z^2} = \frac{S_y}{Kb} \frac{\partial h}{\partial t} \qquad (3.45)$$

3.3.7 Aquifer diffusivity

In problems dealing with surface water and groundwater interaction, it is often required to estimate the transmission of a pressure wave (e.g. due to a flood event) through the aquifer system. For homogeneous, isotropic aquifers, Eqn 3.45 can be rewritten as:

$$\frac{T}{S} \left(\frac{\partial^2 h}{\partial x^2} + \frac{\partial^2 h}{\partial y^2} + \frac{\partial^2 h}{\partial z^2} \right) = \frac{\partial h}{\partial t} \qquad (3.46)$$

where S = storage coefficient (dimensionless) and the ratio of transmissivity to storage coefficient (T/S) is called *aquifer diffusivity*.

The above equation is applicable, for confined and for unconfined conditions, where change in aquifer thickness is insignificant, and demonstrates a direct relationship between dispersal of a groundwater flood wave (and pressure wave) and aquifer diffusivity.

3.3.8 Flow lines and flow nets

Solutions of the Laplace equation (Eqn 3.39) have a unique property and are called harmonic functions. For each harmonic function, $h(x, y)$, there is a corresponding function, $\psi(x, y)$, known as the conjugate harmonic function of h. The function, $\psi(x, y)$ is also a solution of the Laplace equation and is generally called the stream function. The two functions (h and ψ) are related in such a way that their gradients are normal to each other and have the same magnitude. Because of this, it is possible to adopt a simple, flexible graphical technique for estimating the distribution of heads $h(x, y)$ and the streamlines $\psi(x, y)$ for a steady state two-dimensional flow.

Two examples of a flow net are illustrated in Fig. 3.11. There are two sets of curves in both

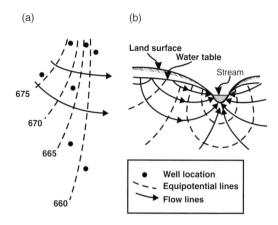

Fig. 3.11 (a) Typical groundwater flow net showing lines of equal hydraulic head (equipotential lines) and orthogonal flow lines in the horizontal plane. The flow lines indicate direction of groundwater flow – from higher potential level to lower level. (b) A typical cross-section with equipotential and flow lines in a vertical plane across an effluent stream. Redrawn from http://www.ees.nmt.edu/vivoni/mst/lectures/ Lecture16.pdf (accessed 7th April 2007).

flow nets: the lines of constant hydraulic head (the equipotential lines) and the *streamlines* (or the *flow lines*). These two sets of curves are mutually orthogonal where they intersect, and form boxes that are roughly square. These geometrical properties make it possible to draw reasonably accurate flow nets by hand and then use the result to analyse the distribution of heads, pressure discharges, and flow paths.

The general rules for flow net construction are:

- Flow nets apply only to two-dimensional steady flow in homogenous domains, where the flow is governed by the Laplace equation. Flow nets are generally used for vertical flow in a given plane (Fig. 3.11b) or for flow in a regionally confined aquifer with zero net recharge/leakage (Fig. 3.11a).
- Streamlines are perpendicular to equipotential lines and do not cross each other.
- Streamlines and equipotential lines intersect to form approximate squares.
- The conductivity K is assumed to be isotropic. In anisotropic aquifers, streamlines and equipotential lines are not orthogonal, except when the flow is parallel to the principal direction (Bear

and Dagen 1965). In order to calculate flow values for this situation, the boundaries of flow section must be transformed so that an equivalent isotropic medium is obtained. For a typical alluvial case of $K_x > K_z$, all horizontal dimensions are reduced in the ratio $\sqrt{K_z/K_x}$. This creates a transformed section with equivalent isotropic hydraulic conductivity K':

$$K' = \sqrt{K_x K_z} \qquad (3.47)$$

- The head difference between adjacent equipotential lines is the same throughout the flow net.
- The discharge [L³/T] through each stream tube (a channel bounded by two adjacent streamlines) is the same throughout the flow net.

The magnitude of discharge through a square ($\Delta s \times \Delta s$) and entire saturated thickness, b, normal to the plane of the flow net, is given by Darcy's law:

$$|\Delta Q| = Kb\Delta s \frac{|\Delta h_s|}{\Delta s} \qquad (3.48)$$

where $b\Delta s$ is the cross-sectional area through which the discharge Q passes; and $|\Delta h_s|$ is the head difference between adjacent equipotential lines in the flow net. If the total head loss $|h|$ is divided into n squares between any two adjacent flow lines, then $|\Delta h_s| = |h|/n$. For a given flow net, the discharge through each stream tube is the same regardless of location within the flow net. Therefore, if the total flow in divided into m stream tubes, one obtains

$$|Q| = |\Delta Q|m = \frac{Kb|h|m}{n} \qquad (3.49)$$

3.4 Dispersion

In saturated flow through porous media, velocities vary widely across any single pore, just as in a capillary tube where the velocity distribution for laminar flow is parabolic (Eqn. 2.43). In addition, pores are of different sizes, shapes, and orientations. As a consequence, when a labelled mass of water is introduced into a flow system, its elements move with different velocities, in different directions and, therefore, mix with other unlabelled water

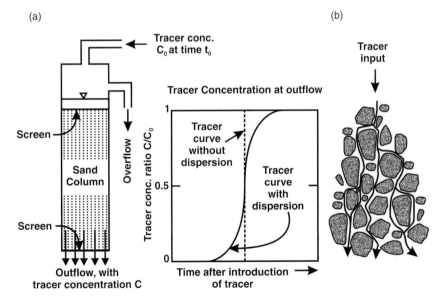

Fig. 3.12 (a) Sand column experiment to demonstrate the effect of longitudinal dispersion on tracer concentration curve at the outflow. (b) Lateral dispersion of a tracer originating from a point source in a porous medium. Redrawn from Todd (2006). © John Wiley & Sons Inc.

elements to progressively occupy an increasing portion of the flow regime. This phenomenon is known as *dispersion* and constitutes a non-steady, irreversible mixing process by which the tracer disperses within the surrounding water mass.

The effect of dispersion in the longitudinal direction can easily be seen and measured using a column packed with sand (Fig. 3.12a), supplied continuously after time, t_0, with water containing tracer of concentration, C_0. Tracer concentration, C, at the outflow follows a typical S-shaped curve. In addition to *longitudinal dispersion*, lateral dispersion also occurs because flowing water is continually bifurcating and reuniting as it follows tortuous flow paths around grains of a medium (Fig. 3.12b).

The equation for dispersion in a homogenous isotropic medium for a two-dimensional case has the form:

$$\frac{\partial c}{\partial t} = D_L \frac{\partial^2 c}{\partial x^2} + D_T \frac{\partial^2 c}{\partial y^2} - v \frac{\partial c}{\partial x} \qquad (3.50)$$

where c is the tracer concentration relative to that in the inflow (C/C_0); D_L and D_T are longitudinal and *transverse dispersion* coefficients; v is fluid velocity in the x-coordinate direction; y is the coordinate normal to the direction of flow; and t is time. Dimensions of the dispersion coefficient is L^2T^{-1}.

Dispersion is essentially a microscopic phenomenon caused by a combination of molecular diffusion and hydrodynamic mixing occurring with laminar flow through porous media.

3.5 Specialized flow conditions

Darcy's Law (Eqn 3.4) is an empirical relationship that is valid only under the assumption of laminar flow for a fluid with constant density. These assumptions are not always valid in Nature. Flow conditions in which Darcy's Law is generally not applicable are discussed below.

3.5.1 Fractured flow

Fractured-rock aquifers occur in environments in which the flow of water is primarily through fractures, joints, faults, or bedding planes, which have not been significantly enlarged by dissolution.

Fracturing adds secondary porosity to a rock/soil medium that already has some original porosity.

Fractures consist of pathways that are much longer than they are wide. These pathways provide conduits for groundwater flow that are much less tortuous than flow paths due to primary porosity in clastic sedimentary rocks/soils. At a local scale, fractured rock can be extremely heterogeneous. However, on a larger scale, regional flow can be treated as a flow through porous media. Effective permeability of crystalline rock typically decreases by two or three orders of magnitude in the first 300 m or so below ground surface, as fractures decrease in number or close completely under increased lithostatic load (Smith and Wheatcraft 1992).

3.5.2 Karst aquifers

Karst aquifers occur in environments where all or most of the flow of water is through joints, faults, bedding planes, pores, cavities, conduits, and caves that have been significantly enlarged by dissolution. Effective porosity in karst environments is mostly tertiary; where secondary porosity is modified by dissolution of carbonate rocks such as limestone/dolomite. Karst aquifers are generally highly anisotropic and heterogeneous. Flow in karst aquifers is often rapid and turbulent, where Darcy's law rarely applies. Solution channels leading to high permeability are favoured in areas where topographic, bedding, or jointing features promote localization of flow, which aids the solvent action of circulating groundwater, or in situations where well-connected pathways exist between recharge and discharge zones, favouring higher groundwater velocities (Smith and Wheatcraft 1992). Karst aquifers are generally found in limestone/dolomite rocks.

3.5.3 Permafrost

Temperatures significantly below $0°C$ produce permafrost, i.e. frozen ground surface. The depth and location of frozen water within the soil depends upon several factors, such as fluid pressure, salt content of the pore water, grain size distribution, mineralogy, and structure of the soil. Presence of frozen or partially frozen groundwater has a tremendous effect upon the flow. As water freezes, it expands to fill pore spaces. Soil that normally conveys water easily behaves like an aquitard or aquiclude when frozen. The flow of water in permafrost regions is analogous to fractured flow-like environments, where flow is confined to conduits in which complete freezing has not taken place.

3.5.4 Variable-density flow

Unlike aquifers containing constant-density water, where flow is controlled only by the hydraulic head gradient and the hydraulic conductivity, variable-density flow is also affected by change in the spatial location within the aquifer. Water density is commonly affected by temperature and by salinity. As water temperature increases and/or salinity decreases, its density decreases. A temperature gradient across an area influences the measurements of the hydraulic head and the corresponding hydraulic gradient. Intrinsic hydraulic conductivity is also a function of water temperature.

3.5.5 Seawater intrusion

Due to different concentrations of dissolved solids, the density of saline water is higher than the density of fresh water. In aquifers hydraulically connected to the sea, a significant density difference occurs, which can discourage mixing of waters and result in a diffuse interface between salt water and sea water. The depth of this interface can be estimated by the *Ghyben-Herzberg relationship* (Fig. 3.13, Eqn 3.51):

$$Z_s = \frac{\rho_f}{\rho_s - \rho_f} Z_w \qquad (3.51)$$

where Z_s = depth of interface below sea level; Z_w = elevation of water table above sea level; ρ_s = seawater density; ρ_w = freshwater density. For the common values of $\rho_w = 1.0$ and $\rho_s = 1.025$:

$$Z_s = 40 Z_w \qquad (3.52)$$

The Ghyben-Herzberg relationship assumes hydrostatic conditions in a homogeneous, unconfined aquifer. In addition, it assumes a near sharp interface between fresh water and salt water. In reality, there tends to be a mixing of salt water and fresh water in a zone of diffusion around the interface. If the aquifer is subjected to hydraulic head fluctuations caused by tides, the zone of mixed water will be broadened.

3.6 Groundwater measurements

A fundamental problem of groundwater systems is that most of the subsurface is inaccessible, therefore most measurements related to groundwater and its flow as well as aquifer characteristics are generally indirect.

3.6.1 Groundwater level measurements

Perhaps the most direct groundwater measurements are groundwater levels measured in wells. Groundwater level data are extremely important in providing information about the overall groundwater flow regime and water budget of an aquifer. Generally, water level data are recorded as X, Y

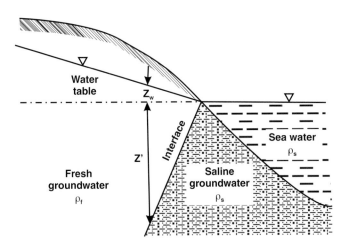

Fig. 3.13 Sea water–fresh water interface in an unconfined coastal aquifer.

(global coordinates) and *Z* (elevation) and provide the basic dataset for constructing contour maps of water table in a given area. In order to reliably interpret field measurements of groundwater level, a few basic ideas need to be clearly understood.

Unlike surface water, groundwater does not always 'run downhill'. Instead, groundwater flows from regions of higher potential head to regions of lower potential head. Hydrologists measure the driving force for groundwater flow in terms of 'head' [L]. As already defined, hydraulic head is a combination of gravitational potential and pressure potential (Eqn 3.1, 3.2 and 3.3). Relationship between hydraulic head, pressure head, and elevation, for both unconfined and confined aquifers, is depicted in Fig. 3.5.

Wells are generally open only at the screen – the bottoms are assumed to be sealed (although this may not be true in practice, particularly in hard rock, where screening is generally omitted). When a well has a significant depth (or screened interval) open to the aquifer, the water level in the well is an average of the head for the entire screened portion. If the well has only a very limited screened portion (approximating a point), the well is known as a *piezometer*, in which water level measured is the head at the open point (Fig. 3.14).

If head measurements from regular wells are available, it is possible to infer the horizontal di-

rection of flow from higher to lower head, but not the vertical flow. If there is a vertical component of the groundwater flow, there must be differences in vertical head. This can be determined if two or more piezometers, designed to probe different depths, are available at the same location. For example, in Fig. 3.14, water elevations (heads) in the monitoring wells W-1 and W-7 clearly indicate that groundwater must flow from W-1 towards W-7 in the unconfined aquifer. However, only by noticing the difference in head between the two piezometers, W-5 and W-6, it is possible to infer that groundwater in the unconfined aquifer is flowing upwards, as well as from left to right.

Vertical and horizontal gradients often become important when dealing with multiple aquifer systems, because it is necessary to know if water is flowing from a deeper aquifer to a shallower aquifer, or vice versa. For example, the piezometric surface in Fig. 3.14, measured in the monitoring wells W-2, W-4, and W-8, slopes from W-8 to W-2 and is higher than the water table. From this observation flow in the confined aquifer is from right to left and there is also some upward flux across the confining aquitard. It is, therefore, important to have knowledge of screen positions when measuring the heads in several wells to properly interpret the data.

Long-term, *in-situ* monitoring of groundwater levels is useful for:

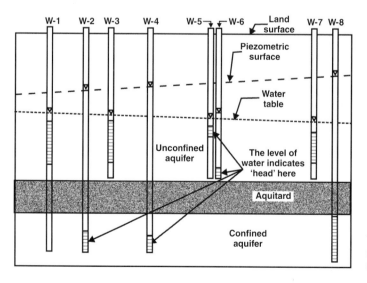

Fig. 3.14 Measuring groundwater levels, water table, and potentiometric surface gradients using wells in a multi-aquifer system.

- developing regional water tables in order to establish sources of water for industry, agriculture, and human consumption;
- estimating groundwater flow velocity and direction for risk assessment;
- analysing pump test data to determine the hydraulic properties of the aquifers;
- highlighting areas of concern for source water protection initiatives;
- determining potential impacts to wetlands, ecology, and freshwater streams;
- identifying groundwater divides to be used as boundary conditions for modelling of groundwater systematics in aquifers;
- providing a source of 'real field' data for model calibration;
- comparing historical data to establish trends in water table changes resulting from overuse or climatic changes.

Measurement of groundwater levels, especially in monitoring wells, is done by several means. The method used is generally chosen based on the accuracy and speed of the desired measurements. Drillers often make an initial measurement with a 'plopper' fixed on one end of a length of twine or measuring tape. Knots in the twine indicate approximate depth intervals. More accurate measurements are obtained by using a steel tape is coated with chalk, which is lowered in the well until the bottom part of the tape gets wet and the chalk marking changes colour. Carpenter's blue marking chalk is commonly used for this purpose. The colour changes from light blue to dark blue when the chalk becomes wet. This method can be accurate, but is slow. Accurate measurements can be made more rapidly with an electronic water level meter (or water level indicator). Simple and reliable water level data loggers are available that can additionally monitor and record groundwater temperature and electrical conductivity variations recorded together with date and time of measurements. In the field, groundwater level measurements are made with reference to a local datum that is often the top of well casing, concrete plinth, drill table, etc. These measurements must be reduced to a regional datum or the mean sea level before plotting and interpretation for regional groundwater flow.

3.6.2 Groundwater temperature

In many regions of the world, groundwater is the primary source of water, sustaining ecologically sensitive freshwater streams, and rivers. Even a small change in water temperature can severely impact the health of a freshwater system. Temperature also plays an important role in the aqueous geochemical processes occurring in groundwater. As groundwater flows through porous media, it dissolves several constituents from the aquifer material, acquiring distinct characteristics. However, as groundwater temperature fluctuates, so does the ionic balance of the dissolved chemicals. Long-term temperature records are useful to:

- monitor fluctuations in geothermal well fields used for industrial activities;
- identify and characterize hydrogeological properties of an aquifer;
- long-term monitoring of seawater intrusion;
- assess the aqueous geochemical properties of groundwater for remediation;
- observe impacts to ecologically sensitive areas;
- highlighting preferential pathways in karstic regions.

3.6.3 Groundwater electrical conductivity

More than 50% of the world's population lives within about 60 km of the coastal regions and this number is growing rapidly. Increasing demand of groundwater in these areas is resulting in seawater intrusion into the coastal aquifers. Electrical conductivity is an effective parameter, which can be measured conveniently to estimate the total dissolved solids (TDS) and hence concentration of dissolved salts in groundwater. Identifying and mapping these water quality parameters provides the base data for developing mitigation measures and to:

- estimate distribution of saltwater in an aquifer;
- assess the life span of an aquifer being affected by seawater intrusion;
- determine suitability of an aquifer for irrigation;
- develop water quality maps for regulatory compliance monitoring.

3.7 Groundwater pollution

Groundwater pollution is a global phenomenon with potentially disastrous consequences.
Prevention of pollution, as far as possible, is an
ideal approach. However, in real life, pollution
cannot be completely prevented. The approach,
therefore, should be to minimize it to the greatest
extent possible. Groundwater quality monitoring is
the main tool for timely detection of pollutants and
protection of groundwater resources. Monitoring
groundwater quality is a specialized task, often requiring expert interpretation of the data. Although
some general prescriptions for management of
both solid and liquid wastes exist, these may not be
applicable in all cases. In the literature, different approaches have identified various sets of pollutants
and pollution indicators. Identifying and quantifying groundwater pollution needs an approach
that is specific to the aquifer, the site, and the type
of pollution present. This involves inputs from
hydrogeologists and pollution control experts. In
this approach, the conceptual hydrogeological
framework, monitoring network, and the parameters that need to be monitored need to be defined.
'Bulk parameters', such as electrical conductivity
and dissolved organic carbon, are extremely useful
for delineating pollution plumes. Electrical conductivity, in particular, can easily be determined

in the field and in the majority of cases, it gives a
reliable indication of the extent of the pollution.

Central to all groundwater measurements and
resource exploitation are groundwater wells.
Therefore, in the next chapter we shall describe
flow of groundwater to wells, as well as the
hydraulics of groundwater wells.

3.8 Composite nature of surface water and groundwater

Groundwater and surface water are fundamentally
interconnected. In fact, it is often difficult to separate the two because they 'feed' upon each other.
One way to visualize this is by having a closer look
at the fate of precipitation on land (Fig. 3.15). As
rain or snow falls on the Earth's surface, some water
runs off the land and joins rivers, lakes, streams, and
oceans (surface water). Water can also move into
these bodies by percolation into the ground. Water
entering the soil can infiltrate deeper to reach
groundwater, which can discharge to surface water or return to the surface through wells, springs,
and marshes, thus becoming surface water again.
Upon evaporation, it completes the hydrologic
cycle.

One of the most commonly used forms of
groundwater comes from shallow, unconfined

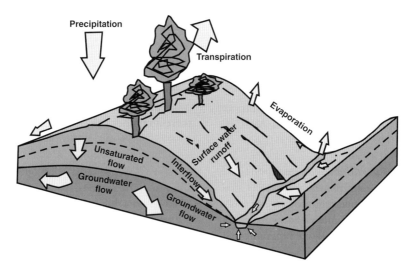

Fig. 3.15 Partitioning of precipitation falling on land and the interconnection between surface- and groundwater. See also
Plate 3.15.

(water table) aquifers, which are major sources of drinking and irrigation water. Unconfined aquifers also interact closely with streams – sometimes discharging water into a stream or lake and sometimes receiving water from surface water bodies. An unconfined aquifer that feeds a stream contributes to the stream's base flow, which sustains it during the lean season. This is called a gaining stream. In fact, groundwater can be an important component in maintaining the hydrologic balance of surface water sources such as streams, springs, lakes, wetlands, and marshes.

The source of groundwater recharge is through precipitation, surface water as well as return seepage of irrigation water that percolates downward under the influence of gravity. Approximately 5–50% (depending on climate, land use, soil type, geology, and many other factors) of annual precipitation results in groundwater recharge. In some areas, particularly in arid regions, streams literally recharge the aquifer through stream bed infiltration, called perched losing streams (see Fig. 5.7c).

An unconfined aquifer often extends a lake, river or estuary and hence to the watershed of the water body concerned. The biggest risk for pollution of an unconfined aquifer is contaminated water moving through the permeable materials directly above it. This area is known as the primary recharge area. Depending on the depth and geological characteristic of the overlying strata, travel time from the surface to the aquifer can be quite short.

Less permeable deposits located at higher elevations than the aquifer form a secondary recharge area. These areas also recharge the aquifer through both overland runoff and groundwater flow. Because they are less permeable and tend to be further away from the aquifer, they often filter out contaminants.

3.9 Conjunctive use of surface water and groundwater

In most parts of the world, precipitation and consequently peak runoff that form a significant part of the total discharge of the rivers often, occur during a particular season of the year that usually coincides with the lowest water demand. The water resource development problem therefore consists of transferring water from the high supply season to the high demand season. The most common solution consists of storing surface water in reservoirs created by dams. But surface reservoirs have many drawbacks, especially:

- *Evaporation*: large open water bodies are exposed to high evaporation rates, leading to water losses sometimes exceeding 20% of the average annual runoff.
- *Sedimentation*: soil erosion in the catchment results in siltation in the surface reservoirs and in the equivalent reduction of the storage capacity. Flushing out part of the mud from the reservoirs is occasionally possible through specially designed pipes placed at the bottom of the dam, but this uses a lot of water and may also be detrimental to the environment of the downstream region.
- *Environmental impact* of surface reservoirs may often be highly undesirable for human health, besides causing flooding of inhabited or good agricultural land. Distribution of water from the reservoir may be expensive and require the construction of an extensive canal network because of the distance between dam and utilization areas.

When these disadvantages are taken into account, underground storage of water in the ground may be a better alternative compared to surface storage systems, although not always given due consideration in planning the development of water resources. Large and concentrated water demand, such as that from large irrigation schemes, is usually supplied from surface water storage, and there a number of reasons for this choice:

1. Groundwater aquifers seldom offer large storage capacity able to absorb large volumes of flood in a short period of time, and are unable to return them as significant discharge per unit production system of a well or a borehole.
2. Surface water storage, because of the large investments involved, is often preferred because it offers a much higher political and social visibility. Yet another reason, which is not

often documented, is that high construction costs give an opportunity for making substantial private profits, opening a way for unethical influence on decision making.

Conjunctive use of surface- and groundwater, combining the use of both sources of water, offers a reasonable option to minimize the undesirable physical, environmental, and economical effects and to optimize the water demand/supply balance water. Usually conjunctive use of surface water and groundwater is considered within a river basin management programme, i.e. both the river and the aquifer belong to the same basin.

For this approach to be a viable part of resource management strategy, several problems need to be carefully studied before selecting the different options:

- underground storage availability to be determined;
- production capacity of the aquifer(s) in term of potential discharge;
- natural recharge of the aquifer(s);
- induced natural recharge of the aquifer(s);
- potential for artificial recharge of the aquifer(s);
- comparative economic and environmental benefits derived from the various possible options.

3.9.1 Underground storage and production capacity of the aquifer

In order to use the underground reservoir to store a significant volume of water – possibly of the same order of magnitude as the annual runoff – with the intent to use it at a later time, the groundwater reservoir should present sufficient free space between the ground surface and the water table to accommodate and retain the water to be recharged, for the period during which water is not needed. The suitability of an aquifer for recharging may be estimated using the following criteria:

- Surface material has to be highly permeable so as to allow water to percolate easily.
- The unsaturated zone should present a high vertical permeability, and vertical flow of water should not be impeded by less permeable clay layers.

- Depth to water table should not be less than 5 m.
- Aquifer transmissivity should be high enough to allow water to move rapidly from the mound created under the recharge basin, but should not be too high (as in karstic channels) so that water moves away from where it is recharged and cannot be recovered subsequently.

A reasonable value of transmissivity for recharging is also a good indicator of the aquifer capacity to produce high well discharge and, therefore, easily yield the stored water when desired. See Section 11.2.3 for a brief description of some of the commonly used artificial groundwater recharge structures.

3.9.2 Natural and induced recharge of the aquifer

Any modification of the natural course of surface water may significantly alter the renewable groundwater renewable resources. Therefore careful estimation of natural recharge and potential for additional recharge must be made, avoiding double counting of water resources as if surface water and groundwater were independent of each other. Induced natural recharge occurs when intensive exploitation of groundwater close to a river results in a significant depression of the groundwater level resulting in dwindling river flow. This phenomenon is well-known in temperate climates where rivers are perennial. However, it may also occur in semiarid climates where a depression of the piezometric level of an aquifer underlying an ephemeral river creates void space in the aquifer, which facilitates its recharge during flooding when a large crosssection of the river floodplain has water.

3.9.3 Conjunctive use with poor-quality water

Practical difficulties and costs involved in disposing off waste water often present new opportunities for conjunctive use. Growing wastewater use in peri-urban agriculture in cities around the world is a case in point. Research in several cities in India, Pakistan, and Mexico points to ingenious practices developed by peri-urban farmers to use urban waste water and groundwater conjunctively for irrigation (Buechler and Devi 2003). However, in water-scarce situations, some industrial waste waters also offer opportunities for creating livelihood through irrigation.

3.9.4 *Conjunctive use with saline groundwater*

In regions with primary salinity, conjunctive use of surface water and groundwater presents unique challenges and opportunities. In such places, the objective of conjunctive management is to maintain both water and salt balances. In this situation, system managers need to exercise effective control and precision in canal water deliveries to different parts of the command areas and to maintain an optimal ratio of fresh and saline water for irrigation (Murray Rust and Vander Velde 1992). In many systems, it makes sense to divide the command areas into surface water irrigation zones and groundwater irrigation zones, depending on the aquifer characteristics and water quality parameters. In others, providing recharge structures within a surface system is often a useful component of a rehabilitation and modernization package. It is a risky business and requires a sound conceptual model of the fate of the salts mobilized, if it is not to cause more problems than it solves.

For additional tools to facilitate conjunctive use of surface water and groundwater, such as artificial groundwater recharge, rainwater harvesting, and watershed management, see Chapter 11.

3.10 Tutorial

Ex 3.1 Data from three piezometers located within a few metres of each other as shown in Fig. 3.16 is as follows:

	A	B	C
Elevation at land surface (m)	102	102	102
Depth of monitoring well (m)	52	40	26
Depth to water (m below surface)	27	25	21

What is the hydraulic head (h) at each piezometer?
[Ans. 75, 77, 81]
What is the pressure head (h_p) at each piezometer?
[Ans. 25, 15, 5]
What is the elevation head (z) at each piezometer?
[Ans. 50, 60, 76]
What is the vertical hydraulic gradient between Well A and Well B?

[Ans. 0.18]

Ex 3.2 Two wells are located 30 m apart in a sand aquifer with a hydraulic conductivity of $1.2 \times 10{-}2\ m\ d^{-1}$ and 35% porosity. The head in well A is

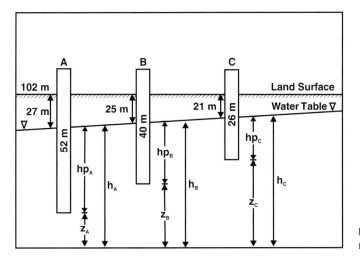

Fig. 3.16 Schematic illustration of the data for use with Ex 3-1.

29 m and the head in well B is 30 m. (a) What is the horizontal hydraulic gradient between the two wells? (b) What is the velocity of ground water between the two wells?

[Ans. 0.03, 1.143×10^{-3} m d^{-1}]

Ex 3.3 Two tensiometers located at elevations of 32 m and 28 m record hydraulic pressure of -1 m and -2 m, respectively. Determine the vertical hydraulic gradient between the two tensiometers and also the capillary gradient between both tensiometer locations.

[Ans. 1.25, 0.25]

Ex 3.4 Estimate the quantity of water discharging from an aquifer 100 m thick and 1 km wide, for a hydraulic gradient of 1 in 1000. The hydraulic conductivity of the aquifer is 50 m d^{-1}.

[Ans. 5000 m^3 d^{-1}]

Ex 3.5 In a certain for stream, the average daily flows measured during the dry season at two gauging stations 5 km apart are 2.485 m^3 d^{-1} and 2.355 m^3 d^{-1}. The average slope of the water table on both sides of the stream, as determined from measurements in the observation wells, is 1:2000. The average thickness of the aquifer contributing to the base flow is 50 m. Determine transmissivity and hydraulic conductivity of the aquifer.

[Hint: Discharge from half the thickness of of an aquifer contributes to half the increase in stream flow between gauging stations. Ans. 2246 m^2 d^{-1}, 45 m d^{-1}]

Ex 3.6 A confined aquifer 100 m thick, having a porosity of 0.2 and containing water at a temperature of $\sim15°C$, releases $\sim9 \times 10^{-7}$ m^3 of water per cubic metre of aquifer per metre of decline in head of the water due to its expansion alone. Determine the storage coefficient of the aquifer.

[Ans. $>9 \times 10^{-7}$]

4

Well hydraulics and test pumping

Well hydraulics is one of the most important topics in hydrogeology. Because wells are in direct hydraulic communication with aquifers, they are used to extract ground water and sometimes for recharging of aquifers as well. By conducting pumping tests on wells, which provide direct access to groundwater conditions, they can also be used to estimate aquifer properties, such as the storage coefficient and transmissivity in the vicinity of the well. The coefficient of storage, S, is a measure of the amount of water obtained from aquifer storage by pumping and is defined as 'the amount of water given out by the aquifer per unit surface area per unit change (decrease) of pressure head'. Equivalently, when an aquifer is recharged, S is defined as 'the amount of water taken into storage by the aquifer per unit surface area per unit change (increase) of pressure head'. The coefficient of transmissivity, T, is a measure of the rate at which water flows through the saturated thickness of an aquifer.

Often the term applied to this type of quantitative method is the *aquifer performance test*, or simply the aquifer test. With the knowledge of these aquifer characteristics, future declines in groundwater levels associated with pumpage can be calculated to help with resource management and environmental conservation when needed.

To understand what constitutes a pumping test, let us first consider an unconfined aquifer. As pumping proceeds, the water table drops and air replaces water in the voids in, say, 20% (or whatever is the porosity of the aquifer) of the dewatered volume of the aquifer. Pumping creates a gradient, or slope of the water table towards the well, causing water in the aquifer to move towards the well. Careful observation of the behaviour of the water levels in nearby wells during pumping enables evaluation of both the coefficient of storage and the coefficient of transmissivity.

In an unconfined aquifer, water is released from storage by dewatering due to a drop in the water table. But in a confined aquifer, i.e. an artesian aquifer – the concept of obtaining water from storage may be slightly more complex to visualize. An artesian aquifer is bounded by confining beds at its top as well as at its bottom and the water in such a well penetrating the aquifer rises above the top of the aquifer. When pumping starts and water is taken from the well, the aquifer still remains saturated (i.e. still full of water). At some distance away from the well the conditions that existed prior to pumping may not have changed. Where, then, does the pumped water come from? First, a very small fraction of it comes from expansion of the water itself, as the pressure is reduced near the well due to pumping. The decrease in pressure also reduces support of the aquifer skeleton due to loss of buoyancy, resulting in compaction of the aquifer due to the weight of the overburden. This is also true of less permeable beds lying within the aquifer, and some water is squeezed out of these beds.

The storage coefficient of an artesian aquifer is very small compared to its value for an unconfined

Modern Hydrology and Sustainable Water Development, 1st edition. © S.K. Gupta
Published 2011 by Blackwell Publishing Ltd.

(water table) aquifer. In contrast to a water table aquifer, the entire thickness of an artesian aquifer within the area of a pressure decline contributes water from aquifer storage. Of course, in an un-confined aquifer some water is also released from storage by compaction, but this amount is so small that it is usually neglected. However, in a very thick unconfined aquifer, it has been shown to be im-portant. As pumping continues in a water-table or an artesian aquifer, the effects continue to spread radially outwards from the pumping well until suf-ficient natural discharge or recharge rejected for-merly due to inability of aquifer to take in more on account of hydraulic pressure is now intercepted to offset the quantity pumped out. Leakage through adjacent beds may also contribute to the recharge.

Pumping tests are carried out periodically, which involve measurements of well discharge and wa-ter levels in a pumped well together with wa-ter levels measured in one or more observation wells. This data is used to obtain particular solu-tions to equations of groundwater flow. There are innumerable conditions under which ground wa-ter can occur, and for each situation a hydrologist must be aware of the limitations of the chosen for-mula. In this chapter, applying Darcy's law and the fundamental equations governing the groundwa-ter movement to particular situations, some simple analytical methods are presented to model flow conditions in the vicinity of pumping wells. The methods employed are based on one- and two- di-mensional geometry of groundwater flow through aquifers. It is also assumed that the aquifer is homo-geneous and isotropic, resistance to vertical flow can be neglected, and only resistance to horizon-tal flow is accounted for. Both steady state and transient flow conditions are considered. Some of these models and equations are used during anal-yses of pumping tests for estimation of aquifer parameters, such as transmissivity and storage coefficient.

4.1 Steady flow

Steady flow implies that no change occurs with time, both in flow rate and head, so that $Q =$ constant and $\partial h / \partial t = 0$. The governing equation

(Eqn. 3.10) in a confined aquifer, therefore, reduces to the *Laplace equation*, and for two-dimensional flow may be written as:

$$\frac{\partial^2 h}{\partial x^2} + \frac{\partial^2 h}{\partial y^2} = 0 \qquad (4.1)$$

The solution to this equation can be derived by observing that one possible set of solutions would have both:

$$\frac{\partial^2 h}{\partial x^2} = 0 \quad and \quad \frac{\partial^2 h}{\partial y^2} = 0 \qquad (4.2)$$

Integration yields:

$$h = Ax + By + C \qquad (4.3)$$

This equation can be used to estimate $h(x, y)$, by knowing h at three locations, enabling evaluation of the three constants A, B, and C.

Example 4.1. Three non-pumping observation wells L(0; 0), M(500; 0), and N(0; -300), with mea-sured head $h_L = 120$ m, $h_M = 116$ m, and $h_N = 122$ m, respectively, are located in a plan view of an aquifer. Determine the mathematical model for uniform flow that fits these observations and pre-dict the head in another observation well located at P(300; 200).

Solution. One can form three equations using Eqn 4.3 for three observation wells at A, B, and C. Thus:

at M: $h_L = A(0) + B(0) + C$; therefore $h_L = C = 120$ m

at N: $h_M = A(500) + B(0) + C$; therefore $h_M = A(500) + 120 = 116$ m

at O: $h_N = A(0) + B(300) + C$; therefore $h_N = B(-300) + 120 = 122$ m

Solving Eqn. 4.3 with the above boundary con-ditions yields $A = -4/500$ and $B = -2/300$ and the mathematical model is:

$$h = (-4/500)x + (-2/300)y + C$$

Hence:

$$h_P = (-4/500)300 + (-2/300)200 + 120$$

$$= 119.1 \ m$$

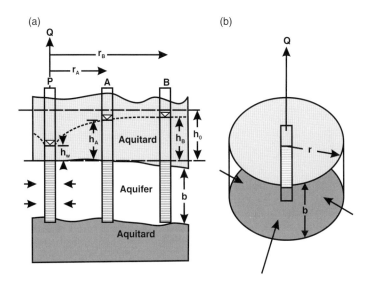

Fig. 4.1 (a) Vertical section through a confined aquifer near a pumping well 'P' and two observation wells 'A' and 'B'. (b) Aquifer top and bottom showing radial flow to the pumping well 'P' of (a). The discharge, Q, of the well must flow through the sides of a cylinder centred on the well 'P'.

For a similar flow situation in an unconfined aquifer, solution of the Laplace equation (Eqn. 3.39) is not possible because the water table in a two-dimensional case represents a flow line and determines the flow distribution that in turn governs the water table. To obtain a solution, Dupuit (1863) assumed: (i) velocity of flow to be proportional to the tangent of hydraulic gradient; and (ii) flow to be horizontal and uniform everywhere in a vertical section. Now, for unidirectional flow, the discharge per unit width, q, at any vertical section can be written as:

$$q = -Kb\frac{db}{dx} \qquad (4.4)$$

Integrating : $qx = \frac{K}{2}b^2 + C \qquad (4.5)$

And if $b = b_0$ where $x = 0$, then the *Dupuit approximation* results in:

$$q = \frac{K}{2x}\left(b_0^2 - b^2\right) \qquad (4.6)$$

This indicates that the shape of the water table for an unconfined aquifer in steady state is parabolic.

4.1.1 Steady radial flow to a well

When water is pumped from an aquifer, the water table or the piezometric surface, depending on the type of aquifer, is lowered. The *drawdown* at a given point is the vertical distance to which the water level is lowered with respect to its pre-pumping level. A drawdown curve shows variation of drawdown with distance from the pumped well (Fig. 4.1a). In three dimensions the drawdown curve is cone shaped and is known as the *cone of depression.* The outer limit of the cone of depression defines the 'area of influence' of the well.

4.1.1.1 Confined aquifer

In a *confined aquifer*, the discharge Q of the well must flow through the sides of a cylinder centred on the well (Fig. 4.1a). Using plane polar coordinates with the well being at the origin; the well discharge, Q, at a radial distance, r, equals:

$$Q = -2\pi rbK\frac{db}{dr} \qquad (4.7)$$

Rearranging and integrating yields:

$$b = \frac{Q}{2\pi bK}\ln r + C \qquad (4.8)$$

This satisfies the Laplace equation (Eqn. 3.42).

Because of the natural-log (*ln r*) in Eqn 4.8, the solution behaves such that the boundary conditions:

$$r \rightarrow 0; \quad b \rightarrow -\infty \quad and \quad r \rightarrow \infty; \quad b \rightarrow \infty \qquad (4.9)$$

Thus, from a theoretical aspect, steady radial flow in an extensive aquifer does not exist because the cone of depression must expand indefinitely. However, from a practical standpoint, b varies with the logarithm of the distance from the well and approaches b_0 at a large enough distance (say r_0) from the pumping well. In the more general case for boundary condition at the well, $b = b_w$ and $r = r_w$, with no external limit on r, Eqn 4.8 yields:

$$b = b_w + \frac{Q}{2\pi bK} \ln \frac{r}{r_w} \qquad (4.10)$$

This equation, known as the equilibrium or *Thiem equation*, enables determination of hydraulic conductivity K (or the transmissivity T) of a confined aquifer employing the drawdown data from a pumping well. Because any two points can define a logarithmic drawdown curve, the method consists of measuring drawdowns in two observation wells at different distances from a well being pumped at a constant rate. Theoretically, b_w at the pumped well itself can serve as one of the observation points. However, well losses caused by flow through the well screen and inside surface of the well introduce errors so that measurement of b_w should be avoided. With reference to Fig. 4.1a, the transmissivity is given by:

$$T = Kb = \frac{Q}{2\pi(b_B - b_A)} \ln \frac{r_B}{r_A} \qquad (4.11)$$

where r_A and r_B are radial distances of observation wells A and B from the pumping well P and b_A and b_B are heads at the respective observation wells. From a practical standpoint, the drawdowns rather than the head, b, are measured, so Eqn 4.11 can be rewritten as:

$$T = \frac{Q}{2\pi(s_A - s_B)} \ln \frac{r_B}{r_A} \qquad (4.12)$$

where s_A and s_B are the drawdowns (i.e. the difference between pre-pumping water level and the pumping water level) in the respective observation wells.

Application of Eqn 4.12 requires that the observation wells are located sufficiently close to the pumping well, so that their drawdowns are appreciable and can be measured reasonably well, and that pumping at a constant rate has been going on for a sufficiently long time for the steady state (implying negligible change in the drawdown with time) to have been achieved.

Example 4.2. An 8 m thick confined aquifer is being pumped by a well at Q = 500 $m^3 \, d^{-1}$. At the steady state, with two observation wells A and B at radial distances 10 m and 25 m, respectively show $b_A = 80$ m and $b_B = 82$ m. Estimate T and K for the aquifer.

Solution. Using Eqn 4.10:

$$b_B = \frac{Q}{2\pi T} \ln \frac{r_B}{r_A} + b_A$$

Solving yields $T = 36.5$ $m^3 d^{-1}$. Hence $K = T/b = 4.6$ $m \, d^{-1}$.

4.1.1.2 Unconfined aquifer

In an *unconfined aquifer*, the saturated thickness varies with distance from the pumping well and is a function of b. If the well penetrates the aquifer completely up to the horizontal base and a concentric boundary of constant head surrounds the well, the well discharge is obtained by rewriting Eqn 4.7 as:

$$Q = -2\pi r Kb \frac{db}{dr} \qquad (4.13)$$

which, after integration between the limits $b = b_1$ at $r = r_1$ and $b = b_2$ at $r = r_2$, yields:

$$Q = \pi K \frac{b_2^2 - b_1^2}{\ln(r_2/r_1)} \qquad (4.14)$$

Rearranging to solve for hydraulic conductivity:

$$K = \frac{Q}{\pi(b_2^2 - b_1^2)} \ln\left(\frac{r_2}{r_1}\right) \qquad (4.15)$$

This equation fails to accurately describe the drawdown curve near the pumping well because of large vertical flow components. Therefore, in practice, drawdowns should be small in relation to the saturated thickness of the aquifer.

4.1.1.2.1 Unconfined aquifer with uniform recharge

In the case when an unconfined aquifer is recharged uniformly by percolation of rainwater, irrigation return flow, or any other surface water source(s) at a rate W (L^3 L^{-2} T^{-1}), the flow, Q, towards the well is no longer limited to being predominantly horizontal from the sides of the cylinder surrounding the well, as in the previous two cases. It increases by addition of recharge water towards the well, reaching a maximum of Q_w at the well face. The incremental inflow, dQ, through the cylinder of thickness, dr, and radius, r, is given by:

$$dQ = -2\pi r dr W \qquad (4.16)$$

Integrating:

$$Q = -\pi r^2 W + C \qquad (4.17)$$

At the centre of the well, $r = 0$ and $Q = Q_w$, so that:

$$Q = -\pi r^2 + Q_w \qquad (4.18)$$

Substituting Q from above in the equation for flow to the well (Eqn 4.13):

$$Q = -\pi r^2 W + Q_w = -2\pi r K b \frac{db}{dr} \qquad (4.19)$$

Integrating and noting that $b = b_0$ at $r = r_0$, the equation for the drawdown curve is obtained as:

$$b_0^2 - b^2 = \frac{W}{2K}\left(r^2 - r_0^2\right) + \frac{Q_w}{\pi k} \ln \frac{r_0}{r} \qquad (4.20)$$

when $r = r_0$ and $Q = 0$, so that from Eqn 4.18:

$$Q_w = \pi r_0^2 W \qquad (4.21)$$

Thus total flow at the well equals the recharge within the circle defined by the radius, r_0. This is the radius at which $b = b_0$ and there is no drawdown ($s_0 = 0$) for a given pumping rate under steady state conditions. This is defined as the *radius of influence*, which is a function of both the pumping rate and the recharge rate. This analysis assumes an idealized circular boundary with constant head and no flow conditions, which are rarely encountered in the field.

4.1.2 Unsteady radial flow

So far, only steady-state (i.e. where discharge and heads, or drawdowns, do not change with time) groundwater flow models to a pumping well have been considered. These models describe the drawdown for a given pumping rate reasonably well, when long-term average flows are considered, but fail at the instant of time when pumping wells are started or shut down.

4.1.2.1 Confined aquifers

When a well penetrating an extensive confined aquifer is pumped at a constant rate, the influence of the discharge extends radially outwards with time. The head continues to decline, because water must come from reduction of storage within the aquifer. The rate of decline, however, continuously decreases as the area of influence expands outwards from the pumping well with time.

In a homogenous and isotropic aquifer, for the case of two-dimensional horizontal flow with a fixed saturated thickness and radially-symmetric groundwater flow to wells, the Laplace equation (Eqn. 3.42) in plane polar coordinates describes the system. Theis (1935) obtained a solution to Eqn. 3.42 based on the analogy between groundwater flow and heat conduction in solids. By assuming that the well is replaced by a mathematical sink of constant strength and imposing the boundary conditions $b = b_0$ for $t = 0$, and $b \rightarrow b_0$ as $r \rightarrow \infty$ for $t \geq 0$, the solution (known as *Theis Equation*) is:

$$s = \frac{Q}{4\pi T} \int_u^\infty \frac{e^{-u}du}{u} \qquad (4.22)$$

where s is drawdown; Q is the constant well discharge; and:

$$u = \frac{r^2 S}{4Tt} \qquad (4.23)$$

The integral in Eqn 4.22 is a function of the lower limit and is known as the exponential integral. It can be expanded in the form of a convergent series as:

$$\int_u^\infty \frac{e^{-u}du}{u} = -0.5772 - \ln u + u - \frac{u^2}{2.2!}$$

$$+ \frac{u^3}{3.3!} - \frac{u^4}{4.4!} + \ldots\ldots$$

$$= W(u) \qquad (4.24)$$

where $W(u)$, termed the *well function*, is a convenient symbolic form to denote the exponential integral. Eqn 4.22 may thus be written as:

$$s = \frac{Q}{4\pi T} W(u) \qquad (4.25)$$

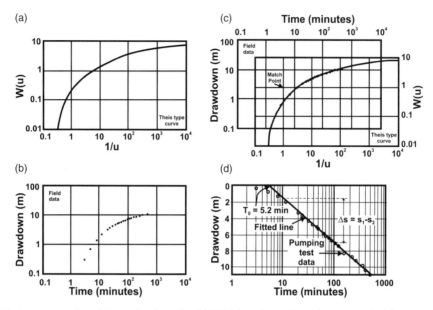

Fig. 4.2 (a) Theis type curve for a fully confined aquifer. (b) Field-data plot on logarithmic paper for Theis curve-matching technique. (c) Matching of field-data plot to Theis type curve. (d) Cooper-Jacob straight-line time-drawdown method for a fully confined aquifer. Modified from Heath (1983). © U.S. Geological Survey.

Rewriting Eqn 4.23 as:

$$t = \left(\frac{r^2 S}{4T} \right) \frac{1}{u} \qquad (4.26)$$

4.1.2.1.1 Theis method of solution

Because the terms in parentheses in Eqn 4.25 and Eqn 4.26 are constants, the relation between $1/u$ and $W(u)$ must be similar to that between t and s. Based on this similarity, Theis (1935) suggested an approximate solution for S and T by employing a graphical method of superposition. This involves the following steps:

- Making a plot of $W(u)$ versus $1/u$ on log-log paper, or using a spreadsheet. This graph has the shape of a cone centred on the pumping well and is referred to as the *Theis type curve*, or the non-equilibrium type curve (Fig. 4.2a). Both $W(u)$ and $1/u$ are dimensionless numbers.
- Plotting the field drawdown at the observation well, $(s = h_o - h)$, versus t, using the same logarithmic scale as the type curve (Fig. 4.2b).
- Superimposing the type curve over the field-data graph, with the axes of both graphs kept par-

allel and adjusting the two graphs until the data points match the type curve (Fig. 4.2c). Selecting the intersection of the line $W(u) = 1$ and the line $1/u = 1$ as the match point, the values of s (= $h_o - h$) and t corresponding to the match point on the field-data graph are determined. Any other convenient point can also be selected as the match point.
- The transmissivity (T) and aquifer storativity (S) are computed with values of $W(u)$, u, s, and t for the selected match point from Eqn 4.25 and Eqn 4.26.

In areas where several wells exist near a well being pump-tested, simultaneous recording of s in the monitored wells enables distance-drawdown data to be fitted to the type curve in a manner identical to the time-drawdown data as above.

4.1.2.1.2 Cooper-Jacob method of solution

It was noted by Cooper and Jacob (1946) that for small values of r and large values of t, u is small, so that the series terms in Eqn 4.24 become negligible

WELL HYDRAULICS AND TEST PUMPING

after the first two terms. As a result, the drawdown can be expressed by an asymptotic relationship:

$$s = \frac{Q}{4\pi T}(-0.5772 - \ln u)$$

$$= \frac{Q}{4\pi T}\left(-0.5772 + \ln \frac{4Tt}{r^2 S}\right) \quad (4.27)$$

For a particular value of r, Eqn 4.27 is the equation of a straight line plot of s versus ln (t) or log (t), as seen by rewriting and changing to decimal logarithms:

$$s = \frac{2.30Q}{4\pi T} \log \frac{1}{u} - \frac{0.5772Q}{4\pi T}$$

$$= \frac{2.30Q}{4\pi T} \log \frac{4Tt}{r^2 S} - \frac{0.5772Q}{4\pi T} \quad (4.28)$$

Similarly, for a particular value of time, t, Eqn 4.27 is also the equation of a straight line plot of s versus ln (r) or log (r). The straight line semilog plots of distance-drawdown and time-drawdown (Fig. 4.2d) may be extrapolated to intersect the zero-drawdown axis. At the axis $(s = 0)$ and from Eqn 4.27:

$$-0.5772 = \ln \frac{1}{u} = \ln \frac{4Tt_0}{r^2 S} = \ln \frac{4Tt}{r_0^2 S} \quad (4.29)$$

where t_0 = intersection of time-drawdown straight line semilog plot with zero-drawdown axis and r_0 = intersection of distance-drawdown straight line semilog plot with zero-drawdown axis.

Converting to decimal logarithms, Eqn 4.29 yields:

$$S = \frac{2.25Tt_0}{r^2} = \frac{2.25Tt}{r_0^2} \quad (4.30)$$

The value of T can be obtained by noting that in Eqn 4.28, if $s = s_1$ at $t = t_1$, $s = s_2$ at $t = t_2$, and $t_1/t_2 = 10$, then log $t_1/t_2 = 1$; therefore, $s_1 - s_2 = \Delta s$ is the drawdown difference per log cycle of t:

$$T = \frac{2.30Q}{4\pi \Delta s} \quad (4.31)$$

Thus the procedure involves first solving for T with Eqn 4.31 and then solving for S with Eqn 4.30. The straight line approximation for this method should be restricted to small values of u (<0.01) to avoid large errors.

Example 4.3. The time-drawdown semilog plot of an observation well located at a distance of 40 m from a pumping well discharging at $Q = 2$ m^3/min was fitted to a straight line that indicated drawdown over one log cycle of time (Δs) as 1.1 m. The fitted straight line intersected the zero-drawdown axis at 28 sec. Estimate T and S, assuming that the aquifer is fully confined.

Solution. A straightforward application of equations for T and S gives:

$$T = \frac{2.30Q}{4\pi \Delta s} = \frac{2.30 \times 2}{4\pi \times 1.1}\left(\frac{min}{60 \sec}\right) = 0.0055 \ m^2 \sec^{-1}$$

$$S \approx \frac{2.25Tt_0}{r^2} = \frac{2.25 \times 0.0055 \times 28}{40^2} = 2.2 \times 10^{-4}$$

4.1.2.1.3 Recovery test

At the end of a pumping test, when pumping is stopped, water levels in the pumping- and observation wells begin to rise or recover towards the original (pre-pumping) static water level. The drawdown during the recovery period is known as *residual drawdown*. A schematic diagram of change in water level with time, during and after the pumping, is shown in Fig. 4.3.

If a well is pumped for some time and then shut down, the drawdown in the observation well thereafter becomes the same as if the pumping had been continued and the influence of a hypothetical recharge with the same flow was superposed on the pumping well at the instant of shutting down. Using this principle, Theis (1935) showed that the residual drawdown s' is given by:

$$s' = \frac{Q}{4\pi T}[W(u) - W(u')] \quad (4.32)$$

where:

$$u = \frac{r^2 S}{4Tt} \quad and \quad u' = \frac{r^2 S}{4Tt'} \quad (4.33)$$

and t and t' are defined in Fig. 4.3. As mentioned above, for small r and large t', the well function can be approximated by the first two terms of Eqn 4.24, so that Eqn 4.32 can be written as:

$$s' = \frac{2.30Q}{4\pi T} \log \frac{t}{t'} \quad (4.34)$$

Thus a plot of residual drawdown s' versus the logarithm of t/t' is a straight line with a slope equal to

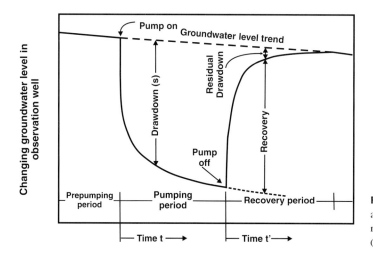

Fig. 4.3 Schematic depiction of drawdown and recovery phases in an observation well near a pumping well. Modified from Heath (1983). © U.S. Geological Survey.

$2.30Q/4\pi T$. Therefore, for $\Delta s'$, the difference of residual drawdown per log cycle of t/t', the transmissivity is given by:

$$T = \frac{2.30Q}{4\pi \Delta s'} \qquad (4.35)$$

No comparable value of S can be determined by the recovery method. Even so, it is good practice to measure residual drawdowns, because analysis of the data enables an independent estimate of transmissivity as a check on the pumping test results. In addition, since Eqn 4.35 is independent of r, the measurement of water levels in the pumped well during the recovery phase provides an estimate of transmissivity, even in the absence of an observation well.

4.1.2.2 Unconfined aquifers

Where drawdowns are significant, the assumption that water from storage is released instantaneously with decline of head, though valid for confined aquifers, is generally not applicable in cases of unconfined aquifers. Pumping test data often show that, as the water table is lowered, gravity drainage of water from the unsaturated zone proceeds at a variable rate (Fig. 4.4a). This is known as *delayed yield* (Boulton 1954; Neuman 1972).

In the initial phase after starting of the pump, the well draws water from elastic storage in the saturated zone by aquifer compression as well as by expansion of pore water. The drawdown response

during this early phase is essentially the same as predicted by the Theis equation (Fig. 4.4b) for the confined aquifer with $S = S_s b$. As pumping is continued, gravity drainage of water at the water table begins to supply significant amounts of water, and the s versus t curve levels off. Later, the rate of drainage at the water table slows and s begins to creep upwards again. In this phase, the s versus t curve approaches the shape of the Theis curve, but with a much larger value of the storage coefficient ($S = S_y$) corresponding to the unconfined aquifer (called the *specific yield*).

From the perspective of water production and pumping test analysis, the storage coefficient obtained from the third segment of the drawdown curve (Fig. 4.4b) is the most important. Therefore, a pumping test should be continued sufficiently long enough to define the third segment of the curve.

Boulton (1963) and Boulton and Streltsova (1975) developed special type-curves for analysing pumping test data of unconfined aquifers, taking into account the delayed yield.

Neuman (1975) developed a solution that allows for anisotropy of $K(K_r \neq K_z)$. Using this solution, the drawdown at a radial distance, r, from the pumping well is given by:

$$s = \frac{Q}{4\pi T} W(u_A, u_B, \eta) \qquad (4.36)$$

where:

$$u_A = \frac{r^2 S}{4Tt} (S = S_s b) \quad \textit{for elastic storage} \qquad (4.37)$$

(a)

(b)

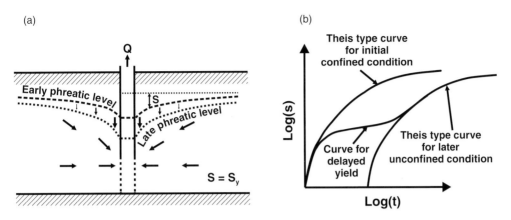

Fig. 4.4 (a) Schematic illustration of non-equilibrium flow to a well pumping an unconfined aquifer. (b) Type curves of drawdown versus time illustrating the effect of delayed yield for pumping tests in an unconfined aquifer. Modified from Bureau of Reclamation (1977) and http://www.scribd.com/doc/3965894/Pump-Test (accessed 11 March 2010).

and:

$$u_B = \frac{r^2 S_y}{4Tt}(S = S_y) \quad \begin{array}{l} for\ unconfined \\ aquifer\ storage \end{array} \qquad (4.38)$$

The dimensionless parameter η is defined as:

$$\eta = \frac{r^2 K_Z}{b^2 K_r} \qquad (4.39)$$

where b is the saturated thickness of the aquifer.

4.1.2.3 Leaky aquifers

When a *leaky aquifer* is pumped, water is drawn both from the aquifer and from the saturated portion of the overlying aquitard or the semi-pervious layer (Fig. 4.5a). As a result of lowering of the piezometric surface, hydraulic gradients are created within the aquitard, leading to vertical movement of ground water across the aquitard layer downwards into the aquifer. Because of the vertical flow, if the discharge rate of the pump equals the vertical flow, a steady state condition is possible in a leaky aquifer.

4.1.2.4 Hantush-Jacob solution for leaky
aquifers

A solution for the steady state situation was provided by Hantush and Jacob (1955). A more general analysis for unsteady state, useful for pumping test

analysis (Hantush 1956, 1964b; Hantush and Jacob 1955), can be given by:

$$s = \frac{Q}{4\pi T}W(u, r/B) \qquad (4.40)$$

where, s, q, r, and u are as defined above. The quantity r/B is given by:

$$\frac{r}{B} = \frac{r}{\sqrt{T/(K'/b')}} \qquad (4.41)$$

where T is the transmissivity of the aquifer; K' is the vertical hydraulic conductivity of the aquitard; and b' is the thickness of the aquitard (Fig. 4.5a). This solution is presented graphically in Fig. 4.5b as a family of curves for varying the values of r/B. The function $W(u,r/B)$ has the form of the Theis equation and reduces to it for a confined aquifer, because as $k' \to 0$, $B \to \infty$ and $r/B \to 0$. Furthermore, all curves converge on the Theis curve for large u (corresponding to early phase of pumping).

Walton (1960) suggested a method similar to the Theis type curve matching solution for finding values of $W(u,r/B)$, $1/u$, s, and t from the match point. T is then found from Eqn 4.40 and S from Eqn 4.23. From the value of r/B belonging to the type curve of the best fit, K'/b' can be calculated from Eqn 4.41. Finally, K' can also be evaluated, if b' is known from the field condition.

Example 4.4. A well in a confined aquifer is to be pumped at a rate of 500 $m^3\ d^{-1}$ for 10 *days*

(a)   (b)

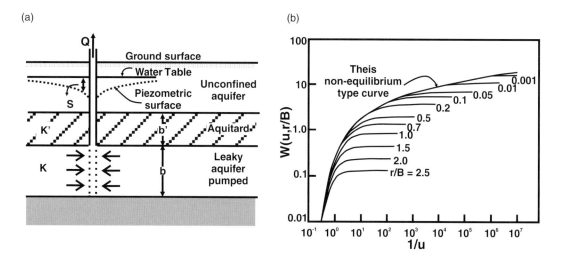

Fig. 4.5 (a) Well pumping a leaky aquifer. (b) Hantush-Jacob solution for leaky aquifer. All curves converge on the Theis curve (upper left for small $1/u$). Redrawn from Walton (1960).

to allow an excavation in the overlying aquitard 5 m thick and with a vertical permeability of 0.01 $m\ d^{-1}$. The aquifer is 10 m thick and its estimated hydraulic properties are: horizontal $K = 5\ m\ d^{-1}$; $S_s = 10^{-4}\ m^{-1}$. Estimate the drawdown in an observation well that is 100 m from the well to be pumped after 10 *days* of pumping.

Solution. Assuming that the confined aquifer is nonleaky, the Theis solution applies:

$$T = Kb = 5 \times 10 = 50\,m^2\,d^{-1}$$
$$S = S_s b = 10^{-4} \times 10 = 10^{-3}$$

To estimate the drawdown in the observation well, we first calculate u (or $1/u$) for this well after 10 days of pumping:

$$\frac{1}{u} = \frac{4Tt}{r^2 S} = \frac{4 \times 50 \times 10}{100^2 \times 10^{-3}} = 200\ or\ u = 5 \times 10^{-3}$$

The corresponding value of $W(u)$ is 4.73. Thus the drawdown is:

$$b_0 - b = \frac{Q}{4\pi T} W(u) = \frac{500}{4\pi \times 50} \times 4.73 = 3.76\ m$$

One of the implicit assumptions in the Theis' solution is that the pumping does not induce any additional flow through the layers above and below that sandwich the aquifer. In reality, the pumped well does induce some amount of leakage. This case is solved by using the Hantush and Jacob formulation,

which has the same form as the Theis solution, but in this case W is a function of u and a new parameter, r/B. In the present case:

$$\frac{r}{B} = \frac{r}{\sqrt{T/(K'/b')}} = \frac{100}{\sqrt{50/(0.01/5)}} = 0.632$$

The corresponding value of $W(u, r/B)$ is approximately 1.55. Thus the drawdown is:

$$b_0 - b = \frac{Q}{4\pi T} W(u, r/B)$$
$$= \frac{500}{4\pi \times 50} \times 1.55 = 0.796\ m$$

This indicates that the drawdown is significantly less due to leakage influx through the overlying aquitard.

4.2 Superposition in space and time

In a linear differential equation, the dependent variable (b in our case), or derivatives of it, occur in linear combinations, as below:

$$Ax, \quad \partial b/\partial x, \quad B\partial^2 b/\partial x^2, \quad \partial^2 b/\partial x \partial y, \quad C\partial b/\partial t$$

with A, B, and C as constants.

Because the governing equations of groundwater flow, viz. the Laplace equation (Eqn. 3.39) and

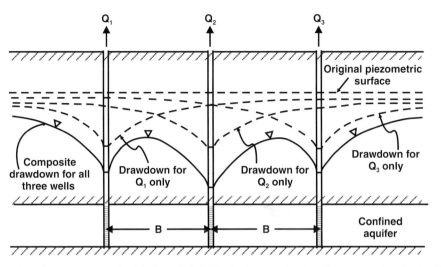

Fig. 4.6 Illustration of superposition principle through individual and composite drawdown curves for three wells in a line. Redrawn from Todd (2006). With permission from John Wiley and Sons Inc.

the Poisson equation (Eqn. 3.40), are linear differential equations, it is possible to add together different solutions to make new, composite solutions of groundwater flow that are more versatile. As an example, if the solutions of three Laplace equations are h_1, h_2, and h_3 and the solution of one Laplace equation is h_4, the superposition principle can be mathematically stated as:

if $\quad \Delta^2 h_1 = \Delta^2 h_2 = \Delta^2 h_3 = 0$ $\qquad$ (4.42)

and $\quad \Delta^2 h_4 = E$ $\qquad$ (4.43)

then $\quad \Delta^2 (h_1 + h_2 + h_3) = 0$ $\qquad$ (4.44)

and $\quad \Delta^2 (h_1 + h_2 + h_3 + h_4) = E$ $\qquad$ (4.45)

where E is a constant. There is no limit to the number of solutions that may be superposed. By superposing a few solutions, some complex problems can be easily solved. For example, *superposition of two, three, or many radial flow solutions gives a solution to the multiple well system problems*. As a result, drawdown at any point within the area of influence caused by the discharge of several wells is equal to the sum of the drawdowns caused by each well individually. Thus from Eqn 4.8 for steady radial flow in a confined aquifer:

$$s_T = \frac{Q_1}{2\pi T} \ln r_1 + \frac{Q_2}{2\pi T} \ln r_2 + \cdots \qquad (4.46)$$

where Q_1 and Q_2 are the well discharges; and r_1 and r_2 are the radial distances from the centres of two different wells to the point where s_T is evaluated. The superposition of drawdowns may be illustrated in a simple way by the well line of Fig. 4.6. The solutions can be based on equilibrium or non-equilibrium equations. Therefore, from the drawdown caused by two pumping wells that are turned on at different times, t_1 and t_2, and pump with different rates, Q_1 and Q, the total drawdown can be predicted by adding together the two Theis solutions (Eqn 4.25) for each well:

$$s_T = \frac{Q_1}{4\pi T} W(u_1) + \frac{Q_1}{4\pi T} W(u_2) \qquad (4.47)$$

where

$$u_1 = \frac{r_1^2 S}{4T(t - t_1)} \qquad (4.48)$$

and

$$u_2 = \frac{r_2^2 S}{4T(t - t_2)} \qquad (4.49)$$

Example 4.5. Define the capture zone of a well with radial flow located in a uniform flow field.

Solution. Because Eqn 4.3 and Eqn 4.8 are solutions of linear differential equations, they can be

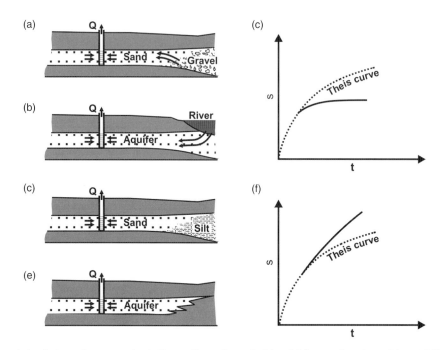

Fig. 4.7 Boundaries that supply more water to the aquifer as shown in (a) and (b) cause drawdown (c) to stabilize earlier than predicted by the Theis solution (dashed line). On the other hand, boundaries that limit water supply to the aquifer (d) and (e) result in greater drawdown (f) than predicted by the Theis solution. Redrawn from Fitts (2002). © Academic Press.

superposed to create an infinite number of useful solutions:

$$b = \frac{Q}{2\pi T} \ln r + Ax + C \qquad (4.50)$$

One of the solutions can be used to mathematically define the capture zone of a pumping well placed, for example, within a contaminant plume. This solution at an arbitrary point, S, requires:

$$\frac{\partial b}{\partial x} = 0, \quad \frac{\partial b}{\partial y} = 0 \qquad (4.51)$$

Placing the centre of the x-y coordinate system at the centre of the well, the location of the point where $\partial b/\partial x = 0$ on the x-axis, is obtained by differentiating Eqn 4.50:

$$\frac{\partial b}{\partial x} = \frac{Q}{2\pi T} \frac{1}{r} \frac{\partial r}{\partial x} + A \qquad (4.52)$$

If we restrict our analysis to the derivative on the x-axis, then $r = x$ and this equation becomes:

$$\frac{\partial b}{\partial x} = \frac{Q}{2\pi T} \frac{1}{r} + A \quad \text{on the x axis} \qquad (4.53)$$

At the point S, $x = x_S$ and $\partial b/\partial x = 0$. Solving for x_S gives:

$$x_S = \frac{Q}{2\pi T A} \qquad (4.54)$$

When $A < 0$, the uniform flow is in the positive x-direction and the point S lies to the right of the well, and vice versa.

4.3 Boundaries and images in flow modelling

Superposition can also be used to model certain aquifer boundary conditions. If the boundaries are close enough to the pumping and observation wells, their effect causes deviation from the drawdown pattern predicted by solutions based on radial symmetry. A typical boundary is a surface water source or a more conductive zone that can supply additional water to the aquifer (Fig. 4.7a and b). In such a case, the actual drawdown at later times

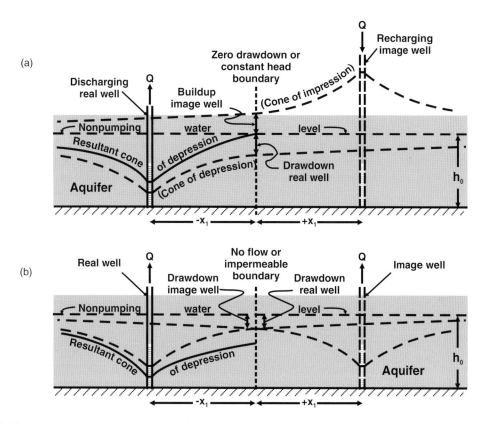

Fig. 4.8 The method of images applied to construct an equivalent hydraulic system (a) for the case of a discharging well near a zero drawdown or constant head boundary (e.g. a stream); (b) Same for the case of a no-flow or impermeable boundary. Modified from Todd (2006). With permission from John Wiley and Sons Inc.

is less than that predicted by the Theis solution (Fig. 4.7c). On the other hand, if the pumping is done from close to the low conductivity boundary (Fig. 4.7d and e), the actual drawdown (Fig. 4.7f) occurs at later times and exceeds the theoretical drawdown due to reduced water supply.

Multiple radial flow solutions can be superposed to solve problems involving wells with nearly constant heads or impermeable boundaries. This method, known as the *method of images*, involves introducing imaginary (or image) wells opposite to the real well, located at an equal distance beyond the boundary in such a way that an aquifer of finite extent (with a boundary close by) can be transformed into an infinite aquifer, so that the solution in the nearby real aquifer domain meets the required conditions at the boundary.

This is achieved in the case of a constant head boundary (e.g. a stream), by an image well as a recharge well, which operates simultaneously and at the same rate as the real well, so that the buildup (increase of head around a recharge well) and drawdown of head along the line of the boundary exactly cancel out (Fig. 4.8a). This provides a constant head along the stream. The resultant asymmetrical drawdown of the real well is obtained using superposition of the drawdown of the real well and the build-up of the image recharge well, as if both were located in an infinite aquifer.

In the case of an impermeable (or no flow) boundary, an image discharging well is placed with the same rate of discharge and at an equal distance from the boundary (Fig. 4.8b). Therefore, along the boundary, the effects of the two wells offset each

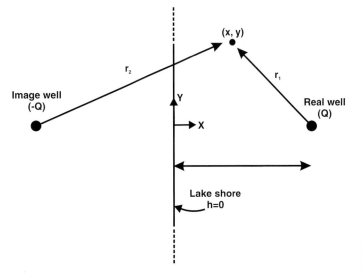

Fig. 4.9 Schematic illustration of the problem in Example 4.6.

other, causing no flow across the boundary, which is the desired condition.

Example 4.6. A well is located 50 m inland from a long, straight lake shore in an aquifer that has an average transmissivity of $T = 200\ m^2\ d^{-1}$. When the well is not pumped, the steady-state water level is 2.0 m above the lake. Define a model to predict the maximum discharge of the well before some water flows back to the well from the lake.

Solution. A schematic illustration of the problem is given in Fig. 4.9.

For a non-pumping condition there is uniform flow given by $h = Ax + C$. At the shore, $x = 0$ and $h = 0$, so that $C = 0$. Again, at the location of the pumping well, $x = 50$ and $h = 2$, so that $A = 2/50 = 0.04$.

When the well has been pumped for a long time, the new steady-state model will have additional terms for the well plus an image well located at $x = 50, y = 0$:

$$h = \frac{Q}{2\pi T} \ln \frac{r_1}{r_2} + Ax$$

If the well draws water from the lake, water flows in the +ve x-direction and at the shore (i.e. $x = y = 0$), $\partial h/\partial x < 0$. On the other hand, if water flows towards the lake, $\partial h/\partial x > 0$ at the shore. So

the maximum discharge of the well before some water will flow from the lake is when $\partial h/\partial x = 0$ at the shore:

$$\frac{\partial h}{\partial x} = \frac{Q}{2\pi T} \frac{\partial}{\partial x}(\ln r_1 - \ln r_2) + A$$

Taking $y = 0$; $r_1 = d - x$; and $r_2 = d + x$:

$$\frac{\partial h}{\partial x} = \frac{Q}{2\pi T} \left(\frac{-1}{d - x} - \frac{1}{d + x} \right) + A$$

Therefore, for $x = y = 0$ and for Q_{max}:

$$\frac{\partial h}{\partial x} = \frac{Q_{max}}{2\pi T} \frac{-2}{d} + A = 0;$$

or

$$Q_{max} = \pi T d A = 1256.6\ m^3\ d^{-1}.$$

4.4 Well flow under special conditions

Several solutions are available for special conditions of the aquifer, pumping wells, and boundaries (Hantush 1964a; Kruseman and Ridder 1970), which include:

1) Constant well drawdown (Bennett and Patten 1962; Hantush 1964b);
2) Varying, cyclic, and intermittent well discharges (Hantush, 1964b; Lennox and Vanden

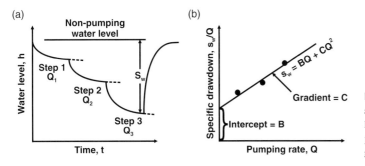

Fig. 4.10 Step-drawdown pumping test analysis to evaluate formation (B) and well loss (C) coefficients. (a) Time drawdown plot. (b) Determination of B and C from the graph of s_w/Q versus Q.

Berg 1962; Moench 1971; Sheahan 1971; Sternberg 1967, 1968);

3) Sloping aquifers (Hantush 1964b);
4) Aquifers of variable thickness (Hantush 1964b);
5) Two-layered aquifers (Hantush 1967; Neuman and Witherspoon 1972);
6) Anisotropic aquifers (Dagan 1967; Hantush 1966; Hantush and Thomas 1966; Neuman 1975; Weeks 1969);
7) Aquifer properties varying with depth (Moench and Prikett 1972; Rushton and Chan 1976);
8) Large-diameter wells (Papadopulos and Cooper 1967; Wigley 1968);
9) Collector wells (Hantush and Papadopulos 1962);
10) Wells with multiple well screens (Selim and Kirkham 1974).

These solutions involve more comprehensive mathematical treatment; their derivation is beyond the scope of this book.

4.5 Well losses

The total drawdown (s_w) in most, if not all, pumping wells consists of two components: (i) the drawdown (s_a) in the aquifer; and (ii) the drawdown (s_c) that occurs as water moves from the aquifer into the well and up to the pump intake. Thus, the drawdown in most pumping wells is greater at the radius of the pumping well than the drawdown in the aquifer. The total drawdown (s_w) in a pumping well can be expressed in the form of the following equations:

For transient flow in a confined aquifer:

$$s_w = \frac{Q}{4\pi T} W(u) + CQ^n \tag{4.55}$$

For steady-state flow in a confined aquifer:

$$s_w = \frac{Q}{2\pi T} \ln \frac{r_0}{r_w} + CQ^n \tag{4.56}$$

where C is a constant governed by the radius, construction, and condition of the well, The first term, for simplicity, is written as BQ. Transient flow, B, involving $w(u)$, is not constant in the strict sense but approaches a constant value for large t:

$$s_w = BQ + CQ^n \tag{4.57}$$

The term BQ is known as the *formation loss* ($= s_a$) and the term CQ^n as the *well loss* ($= s_c$). Since the factor C is normally considered to be constant, in a constant pumping rate test, CQ^n is also constant. As a result, the well loss (s_c) increases the total drawdown in the pumping well but does not affect the rate of change of drawdown with time. It is, therefore, possible to analyse drawdowns in the pumping well with the methods discussed earlier, such as the Jacob time-drawdown method employing semi-logarithmic graph paper.

The well losses are evaluated by conducting a *step-drawdown pumping test.* This consists of pumping a well initially at a low discharge rate until the drawdown within the well essentially stabilizes. The discharge is then increased in successive steps, generally of two hours duration each, as shown by the time-drawdown plot in Fig. 4.10.

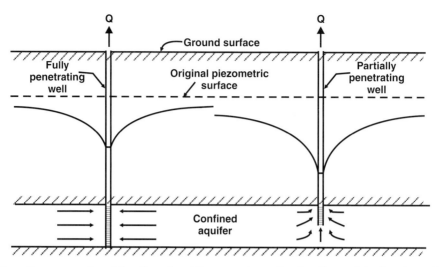

Fig. 4.11 Schematic representation of the effect of partially penetrating well on flow lines and the drawdown. Modified from Todd (2006). With permission from John Wiley and Sons Inc.

From Eqn 4.57, following Jacob (1947), and letting $n = 2$:

$$\frac{s_w}{Q} = B + CQ \qquad (4.58)$$

This is the equation of a straight line between s_w/Q versus Q with the slope of the line being equal to C, the well loss coefficient, and the intercept equal to the formation loss coefficient, B.

4.5.1 Specific capacity and well efficiency

If well discharge is divided by drawdown in the pumping well, the *specific capacity*, a measure of the productive capacity of the well, is obtained. Clearly, the larger the specific capacity, the better the condition of the well.

Substituting for B in Eqn 4.57 and inverting, one obtains an expression for the specific capacity, Q/s_w as:

$$\frac{Q}{s_w} = \frac{1}{W(u)/4\pi T + CQ^{n-1}} \qquad (4.59)$$

This indicates that the specific capacity decreases with increasing values of Q and t. But since $W(u)$ approaches asymptotic value for increasing t, the decrease of specific capacity with t is small and for a given discharge the specific capacity of the well can often be assumed as constant.

Theoretical value of the specific capacity of the well is obtained by neglecting the drawdown due to well loss (CQ^n) and is given by $Q/(BQ)$, i.e. $1/B$. The computed specific capacity, when compared with the value measured in the field (Q/s_w), defines the *well efficiency*. Thus, for a specified duration of pumping, the well efficiency, E_w, expressed as percentage, is given by:

$$E_w = 100\frac{Q/s_w}{Q/BQ} = 100\frac{BQ}{s_w}$$

$$\qquad (4.60)$$

$$= 100\frac{Drawdown\ due\ to\ formation\ loss}{Total\ drawdown}$$

4.5.2 Partial penetration and associated hydraulics

Sometimes the screened or permeable section of the pumped well penetrates only part of the saturated thickness of the aquifer. In the vicinity of a *partially penetrating well*, there are vertical components of flow (Fig. 4.11), which the pumping test solutions considered above do not take into consideration. As a result of an increase in the average length of flow line and generally lower permeability to vertical flow, the actual drawdown at the pumping well is greater than predicted. Detailed methods for analysing effects of partial penetration on well flow for steady and unsteady conditions in confined, unconfined, leaky, and anisotropic aquifers have been outlined by Hantush (1966), Kipp (1973), Lakshminarayana and Rajagopalan (1978) and Sternberg (1973).

Although evaluating the effects of partial penetration is complicated, except in simple cases, common field situations often reduce its practical importance. Any well with 85% or more open or screened borehole area in a saturated thickness may be considered fully penetrating (Todd 2006). Hantush (1966) gives a general guidance that when the radial distance, r, from the pumping well is as large as given by Eqn 4.61, the effect of partial penetration on drawdown is minimal and standard pumping test analysis may be applied, yielding:

$$r \geq 1.5b\sqrt{\frac{K_b}{K_v}} \tag{4.61}$$

where b is the saturated thickness of the aquifer; and K_b and K_v are horizontal and vertical conductivities, respectively.

4.6 Tutorial

Ex. 4.1 A new borewell is to be installed in a regionally confined aquifer with a transmissivity of $150 \, m^2 \, d^{-1}$ and a storativity of 0.0005. The planned pumping rate is $1.5 \, m^3 \, min^{-1}$. As there are several nearby wells tapping the same aquifer, it is required to know if the new well will interfere significantly with these wells:

(a) Compute the theoretical drawdown caused by the new well after 30 *days* of continuous pumping at the following distances: 10, 50, 150, 250, 500, 1000, 3000, and 6000 *m*.
(b) Plot distance-drawdown data on semi-log paper.
(c) If the pumping well has a radius of 0.3 *m*, and the observed drawdown in the pumping well is 39 *m*, what is the efficiency of the well?
(d) If the aquifer is not fully confined but is overlain by a 4 *m* thick confining layer with a vertical hydraulic conductivity of $0.04 \, m \, d^{-1}$, what would be the drawdown after 30 *days* of pumping at

$1.5 \, m^3 \, min^{-1}$ at the indicated distances, neglecting storativity of the aquifer?
(e) For the new well and aquifer system, plot the drawdown at a distance of 6 *m* at the following times: 1, 2, 5, 10, 15, 30, and 60 *min*; 2, 5, and 12 *hr*; and 1, 5, 10, 20, and 30 *days*.

Ex. 4.2 During a pump test data analysis, using the Theis curve matching procedure, the following values were recorded: $Q = 1.9 \, m^3 \, min^{-1}$; $r = 187 \, m$; match-point coordinates: $W(u) = 1$, $1/u = 1$; $s = 2.20 \, m$, 1.8 *min*. Compute T and S.

Ex. 4.3 The data in Table 4.1 are from a pumping test where a well was pumped at a rate of $0.7 \, m^3 \, min^{-1}$. Drawdown, as shown, was measured in an observation well 80 *m* away from the pumped well.

(a) Plot the time-drawdown data on a log-log paper. Use the Theis-type curve to find the aquifer transmissivity and storativity. Compute the average hydraulic conductivity.
(b) Re-plot the data on a semi-log paper. Use the straight-line method to estimate the aquifer transmissivity and storativity.

Ex. 4.4 A well in a water-table aquifer was pumped at a rate of $873 \, m^3 \, min^{-1}$. Drawdown was measured in a fully penetrating observation well located 90 *m* away. The following data (Table 4.2) were obtained:

Table 4.1

Time (min)	Drawdown (m)	Time (min)	Drawdown (m)
0	0.00	24	0.72
1	0.20	30	0.76
1.5	0.27	40	0.81
2	0.30	50	0.85
3	0.34	60	0.88
4	0.41	80	0.93
5	0.45	100	0.96
6	0.48	120	1.00
8	0.53	150	1.04
10	0.57	180	1.07
12	0.60	210	1.10
14	0.63	240	1.12
18	0.67		

Table 4.2

Time (min)	Drawdown (m)	Time (min)	Drawdown (m)
0	0.0	10	1.69
1	0.27	15	1.74
2	0.66	20	1.82
3	0.93	30	1.87
4	1.11	40	1.89
5	1.24	50	1.90
6	1.38	60	1.91
7	1.44	90	1.92
8	1.53	120	1.92
9	1.59		

Find the transmissivity, storativity, and specific yield of the aquifer.

Ex. 4.5 Results of a step-drawdown test are given in Table 4.3. Calculate coefficients of formation loss and well loss. Also estimate the well efficiency at each pumping step.

Table 4.3

	Results of a step-drawdown pumping test			
Step	Discharge $(m^3\ min^{-1})$	Drawdown (m)	Cumulative drawdown (m)	Efficiency (%)
1	0.39	3.26		
2	0.63	2.29		
3	0.81	2.50		
4	0.95	2.19		

Appendix 4.1 Values of the function $W(u)$ for various values of u.

u	W (u)	u	W (u)	u	W (u)	u	W (u)	u	W (u)
1×10^{-10}	22.45	7×10^{-8}	15.90	4×10^{-5}	9.55	1×10^{-2}	4.04		
2	21.76	8	15.76	5	9.33	2	3.35		
3	21.35	9	15.65	6	9.14	3	2.96		
4	21.06	1×10^{-7}	15.54	7	8.99	4	2.68		
5	20.84	2	14.85	8	8.86	5	2.47		
6	20.66	3	14.44	9	8.74	6	2.30		
7	20.50	4	14.15	1×10^{-4}	8.63	7	2.15		
8	20.37	5	13.93	2	7.94	8	2.03		
9	20.25	6	13.75	3	7.53	9	1.92		
1×10^{-9}	20.15	7	13.60	4	7.25	1×10^{-1}	1.823		
2	19.45	8	13.46	5	7.02	2	1.223		
3	19.05	9	13.34	6	6.84	3	0.906		
4	18.76	1×10^{-6}	13.24	7	6.69	4	0.702		
5	18.54	2	12.55	8	6.55	5	0.560		
6	18.35	3	12.14	9	6.44	6	0.454		
7	18.20	4	11.85	1×10^{-3}	6.33	7	0.374		
8	18.07	5	11.63	2	5.64	8	0.311		
9	17.95	6	11.45	3	5.23	9	0.260		
1×10^{-8}	17.84	7	11.29	4	4.95	1×10^{-0}	0.219		
2	17.15	8	11.16	5	4.73	2	0.049		
3	16.74	9	11.04	6	4.54	3	0.013		
4	16.46	1×10^{-5}	10.94	7	4.39	4	0.004		
5	16.23	2	10.24	8	4.26	5	0.001		
6	16.05	3	9.84	9	4.14				

Source: Wenzel (1942).

Appendix 4.2 Values of the function $W(u, r/B)$ for various values of u and r/B. Source: Hantush (1956).

u \ r/B	0.002	0.004	0.006	0.008	0.01	0.02	0.04	0.06	0.08	0.1	0.2	0.4	0.6	0.8	1	2	4	6	8
0	12.7	11.3	10.5	9.89	9.44	8.06	6.67	5.87	5.29	4.85	3.51	2.23	1.55	1.13	0.842	0.228	0.0223	0.0025	0.0003
0.000002	12.1	11.2	10.5	9.89	9.44														
0.000004	11.6	11.1	10.4	9.88	9.44														
0.000006	11.3	10.9	10.4	9.87	9.44														
0.000008	11.0	10.7	10.3	9.84	9.43	8.06													
0.00001	10.8	10.6	10.2	9.80	9.42	8.06													
0.00002	10.2	10.1	9.84	9.58	9.30	8.06													
0.00004	9.52	9.45	9.34	9.19	9.01	8.03													
0.00006	9.13	9.08	9.00	8.89	8.77	7.98	6.67												
0.00008	8.84	8.81	8.75	8.67	8.57	7.91	6.67												
0.0001	8.62	8.59	8.55	8.48	8.40	7.84	6.67	5.87	5.29										
0.0002	7.94	7.92	7.90	7.86	7.82	7.50	6.62	5.86	5.29										
0.0004	7.24	7.24	7.22	7.21	7.19	7.01	6.45	5.83	5.29										
0.0006	6.84	6.84	6.83	6.82	6.80	6.68	6.27	5.77	5.27	4.85									
0.0008	6.55	6.55	6.54	6.53	6.52	6.43	6.11	5.69	5.25	4.85									
0.001	6.33	6.33	6.32	6.32	6.31	6.23	5.97	5.61	5.21	4.83	3.51								
0.002	5.64	5.64	5.63	5.63	5.63	5.59	5.45	5.24	4.98	4.71	3.50								
0.004	4.95	4.95	4.95	4.94	4.94	4.92	4.85	4.74	4.59	4.42	3.48								
0.006	4.54				4.54	4.53	4.48	4.41	4.30	4.18	3.43	2.23							
0.008	4.26				4.26	4.25	4.21	4.15	4.08	3.98	3.36	2.23							
0.01	4.04				4.04	4.03	4.00	3.95	3.89	3.81	3.29	2.23	1.55						
0.02	3.35				3.35	3.35	3.34	3.31	3.28	3.24	2.95	2.18	1.55	1.13					
0.04	2.68				2.68	2.68	2.67	2.66	2.65	2.63	2.48	2.02	1.52	1.13	0.842				
0.06	2.30				2.30	2.29	2.29	2.28	2.27	2.26	2.17	1.85	1.46	1.11	0.839				
0.08	2.03					2.03	2.02	2.02	2.01	2.00	1.94	1.69	1.39	1.08	0.832				
0.1	1.82						1.82	1.82	1.81	1.80	1.75	1.56	1.31	1.05	0.819	0.228			
0.2	1.22						1.22	1.22	1.22	1.22	1.19	1.11	0.996	0.857	0.715	0.227			
0.4	0.702						0.702	0.702	0.701	0.700	0.693	0.665	0.621	0.565	0.502	0.210			
0.6	0.454						0.454	0.454	0.454	0.453	0.450	0.436	0.415	0.387	0.354	0.177	0.0222		
0.8	0.311						0.311	0.310	0.310	0.310	0.308	0.301	0.289	0.273	0.254	0.144	0.0218		
1	0.219									0.219	0.218	0.213	0.206	0.197	0.185	0.114	0.0207		
2	0.049										0.049	0.048	0.047	0.046	0.044	0.034	0.011	0.0021	
4	0.0038											0.0038	0.0037	0.0037	0.0036	0.0031	0.0016	0.0006	0.0002
6	0.0004														0.0004	0.0003	0.0002	0.0001	
8	0																	0	0

5

Surface and groundwater flow modelling

Management of any system means making decisions aimed at achieving desired goals, taking into account the specified technical as well as non-technical constraints on the system. Models are tools to describe and evaluate performance of relevant systems under various real or hypothetical constraints and situations. A hydrologic model may be defined as a simplified conceptual representation of a part of the hydrologic cycle of a real-world system (here a surface water, a groundwater or a combined system) that approximately simulates the relevant excitation-response relations of a real-world system. Hydrologic models are primarily used for hydrologic prediction and for understanding hydrologic processes. Since real-world systems are usually complex, there is a need for simplifying the the process of making planning and management decisions. The simplification is introduced as a set of assumptions that expresses the nature of the system and features of its behaviour that are relevant to the problem under investigation. These assumptions relate, amongst other factors, to geometry of the investigated domain, the way various heterogeneities are smoothed out, nature of the porous medium (e.g. its homogeneity, isotropy), properties of the fluid (or fluids) involved, and type of flow regime under investigation. Because a model is a simplified version of a real-world system, no model is unique to a given hydrologic system. Different sets of simplifying assumptions result in different models, each approximating the investigated system in a different way. Two major types of hydrologic models can be distinguished:

- *Stochastic Models*. These models are essentially black box systems, based on data and using mathematical and statistical concepts to link a certain input (e.g. rainfall) to the model output (e.g. runoff). Commonly used techniques are regression, transfer functions, neural networks, and system identification. These models are known as stochastic hydrologic models and are discussed in Chapter 8.
- *Process-Based Models*. These models aim to represent the physical processes observed in the real world. Typically, such models contain representations of surface runoff, subsurface flow, evapotranspiration, and channel flow, but they can be far more complicated. These models are known as deterministic hydrologic models. A deterministic model can be subdivided into a single-event model and a continuous simulation model.

Recent research in hydrologic modelling aims to adopt a global approach to understanding behaviour of hydrologic systems to make better predictions to meet major challenges in water resources management. While the objectives of hydrologic modelling can encompass the entire hydrologic cycle and how it affects life on Earth, only flow of surface- and groundwater are considered in this chapter.

Modern Hydrology and Sustainable Water Development, 1st edition. © S.K. Gupta
Published 2011 by Blackwell Publishing Ltd.

5.1 Surface water flow modelling

Hydrologic systems are complex, with processes occurring over different geographic areas, characterized by highly variable parameters. In this situation mathematical models are valuable tools that allow one to make assessments, investigate alternative scenarios, and assist in developing effective management strategies. Modelling allows one to interpret real-life information and pose questions such as:

- Where do drainage deficiencies exist in the catchment?
- What effect does an in-stream structure have on flooding?
- How much water can be harvested from a storm water outfall for reuse?
- How does a wetland improve water quality of the watercourse?

Regardless of the phenomena being modelled, the basic principles of mathematical modelling remain the same. Whether the question at hand concerns behaviour of a natural system (e.g. the rainfall-runoff relationship) or an artificial one (e.g. a complex assembly of machines), the modeller can follow one of two approaches:

1) Constructing the model from first principles – which is possible in some cases. In this approach, starting from the physical laws known to govern the system, the laws are represented as equations that describe the relationships amongst variables, where changes in these variables denote changes in the state of the system. The accuracy of the model is then tested by comparing its predictions with the observed behaviour of the actual system.
2) In other instances, however, it is necessary to approach the problem from the opposite direction – that is, by inferring the system behaviour from the observed regularities in it. This behaviour can be included in the model by means of mathematical equations. In this type of approach, the model essentially describes how the system performs without explicitly explaining its *modus operandi*. As with numerical mod-

els in other disciplines, models used for surface water studies simulate the processes of interest through equations.

Hydrologic modelling can be undertaken as distributed or lumped-parameter simulation, differentiated by whether or not spatial variation of hydrological parameters is accounted for. *Lumped parameter modelling* assumes that hydrologic parameters and inputs, such as topography, soil type, vegetation, and rainfall do not vary over the catchment or other modelling scales. Thus, lumped parameter models combine all the data for a sub-basin into a single parameter, or a set of numbers, that define the response of the basin to a storm event. Distributed modelling, on the other hand, attempts to incorporate variability of the relevant hydrologic parameters, processes, and inputs across the catchment, as well as to provide for output of hydrologic information from potential contributing areas within the catchment. The term 'distributed model' implies any hydrologic model that has spatial differentiation of hydrologic parameters, inputs, or processes. Thus distributed or spatial models divide a drainage basin into a set of grid cells. These models take the data for each grid cell and use it to compute flow from one cell to the other for the entire drainage basin. The combined flow at the outlet cell of the watershed can thus be determined.

Surface water hydrologic modelling is perceived to meet two basic needs: (1) to determine the magnitude and frequency of flood flows; and (2) to determine the long-term availability of water for consumptive uses. The two needs require different modelling approaches.

5.1.1 Flood runoff

If only flood runoff is to be considered (e.g. to design storm water drainage systems, culverts, detention basins, and other storm water harvesting facilities), only rainfall and runoff need to be modelled. In this case, relatively simple lumped-parameter models can be used to estimate a design discharge (i.e. the stream flow to be used for engineering design) and the runoff hydrograph (i.e. a plot of stream flow versus time).

The various runoff models differ mainly in the methods used to generate runoff and to route it through a basin. They also differ in the control options available, data handling, and user interface, but these differences have virtually no effect on how the model computes runoff. The various models calculate runoff (excess precipitation) by one of the following: (i) SCS curve number method; (ii) Horton's equation; or (iii) continuous soil moisture balance. The SCS curve number is the most widely used method because of its relative simplicity; it defines the watershed storage and is determined for a watershed or sub-watershed predominantly from the types of soils, vegetative cover, and land-use characteristics (SCS 1986). Horton's equation assumes that the soil infiltration rate decreases exponentially as a function of time reckoned from the commencement of a storm. Some models take into account soil-moisture storage and infiltration using either the Green-Ampt or Phillips equation, or a variation thereof. The Pennsylvania State Runoff Model (PSRM) model uses the SCS curve number for determining soil infiltration, but uses soil moisture balance to estimate available storage. These models are either continuous or quasi-continuous (the process of soil-moisture accounting is continuous, but routing is performed only for a specified storm period). Soil moisture accounting and infiltration procedures are generally more data-intensive than the SCS curve number and Horton methods, and require a number of parameters corresponding to physical soil-water storage and infiltration characteristics. Important characteristics of some of the more commonly used rainfall-runoff models are given in Table 5.1.

After estimating the excess precipitation, surface runoff is calculated for overland flow and channel flow by one of the following methods: (i) unit hydrograph; (ii) SCS unit hydrograph; or (iii) by solving equations of flow. The unit hydrograph procedure derives a hydrograph by assuming a specific shape that represents land-use, soil, and geometric characteristics of a watershed. Techniques are also available to derive the unit hydrograph from observed rainfall-runoff data. The SCS unit hydrograph is an approximation of a nonlinear runoff

Table 5.1 Characteristics of some of the commonly used rainfall-runoff models. Some models have multiple options for runoff generation and flow routing.

Model	Simulation Type	Runoff Generation	Overland Flow	Channel Flow	Watershed Representation
CASC2D[1]	Event	Soil moisture accounting	Cascade	Diffusive wave	Distributed
CUHP[2]	Event	Horton	Unit hydrograph	Unit hydrograph	Lumped
CUHP/SWMM[3]	Event	Horton	Unit hydrograph	Unit hydrograph	Distributed
DR3M[4]		Soil moisture accounting	Kinematic wave	Kinematic wave	Distributed
HEC-1[5]	Event	SCS curve no.	Unit hydrograph	Muskingum	Distributed
HSPF[6]	Continuous	Soil moisture accounting	Kinematic wave	Kinematic wave	Distributed
PSRM[7]	Quasi-continuous	Soil moisture accounting and Soil moisture	Cascade	Kinematic wave	Distributed
SWMM[8]	Event	Horton	Kinematic wave	Kinematic wave	Distributed
TR-20[9]	Event	SCS curve no.	SCS Unit hydrograph		Lumped

[1]CASC2D   Cascade two-dimensional (Julian and Saghafian 1991)
[2]CUHP   Colorado Unit Hydrograph Procedure (UDFCD 1984)
[3]CUHP/SWMM   Sub-basin application of CUHP linked to SWMM
[4]DR3M   Distributed Rainfall Routing Runoff Model (Alley and Smith 1982)
[5]HEC-1   Hydrologic Engineering Centre (HEC 1990)
[6]HSPF   Hydrologic Simulation Program Fortran (Bicknell et al. 1993)
[7]PSRM   Penn State Runoff Model (Aron et al. 1996)
[8]SWMM   Storm Water Management Model (Huber and Dickinson 1988)
[9]TR20   Technical Release No. 20 (SCS 1983)

distribution that is assumed to be constant in a unit hydrograph method. A number of methods exist for solving equations of flow. The Muskingum method is used for channel routing by determination of wedge-shaped channel storage in relation to inflow and outflow channel volume. Overland flow and channel routing is performed in some models by kinematic waves to solve the continuity equation for flow or by diffusive waves, which include an additional pressure-differential term (Miller 1984). The cascade method is a two-dimensional kinematic wave approximation for routing overland flow (Julian et al. 1995). Moreover, modelling of flood flow focuses on the maximum flow (discharge), or peak discharge, for an event with a particular exceedance probability. The exceedance probability is the probability of a particular event being equalled or exceeded over a given period of time, usually one year. The model designer chooses the exceedance probability based on the perceived risk to human life or damage to property if the magnitude of the event is exceeded.

A comparison of the simulated and observed flows, using the nine uncalibrated models in Table 5.1 in two small urban watersheds, indicates that simulated peak flows differed from observed peak flows by as much as 260% and simulated storm volumes differ from the observed storm volumes by as much as 240% (Zarriello 1998). Because in most situations model results are difficult to verify and often one has to rely on the experience and judgment of the analyst, it is the responsibility of the analyst to ensure that the model used produces results that are reasonable for a given watershed. The analyst is also faced with decisions regarding the length of time to be simulated. These modelling decisions require considerable experience, not only as a hydrologist, but also as user of a particular model.

5.1.2 Water supply

If water supply is the important goal of a modelling exercise, the analyst must focus on longer time periods (e.g. years or decades) compared to the short intervals used for flood analysis. Because a specific storm event is not the focus, all hydrologic processes come into play, and the models used to study long-term water volumes must reflect this reality.

The physical processes are represented by simplified sub-models, yet the entire model must yield a reasonable solution.

The basic concept is the hydrologic budget, an accounting of the water as it moves along various pathways at the Earth's surface (in particular on the watershed of interest) in a manner that is consistent with the principle of conservation of mass of water. The components involved are precipitation, runoff, evapotranspiration, and movement of groundwater. The governing equation is:

$$P - E - R - G = \Delta S$$

where P is the precipitation input; E is evapotranspiration; R is runoff from the watershed; G is percolation to groundwater; and ΔS is the change in water storage in the active part of the soil profile. The terms in this equation can be expressed either as time rates or by depth or volume over some convenient period (i.e. 1 day). Because there are many terms involved in the hydrologic budget, computer models are most often used for these analyses.

The river basin scale model SWAT has been developed to quantify the impact of land management practices on water, sediment, and agricultural chemical yields in large complex watersheds with varying types of soils, land use, and management conditions over long periods of time. The main components of SWAT include weather, surface runoff, return flow, percolation, evapotranspiration, transmission losses, pond and reservoir storage, crop growth and irrigation, groundwater flow, reach routing, nutrient and pesticide loading, and water transfer into and out of the basin.

SWAT is a public domain model actively supported by the USDA Agricultural Research Service at the Grassland, Soil and Water Research Laboratory in Temple, Texas.

The use of hydrologic modelling systems for water resources planning and management is becoming increasingly popular. Because these hydrologic models generally deal with the land phase of the hydrologic cycle and use data related to topography and physical parameters of watershed, computer-based Geographic Information Systems (GIS) meet this requirement efficiently. These systems link land cover data to topographic data as well as to other information related to geographic locations. When applied to hydrologic systems, non-topographic information can also be included as description of soils, land use, ground cover, groundwater conditions, as well as man-made systems and their characteristics on or below the land surface. Therefore, integration of GIS with hydrologic modelling holds the promise of a cost-effective alternative for studying watersheds. The ability to perform spatial analysis for development of lumped and distributed hydrologic parameters not only saves time and effort, but also improves accuracy over traditional methods.

A number of catchment models have been developed worldwide, the important ones amongst them are RORB (Australia), Xinanjiang (China), Tank model (Japan), ARNO (Italy), TOPMODEL (Europe), UBC (Canada), HBV (Scandinavia), and MO-HID Land (Portugal). However, not all these models include any chemistry component. Generally speaking, SWM, SHE, and TOPMODEL have the most comprehensive stream chemistry treatment (component) and have evolved to accommodate the latest data sources, including remote sensing and GIS data.

In the United States, the US Army Corps of Engineers, in collaboration with researchers at a number of universities, have developed the Gridded Surface/Subsurface Hydrologic Analysis (GSSHA) model. GSSHA is widely used in the United States for research and analysis by US Army Corps of Engineers, district level and larger consulting companies to compute flow, water levels, distributed erosion, and sediment delivery in complex engineering designs. A distributed nutrient and contaminant fate and transport component is undergoing testing. GSSHA input/output processing and interfacing with GIS is facilitated by the Watershed Modelling System (WMS).

5.2 Groundwater flow modelling

As groundwater is essentially a hidden resource with an enormously complex distribution of material and transient fluxes of water, models are needed first to characterize the flow system and then to make quantitative estimates of flow across boundaries, storages, and drawdowns to devise resource exploitation strategies and estimate cost of exploitation and finally to ensure its sustainable development and management. Only a small fraction of the subsurface can ever be sampled and tested hydraulically. This, undoubtedly, provides an important yet incomplete picture of the actual system. Therefore, models – inherently simpler than the real flow systems, yet capturing their important overall features – are formulated and used for quantitatively describing and solving flow problems. The International Groundwater Modeling Center, located at ----, defines a model as 'a non-unique, simplified, mathematical description of an existing groundwater system, coded in a programming language, together with quantification of the groundwater system, the code simulates in the form of boundary conditions, system parameters, and system stresses. Models of groundwater flow are widely used for a variety of applications ranging from designing of water supply–to contaminant remediation – systems. A large number of computer programs are available to carry out the involved computations in groundwater flow models.

The process begins with developing a conceptual model of the flow problem, followed by constructing a mathematical model to simulate the system employing a conceptual model. For simple conceptual models, flow solutions developed in Chapter 3 may suffice. But with more complex conceptual models, the only practical approach may be to use a computer program to carry out the computations. The complete process of model application comprises several steps, as indicated in Fig. 5.1. These steps are described in the following, with emphasis on basic mathematical formulation in the numerical modelling process.

In the past, complex groundwater flow problems were modelled using physical models or analogs. The physical model of the conceptualized flow domain comprised a miniature scaled laboratory model employing a tank. The most common analog models used the analogy between water flow and flow of electricity through a network of wires and resistors, where voltage is analogous to head, resistance is analogous to inverse of permeability or transmissivity ($1/K$ or $1/T$), current is analogous to discharge, and capacitance is analogous to storage. In the course of time, computer simulation

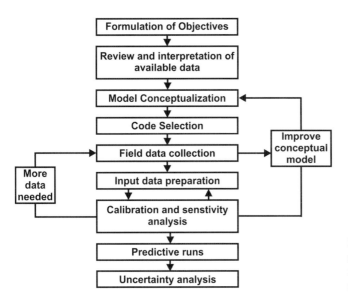

Fig. 5.1 Various steps involved in a groundwater modelling application process. Redrawn from Bear (1992). With permission from John Wiley and Sons Inc.

methods have essentially replaced physical and analog models, so these methods are not discussed any further. The interested reader is, however, referred to Walton (1970) and Prickett (1975).

The problem of finding a function that solves a specified partial differential equation (PDE) in the interior of a given region, which takes prescribed values on the boundary of the region, is known as a Dirichlet problem. The Dirichlet problem in groundwater is to find the head distribution $(h(x, y, z, t) = \ldots)$ in an aquifer domain consistent with Darcy's law and mass balance for given initial and boundary conditions. At present, most computer programs for groundwater flow modelling are based on solutions of partial differential equations of groundwater flow (Eqn. 3.35 or Eqn. 3.36) using one of the three methods: (i) finite difference method (FDM); (ii) finite element method (FEM); or (iii) analytic elements method (AEM). A fourth, less commonly employed method, the boundary integral equation method (BIEM), is conceptually similar to the analytic element method but is not discussed any further. The interested reader is referred to Liggett and Liu (1983). In the following, essentials of the three methods (FDM, FEM, and AEM) are outlined.

Flow in aquifers is often modelled as two-dimensional in the horizontal plane. This can be done because aquifers have horizontal dimensions that are hundreds or thousands times larger than their vertical thickness. In most aquifers, the dominant component of the saturated flow is also horizontal, which gives rise to the total resistance encountered along a typical flow path. Because of this, the real three-dimensional flow system can still be modelled reasonably by employing two-dimensional analysis. This is achieved by assuming that h varies with x and y, but not with z. This simplifying assumption is called the Dupuit–Forchheimer approximation after the French and the German hydrologists (Dupuit 1863; Forchheimer 1886).

5.2.1 Finite difference method

The *finite difference method* (FDM) is a computational procedure based on dividing an aquifer into a grid and analysing the flows associated within a single zone of the aquifer.

5.2.1.1 Finite difference equations

Finite differences are a way of representing continuous differential operators using discrete intervals (e.g. Δx and Δt). The differentials are derived by using Taylor series expansion of the variable around a point. For example, the first-order time derivative

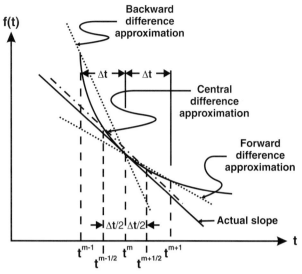

Fig. 5.2 Schematic representation of the three commonly used approaches for estimating the slope of a function in time domain.

of the head variable, h, is often approximated using the following forward finite difference scheme (Fig. 5.2), where the subscripts indicate a discrete time location:

$$\frac{\partial h}{\partial t} = h'(t_i) \approx \frac{h_i(t) - h_{i-1}(t)}{\Delta t} \quad (5.1)$$

The first- and second-order space derivatives of h around x may be written as:

$$\frac{\partial h}{\partial x} = h'(x_i) \approx \frac{h_i(x) - h_{i-1}(x)}{\Delta x} \quad (5.2)$$

$$\frac{\partial^2 h}{\partial x^2} = h''(x_i) \approx \frac{h_{i+1}(x) - 2h_i(x) + h_{i-1}(x)}{(\Delta x)^2} \quad (5.3)$$

Similarly, for y and z, rewriting Eqn. 3.36 as:

$$\frac{\partial}{\partial x}\left(K_x \frac{\partial h}{\partial x}\right) + \frac{\partial}{\partial y}\left(K_y \frac{\partial h}{\partial y}\right) + \frac{\partial}{\partial z}\left(K_z \frac{\partial h}{\partial z}\right) - W$$

$$= S_s \frac{\partial h}{\partial t} \quad (5.4)$$

and recalling that K_x, K_y, and K_z are values of hydraulic conductivity along the x, y, and z coordinate axes, respectively [L T^{-1}]; h is the potentiometric head [L]; W is the volumetric flux per unit volume representing sources and/or sinks of water, where the negative sign indicates extraction [T^{-1}]; S_S is specific storage of the porous material [L^{-1}]; and t is time [T].

Let us examine the first term in Eqn 5.4. In the control volume finite difference method, the nodes coincide with the boundary of the domain. Let (i, j, k) denote the particular node under investigation and $(i + \frac{1}{2})$ be the interface between node i and $(i + 1)$ (similarly for j and k). Thus for node (i, j, k):

$$\frac{\partial}{\partial x}\left(K_x \frac{\partial h}{\partial x}\right) \approx \frac{\left[K\frac{\partial h}{\partial x}\right]_{i+1/2} - \left[K\frac{\partial h}{\partial x}\right]_{i-1/2}}{\Delta x_i} \quad (5.5)$$

A similar relation holds for the other two terms in Eqn 5.4. This is called the backward spatial differencing scheme (BSDS). The finite-difference approximation for the time derivative of head, $\partial h/\partial t$, must next be expressed in terms of specific head and time.

An approximation to the time derivative of the head at time instant, t_m, is obtained by dividing the head difference, $h_{i,j,k}^m - h_{i,j,k}^{m-1}$, by the time interval, $t_m - t_{m-1}$, that is:

$$\frac{\partial h}{\partial t} \approx \frac{h_{i,j,k}^m - h_{i,j,k}^{m-1}}{t^m - t^{m-i}} \quad (5.6)$$

Thus the time derivative is approximated using the change in head at the node over the preceding time interval and ends with the time instant at which flow is evaluated. This is termed a backward-difference approach in that $\partial h/\partial t$ is approximated

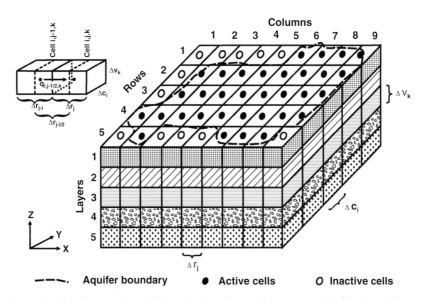

Fig. 5.3 Three-dimensional block centred spatial discretization of an aquifer system used in the finite difference model MODFLOW. A grid of blocks called cells, whose locations are described in terms of rows, columns, and layers and an 'i, j, k' indexing system is used. In terms of Cartesian coordinates, the k index denotes changes along the vertical direction z; because the convention followed in this model is to number layers from the top down, an increment in the k index corresponds to a decrease in elevation. Similarly, rows are parallel to the x-axis, so that increments in the row index, i, would correspond to decreases in y. Columns are considered parallel to the y-axis, so that increments in the column index, j, correspond to increase in x. The flow from cell (i, j^{-1}, k) into the cell (i, j, k) is indicated in the top left corner. Redrawn from Harbaugh (2005). © U.S. Geological Survey.

over a time interval that extends backwards in time from t^m, the time at which the flow terms are calculated. The backward-difference approach is preferred over other approaches of differencing, as it is unconditionally numerically stable. The three commonly used approaches, namely the forward-, backward-, and central-difference in time domain are schematically depicted in Fig. 5.2.

Finally, following Harbaugh (2005) as in MODFLOW – a popular finite difference computer code – the partial differential equation (Eqn 5.4) in a block-centred discrete aquifer domain can be written as:

$$CR_{i,j-\frac{1}{2},k} \left(b^m_{i,j-1,k} - b^m_{i,j,k} \right)$$

$$+ CR_{i,j+\frac{1}{2},k} \left(b^m_{i,j-1,k} - b^m_{i,j,k} \right)$$

$$+ CC_{i-\frac{1}{2},j,k} \left(b^m_{i-1,j,k} - b^m_{i,j,k} \right)$$

$$+ CC_{i+\frac{1}{2},j,k} \left(b^m_{i+1,j,k} - b^m_{i,j,k} \right)$$

$$+ CV_{i,j,k-\frac{1}{2}} \left(b^m_{i,j,k-1} - b^m_{i,j,k} \right)$$

$$+ CV_{i,j,k+\frac{1}{2}} \left(b^m_{i,j,k+1} - b^m_{i,j,k} \right)$$

$$+ P_{i,j,k} b^m_{i,j,k} + W_{i,j,k}$$

$$= SS_{i,j,k} \left(\Delta r_j . \Delta c_i . \Delta v_k \right) \frac{b^m_{i,j,k} - b^{m-1}_{i,j,k}}{t^m - t^{m-1}} \qquad (5.7)$$

where $b^m_{i,j,k}$ is the hydraulic head at the cell (i, j, k) at time step m. This equation (Eqn 5.7) is obtained by using block-centred nodes of rows, columns, and layers, as in Fig. 5.3 for the cell (i, j, k).

CV, CR, and CC are the hydraulic conductances of layers, rows, and columns, or branch conductances between node (i, j, k) and a neighbouring node. Conductance is product of hydraulic conductivity and cross-sectional area of flow divided by the length of the flow path (in this case, the total distance between the nodes). $CR_{i,j-1/2,k}$ is the conductance in row i and layer k between nodes (i, j^{-1}, k) and (i, j, k) [L^2T^{-1}]:

$P_{i,j,k}$ is the sum of coefficients of head from source and sink terms;

$W_{i,j,k}$ is the sum of constants from source and sink terms, where $W_{i,j,k} < 0.0$ is flow out of the groundwater system (such as pumping) and $W_{i,j,k} > 0.0$ is the inflow (such as injection or natural groundwater recharge);

$SS_{i,j,k}$ is the specific storage;

Δr_j, Δc_i, and Δv_k are the three dimensions of cell (i, j, k) which, when multiplied together, represent the volume of the cell;

t^m is the time at time step m.

The finite-difference flow equation for a cell is a representation of the volumetric flow from all sources in dimensional form ($L^3\ T^{-1}$), where L and T denote dimensions of length and time, respectively, provided that length and time units are used consistently for all the terms, in any specific units.

The dimensions of the blocks are made differently for smaller blocks in areas where more accuracy is required. The spacing of the grid boundaries in the x and y directions is fixed throughout the grid, so that there are bands of narrower blocks that are spread over the entire grid. The horizontal boundaries that form the top and bottom of the blocks can be staggered to form irregular surfaces so that model layers correspond to stratigraphic layers.

An equation of this form (Eqn 5.7) is written for every cell in the grid in which the head is free to vary with time (variable head cells), and the system of equations is solved simultaneously for the heads at time t^1, taking $(m^{-1}) = 0$. When these solutions have been obtained, the process is repeated to obtain heads at time t^2, the end of the second time step. To do this, Eqn 5.7 is applied again, using 1 as the time superscript (m^{-1}) and 2 as time superscript m. In the resulting system of equations, heads at time t^2 are the unknowns and this set of equations is solved simultaneously to obtain the head distribution at time t^2. This process is continued for as many time steps as required to cover the time range of interest.

It is convenient to rearrange Eqn 5.7 so that all terms containing heads at the end of the current time step are taken to the left-hand side of the equation, and all terms that are independent of head at the end of the current time step are taken to the right-hand side. All coefficients of $b^m_{i,j,k}$ that do not include conductance between nodes are combined into a single term, $HCOF$, and similarly all right-hand-side terms are combined. Furthermore, the complexity can be reduced by assuming that the time superscript is denoted by m, unless otherwise shown. The resulting equation is:

$$CV_{i,j,k-1/2}b_{i,j,k-1} + CC_{i-1/2,j,k}b_{i-1,j,k}$$

$$+ CR_{i,j-1/2,k}b_{i,j-1,k} + (-CV_{i,j,k-1/2} - CC_{i-1/2,j,k}$$

$$- CR_{i,j,k-1/2} - CV_{i,j,k+1/2} - CC_{i+1/2,j,k}$$

$$- CR_{i,j,k+1/2} + HCOF_{i,j,k})b_{i,j,k}$$

$$+ CV_{i,j,k+1/2}b_{i,j,k+1} + CC_{i+1/2,j,k}b_{i+1,j,k}$$

$$+ CR_{i,j+1/2,k}b_{i,j+1,k} = RHS_{i,j,k} \qquad (5.8)$$

where

$$HCOF_{i,j,k} = P_{i,j,k} - \frac{SS_{i,j,k}\Delta r_j \Delta c_i \Delta v_k}{t - t^{m-1}} \qquad [L^2 T^{-1}]$$

and

$$RHS_{i,j,k} = -Q_{i,j,k} - SS_{i,j,k}\Delta r_j \Delta c_i \Delta v_k \frac{b^{m-1}_{i,j,k}}{t - t^{m-1}}$$

$$[L^3 T^{-1}]$$

The entire system of equations of the type as Eqn 5.8, which includes one equation for each variable head cell in the grid, may be written in matrix notation as:

$$[A]\{b\} = \{q\}$$

where

[A] is a matrix of the coefficients of head on the left-hand side of Eqn 5.8 for all active nodes in the grid;

$\{b\}$ is a vector of head values at the end of time step m for all nodes in the grid; and

$\{q\}$ is a vector of the constant terms on the right-hand side for all nodes of the grid.

The computer code MODFLOW assembles the vector $\{q\}$ and the terms that comprise [A] through a series of subroutines. The vector $\{q\}$ and the terms comprising [A] are then transferred to subroutines that actually solve the matrix equations for the vector $\{b\}$.

MODFLOW gained widespread acceptance in the 1980s, because it is versatile, well tested, well documented, and available in the public domain (URL: http://pubs.water.usgs.gov/tm6a16). The formatted text input and output files associated with *MODFLOW* are cumbersome, but graphical user interface software makes it user friendly by shielding the user from worrying about the cumbersome details.

MODFLOW was programmed in a modular way, so that additional capabilities could be added subsequently. Some of the more important additions include: variable-density flow (Sanford and Konikow 1985); MODPATH for tracing flow path lines (Pollock 1989); an improved matrix solution procedure (Hill 1990); better handling of water-table boundary conditions (McDonald *et al.* 1991); a parameter estimation capability (Hill 1992); and thin barriers to flow (Hsieh and Freckleton 1993).

5.2.1.2 Boundary conditions

All typical *boundary conditions* in flow models, including no-flow boundaries, specified head boundaries, specified flux boundaries, and leakage boundaries constitute important elements in FDMs.

In MODFLOW, the entire external boundary is by default assumed to be a no-flow (impermeable) boundary. For a block at the edge of the grid, the no-flow condition is accomplished simply by omitting the flux term that is in the direction of the boundary. For example, the node equation for a block at the negative x edge of the grid would have $Q_{x-} = 0$.

Similarly, where an impermeable (termed as 'inactive' in MODFLOW terminology) block abuts an active block, the nodal equation for the block omits the flux term coming from it to simulate the internal impermeable block. To simulate an irregularly shaped no-flow boundary, the entire region of blocks may be made inactive by omitting generation of nodal equations for them.

At a specified head boundary, the head at the node is known while the internal discharge through the block Q_s is unknown. The model determines the Q_s needed to maintain the specified head and yields Q_s as output after solving the finite difference equations. There is no limit on the magnitude of Q_s, so a constant head boundary represents a boundary with unlimited capacity to supply or remove water. Therefore, it is important to check the output values of Q_s to see if the discharge is consistent with the actual field situation.

In case of a specified flux boundary condition, the internal discharge term Q_s is specified, in contrast to the default value of $Q_s = 0$. Pumping wells, recharge, and evapotranspiration are typically represented using specified flux boundaries. An extraction well with discharge Q contributes $-Q$ to Q_s and an injection well contributes $+Q$ to Q_s. When representing recharge at a rate N [LT^{-1}], a discharge equal to $N\Delta x\Delta y$ is added to Q_s in the uppermost saturated block of a model. Evapotranspiration flux is defined in the same way as recharge, but its contribution to Q_s is negative. In MODFLOW, evapotranspiration occurs only when the simulated water table rises above a specified threshold level.

Discharges from a semi-permeable river bed to an underlying aquifer are represented by leakage boundaries. These also include discharge from resistant drains and other similar features not in direct contact with the aquifer. The leakage is represented by a discharge, Q, that is added to Q_s. The magnitude of Q is proportional to the leakage factor, L, and the difference between the head at the node, h, and a specified reference head, h_r is given as:

$$Q = L\left(h_r - h\right) \tag{5.9}$$

The leakage factor is defined by Darcy's law for vertical flow through the river bed as:

$$L = \frac{K_Z^* A}{b^*} \tag{5.10}$$

where K_Z^* is the vertical hydraulic conductivity of the semi-permeable river bed material; b^* is the vertical thickness of this material; and A is the surface area of river bed lying within the limits of the block. When the entire surface area of a block underlies the river bed, $A = \Delta x\Delta y$.

5.2.2 Finite element method

The *finite-element method* (FEM) uses a concept termed as a 'piecewise approximation'. The domain of the aquifer to be simulated is divided into

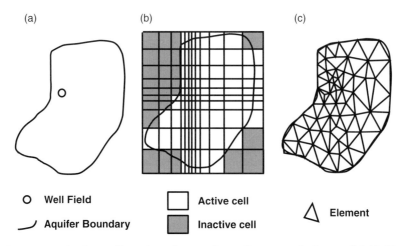

Fig. 5.4 Schematic representation in two dimensions of a groundwater flow system having a well field of interest and impermeable aquifer boundary, as in (a). The aquifer has been discretized using finite-difference model (FDM) in (b) and finite-element model (FEM) in (c). Note that rectangular elements in FDM are smaller at the well field to improve accuracy of simulation. Similarly, the elements are smaller in the area of the well field in FEM. Redrawn from http://www.iaea.org/programmes/ ripc/ ih/volumes/vol_six/chvi_04.pdf.

a set of elements. Point values of the dependent variable (head) are calculated at the nodes that are corners or vertices of the elements, and a simple equation is used to describe the value of the dependent variable within the element. This simple equation is called a basis function and each node that is part of an element has an associated basis function. The simplest form amongst the basis functions that are generally used is linear function.

A major advantage of the FEM is the flexibility of the finite-element grid, which allows a close spatial approximation of irregular boundaries of the aquifer and/or of parameter zones within the aquifer. The most commonly used shapes in the FEM are triangles and trapezoids for two-dimensional flows and triangular and trapezoidal prisms for three-dimensional flows. This difference between the FDM and FEM for a two-dimensional hypothetical aquifer system using triangular elements is schematically shown in Fig. 5.4c. The aquifer system is assumed to have impermeable boundaries and a well field of interest (Fig. 5.4a). The aquifer has been discretized using finite-difference (Fig. 5.4b) and finite-element (Fig. 5.4c) grids. Fig. 5.4b and Fig. 5.4c also conceptually illustrate how their respective grids can be adjusted to use finer mesh spacing in selected areas

of interest. The rectangular finite-difference grid approximates the aquifer boundaries in a step-wise manner, resulting in some nodes or cells falling outside the aquifer, whereas sides of the triangular elements of the finite-element grid can closely follow the outer boundary using a minimal number of nodes. Because of the flexibility in choosing the shape of elements, the boundary conditions can also be specified more realistically in an FEM than in an FDM.

Furthermore, in a FDM the head is defined only at the nodes, but in a FEM the head is defined both at the nodes as well as throughout the entire spatial domain of the element. This is achieved by assuming the head to vary across an element in a simple manner. A linear variation is commonly used, although some higher-order distributions can also be used. The head surface has a constant slope within an element, but the slope abruptly changes across an element boundary. Such an assemblage of linear panels (Fig. 5.5a) is often an acceptable approximation of an actual smoothly varying head distribution.

The third difference between FDM and FEM is in the method of determining the solution of the groundwater flow equation (Eqn. 3.36 or Eqn 5.4) for a restricted type of head distribution (in this

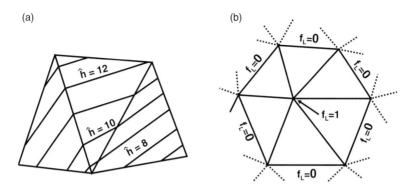

Fig. 5.5 (a) Contours of $\hat{h}$ across three elements in a finite element model. Within any element the gradient of $\hat{h}$ is constant but across element boundaries abrupt changes can be seen. (b) Basis function f_L for the central node L. It varies from $f_L = 1$ at the node to $f_L = 0$ at the opposite edge of any triangle and varies linearly in between. Redrawn from Fitts (2002). © Academic Press.

case linear) allowed for the elements. There are several methods for honing in on the best solution. In the variational method, the solution is obtained by minimizing a variational function at each node. This approach leads ultimately to the same set of algebraic equations and the same solution, as given by the Galerkin method that determines the solution by minimizing a parameter called the residual, which is a measure of how much the approximate solution differs from that obtained from the general equations.

5.2.2.1 Finite element equations

To understand FEM equations, let us consider linear triangular elements and two-dimensional (x and y) flow in a homogenous aquifer governed by Eqn 5.4. The solution to this equation is equivalent to finding a solution for $h(x, y, t)$ that minimizes the variational function F (Todd 2006):

$$F = \iint \left[\frac{K_x}{2} \left(\frac{\partial h}{\partial x} \right)^2 + \frac{K_y}{2} \left(\frac{\partial h}{\partial y} \right)^2 \right.$$

$$\left. + \left(S\frac{\partial h}{\partial t} + W \right) h \right] dxdy \qquad (5.11)$$

To minimize Eqn 5.11, the differential $\partial F / \partial h$ is evaluated for each node and equated to zero. The resulting system of simultaneous equations can be solved on a digital computer.

The Galerkin method starts by defining the variable $\hat{h}(x, y)$ for each node (corners) as the approx-

imate FEM solution. The mathematical expression for h can be written as sum of the contributions associated with various nodes:

$$\hat{h}(x, y) = \sum_{L=1}^{NNODE} \hat{h}_L f_L(x, y) \qquad (5.12)$$

where L = nodal number; $NNODE$ = total number of nodes in the domain under consideration; and f_L is called the basis function for the node L. The basis function, also called interpolation function, for one of the nodes (numbered as L) is shown graphically in Fig. 5.5b. It varies linearly from $f_L = 1$ at the node L to $f_L = 0$ at the opposite edge of any triangle that shares the node L. In all other triangles, $f_L = 0$. At any point (x, y), Eqn 5.12 contains only three non-zero terms, namely those associated with the three nodes of the triangular element containing (x, y). Therefore:

$$\hat{h}(x, y) = \hat{h}_i f_i + \hat{h}_j f_j + \hat{h}_k f_k \qquad (5.13)$$

where i, j, k are the three nodes of the triangle containing (x, y) and $\hat{h}_i, \hat{h}_j, \hat{h}_k$ are the respective heads at these nodes.

With the Galerkin method, the initial value for the residuals is assumed to be zero when integrated over the entire problem domain:

$$\iint_D \left[\frac{K_x}{2} \left(\frac{\partial h}{\partial x} \right)^2 + \frac{K_y}{2} \left(\frac{\partial h}{\partial y} \right)^2 \right.$$

$$\left. + \left(S\frac{\partial h}{\partial t} + W \right) h \right] f_L(x, y) dxdy = 0 \qquad (5.14)$$

where D denotes the entire problem domain. The terms inside the square brackets represent the residual, which is a measure of how poorly (or well) the trial solution satisfies Eqn 5.4. Thus $f_L(x, y)$ becomes a weighting function.

In this manner, a series of algebraic equations represented as basis functions of interpolated values of trial hydraulic heads are obtained. Because each nodal equation involves only the central node and its immediate neighbours, it contains a limited number of terms. If the element mesh is designed efficiently, the resulting system of linear equations forms a narrowly banded matrix of coefficients. The system of resulting equations is usually solved by iterative methods and gives nodal heads $\hat{b}_L$ ($L = 1$ to $NNODE$). With $\hat{b}_L$ known, heads and discharges can be calculated throughout the domain. FEMWATER uses the Galerkin finite element method.

Like the FDM, the FEM can handle all forms of boundary conditions, including no-flow-, specified head-, specified flux-, and leakage boundaries.

A disadvantage of FEM grids, however, is the larger amount of data required for the grid compared to the FDM. The nodes and elements must be numbered consecutively. The coordinates of each node and element must be specified. The numbering of nodes must be specified systematically from top to bottom and from left to right, sequentially across the shortest dimension of the problem domain.

5.2.3 Analytic element method

The *analytic element method* (AEM) is essentially a computer-generated version of the superposition technique. Instead of superposing just a few functions manually, a computer can superpose thousands of functions. The idea of using computers to superpose large numbers of solutions in groundwater flow models was pioneered by Otto Strack and a detailed description of this can be found in Strack (1989) and Haitjema (1995). AEM is a grid-independent method that discretizes the external and internal system boundaries and not the entire domain. Therefore, AEM models are limited by the amount of detail included and not by the spatial extent. As shown subsequently, it uses exact solution

of the governing PDE of the constituent elements and is approximate only in how well the boundary conditions are satisfied.

Unlike the finite element, an analytic element is a mathematical function that is associated with a particular boundary condition associated with the flow domain. For example, one kind of element represents flow to a pumping well, a second kind represents discharge to a stream, and a third kind represents an area of recharge, and so on. These mathematical functions in the modelling domain are summed up (corresponding to superposition) to create a single equation that may have hundreds of terms. This equation is then used to predict the head distribution, $b(x, y, t)$, and discharges within the modelled domain. Most current AEMs are based on two-dimensional aquifer flow, as described in the following. By superimposing several two-dimensional models, it is possible to develop three-dimensional models for representing a layered aquifer system with coupled leakage occurring between adjacent layers (Strack 1999).

5.2.3.1 AFM functions for two-dimensional aquifer flow

As already mentioned, a real three-dimensional flow system can be modelled in a reasonable way by using a two-dimensional analysis (the Dupuit-Forchheimer approximation). However, this is reasonable only when the resistance to vertical flow is a small fraction of the total resistance encountered along the entire flow path.

The general equation for two-dimensional aquifer flow (Eqn 3.36 or Eqn 5.4) is recapitulated below for the case of isotropic transmissivity T:

$$\frac{\partial}{\partial x}\left(T\frac{\partial b}{\partial x}\right) + \frac{\partial}{\partial y}\left(T\frac{\partial b}{\partial y}\right) - W = S\frac{\partial b}{\partial t} \quad (5.15)$$

In the analytic element method, this and other flow equations are written in terms of an aquifer discharge potential φ [L^3/T]. This permits AEM to efficiently model a variety of aquifer conditions (confined, unconfined, etc.). The discharge potential is defined in terms of aquifer parameters and head, ensuring validity of the following equations:

$$\frac{\partial \varphi}{\partial x} = T\frac{\partial b}{\partial x} = -Q_x, \quad \frac{\partial \varphi}{\partial y} = T\frac{\partial b}{\partial y} = -Q_y \quad (5.16)$$

It is thus seen that discharge potential φ is derivative of discharge with a negative sign. Substituting Eqn 5.16 into Eqn 5.15 gives:

$$\nabla^2 \varphi - W = S\frac{\partial h}{\partial t} \qquad (5.17)$$

For steady state flow with leakage/recharge/discharge, the governing equation is the Poisson equation (Eqn 3.40), which is linear in φ:

$$\nabla^2 \varphi = W \qquad (5.18)$$

And for steady state flow without any leakage/recharge/discharge, the governing equation is the Laplace equation (Eqn 3.39), which is also linear in ϕ:

$$\nabla^2 \varphi = 0 \qquad (5.19)$$

As already mentioned in Section 4.2, solutions to these linear equations can be superposed. This is the basis of AFM.

One can examine how φ must be defined in order to be consistent with Eqn 5.16. For uniform transmissivity, each relation of Eqn 5.16 can be integrated to give:

$$\varphi = Th + Const \qquad (5.20)$$

where $Const$ is the constant of integration. For an unconfined aquifer with horizontal base and uniform horizontal hydraulic conductivity, $T = K\,h$, with h measured from the aquifer base. Eqn 5.16 for this case becomes:

$$\frac{\partial \varphi}{\partial x} = Kh\frac{\partial h}{\partial x}, \quad \frac{\partial \varphi}{\partial y} = Kh\frac{\partial h}{\partial y} \qquad (5.21)$$

Integrating either of these equations yields the discharge potential for this type of unconfined flow:

$$\varphi = \frac{1}{2}Kh^2 + Const \quad (unconfined\ aquifer$$

$$with\ horizontal\ base) \quad (5.22)$$

where $Const$ is the constant of integration. An equation such as Eqn 5.20 or Eqn 5.22 relates discharge potential to head, $\varphi(h)$, and another to location and time, $\varphi(x,y,t)$. In fact, φ can be thought of as an intermediate variable between h on the one hand

and (x,y,t) on the other. The $\varphi(h)$ relation depends only on the aquifer geometry and properties, and the $\varphi(x,y,t)$ relation depends on discharges and the boundary conditions in the aquifer.

To verify that the use of φ indeed leads to economy, let us consider the solution for steady radial flow to a well in an aquifer with constant T (Eqn. 4.8):

$$h = \frac{Q}{2\pi T}\ln r + C \qquad (5.23)$$

The corresponding solution for radial flow in an unconfined aquifer with a horizontal base is (from Eqn. 4.13):

$$h^2 = \frac{Q}{\pi K}\ln r + C \qquad (5.24)$$

Both solutions, Eqn 5.23 and Eqn 5.24, can be represented by one discharge potential function for radial flow, φ, as below:

$$\varphi = \frac{Q}{2\pi}\ln r + D \qquad (5.25)$$

where D is a constant and the $\varphi(h)$ relation is either Eqn 5.20 for constant T aquifer, or Eqn 5.22 for an unconfined aquifer.

Several other types of aquifers can be modelled with AEM. All that is required is a definition of discharge potential – head relation $\varphi(h)$ that satisfies Eqn 5.16. The domain of AEM model is infinite in the $X-Y$ plane and all boundaries are internal. The various definitions of $\varphi(h)$ that satisfy Eqn 5.16 (e.g. Eqn 5.20 and Eqn 5.22), lead to the same general equation in terms of φ. Since the steady state general equations and in some cases transient general equations are linear, any number of solutions may be superposed. Any specific problem can be solved by discharge potential function $\varphi(x, y, t)$, which satisfies the general equation(s) exactly and approximates the boundary conditions of the conceptual model. The total discharge potential, φ_t, is the sum of all the discharge potential functions, each associated with particular elements representing aquifer boundary conditions that generally correspond to hydrogeologic features of the area under investigation:

$$\varphi_t(x, y, t) = \varphi_{well}(x, y, t) + \varphi_{river}(x, y, t)$$

$$+ \varphi_{lake}(x, y, t) + \ldots + C \qquad (5.26)$$

where φ_{well}, φ_{river}, φ_{lake}, etc. are analytic functions associated with specific elements such as well, river, lake, etc. and C is a constant. The most commonly used simple elements are: well, river, line sink, lake, and inhomogeneity (e.g. change in conductivity).

As an example, let us consider a model of steady flow with a uniform recharge rate W. One of the functions φ_1, φ_2, ... would be the solution of Poisson's equation and the remaining would be solutions of Laplace's equation to accommodate boundary conditions such as discharging wells, discharge along stream segments, or boundaries arising from aquifer heterogeneities.

Some of the functions φ_1, φ_2, ... do not contain any unknown parameters. For example, the function of a well of known discharge is completely defined at the outset. Other functions contain unknown parameters that are determined by specifying the boundary conditions at or near the element that the function represents. For example, an element representing a well of unknown discharge but known head would have the head specified at the well screen radius. The unknown discharge is determined by writing an equation for the head at the location of the well screen.

Each specified boundary condition yields an equation with adjustable coefficients that can be used to satisfy boundary conditions along its boundary. This results in a system of linear equations. This is solved by standard methods for the unknown parameters. If the number of unknowns, U, equals number of equations, E, the system is solved for the generated square matrix. If $E > U$, the over specified system of equations is solved using least square techniques.

With any AEM model, the head may be evaluated anywhere in the x, y domain by first evaluating the discharge potential, $\varphi_t(x, y, t)$, and the head from this discharge potential is then determined using the appropriate $\varphi(b)$ relation. Surface plots of the head can be made by evaluating b at an array of points within the area of interest, followed by then contouring the values at various points.

Analytic expressions of the gradient of the discharge potential may be derived by summing up the derivatives of the discharge potential function

for each element:

$$\frac{\partial \varphi_t}{\partial x} = \frac{\partial \varphi_1}{\partial x} + \frac{\partial \varphi_2}{\partial x} + \frac{\partial \varphi_3}{\partial x} + \cdots$$

$$\frac{\partial \varphi_t}{\partial y} = \frac{\partial \varphi_1}{\partial y} + \frac{\partial \varphi_2}{\partial y} + \frac{\partial \varphi_3}{\partial y} + \cdots \quad (5.27)$$

These analytic expressions are programmed into the modelling software and are used when specifying certain boundary conditions and also for tracing flow path lines.

Most of the two-dimensional analytic elements are actually expressed in terms of a complex potential comprising the real part, φ (discharge function), and an imaginary part, Ψ (stream function):

$$\Omega(z) = \varphi(z) + i\Psi(z) \quad (5.28)$$

where $z = x + iy$ $(i = \sqrt{-1})$. The stream function, Ψ, is constant along streamlines and its gradient is of the same magnitude as that of φ. This is because any infinitely differentiable analytic complex function instantly has real and imaginary parts, both satisfying the Laplace equation. The expression for the complex discharge, W, is:

$$W = -\frac{\partial \Omega}{\partial z} = Q_x - iQ_y \quad (5.29)$$

When φ and Ψ are contoured at the same contour interval, the resulting plot constitutes a flow net. A flow net has two sets of curves that are mutually perpendicular at the point of their intersection. Any complex function, $\Omega = \varphi + i\Psi$, has this property. With this formulation, amount of discharge between adjacent stream function contours is equal to the contour interval, $\Delta\Psi$.

5.2.4 A comparative account of FDM, FEM, and AEM models

As already noted, FDM and FEM have similar structure, which is distinct from the AEM. The differences stem from the use of discretized domains versus exact solutions. In an AEM, hydraulic properties and elevations are input for the modelled polygonal regions, and boundary conditions are specified on the internal boundaries.

The domain discretization of FDMs is limited to an orthogonal grid of rectangular blocks, which is less flexible than the triangular or quadrilateral elements of FEMs. Because of this, the boundary conditions can be specified more precisely in a FEM than in a FDM. The approximation of boundary conditions in an AEM model can be quite good, due to flexibility in assigning spatial positions to analytic elements.

Heads are defined throughout the spatial domain in AEMs and FEMs, but only at nodes in FDMs.

The domain of AEMs is infinite in the x, y plane and all boundaries are internal. FDMs and FEMs, on the other hand, require that a finite domain be chosen and boundary conditions be assigned on the entire boundary of this finite domain. The infinite domain of an AEM means that these models can be expanded in spatial domain as well as in complexity by simply adding more elements. However, solutions typically become meaningless at some distance away from the modelled area, as there is usually only discharge outflow to or inflow from infinity. The finite discretized domain of the FDMs and FEMs means that a larger amount of input is required than in a comparable AEM.

In general, AEM has higher accuracy at the cost of limited capability. The FDM and FEM have been adapted to a greater variety of flow situations, including transient flow, three-dimensional flow, heterogeneous and anisotropic domains, and also for unsaturated condition.

The accuracy issues are significant only in certain situations. If discretization is fine enough, both FDMs and FEMs can usually provide sufficient accuracy. Near wells, drains, barriers, and other features, that cause non-linear head distributions, the accuracy of AEM is an important advantage.

At present, there are no versatile fully three-dimensional AEM computer programs. On the other hand, fully three-dimensional FDMs and FEMs are already available. At present, FDMs and FEMs have stronger capabilities for simulating heterogeneous and anisotropic domains. The AEM currently has limited transient simulation capabilities compared to FDMs and FEMs.

5.2.5 Model calibration, parameter estimation, and sensitivity analysis

Model calibration is a process of adjusting the input properties and boundary conditions of a model to achieve the desired degree of correspondence between the model simulations and the natural groundwater flow system. A flow model is considered calibrated when it can reproduce, to an acceptable degree, hydraulic heads and groundwater fluxes in a natural system being modelled. Calibration involves adjusting the unknown parameters (hydraulic conductivities, boundary fluxes, etc.) iteratively until the simulated results match the known values (usually the hydraulic heads) to an acceptable degree. Multiple calibrations of the same system are possible using different boundary conditions and aquifer properties. There is no unique calibration that is applicable to all the models, because exact solutions cannot be computed with a multi-variable approach.

If a model is properly calibrated, there would be some random deviations. If there are systematic deviations, such as most of the simulated heads exceeding the observed heads, the calibration is considered poor and appropriate adjustments must be made. Darcy's law:

$$q_x = -K_x \left(\partial h / \partial x \right)$$

can be made use of advantageously in the calibration process. This is readily seen from a simple example taken from Fitts (2002).

Consider an unconfined aquifer where water enters as recharge and leaves as discharge to a local stream, as shown schematically in Fig. 5.6. If the constant heads assigned at the stream are correct and the simulated heads in a regional flow model are systematically higher than the corresponding observed heads, two possible adjustments to the model could be made:

1) increasing the value of the hydraulic conductivity of the aquifer; or
2) decreasing the rate of recharge used.

If the discharges to streams and/or wells are known to be correct, then conductivity values should be increased, as increasing the value of K allows the same flux to be transmitted for a smaller gradient (*cf.* Darcy's Law). On the other hand, if conductivities are known to be correct and the discharges to streams are too high, the recharge rate needs to be lowered. A recent technique in flow model

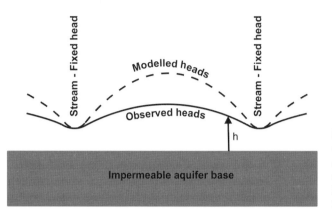

Fig. 5.6 Vertical cross-section through an unconfined aquifer with recharge input and discharge to local streams. As shown, unsatisfactory modelling yields most simulated heads to be higher than the observed heads and hence requires adjustments in model calibration.

calibration relates to automatic optimization of parameters (Doherty 2000; Hill 1992). It is based on minimizing an objective function that is defined to be a measure of the goodness of fit between the simulated and actual results. For example, higher values of the objective function indicate greater discrepancy between simulated and observed heads. One simple and commonly used definition of the objective function, F, is the sum of the squared differences between simulated and observed heads:

$$F = \sum_{i=1}^{n} (b_{oi} - b_{si})^2 \qquad (5.30)$$

where n is the number of observed heads; b_{oi} is the ith observed head; and b_{si} is the model simulated head corresponding to the ith observed head. This equation can be modified by multiplying an appropriate weighting factor to each one of the terms in the sum, to account for the relative importance of an observation to the overall validity and confidence. The automated calibration techniques can save considerable time in the calibration process but are fraught with the possibility of yielding unreasonable results if sufficient constraints are not imposed. Some important aspects of calibration, such as discharges and gradients, are neglected by many automated calibration schemes and hence the need for caution in applying them.

Sensitivity analysis involves quantitative evaluation of the influence on model outputs arising from variation of model inputs. During this process the attempt is to identify parameters that are most important in determining the accuracy and precision of model predictions (Kerr 1992). Sensitivity analysis involves numerous model runs; for each run only one of the parameters is varied by a specified percentage. Both positive and negative variances are tested. Sensitivity analyses can be used to aid model building by identifying inputs that need to be defined more precisely. For example, sensitivity analysis may show that existing hydraulic conductivity data has such a large range that additional pumping tests are needed to achieve the desired level of accuracy in modelling results. Sensitivity analyses also help in interpreting the results. For example, uncertainty about the head values at a boundary may not be of concern if the analysis shows that the output of interest is insensitive to these head values. The analysis shows that sensitivity of groundwater flow to variation in hydraulic conductivity is relatively high.

After calibration and sensitivity analysis, the model application should be tested by reproducing the historical data. This involves comparing model predictions with a dataset independent of the calibration data. If the comparison is unsatisfactory, the model needs further calibration. If the comparison is satisfactory, it gives credence to the argument that the model can be used for predictive purposes with a reasonable degree of accuracy. This confidence does not, however, extend to conditions other than those tested and, therefore, does not account for any unforeseen situations. Matching with historical data shows how well the model application can simulate past conditions. It does not, however, necessarily indicate accuracy for predictive simulations into the future.

5.2.6 Interpreting model results

Models are often calibrated to simulate existing conditions in a flow system and then used to predict future condition for different values of boundary conditions.

Model output is usually given in the form of hydraulic heads and flow vectors at grid nodes. From these, contour map of heads, flow field vector maps, groundwater path line maps, and water balance calculations can be made. Although a well-calibrated model may simulate the conceptual model with reasonable accuracy, it still represents a simplified approximation of the real flow system. It is important to remember this fact when interpreting the model results.

It may also be noted from the general equation for steady two-dimensional flow with recharge (obtained from Eqn 5.4 by substituting $\partial h / \partial t = 0$):

$$\nabla^2(h) = W/K \qquad (5.31)$$

that the same pattern of h can be achieved with any number of models that maintain fixed W/K, that is, recharge to aquifer conductivity ratio. Therefore, in some cases, there is no unique solution to the calibration problem. The problems of non-uniqueness are more prevalent when observational data are limited.

One key issue is how to optimize the modelling runs to account for the uncertainty, while still meeting the modelling objectives. This can be accomplished by using one of the following approaches:

- Producing 'best estimate' results by using the most representative input values usually provides a useful indicator of groundwater velocities, heads, and fluxes. However, single value modelling results do not, by themselves, ensure desired accuracy of the model output.
- One possible modelling objective is to determine if a certain groundwater level or flow rate may arise for the most unfavourable condition that can be anticipated. In this case, the model is calibrated using the best estimates obtained from the range of input values, but simulations are performed using input values from the most unfavourable ones of the range of input values. For example, field estimates for transmissivity are identified at 600–800 $m^2\ d^{-1}$. However, if model

simulations predict that a certain well field design will meet its production goals, even when using transmissivity value as low as 500 $m^2\ d^{-1}$, there is reasonable assurance that the design is adequate.

- Best estimate results can be coupled with results from the sensitivity analysis to provide the extreme scenarios of expected aquifer performance. For example, if the modelling objective is to predict whether head fluctuations at a location could exceed 5 m and, if the best estimate results plus additional adjustment from the sensitivity analysis predict only an expected fluctuation of 2 m, then further analysis may not be necessary.
- In such a case, two or more calibrations of the model are made with different sets of values for key input parameters. The results should bracket the expected possible range of results that gives better understanding of how overall model performance varies in response to the input uncertainty.
- Various other methods, such as inverse modelling and Monte Carlo analysis, can be used to more fully analyse the effects of uncertainty. Usually these methods require specifying an upper bound to the range of uncertainty or define a probability distribution for the associated variable. Results are obtained in bounded or distributed form.

5.2.7 An overview of available groundwater modelling software

The most important code for groundwater modelling is MODFLOW. The origin of MODFLOW can be traced back to the early part of the 1980s. An overview of the history of MODFLOW is given by McDonald and Harbaugh (2003). Today there are several graphical user interfaces (GUI) such as Processing MODFLOW, Visual MODFLOW, GMS, and MODFLOW-GUI, that are 'built around' MODFLOW. These programs, employing MODFLOW for groundwater flow models or other codes for transport simulations or other tasks, enable the user with various pre- and post-processing tasks. Moreover, it is possible to use the GUI for direct and inverse modelling. A select list of groundwater modelling codes is given in Table 5.2. The second column of

Table 5.2 A select list of groundwater modelling codes. 2nd Column: if the code solves the resulting systems of equations itself (S) or if it is built around another solver code as a graphical user interface (GUI). 3rd column: numerical method used for discretization (FD = Finite Differences; FE = Finite Elements; FV = Finite Volumes). 4th Column: PD – Public domain (freeware), Com. – Commercial. Reproduced from Holzbecher and Sorek (2005).

Code name	Solver GUI Tool	FD/FE/FV	Com./PD	Short Description	Internet Link
CFEST	S	FE		Coupled fluid-energy-solute transport	http://db.nea.fr/abs/html/nesc9537.html
CHAIN-2D	S	FE	PD	2D transport of decay chain	http://www.ussl.ars.usda.gov/MODELS/CHAIN2D.HTM
FAST	S, GUI	FV		3D flow, 2D transport, 2D density driven	http://www.igbberlin.de/abt1/mitarbeiter/holzbecher/index—e.shtml
FEFLOW	S, GUI	FE	Com.	3D flow, solute, and heat transport	http://www.wasy.de/english/produkte/feflow/index.html
FEMWATER	S, GUI	FE		3D groundwater flow	http://www.scisoftware.com/products/gms—fem/gms—fem.html
GMS	GUI	FE, FV	Com.	For FEMWATER, MODFLOW, MT3D, RT3D, SEEP2D, SEAM3D	http://www.ems-i.com
HST3D	S	FD	PD	3D flow, solute, and heat transport	http://water.usgs.gov/software/hst3d.html
MOC3D	S		PD	3D method of characteristics (flow and transport)	http://water.usgs.gov/software/moc3d.html
MODFLOW	S	FV	PD	3D flow	http://wa:er.usgs.gov/software/modflow.html
MODFLOW-GUI	GUI	FV	PD	for MODFLOW and MOC3D, works under ARGUS ONE only	http://wa:er.usgs.gov/nrp/gwsoftware/mfgui4/modflowgui.html
MODPATH	Tool	–	PD	Particle tracking for MODFLOW	http://water.usgs.gov/software/modpath.html
Model Viewer	Tool	–	PD	Visualization of 3D model results	http://water.usgs.gov/nrp/gwsoftware/modelviewer/ModelViewer.html
MT3D	S	FV	PD	3D transport	http://hydro.geo.ua.edu/
MT3D-MS	S	FV	PD	3D multiple species transport	http://hydro.geo.ua.edu/
PATH3D	Tool	–		Particle tracking for MODFLOW	http://hydro.geo.ua.edu/mt3d/path3d.htm
PEST	Tool		Com.[b]	Parameter Estimation	http://www.parameterestimation.com
PHREEQC	S	Cells[a]	PD	Geochemistry and 1D Transport	http://water.usgs.gov/software/phreeqc.html
PMWIN	GUI	FV	Com.[b]	for MODFLOW, MOC, MT3D, PEST and UCODE	http://www.scisoftware.com/products/pmwin—details/pmwin—details.html
PORFLOW		FD	Com.	3D flow, solute, and heat transport	http://www.acri.fr/English/Products/PORFLOW/porflow.html
ROCKFLOW		FE		3D flow, solute, and heat transport	http://www.hydromech.uni-hannover.de/Projekte/Grundwasser/misc/news.html
RT3D		FV	PD	Reactive transport based on MT3D-MS	http://bioprocess.pnl.gov/rt3d.htm
SEAM3D	GUI	FV		Reactive transport based on MT3D-MS	http://modflow.bossintl.com/html/seam3d.html
SUTRA	S, GUI	FE	PD	Flow and transport	http://water.usgs.gov/software/sutra.html
SWIFT	S	FV	Com.	3D fluid, solute, and heat transport	http://www.scisoftware.com/products/swift—overview/swift—overview.html
TBC	S	FE, FV	PD	Transport, Biochemistry, and Chemistry	http://www.iwr.uniheidelberg.de/~Wolfgang Schafer/tbc201.pdf
UCODE	Tool	–	PD	Parameter Estimation	http://water.usgs.gov/software/ucode.html
Visual MODFLOW	GUI	FV	Com.	for MODFLOW, MT3D, RT3D and PEST	http://www.flowpath.com/software/visualmodflow/visualmodflow.html

[a] Can be regarded as a special type of Finite Volumes, for 1D only.
[b] Limited version is freeware.

the table explains if the code solves the resulting system of equations (S) itself, or if it is built around another solver code as a graphical user interface (GUI). Some programs function simply as tools for pre- and post-processing. GUIs also include several tools. The third column indicates the numerical method used for the discretization (FD = Finite Differences; FE = Finite Elements; FV = Finite Volumes). The method of Finite Volumes (FV) is derived from a mass or volume balance for all blocks of the model region, as in the case of MODFLOW. GUIs, which are connected to different solvers, can have more than one entry point. For the tools, the discretization method does not have to be specified, as it depends on the solver.

5.3 Surface and groundwater interactions and coupled/integrated modelling

Aquifers are often partially fed by seepage from streams and lakes. In other locations, the same aquifers may discharge through seeps and springs to feed streams, rivers, and lakes.

The Big Lost River and Snake River (USA) provide excellent examples of a losing stream feeding an aquifer and a gaining stream fed by groundwater. The Big Lost River flows out of a mountain valley on the northwest margin of the Snake River Plain and completely disappears through seepage into the permeable lava of the plain.

The groundwater flow in the aquifer underlying Snake River Plain is towards the southwest, ultimately discharging in the form of springs along the wall of the Snake River canyon. As the Snake River flows across southern Idaho, much of the flow is diverted for irrigation. At Shoshone Falls, about 50 km downstream of the Milner Dam, the river may nearly dry up due to irrigation diversions. A further 60 km downstream, the river is 'reborn' in the impressive Thousand Springs area, where springs collectively discharge more than 140 m^3 s^{-1}. Niagara Springs is an example of the many scenic springs in the Thousand Springs area. These river gains provide much of the downstream flow during summer.

When the aquifer water level is near the land surface, seepage from the river is partially con-

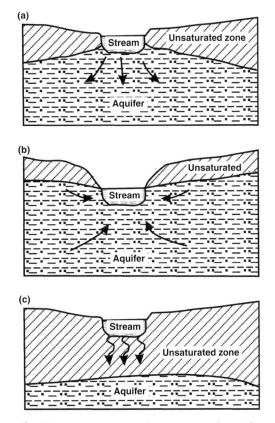

Fig. 5.7 (a) *Losing Streams*: losing water to the aquifer when the stream stage is higher than the water table. (b) *Gaining Streams*: gaining water when the stream stage is below the surrounding water table. (c) *Perched Streams*: when the stream beds are well above the water table.

trolled by the height of the aquifer water level (Fig. 5.7a). Activities or events, such as groundwater pumping, that result in lowering of the water table induce more seepage from the river bed. Conversely, events that cause the aquifer water level to rise (recharge events) result in a decrease in river seepage. If aquifer water levels rise above the level of the river, a losing river reach gains water from the aquifer (Fig. 5.7b). Another hydrologic situation, which is very important in understanding surface water and groundwater interaction, is a surface water body 'perched' above an aquifer when the aquifer water level is well below the bed of the river, stream, or lake (Fig. 5.7c). Under this condition, water seeps from the surface water body to the groundwater, but the surface water body is not affected by aquifer water level and consequently does not change in response to groundwater pumping. Nearby groundwater pumping will cause a lowering of the water table, but will not affect surface water supplies.

So far, the focus on hydrologic modelling in this book has been on surface or subsurface hydrology, treating the two systems separately. To some extent this can be justified because of different hydrologic, temporal, and spatial characteristics of surface water and groundwater. However, with the realization that surface water and groundwater are intimately linked, modellers have developed closely coupled and/or integrated models to better represent the interaction between these two hydrologic regimes. Usually the focus of such hydrologic models is to plan and manage water resources on a larger scale.

Combining surface water and groundwater models to better represent real-life situations is a difficult undertaking. First of all, they tend to operate on very different temporal scales. In various numerical models, not only is the space discretized, the MODFLOW with the layers and grid, and the computations are performed at discrete time intervals – being the operating time-step. The operating time-steps in a typical groundwater model and a typical surface model are extremely different. Groundwater models tend to run on time-steps of the order of months to a year as groundwater movement is quite slow. With a surface water system, for example, in flood or reservoir routing, a time step of a few minutes to a few hours would be more appropriate. It may take several days for the flow to make it through the section, but one has to track it in much shorter time-steps. Because of these and other difficulties in solving groundwater, surface-water, and unsaturated zone flow in an integrated manner, existing individual sectoral models are often coupled.

Model coupling conceptually requires more complete description of the hydrologic process and joint calibration, for example, measured discharges in addition to measured groundwater parameters, such as water levels and flow velocities. Coupling therefore provides a means to better identify groundwater in-flow and out-flow from and to the catchment and cross system processes, for example, groundwater recharge calculations. In practice, however, these potential benefits of coupled models are often offset for by high demand on computation time and required storage capacity. Quite often, results obtained from coupled models might be worse than those obtained from the stand alone sectoral models, the latter requiring inputs from hydrochemistry, natural tracers, and isotopes to achieve more meaningful results.

One of the commonly used codes is GSFLOW, which is a coupled groundwater and surface-water FLOW model based on the integration of the US Geological Survey Precipitation-Runoff Modelling System (PRMS) (Leavesley *et al.* 1983) and the US Geological Survey Modular Ground-Water Flow Model (MODFLOW-2005) (Harbaugh 2005). There are several other coupled/integrated surface-groundwater models suited for various applications appropriate for the varying scales of modelling. Detailed description of the various codes is, however, beyond the scope of this book.

6

Aqueous chemistry and human impacts on water quality

Why is water chemistry, in general, and groundwater chemistry, in particular, important? Water quality is an important aspect for all uses of water, be it for drinking, irrigation, industrial, or other purposes. Some of the quality considerations were discussed in Section 1.2. Aqueous chemistry is central to understanding: (i) sources of chemical constituents in water; (ii) important natural chemical processes occurring in groundwater; and (iii) variations in chemical composition of groundwater in space and time. Aqueous chemistry is also important to estimate the fate of contaminants, both in surface waters and groundwaters, and for remediation of contamination.

As mentioned in Chapter 1, water is called the 'universal solvent' because it dissolves more substances than any other liquid. This is due to the polar nature of the water molecule, which makes it a good solvent of ionic as well as polar molecules. As a result, chemical composition of natural water is derived from multiple sources of solutes, including gases and aerosols from the atmosphere, weathering and erosion of rocks and soil, solution or precipitation reactions occurring below the land surface, and those resulting from human activities. Some of the common inorganic solutes in water are listed in Table 6.1. In this chapter, the focus is on understanding the broad interrelationships between the various processes and their effects that make up the chemical character of natural waters. These include

principles of chemical thermodynamics, processes of dissolution and/or precipitation of minerals, and the laws of chemical equilibrium including the law of mass action, etc.

Basic data used in the determination of water quality are obtained by the chemical analysis of water samples in the laboratory or onsite measurement/sensing in the field. Most of the measured constituents are reported in gravimetric units, in milligrams per litre (mg/L or $mg\ l^{-1}$) or milliequivalents per litre (meq/L or $meq\ l^{-1}$). However, in chemical calculations, it is common to use molar concentration units, which give moles of solute per unit volume of solution (mol/L or $mol\ l^{-1}$), denoted by M (e.g. a 2.5 $M\ Ca^{2+}$ solution contains 2.5 moles of Ca^{2+} per litre. A *mole* is the amount of substance containing N atoms or molecules, where $N = 6.022 \times 10^{23}$ is the *Avogadro's number*, rounded off to 4 significant digits. The mass/weight of one mole of atoms is called the *formula mass*/weight (also called *molecular mass*/weight). For example, the atomic mass of oxygen is 16.00 g, and the formula mass of CO_2 is ($12.01 + 2 \times 16.00$) $= 44.01\ g$. For calculations involving chemical reactions, it is useful to employ moles and molar concentrations, because chemicals react in direct proportion to the number of respective molecules present. Conversion from molar mass per volume to units (i.e. $mg\ l^{-1}$ or $\mu g\ l^{-1}$) can be done by using the formula weight of the solute below:

Modern Hydrology and Sustainable Water Development, 1st edition. © S.K. Gupta
Published 2011 by Blackwell Publishing Ltd.

Table 6.1 Some common inorganic solutes in water.

Cations	Anions	Other
	Major Constituents	
Calcium (Ca^{2+})	Bicarbonate (HCO_3^-)	Dissolved CO_2 (H_2CO_3)
Magnesium (Mg^{2+})	Chloride (Cl^-)	Silica ($SiO_2(aq)$)
Sodium (Na^+) Potassium (K^+)	Sulphate (SO_4^{2-})	
	Minor Constituents	
Iron (Fe^{2+}, Fe^{3+})	Carbonate (CO_3^{2-})	Boron (B)
Strontium (Sr^{2+})	Fluoride (F^-) Nitrate (NO_3^-)	

$$\frac{mol}{L} \times formula\ weight \left(\frac{g}{mol}\right) \times \frac{1000\ mg}{g} = \frac{mg}{L}$$

$$(6.1)$$

Chemical analyses may be grouped and statistically evaluated by parameters such as mean, median, frequency distribution, or ion correlations to characterize large volumes of data. Drawing graphs of analyses, or of groups of analyses, aids in showing chemical relationships between various dissolved constituents of water and identifying their probable sources, regional water-quality relationships, temporal and spatial variation of water quality, and assessment of water resources potential. Graphs may show water types based on chemical composition, relationships between ions (or groups of ions) in individual waters, or water from many sources considered simultaneously. Relationship between water quality and hydrogeological characteristics, such as the stream discharge rate or groundwater flow patterns, can be shown by appropriate mathematical equations, graphs, and maps.

6.1 Principles and processes controlling composition of natural waters

The fundamental concepts relating to chemical processes, which are useful in developing a unified approach to understand the chemistry of natural waters, are mainly related to chemical thermody-

namics as well as to rates of reactions and their causative mechanisms.

Thermodynamic principles, applied to chemical energy transfers, form a basis for quantitatively evaluating the feasibility of various possible chemical processes occurring in natural water systems, for predicting the direction in which chemical reactions may proceed, and in many cases for predicting the actual dissolved concentrations of reaction products that may be present in a given water body.

6.1.1 Thermodynamics of aqueous systems

A review of fundamental relationships and principles relating to chemical energy is helpful in understanding how *thermodynamic* concepts may be used.

The law of conservation of energy states that although its form may change, the total amount of energy in any system remains constant. This principle is also known as the *first law of thermodynamics*.

The chemical energy stored in a substance at constant temperature and pressure is termed 'enthalpy' and is represented by ΔH. The 'delta' indicates that it represents a departure from an arbitrary standard state taken as zero point. For chemical elements, this standard reference state is that of 1 mole (an amount equal to the molecular weight of the constituent expressed in grams) at $25°C$ temperature and 1 atmosphere pressure. Enthalpy may be thought of as having two components: an internal component that is termed 'entropy' (ΔS) and a component that is or can become available externally, termed 'free energy' (ΔG) or Gibbs free energy.

The concept of entropy is implicit in the *second law of thermodynamics*, which can be stated as a spontaneously occurring process in an isolated system that will tend to convert a less probable state to one or more probable states. There is a finite probability that such a system tends to favour a generally random or disordered condition, or eventually a state of relative chaos. Entropy may thus be considered a measure of the degree of disorganization or disorder within a system. However, entropy is more difficult to estimate quantitatively than its corollary, free energy, which is released in a spontaneous chemical process.

The second law of thermodynamics can also be rephrased to the effect that in a closed system the reaction affinities tend to reach their minimum values. At the point of equilibrium for a specific reaction, the value of reaction affinity is zero. The relationship governing these chemical energy manifestations is:

$$\Delta H = \Delta G + T \Delta S \qquad (6.2)$$

where T is temperature on the Kelvin scale. This is a general statement of the *third law of thermodynamics*, which also may be paraphrased 'the entropy of a substance at absolute zero temperature $(T = 0K)$ is zero'.

Enthalpy, entropy, and free energy values are expressed in units of heat.

6.1.2 Chemical reactions

Chemical reactions in which various elements participate involve changes in the arrangement and association of atoms and molecules and interactions amongst electrons that surround the atomic nuclei. The field of natural water chemistry is concerned principally with reactions that occur in relatively dilute aqueous solutions, although some natural waters have rather high solute concentrations. The reacting systems of interest are generally heterogeneous, that is, they involve either a liquid phase, a solid or a gaseous phase, or all the three phases coexist together.

6.1.2.1 Reversible and irreversible reactions in water chemistry

In the strict sense, an irreversible process is one in which reactants are completely converted into products. Thus a *reversible* process is one in which both reactants as well as products can be present when the reaction affinity is zero or nearly zero. It is inferred that to achieve and sustain this condition, both the forward and reverse reactions occur simultaneously, at least on the micro-scale and at comparable rates when reaction affinities are small.

Some reactions, even though favoured thermodynamically, do not take place to a significant extent owing to energy barriers in some of the reaction pathways. If such a condition applies to one of the reactions in a reversible process as defined above, the process apparently behaves as irreversible. Similarly, in open systems in which reactants and/or products may enter and leave, irreversible behaviour may be expected. Therefore, in natural water systems, the reversible or irreversible nature of a chemical reaction is dependent on kinetic factors and on some of the physical features of the system of interest as well as on thermodynamic considerations. Therefore, it has been found more convenient in natural-water systems to consider chemical reactions of interest on the basis of the ease with which these can be reversed. On this basis, three general types of processes are: (1) readily reversible processes; (2) processes whose reversibility is retarded; and (3) processes that are irreversible in a fundamental (thermodynamic) sense. Specific processes in natural-water systems represent a continuum from type 1 to type 3. An example of an easily reversible process is the formation of complex ions or similar homogeneous (single-phase) reactions. Dissolved carbon dioxide, represented as H_2CO_3, dissociates reversibly as:

$$H_2CO_3(aq) \leftrightarrow HCO_3^- + H^+ \leftrightarrow CO_3^{2-} + 2H^+$$

$$(6.3)$$

The symbol (aq) indicates the aqueous phase. Other symbols commonly used to indicate phase of reactants or products on chemical reactions are (s) for solid, (l) for liquid, and (g) for gas. An example of reactions in which reversibility is severely affected can be seen in the weathering of albite, a common feldspar mineral, in which a solid product kaolinite is formed:

$$2Na\,Al\,Si_3\,O_8 + 2H^+ + 9\,H_2O \rightarrow 2Na^+$$
$$+ 4\,H_2SiO_4\,(aq) + Al_2Si_2O_5\,(OH)_4(s) \qquad (6.4)$$

Kaolinite dissolves reversibly as:

$$Al_2Si_2O_5\,(OH)_4(s) + 7H_2O \leftrightarrow 2\,Al(OH)_4$$
$$+ 2Si\,(OH)_4(aq) + 2H^+ \qquad (6.5)$$

Here dissolution of kaolinite is essentially irreversible because it cannot be reconstituted to a significant extent without subjecting it to temperatures and pressures that differ greatly from those prevailing in normal weathering regimes.

A process that is thermodynamically irreversible is that of altering the crystal structure of a solid to a more stable form during its aging. For example, when a ferric hydroxide amorphous precipitate changes to goethite with aging:

$$Fe(OH)_3(s) \rightarrow FeOOH(c) + H_2O \qquad (6.6)$$

the reaction affinity will be greater than zero as long as any $Fe(OH)_3$ is available.

6.1.3 Chemical equilibrium – the law of mass action

Study of chemical equilibria is based on the *law of mass action*, according to which the rate of a chemical reaction is proportional to the active masses of the participating substances. A hypothetical reaction between two substances *A* and *B* producing products *C* and *D*, in a closed system, can be written as:

$$aA + bB \leftrightarrow cC + dD \qquad (6.7)$$

where the lower case letters represent coefficients required to balance the equation. The rates of forward and reverse reaction, according to the law of mass action are, respectively:

$$R_1 = k_1'[A]^a[B]^b \qquad (6.8)$$

and

$$R_2 = k_2'[C]^c[D]^d \qquad (6.9)$$

where bracketed terms represent active masses or activities of reactants/products. The quantities k_1' and k_2' are proportionality constants for the forward and reverse reactions, respectively. When $R_1 = R_2$, the system is in a state of dynamic equilibrium and no change in active concentrations (represented by quantities in the square brackets) occurs. This leads to the equation:

$$\frac{[C]^c[D]^d}{[A]^a[B]^b} = \frac{k_1'}{k_2'} = K \quad (at\ equilibrium) \qquad (6.10)$$

The quantity, K, is referred to as the *equilibrium constant*. Activities are dimensionless quantities and, therefore, the equilibrium constant is also dimensionless. It has a characteristic value for any given set of reactants and products and many experimentally determined values are available in

published chemical literature. The value of the equilibrium constant is influenced by temperature and pressure. Standard thermodynamic conditions ($25°C$ temperature and 1 atmosphere pressure) are usually specified, but K values for many reactions have been determined over a wide temperature range. This form of the mass law (Eqn 6.10) is a statement of final conditions in a system at equilibrium. If a reaction is not in equilibrium, the right-hand side of Eqn 6.10 is called the *reaction quotient* or the *ion activity product* (IAP):

$$Q = IAP = \frac{[C]^c[D]^d}{[A]^a[B]^b} \qquad (6.11)$$

When IAP $<K$, the reaction proceeds to the right and the concentrations of A and B fall while those of C and D rise; when IAP $>K$, the reaction proceeds to the left and concentrations change in the opposite direction.

Example 6.1. The following reaction describes dissociation of bicarbonate ion in water (see also Eqn 6.3):

$$HCO_3^- \leftrightarrow CO_3^{2-} + H^+$$

The equilibrium constant for this reaction at $15°C$ is $10^{-10.43}$, i.e.:

$$K_{HCO_3^-} = \frac{[H^+]\left[CO_3^{2-}\right]}{\left[HCO_3^-\right]} = 10^{-10.43}$$

If the water has a pH of 5.9 (i.e. $[H+] = 10^{-5.9}$ (see Section 1.2.2.5 in Chapter 1) and $\left[HCO_3^-\right] = 2.43 \times 10^{-3}$, assuming equilibrium, the $\left[CO_3^{2-}\right]$ is given by:

$$[CO_3^{2-}] = \frac{K_{HCO_3^-}\left[HCO_3^-\right]}{[H^+]}$$

$$= \frac{\left[10^{-10.43}\right]\left[2.43 \times 10^{-3}\right]}{\left[10^{-5.9}\right]} = 7.2 \times 10^{-8}$$

The equilibration equations such as Eqn 6.10 can be written for reactions involving a variety of phases associated with groundwater: solute-solute, solute-water, solute-solid, and solute-sorbed. By convention activities of water, solid phases or non-aqueous liquid phases in contact with water are set equal to 1.0 in the equilibrium equation. This is because there is essentially an unlimited supply of these substances in contact with reactants/products and equilibrium concentrations are independent of the amount of these substances. For example, dissolution reaction of the mineral calcite (calcium carbonate) in groundwater is:

$$CaCO_3(s) = Ca^{2+} + CO_3^{2-} \qquad (6.12)$$

The corresponding equilibrium equation is written as:

$$K_{CaCO_3} = \frac{[Ca^{2+}][CO_3^{2-}]}{[CaCO_3]} = [Ca^{2+}][CO_3^{2-}]$$

$$(6.13)$$

Thus, activity of solid phases such as calcite is usually omitted from the equilibrium equation, as in Eqn 6.13.

In dilute solutions with low ion concentrations, the electrostatic repulsions between different ions and their ability to collide and react with each other are not drastically hampered, but this is not the case with water having a high ion concentration. In the latter case, it becomes necessary to use a correction factor known as the *activity coefficient* to get the effective activity for use in the equilibrium equation (Eqn 6.10). For a chemical, the activity $[D]$, concentration (D), and the activity coefficient, γ_D, are related as:

$$[D] = \gamma_D(D) \qquad (6.14)$$

In Eqn 6.14, activity $[D]$ is dimensionless and concentration (D) is in molar $(M,\ mol\ l^{-1})$ units. For dilute solutions $\gamma \approx 1$ in magnitude, and the activity is essentially equal to that corresponding to the magnitude of the molar concentration. In more concentrated solutions, such as $\gamma \neq 1$, the activity deviates from that corresponding to the magnitude of the molar concentration.

Several mathematical models for activity coefficients have been developed based on electrostatic theory and empirical observations. The most widely used model for ions in low- to moderately-concentrated- solutions is the extended *Debye-Hückle equation*:

$$\log \gamma_i = \frac{-A\,Z_i^2\,\sqrt{I}}{1 + B\,a\,\sqrt{I}} \qquad (I < 0.1) \qquad (6.15)$$

where A and B are constants for a given solvent that depend on pressure and temperature and are

Table 6.2 Values of constants in the Debye-Hückle equation for activity coefficients A and B (Eqn 6.15). After Manov *et al.* (1943).

$T\ (^\circ C)$	A	B
0	0.4883	0.3241
10	0.4960	0.3258
20	0.5042	0.3272
25	0.5085	0.3281
30	0.5130	0.3290
40	0.5221	0.3302

given in Table 6.2 for water. Z_i is the charge of the ion under consideration; I is the ionic strength of the solution in moles; and a is related to the size of the hydrated ion. Its values for various ions of interest are given in Table 6.3.

For brackish water with ionic strength $I > 0.1M$, the Davies equation (Eqn 6.16) is a better approximation of an ion's activity coefficient (Stumm and Morgan 1996):

$$\log \gamma_i = -A\,Z_i^2 \left(\frac{\sqrt{I}}{1 + \sqrt{I}} - 0.3I \right) \qquad (I < 0.5)$$

$$(6.16)$$

Table 6.3 Values of the parameter 'a' in the Debye-Hückle equation (Eqn 6.15). After Kielland (1937) and Butler (1998).

Ions	Charge	a
$Th^{4+},\ Sn^{4+}$	4	11
$Fe(CN)_6^{4-}$	4	5
$Al^{3+},\ Fe^{3+},\ Cr^{3+}$	3	9
PO_4^{3-}	3	4
$Mg^{2+},\ Be^{2+}$	2	8
$Ca^{2+},\ Cu^{2+},\ Zn^{2+},\ Sn^{2+},\ Mn^{2+},\ Fe^{2+},$ $Ni^{2+},\ CO^{2+}$	2	6
$Sr^{2+},\ Ba^{2+},\ Cd^{2+},\ Hg^{2+},\ S^{2-},\ CO_3^{2-},$ $SO_3^{2-},\ MoO_4^{2-},\ Pb^{2+},\ Ra^{2+}$	2	5
$CrO_4^{2-},\ HPO_4^{2-},\ SO_4^{2-},\ Hg_2^{2+},\ SeO_4^{2-}$	2	4
H^+	1	9
Li^+	1	6
$HCO_3^-,\ H_2PO_4^-, HSO_3^-,\ Na^+$	1	4
$Cl^-,\ Br^-,\ I^-,\ F^-,\ OH^-,\ HS^-,\ NO_3^-,$ $NH_4^+,\ K^-,\ Ag^+,\ CNS,\ CNO^-,\ ClO_4^-,$ $NO_2^-,\ Rb^+,\ Cs^+,\ CN^-$	1	3

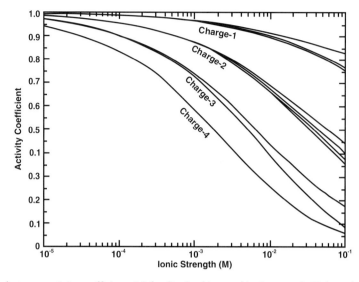

Fig. 6.1 Relationship between activity coefficients (γ) for dissolved ions and ionic strength (I), in molar units, of a solution at one atmospheric pressure and at $25°C$ temperature. Redrawn from Hem (1985). © U.S. Geological Survey.

where A is the same constant that appears in the previous equation.

Activity coefficients versus ionic strength curves, obtained using the Debye-Hückle and Davies model, are shown in Fig. 6.1. In general, the activity coefficient decreases with increasing ionic strength and is smaller for ions with higher charge.

6.1.4 Reaction rates and deviation from the equilibrium

A chemical reaction can occur spontaneously in a closed system when the total free energy of reactants exceeds that of the reaction products. The chemical equation representing such a process and standard free energy data for participating species can be used to determine if the reaction can be spontaneous. However, the knowledge that a given reaction is thermodynamically favourable gives only a limited amount of information that can be used to predict how fast the reaction will proceed. In fact, many reactions that are feasible do not occur at significant rates and some consideration of reaction rate theory is, therefore, necessary to study natural-water chemistry.

In an irreversible chemical process:

$$\alpha A \rightarrow products \tag{6.17}$$

The rate of reaction of A can be written as:

$$-\frac{d[A]}{dt} = k[A]^\alpha \tag{6.18}$$

The rate constant, k, and the exponent, α, are proportionality factors that must be determined experimentally. When the value of the exponent, α, is unity, the process is termed 'first order' in A, implying that the rate of the process is a function only of the concentration of A. It is implicitly assumed that temperature and pressure remain constant during the process and that the ions or molecules of A do not interact with each other. It is possible for a process to be of second order in A (for $\alpha = 2$), implying that two ions of A must interact to create a product. Another type of second-order process is one involving two reactants:

$$\alpha A + \beta B \rightarrow products \tag{6.19}$$

The rate of such a reaction, where α and β are both unity, would be:

$$-\frac{d[A]}{dt} = -\frac{d[B]}{dt} = k[A][B] \tag{6.20}$$

Higher-order reactions occur when α and/or β in the above schematic reaction is greater than 1.

Zero-order reactions occur when the rate is independent of concentration of the reactant, i.e.:

$$d[A]/dt = k$$

Such reactions are of considerable interest in some types of natural water systems and might occur when concentration of A is much smaller compared to that of another reacting substance. Similarly, for processes whose rate is controlled by availability of reaction sites on a solid surface, zero-order kinetics will be applicable if the number of reaction sites is large compared to the concentration of A. Reactions of fractional orders can also be observed for some processes. Generally these involve combinations of several reactions or effects, such as diffusion or mixing, that are not entirely chemical in nature. In evaluating the kinetic properties of any chemical process it is important to consider the effects of processes like these. Generally, only one step in the process controls the rate and determines the order of the whole reaction. This is known as the rate-determining step.

Integration of the first-order rate equation leads to:

$$[A_t] = [A_0] \ e^{-kt} \qquad (6.21)$$

where $[A_0]$ is the concentration of the reactant A at $t = 0$ and $[A_t]$ is the concentration at time t. Note that Eqn 6.21 is similar to the radioactive decay equation (Eqn. 7.1). Therefore, one can define the half-life of the reaction as the time required for half the amount of A present at any moment to disappear:

$$t_{1/2} = \frac{\ln 2}{k} = \frac{0.693}{k} \qquad (6.22)$$

Reaction rates vary widely; some reactions are so rapid that they can result in runaway situations leading to explosions, while others are so slow that it requires geological times to measure their kinetics. A general range of half-lives for several common types of aqueous reactions is shown in Table 6.4.

Rates of some reactions are limited by factors other than their corresponding theoretical chemical reaction rate. Some sorption-desorption reactions are limited by molecular diffusion that transports solute to sorption sites that are not in

Table 6.4 Approximate range of reaction half-lives for some common types of aqueous reactions. After Langmuir and Mahoney (1984).

Type of Reaction	Typical half-life
Solute-solute	Fraction of seconds to minutes
Sorption-desorption	Fraction of seconds to days
Gas-solute	Minutes to days
Mineral-solute	Hours to millions of years

direct contact with flowing water. Other reactions, including many redox reactions, involve micro-organisms and are, therefore, affected by the population density of micro-organisms as well as concentration of various nutrients used by organisms.

The use of equilibrium calculations is valid only if a chemical system remains essentially closed long enough for the reaction of interest to approach equilibrium. In this sense, it is reasonable to apply equilibrium calculations for many solute-solute reactions (Table 6.4). Disequilibrium prevails when transport and other agents of change are rapid compared to the reaction rate, as for example in some mineral-solute reactions. Even in such cases, equilibrium calculations may still be useful to show where the system is heading in the long run.

A strictly chemical parameter for evaluating departure from equilibrium is the saturation index, SI. This is the difference between the logarithms of the activity quotient, Q, and the equilibrium constant, K:

$$SI = \log Q - \log K = \log(Q/K)) \qquad (6.23)$$

where Q is the reaction quotient or ionic activity product, defined by Eqn 6.11, obtained by using observed activities of participating substances in an actual system; K being defined by Eqn 6.10 for equilibrium situation.

6.1.5 Mineral dissolution and precipitation

For a mineral–water system, a positive value of SI indicates that the solution is supersaturated and the reaction should proceed in the direction that will cause more solute to precipitate out of the solution. A negative value of SI indicates under-saturation of

Table 6.5 Reactions and solubility products for some common minerals for standard conditions (25°C *temperature*; 1 atmosphere pressure). After Morel and Hering (1993); Nordstrom *et al.* (1990).

Mineral	Reaction	$Log(K_{so})$	Ref.
Salts:			
Halite	$NaCl = Na^+ + Cl^-$	1.54	(1)
Sylvite	$KCl = K^+ + Cl^-$	0.98	(1)
Fluorite	$CaF_2 = Ca^{2+} + 2F^-$	−10.6	(2)
Sulphates:			
Gypsum	$CaSO_4 . 2H_2O = Ca^{2+} + SO_4^{2-} + 2H_2O$	−4.58	(2)
Anhydride	$CaSO_4 = Ca^{2+} + SO_4^{2-}$	−4.36	(2)
Barite	$BaSO_4 = Ba^{2+} + SO_4^{2-}$	−9.97	(2)
Carbonates:			
Calcite	$CaCO_3 = Ca^{2+} + CO_3^{2-}$	−8.48	(2)
Aragonite	$CaCO_3 = Ca^{2+} + CO_3^{2-}$	−8.34	(2)
Dolomite	$CaMg(CO_3)_2 = Ca^{2+} + Mg^{2+} + 2CO_3^{2-}$	−17.1	(2)
Siderite	$FeCO_3 = Fe^{2+} + CO_3^{2-}$	−10.9	
Hydroxides:			
Gibbsite	$Al(OH)_3 = Al^{3+} + 3OH^-$	−33.5	(1)
Goethite	$\alpha \cdot FeOOH + H_2O = Fe^{3+} + 3OH^-$	−41.5	(1)
Silicates:			
Quartz	$SiO_2 + 2H_2O = Si(OH)_4$	−3.98	(2)
Chalcedony	$SiO_2 + 2H_2O = Si(OH)_4$	−3.55	(2)
Amorphous silica	$SiO_2 + 2H_2O = Si(OH)_4$	−2.71	(2)

the solution, and a zero value of *SI* indicates that the system is in equilibrium.

The equilibrium constant for a mineral–solute reaction is called the solubility product and is denoted by K_{so}. Like all equilibrium constants, the solubility product is defined by an equation such as Eqn 6.10; for example, for the dissolution/precipitation reaction for the mineral anhydrite:

$$CaSO_4 \leftrightarrow Ca^{2+} + SO_4^{2-}$$

would be:

$$K_{so} = \frac{[Ca^{2+}][SO_4^{2-}]}{[CaSO_4]} = [Ca^{2+}][SO_4^{2-}] \quad (6.24)$$

where

$$[Ca^{2+}] \quad \text{and} \quad [SO_4^{2-}]$$

are activities of the two ions. Under equilibrium conditions the activity of the solid phase equals 1 so that:

$$[CaSO_4] = 1$$

Solubility products for standard conditions of temperature and pressure for some common minerals are given in Table 6.5.

The amount of mineral that can be dissolved in a given volume of water depends on the initial concentration of dissolution products in water. More dissolution occurs if water has a low initial concentration than if it has a high initial concentration of the minerals to be dissolved. This explains why caves develop at shallow depths in recharge areas in limestone (calcite) country. The freshly infiltrated pore waters in these areas have low concentrations of dissolution products, namely:

$$Ca^{2+}, CO_3^{2-}, \quad \text{and} \quad HCO_3^-$$

Therefore, limestone dissolves rapidly creating voids that eventually give rise to caves along fractures through which water seepage occurs. As water seeps down, concentration of the dissolution products keeps increasing, which results in progressively less dissolution of the limestone. It is also interesting to note that solubility of solid salts in water, and in most other solvents, increases with temperature while that of gases decreases.

6.1.6 Gas–water partitioning

Due to surface tension of water, the surface of a water body in contact with the atmosphere rather tends to be maintained. However, water molecules are able to cross the surface and escape into the atmosphere and gas molecules from the air can diffuse into the water body. Both processes operating simultaneously tend to produce saturation near the air–water interface. Rates of absorption of gases by water or rates of evaporation of water are functions of: (i) the total surface area of the air–water interface; (ii) the degree of departure from saturation in the layer immediately below the interface (humidity in the case of water vapour); and (iii) the rate at which the molecules of the dissolved or vapour phase are transported away from the interface. The transport rate would be slow if it were controlled solely by molecular diffusion. In most natural systems, however, motion of the gas or liquid phase helps to transport the evaporated or dissolved material away from the interface.

The processes by which gases from the atmosphere dissolve in water are of direct relevance to water quality. As the dissolved oxygen is essential to aquatic organisms, the occurrence, dissolution, and transport of oxygen are important in the study of biochemical processes relating to water pollution. The process of photosynthesis is a major source of atmospheric oxygen. Langbein and Durum (1967) reviewed some properties of stream-channel geometry and stream flow rates as applicable to the role of rivers in the uptake of oxygen from the atmosphere. Understanding systems of this kind entails consideration of rates and the manner in which the rate of one process may affect the rate of another. Some gases, notably carbon dioxide, react with water and their rate of assimilation is affected by subsequent changes in their form. Carbon dioxide is an important constituent in many geochemical processes.

Gas partitioning at the free air–water interface can be reasonably well described by *Henry's law for solubility* of a gas into a liquid. According to this law, concentrations in the two phases are directly proportional to each other (see Section 7.5 in Chapter 7; Eqn. 7.18). If same concentration units (e.g. $mol\ l^{-1}$) are used for both phases, the *Henry coefficient*, k_i, is 'dimensionless'. It depends on temperature, T, and concentrations of all dissolved species, c_j (Ballentine and Burnard 2002). In this form, Henry coefficients (k_i) are the inverse of 'solubility of a gas in water'. Poorly soluble gases have large values of k_i and vice versa.

Generally Eqn. 7.18 is formulated for each gas in terms of its partial pressure, p_i, (expressed in the units of *atmosphere, bar, ...*) and corresponding equilibrium concentration of gas $C_{i,eq}$ ($cm^3\ STP\ g^{-1}$ in water, $mol\ l^{-1}$, $mol\ kg^{-1}$, $mol\ mol^{-1}$). STP denotes Standard Temperature ($T = 0°C = 273.15\ K$) and Pressure ($P = 760\ mmHg = 1\ atm$). To differentiate between different units that are in use, Henry's law is written with coefficient, H_i, (in appropriate units) as:

$$H_i = \frac{p_i}{C_{i,eq}} \qquad (6.25)$$

Partial pressure of a gas A in a gas mixture is defined as the pressure that gas A would exert, if it was the only gas occupying the same volume as the total gas mixture. The sum of partial pressures of each component gas equals the total pressure of a gas mixture (Dalton's Law). For example, in the Earth's atmosphere, about 21% of molecules are O_2 and 0.036% are CO_2. Assuming a sample of the atmosphere at 1 *atm* pressure, the partial pressure of O_2 is 0.21 *atm* and the partial pressure of CO_2 is 0.00036 *atm*. The atmospheric CO_2 partial pressure is, therefore, equivalent to 360 *ppmv*.

Sometimes, instead of Henry's coefficient, the equilibrium concentration ($C_{i,eq}$) for $p_i = 1$ *atm* is also reported. This last formulation of Henry's Law constant is in a way the inverse of Eqn 6.25 (or Eqn. 7.18), and is related to solubility of a given gas expressed in appropriate units (e.g. $mol\ l^{-1}\ atm^{-1}$). To differentiate this formulation from the previous two formulations of Henry's constant, the coefficient, K_H, is used:

$$K_H = \frac{C_{i,eq}}{p_i} \qquad (6.26)$$

Equilibrium constants (K_H) for dissolution of some important natural gases are listed in Table 6.6.

Table 6.6 Dissolution equilibrium constants of some natural gases at 25°C. After Stumm and Morgan (1996).

Dissolution Reaction	$Log(K_H)$ $(mol\ l^{-1}\ atm^{-1})$
$CO_2(g) + H_2O \leftrightarrow H_2CO_3^*$	−1.47
$CO(g) \leftrightarrow CO(aq)$	−3.02
$O_2(g) \leftrightarrow O_2(aq)$	−2.90
$O_3(g) \leftrightarrow O_3(aq)$	−2.03
$N_2(g) \leftrightarrow N_2(aq)$	−3.18
$CH_4(g) \leftrightarrow CH_4(aq)$	−2.98

* Represents the sum of $[CO_2(aq)]$ and $[H_2CO_3]$ formed due to hydration of a small part of $[CO_2(aq)]$. For all values of pH, $[CO_2(aq)] \gg [H_2CO_3]$.

Example 6.2. Calculate the dissolved concentration of O_2 in water that is in equilibrium with the atmosphere at 25°C.

Solution. Partial pressure of oxygen is 0.21 *atm* (21 000 *ppm*). Using Eqn 6.26, and dissolution coefficient of O_2 from Table 6.6:

$$O_2(aq) = p_{O_2}.K_H$$

$$= (0.21\ atm)\left(10^{-2.90}\ M.l^{-1}\ atm^{-1}\right)$$

$$= 2.6 \times 10^{-4}\ M.l^{-1}$$

6.1.7 Carbonate reactions, alkalinity, and hardness

In most natural waters, acid–base reactions and pH are dominated by the interaction of carbon dioxide and the aqueous carbonate compounds $H_2CO_3^*$ (dissolved CO_2), HCO_3^- (bicarbonate), and CO_3^{2-} (carbonate). The dissolution reaction of atmospheric $CO_2(g)$ and the associated acid–base reactions between the carbonate compounds are:

$$CO_2(g) + H_2O \leftrightarrow H_2CO_3^* \qquad (6.27)$$

$$H_2CO_3^* + H_2O \leftrightarrow H_3O^+ + HCO_3^- \qquad (6.28)$$

$$HCO_3^- + H_2O \leftrightarrow H_3O^+ + CO_3^{2-} \qquad (6.29)$$

The equilibrium equations and the constants for these reactions at 1 *atm* pressure and 25°C tem-

perature are:

$$K_{CO_2} = \frac{[H_2CO_3^*]}{p_{CO_2}} = 10^{-1.47}\ atm^{-1} \qquad (6.30)$$

$$K_{H_2CO_3} = \frac{[H^+][HCO_3^-]}{[H_2CO_3^*]} = 10^{-6.35} \qquad (6.31)$$

$$K_{HCO_3^-} = \frac{[H^+][CO_3^{2-}]}{[HCO_3^-]} = 10^{-10.38} \qquad (6.32)$$

The constant defined in Eqn 6.30 is the same as the first constant in Table 6.6, except that the constant here is defined with respect to activity $[H_2CO_3^*]$ instead of concentration $(H_2CO_3^*)$. Eqn 6.31 and Eqn 6.32 represent two equations with four unknowns, namely:

$$[H^+], [H_2CO_3^*], [HCO_3^-], \quad \text{and} \quad [CO_3^{2-}]$$

When pH (the negative logarithm of hydrogen ion concentration) is known, $[H^+]$ is known and, therefore, the ratios:

$$[HCO_3^-]/[H_2CO_3^*] \quad \text{and} \quad [HCO_3^-]/[CO_3^{2-}]$$

are also known, without knowing the magnitude of individual carbonate activities. The relative distribution of dissolved carbonate species as a function of pH is shown in Fig. 6.2. It is seen that $H_2CO_3^*$ is the dominant species when $pH < 6.35$; HCO_3^- is dominant when $6.35 < pH < 10.38$; and CO_3^{2-} is dominant when $pH > 10.38$. The pH of most natural waters falls in the range $6.5 < pH < 10$, in which HCO_3^- is the dominant carbonate species.

In the saturated zone, the carbonate system is closed as there is no direct contact with the gas phase and Eqn 6.30 does not apply. The assumption of equilibrium with atmospheric CO_2 is reasonable in open systems such as streams and some unsaturated pore waters. The partial pressure of CO_2 in soil gas is often higher than in the atmosphere due to plant decay and root respiration. In open systems, $H_2CO_3^*$ is independent of pH, fixed by the atmospheric p_{CO_2}. Given the pH of water and p_{CO_2}, the equilibrium activities of all carbonate species may be calculated for an open system, as shown below.

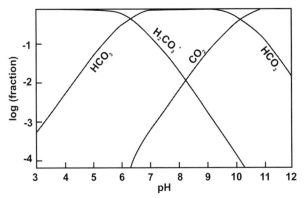

Fig. 6.2 Distribution of the dissolved carbonate species as function of pH for a closed natural water system. Adopted from Fitts (2002). © Academic Press.

Example 6.3. Determine the activities of all carbonate species and the molar concentration of bicarbonate for an open equilibrium system of water with ionic strength $I = 0.01\ M$, $pH = 5.7$, and an atmospheric CO_2 concentration of 360 ppm.

Solution. Atmospheric CO_2 concentration of 360 $ppm = p_{CO_2} = 3.60 \times 10^{-4}\ atm$. Using Eqn 6.30, the activity of $H_2CO_3^*$ is calculated as:

$$\left[H_2CO_3^*\right] = K_{CO_2}\, p_{CO_2} = \left(10^{-1.47}\ atm^{-1}\right)$$

$$\times \left(3.60 \times 10^{-4}\ atm\right) = 1.22 \times 10^{-5}$$

Next, HCO_3^- is now calculated using Eqn 6.31:

$$\left[HCO_3^-\right] = \frac{K_{H_2CO_3}\left[H_2CO_3^*\right]}{[H^+]}$$

$$= \frac{(10^{-6.35})\,(1.22 \times 10^{-5})}{10^{-5.7}} = 2.73 \times 10^{-6}$$

Finally, using Eqn 6.32, $\left[CO_3^{2-}\right]$ is calculated as:

$$\left[CO_3^{2-}\right] = \frac{K_{HCO_3^-}\left[HCO_3^-\right]}{[H^+]}$$

$$= \frac{(10^{-10.38})\,(2.73 \times 10^{-6})}{10^{-5.7}} = 5.71 \times 10^{-11}$$

In this case, about 80% of the dissolved carbonate is $H_2CO_3^*$; about 20% is HCO_3^-; and a negligible fraction is CO_3^{2-}. This is consistent with Fig. 6.2 for pH 5.7.

The bicarbonate molar concentration is calculated from its activity, and activity coefficient for

ionic strength $I = 0.01\ M$:

$$\left(HCO_3^-\right) = \frac{\left[HCO_3^-\right]}{\gamma_{HCO_3^-}} = \frac{2.73 \times 10^{-6}}{0.90}$$

$$= 3.03 \times 10^{-6}\ M$$

6.1.7.1 Alkalinity

Total alkalinity is a parameter that defines the capacity of water to neutralize acid added to it. It is defined as:

$$Alk = (HCO_3^-) + 2\left(CO_3^{2-}\right) + (OH^-) - (H^+)$$

$$(6.33)$$

Total alkalinity is measured in units of equivalent charge, generally expressed as milli-equivalents per litre ($meq\ l^{-1}$), which is related to concentration as:

$$\frac{meq}{L} = \frac{mg}{L} \cdot \frac{Z}{gram\ formula\ weight} = \frac{mmol}{L} \cdot Z$$

$$(6.34)$$

where Z is the charge on the ion, which is also its valency. Sometimes alkalinity is also expressed as $mg\ CaCO_3\ l^{-1}$, $mg\ HCO_3^-\ l^{-1}$, or $mg\ Ca\ l^{-1}$.

Equivalent charge units are used because hydrogen ions are neutralized by their charge rather than their mass. Thus a carbonate ion CO_3^{2-} can neutralize two hydrogen ions and, therefore, $[H^+] = 2[CO_3^{2-}]$, with concentration expressed in molar units. The standard adopted for alkalinity is pure water in which only CO_2 is dissolved. In this state, only ions on the right-hand side of Eqn 6.33 are in solution and the charge balance requires $Alk = 0$. There are other definitions of alkalinity, depending on the reference state chosen, but pure water is commonly used as the standard for natural waters.

Adding a base to water is equivalent to adding cations as well as OH^- ions to the solution, while adding acid generally adds anions plus H^+ ions to the solution. Therefore, adding a base increases $[Alk]$ and adding an acid decreases $[Alk]$. Water with high level of alkalinity is more effective in neutralizing acid that is added to it. In acidic waters, $Alk < 0$. Dissolution of basic minerals, especially carbonate minerals, in water tends to increase its alkalinity. Therefore, as groundwater moves from a recharge area to a discharge area, its alkalinity typically changes from low to high.

Caustic alkalinity is the amount of H^+ required to decrease the pH to ~ 10.8, which equals the stoichiometric amount of H^+ required to complete only the reaction:

$$H^+ + OH^- = H_2O$$

In this case, all the carbonate species will be present entirely as CO_3^{2-}. The amount of acid required to reach this end point cannot be determined readily because of the poorly defined end point, caused by the masking effect of the buffering of water. But this can be calculated if the amounts of carbonate and the total alkalinity are known.

Carbonate alkalinity, also called *phenolphthalein alkalinity*, is the amount of H^+ required to lower the pH to 8.3 (phenolphthalein end point). It is often determined by the colour change of the phenolphthalein indicator. It equals the stoichiometric amount of H^+ required to complete the two reactions: (i) $H^+ + OH^- = H_2O$; and (ii) $H^+ + CO_3^{2-} = HCO_3^-$.

Therefore, *total alkalinity*, sometimes also called *methyl orange alkalinity*, is the amount of H^+ required to decrease pH to about 4.5 (methyl orange end point). In a system dominated by carbonate, the H^+ added is the stoichiometric amount required for the 3 reactions to take place: (i) $H^+ + OH^- = H_2O$; (ii) $H^+ + CO_3^{2-} = HCO_3^-$; and (iii) $H^+ + HCO_3^- = H_2CO_3$.

Total acidity is the number of moles/litre of OH^- that must be added to raise the pH of the solution to approximately 10.8, or to whatever pH is con-

sidered to represent a solution of pure Na_2CO_3 in water at the concentration of interest.

CO_2 *acidity* is the amount of OH^- required to titrate a solution to a pH of 8.3. This assumes, of course, that the pH of the solution is initially below 8.3. Such a solution contains H_2CO_3 as a major component, and the titration consists of converting the H_2CO_3 into HCO_3^-.

If we assume that the alkalinity of water is due solely to the carbonate species and OH^-, the following deductions about the initial composition of a solution from an alkalinity titration can be made to a reasonably good approximation. These are based on the assumption that the titration is a closed system, (which is reasonable if the solution is not shaken and if the titration is conducted). Under these conditions each mole of CO_3^{2-} present will consume one mole of H^+ when the solution is titrated to pH 8.3 and another mole of H^+ as it is titrated from pH 8.3 to pH ~ 4.5.

- If the volume of acid to reach pH 8.3 (V_p) is equal to the volume of acid required to proceed from pH 8.3 to 4.5 (V_{mo}), then the original solution contains only CO_3^{2-} as the major alkalinity causing species.
- If, on adding a phenolphthalein indicator to the solution, it immediately becomes colourless (i.e. if the initial pH is below 8.3) and a volume V_{mo} is required to reach pH of about 4.5, the original solution must have contained only HCO_3^- as the major anion species contributing to alkalinity.
- If the original solution requires V_p ml of acid to reach pH 8.3, but no further acid addition is required to reach pH 4.5, the alkalinity is due to OH^- alone.
- By similar reasoning, it can be shown that if $V_p > V_{mo}$, then the major alkalinity causning species are OH^- and CO_3^{2-}.
- If $V_{mo} > V_p$, then the major species are CO_3^{2-} and HCO_3^-.
- The other possible combination of major alkalinity-causing species, OH^- and HCO_3^-, cannot exist because there is no pH range over which these two species are concurrently the major alkalinity causing species (Fig. 6.2).

Table 6.7 is a summary of the alkalinity titration relationships as explained above.

Table 6.7 Approximate relations between the results of an alkalinity titration and the concentrations of predominant species in carbonate solutions.

Condition	Predominant form of alkalinity	Approximate Molar Concentration (M)
$V_p = V_{mo}$	$CO_3{}^{2-}$	$[CO_3{}^{2-}] = V_pN/V$
$V_p = 0$	$HCO_3{}^-$	$[HCO_3{}^-] = V_{mo}N/V$
$V_{mo} = 0$	OH^-	$[OH^-] = V_pN/V$
$V_{mo} > V_p$	$CO_3{}^{2-}$ and $HCO_3{}^-$	$[CO_3{}^{2-}] = V_pN/V$; $[HCO_3{}^-] = (V_{mo} - V_p) N/V$
$V_p > V_{mo}$	OH^- and $CO_3{}^{2-}$	$[CO_3{}^{2-}] = V_{mo}N/V$; $[OH^-] = (V_p - V_{mo})N/V$

Source: http://www.cheml.com/acad/pdf/c3carb.pdf (accessed on 18th April, 2009)
V_{mo} and V_p are the volumes of strong acid of normality, N, required to reach the end points at pH 4.5 and 8.3, respectively. V is the initial volume of the solution.

It may be noted that the stated end points are approximate values, as the actual end points are not precisely fixed values and vary with the total carbonate concentration in solution.

Some important considerations for understanding the behaviour of natural waters in which carbonate species are the principal buffering agents are:

- Alkalinity and acidity are conservative parameters, unaffected by temperature, pressure, and activity coefficients. Note that this is *not* true for the *pH*, which varies with all of these factors.
- Water can simultaneously possess both alkalinity and acidity. Indeed, this is the usual case over the *pH* range in which $HCO_3{}^-$ predominates.
- Addition or removal of CO_2 (e.g. by the action of organisms) has no effect on [*Alk*]. However, this affects both the acidity and total carbonate concentration; an increase in $[H_2CO_3^*]$ raises both of these quantities.
- Addition or removal of solid $CaCO_3$ or other carbonates has no effect on the acidity. Thus acidity is conserved in solutions that are brought into contact with calcite and similar sediments.
- In a system that is closed to the atmosphere and is not in contact with solid carbonates, the total carbonate concentration is unchanged by the addition of a strong acid or a strong base.
- The presence of mineral acidity or caustic alkalinity in natural water is indicative of a source related to industrial pollution. The limits of *pH* represented by these two conditions correspond roughly to those that are well tolerated by most living organisms.

6.1.7.2 Hardness

Another parameter associated with the carbonate chemistry of water is known as the hardness of the water. Water *hardness* is the sum of the bivalent cations in water, expressed as equivalent $CaCO_3$. The major bivalent cations are calcium, Ca^{2+} and magnesium, Mg^{2+}, though there may also be a minor contribution from iron, Fe^{2+}, and divalent manganese, Mn^{2+}. These bivalent cations react with soap to form a soft precipitate or form solid precipitates that form scale on the inner surface of boilers.

$$Hardness = (Ca^{2+} + Mg^{2+}) \text{ as mg } CaCO_3$$

$$\text{per litre of water} \tag{6.35}$$

Sometimes *hardness* is also expressed as *mg Ca* per litre of water. Common descriptions of water hardness, in terms of *mg CaCO_3* per litre of water, are given in Table 6.8.

Table 6.8 Common descriptions of water hardness based on WHO (2009).

Hardness (as *mg CaCO_3.l^{-1}*)	Description
0–50	Soft
50–100	Moderately soft
100–150	Slightly hard
150–200	Moderately hard
200–300	Hard
>300	Very hard

Example 6.4. What are the alkalinity and hardness of a groundwater sample from a limestone area with $HCO_3^- = 270\ mg\ l^{-1}$; $Ca = 55\ mg\ l^{-1}$; and $Mg = 30\ mg\ l^{-1}$?

Solution. Atomic and molecular weights of the atoms and molecules involved are:

Atomic weights:
  $H = 1; C = 12; O = 16; Mg = 24.3; Ca = 40$
Molecular weights:
  $HCO_3^- = 1 + 12 + 16 + 16 + 16 = 61$
  $CaCO_3 = 40 + 12 + 16 + 16 + 16 = 100$

Alkalinity:

$$270\ mg\ l^{-1}\ HCO_3^- = \frac{270}{61} = 4.43\ mmol\ l^{-1}\ HCO_3^-$$

$$= 4.43 \times 100\ mg\ l^{-1}\ CaCO_3$$

Since the valency of HCO_3^- is 1, the alkalinity of groundwater in various units is $4.43\ meq\ l^{-1}$; $HCO_3^- = 270\ mg\ l^{-1}$; $HCO_3^- = 443\ mg\ l^{-1}\ CaCO_3$.
  Hardness:

$$55\ mg\ l^{-1}\ Ca = \frac{55}{40} = 1.375\ mmol\ l^{-1}\ Ca$$

$$30\ mg\ l^{-1}\ mg = \frac{30}{24.3} = 1.235\ mmol\ l^{-1}\ mg$$

Therefore, $Hardness = 1.375 + 1.235 = 2.61\ mmol\ l^{-1} = 261\ mg\ l^{-1}\ CaCO_3$.

6.1.8 Electro-neutrality

All solutions are required to be electrically neutral, that is, in any given volume of water the sum of electrical charges of all cations must equal the sum of charges of all anions. If the number of cations and anions present in a solution are j and k, respectively, the electro-neutrality condition requires:

$$\sum_{i=1}^{j} ce_i^+ = \sum_{i=1}^{k} ce_i^- \qquad (6.36)$$

where ce_i^+ and ce_i^- are the charge concentration of the ith cation and i^{th} anion, respectively, expressed in equivalents or milli-equivalents per litre units. When results of water analysis are plugged into Eqn 6.36, they should be approximately equal, otherwise either the analysis is erroneous or one or more significant ions are being inadvertently missed from the analysis.

Example 6.5. Refer to Table 6.9 for calculation.

6.1.9 Oxidation and reduction reactions

Oxidation and reduction (redox) reactions transfer energy to many inorganic and life processes. Many common processes such as photosynthesis,

Table 6.9 Calculation of ionic strengths and electro-neutrality – an illustrative example.

	Conc.	Mol. wt.	Valency	Molar Conc.	Ionic strength	%
Cations	$(mg\ l^{-1})$	(g)		$mM\ l^{-1}$	$(meq\ l^{-1})$	
Na^+	8	23	1	0.35	0.35	17
K^+	10	39	1	0.26	0.26	13
Ca^{2+}	12	40	2	0.30	0.60	30
Mg^{2+}	10	24.3	2	0.41	0.82	40
				Sum	2.03	
Anions						
Cl^-	10	35.5	1	0.28	0.28	16
SO_4^{2-}	10	96	2	0.10	0.21	11
HCO_3^-	80	61	1	1.31	1.31	73
				Sum	1.80	

Electro-neutrality, $EN\ (\%) = \dfrac{\sum Cations\ (meq.l^{-1}) - \sum Anions\ (meq.l^{-1})}{\sum Cations\ (meq.l^{-1}) + \sum Anions\ (meq.l^{-1})} \times 100$

$= 6.0\%$

respiration, corrosion, combustion, and even batteries involve redox reactions. Usually redox reactions usually proceed only in one direction, slowly towards completion, but seldom reach there.

In a chemical reaction when an electron is transferred from one atom to another, the processes of oxidation and reduction occur simultaneously; the atom gaining the electron is reduced and that losing the electron is oxidized. The term 'oxidation' is used because oxygen has a strong tendency to accept electrons, which itself gets reduced while oxidizing other atoms that donate electrons to it.

The *oxidation number*, or *oxidation state*, refers to a hypothetical charge that an atom would have if it were to dissociate from the compound of which it is a part. It provides a useful method to study redox reactions. Oxidation numbers are denoted with Roman numerals and are calculated with the following set of rules (Stumm and Morgan 1996):

- The oxidation state of a monoatomic substance is given by its electric charge.
- In a compound formed by covalent bonds between various atoms, the oxidation state of each atom is the charge remaining on the atom when each shared pair of electrons is assigned completely to the more electro-negative of the two atoms sharing a covalent bond.
- The sum of oxidation states of various atoms of a molecule equals zero for neutral molecules, and for ions it equals the charge on the ion.

Electro-negativity of an element is the measure of its affinity for electrons; the higher the electro-negativity, the more it tends to attract and gain electrons. Bonds between different atoms with similar electro-negativity tend to be covalent, while bonds between atoms with very different electro-negativity tend to be ionic. The following is a list of some common elements in order of decreasing electro-negativity:

$$F, O, Cl, N, C = S, H, Cu, Si, Fe, Cr, Mn$$

$$= Al, Mg, Ca, Na, K$$

In a compound, the Group *1A* elements (*H, Li, Na,* ...) are usually with oxidation number (*+I*);

the Group *2A* elements (*Be, Mg, Ca,* ...) are usually (*+II*); and for oxygen it is usually (*-II*). The oxidation number is reduced to a lower value on reduction of an element and is increased on its oxidation. Several common substances and their oxidation states are listed in Table 6.10.

In the following redox reaction involving oxidation of iron, oxygen is reduced from *O(0)* to *O(-II)*:

$$O_2 + 4Fe^{2+} + 4H^+ = 4Fe^{3+} + 2H_2O \qquad (6.37)$$

while iron is oxidized from *Fe(+II)* to *Fe(+III)*; hydrogen remains *H(+I)* and is neither oxidized nor reduced. This redox reaction can also be thought of as the sum of two half reactions as below:

$$O_2 + 4H^+ + 4e^- = 2H_2O \qquad (6.38)$$

$$4Fe^{2+} = 4Fe^{3+} + 4e^- \qquad (6.39)$$

It is clear from the above-mentioned reactions that O_2 is an electron acceptor and Fe^{3+} is an electron donor. Different waters vary in their tendency to oxidize or reduce, depending on the relative concentration of electron acceptors (i.e. O_2) and donors (i.e. *Fe(0)*), respectively.

Just as *pH* is a measure of the hydrogen ion concentration $[H^+]$ of a solution, a measure of the oxidizing or reducing tendency of water is the electron activity $[e^-]$. Similar to *pH*, the parameter *pE* is

Table 6.10 Oxidation states of some common substances. After Fitts (2002).

Substance	Element and its Oxidation state		
H_2O	$H(+I)$	$O(-II)$	
O_2	$O(0)$		
NO_3^-	$N(+V)$	$O(-II)$	
N_2	$N(0)$		
NH_3, NH_4^+	$N(-III)$	$H(+I)$	
HCO_3^-	$H(+I)$	$C(+IV)$	$O(-II)$
CO_2, CO_3^-	$C(+IV)$	$O(-II)$	
CH_2O	$C(0)$	$H(+I)$	$O(-II)$
CH_4	$C(-IV)$	$H(+I)$	
SO_4^{2-}	$S(+VI)$	$O(-II)$	
H_2S, HS^-	$(H+I)$	$S(-II)$	
Fe^{2+}	$Fe(+II)$		
$Fe(OH)_3$	$Fe(+III)$	$O(-II)$	$H(+I)$
$Al(OH)_3$	$Al(+III)$	$O(-II)$	$H(+I)$
$Cr(OH)_3$	$Cr(+III)$	$O(-II)$	$H(+I)$
CrO_4^{2-}	$Cr(+VI)$	$O(-II)$	

defined as the negative logarithm of electron activity, i.e.:

$$pE = -\log[e^-] \qquad (6.40)$$

The electron activity is a measure of the relative activity of electron donors and electron acceptors that are in solution. The actual electron concentration in a solution is likely to be quite low as electrons are not 'free', except very briefly during redox reactions. High pE (i.e. low $[e^-]$) has fewer reducing species (electron donors) than oxidizing species (electron acceptors). Conversely, low pE waters have an excess of reducing species compared to oxidizing species. The pE scale is established by the convention of assigning $K = 1$ for the equilibrium reduction of hydrogen at standard pressure-temperature condition (Morel and Hering 1993):

$$H^+ + e^- = \frac{1}{2}H_2 \qquad (6.41)$$

For a general half-reaction involving reduction of OX to RED:

$$OX + ne^- = RED \qquad (6.42)$$

the pE at equilibrium can be estimated as:

$$K = \frac{[RED]}{[OX][e^-]} \quad \text{or} \quad pE = -\log[e^-]$$

$$= \frac{1}{n}\left[\log K + \log \frac{[OX]}{[RED]}\right] \qquad (6.43)$$

where K is the equilibrium constant for the half reaction OX reduced to RED as above. Any redox pair that is in equilibrium should have the same pE value based on Eqn 6.43. Thus there is a unique value of pE for a solution at redox equilibrium. If there is a consistent computed value of pE for several redox pairs but a different computed value for one of the redox pairs, that particular pair is probably not in equilibrium.

A parameter closely related to pE is *redox potential*, Eh, defined as:

$$Eh = \frac{2.3\,RT}{F}pE = (0.059V)pE \qquad (6.44)$$

where R is the universal gas constant; T is temperature (in *Kelvin*); and F is Faraday's Constant. Eh has units of volt and is essentially proportional to

pE. Eh can be measured through electrochemical reactions.

Elements that are involved in redox reactions are sensitive to pE in the same way that acids and bases are sensitive to pH. A pE-pH or Eh-pH diagram shows which species of a particular element would be stable at equilibrium under a given range of pE and pH conditions. Such diagrams are useful for predicting changes that may be expected with change of pE and/or pH. For example, precipitates and scales/stains often form when low-pE anoxic groundwaters are pumped to the surface and get exposed to the atmospheric oxygen, which suddenly raises the pE of water. Dissolved Fe^{2+} forms a red-brown precipitate of ferric hydroxide, $Fe(OH)_3$, on bathroom showers, sinks, laundry, etc. Similarly, manganese-rich water forms a dark brown to black precipitate and copper-rich water forms a bluish precipitate.

6.1.9.1 Biogeochemical redox reactions

Microbially mediated redox processes are regulated by the ratio of accessible oxidizing agents to the amount of available and degradable organic substances in the water. The resulting biogeochemical conditions induce further chemically and microbially mediated transformations.

During photosynthesis, carbon is reduced, as $C(IV) \rightarrow C(0)$, while some of the oxygen is oxidized as $O(-II) \rightarrow O(0)$. Photosynthesis occurs only with the availability of additional energy from the sunlight. There are other reactions involving sulphur and nitrogen but they produce far less organic matter than photosynthesis. Hydrocarbons store chemical energy, which can be released subsequently through a variety of redox reactions.

During aerobic degradation of organic carbon, bacteria present use oxygen directly as an oxidizing agent. Once the available oxygen is consumed, other bacteria take over and use the remaining oxidants that are present in water during the course of a characteristic sequence of reactions (Fig. 6.3). The important ones amongst these oxidizing agents are nitrates, sulphates, and carbon dioxide dissolved in water as well as solid oxides/hydroxides of manganese and iron. In chemical parlance, microbial decomposition involves transfer of

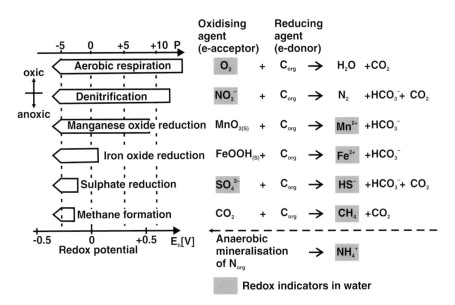

Fig. 6.3 Sequence of the major microbially mediated redox processes in aquatic systems. Redrawn from Zobrist *et al.* (2000). © EAWAG.

electrons mediated by bacteria from organic carbon to the various oxidizing agents. These reactions take place in a typical sequence that is governed by the chemical energy released during each reaction. Energy output is the highest during the first reaction and diminishes steadily thereafter (Fig. 6.3). The energy liberated is used by bacteria for sustaining their metabolism and growth. The oxidizing agents and products of the redox processes can serve as redox indicators in water. From their presence or absence, the status of redox processes and conditions in the aquifer can be inferred.

When all the oxygen has been consumed in an aquatic system, its biogeochemical environment changes drastically, because products of the redox processes initiate secondary geochemical reactions (von Gunten and Zobrist 1993) as:

- CO_2 reacts with the rocks and mineral grains present in an aquifer.
- Dissolved iron and manganese, *Fe(II)* and *Mn(II)*, are precipitated by sulphides and carbonates.
- Any surplus sulphide reductively dissolves iron and manganese (hydr)oxide. It is now known that there are certain additional chemical reactions that do not take place under oxidizing

conditions. For example, *Fe(II)* is also a reactive reducing agent that can react both with inorganic and organic pollutants (Haderlein 2000). Anoxic water must be treated before it can be used as drinking water, since dissolved manganese and iron are just as undesirable as sulphides, from health as well as taste considerations. From a microbiological point of view, anoxic groundwater may contain anaerobic bacteria that decompose organic pollutants that are not persistent in the presence of oxygen (Van der Meer and Kohler 2000).

Rainwater, due to its contact with the atmosphere, is saturated with dissolved O_2, resulting in O_2 concentration close to 10 $mg\ l^{-1}$. As water infiltrates into the ground, oxidation of soil organic matter tends to decrease the amount of dissolved O_2 and increase the level of dissolved CO_2. Further, oxidation reactions in the saturated zone that is isolated from the atmosphere explain progressively lower O_2 levels in the infiltrated water as it moves down into the aquifer. Some groundwaters are anoxic and other oxidation reactions may be important (Fig. 6.3). The unpleasant 'rotten egg' smell of hydrogen sulphide occurs as a by-product of bacterially mediated sulphate reduction in some anoxic groundwaters. The odour of H_2S can be detected at concentrations as low as 0.1 mg l^{-1} (Tate and Arnold 1990).

6.1.10 Sorption

Sorption is a combination of two processes, namely, *adsorption* (implying attachment onto the particle surfaces), and *absorption* (implying incorporation into something). Many of the solutes, particularly nonpolar organic molecules and certain metals, get sorbed onto the surfaces of solids in the aquifer matrix, and onto the rocks, minerals, and sediments. Some of these surfaces are internal to a grain or a rock mass, so technically these cases can be called absorption. Sorption slows down migration of solutes and is, therefore, a key process in the fate and transport of dissolved contaminants. For instance, certain compounds such as methyl tertiary butyl ether (MTBE), a constituent of gasoline, might sorb negligibly and migrate at about the same rate as the average water molecules, while other compounds such as toluene, also a constituent of gasoline, might sorb strongly and migrate at a much slower rate. As a plume of dissolved gasoline migrates, the constituents get segregated, depending on the degree of their sorption.

Most sorption reactions are relatively rapid, approaching equilibrium on time scales of minutes or hours (Morel and Hering 1993). Reaching equilibrium can take much longer if the process

is diffusion-limited in either the liquid or the solid phase.

6.1.11 Metal complexes and surface complexation of ions

In water, free metal cations are actually surrounded by a layer of water molecules. This coordinated rind of molecules closest to the metal cation usually consists of two, four, or six H_2O molecules, six being the most common (Morel and Hering 1993). For example, dissolved *Cr(III)* tends to be coordinated as $Cr(H_2O)_6{}^{3+}$. Beyond the coordinated rind, the water molecules are not chemically bonded but are oriented due to the electrostatic charge on the cation, the degree of orientation decreasing with distance from the cation.

Anions, also called *ligands*, may displace some or all of the coordinated water molecules and bond with the central metal cation. Common ligands in water include the major anions (Table 6.1), other inorganic anions, as well as a variety of organic molecules. All ligands have electron pairs available for sharing to form bonds with central metal cation.

If the metal and ligands bond directly to each other, displacing the coordinated water, one has an *inner sphere complex* or *ion complex*. If, on the other hand, a metal cation along with its coordinated water molecule combines with a ligand anion and its coordinated water molecules through electrostatic bonding, it leads to the formation of

Table 6.11 Dominant species of trace metals in natural waters. After Morel and Hering (1993).

Metal	Aqueous Species	Solid Species
Al	$Al(OH)_3$, $Al(OH)_4{}^-$, AlF^{2+}, $AlF_2{}^+$	$Al(OH)_3$, Al_2O_3, $Al_2Si_2O_5(OH)_4$, Al-silicates
Cr	$Cr(OH)_2{}^+$, $Cr(OH)_3$, $Cr(OH)_4{}^- HCrO_4{}^-$, $CrO_4{}^{2-}$	$Cr(OH)_3$
Cu	Cu^{2+}, $CuCO_3$, $CuOH^+$	CuS, $CuFeS_2$, $Cu(OH)_2$, $Cu_2CO_3(OH)_2$, $Cu_2(CO_3)_2(OH)_2$, CuO
Fe	Fe^{2+}, $FeCl^+$, Fe_2SO_4, $Fe(OH)_2{}^+$, $Fe(OH)_4{}^-$	FeS, FeS_2, $FeCO_3$, $Fe(OH)_3$, Fe_2O_3, Fe_3O_4, $FePO_4$, $Fe_3(PO_4)_2$, Fe-silicates
Hg	Hg^{2+}, $HgCl^+$, $HgCl_2$, $HgCl_3{}^-$, $HgOHCl$, $Hg(OH)_2$, $HgS_2{}^{2-}$, $HgOHS^-$	HgS, $Hg(OH)_2$
Mn	Mn^{2+}, $MnCl^+$	MnS, $MnCO_3$, $Mn(OH)_2$, MnO_2
Pb	Pb^{2+}, $PbCl^+$, $PbCl_2$, $Pb(OH)^+$, $PbCO_3$	PbS, $PbCO_3$, $Pb(OH)_2$, PbO_2

an outer *sphere complex* or *ion pair* and the bonding tends to be weak and ephemeral. The number of sites where the central metal atom bonds to ligands is its *coordination number*. Some ligands form bond with metal atom at only one site and are called *unidentate ligands*. When the ligands bond to metal atoms at multiple sites, these are called *multidentate ligands*. *Chelates* are complexes with multidentate ligands and one central metal atom. Chelates tend to be more stable than unidentate complexes. The major inorganic cations in fresh waters do not typically form complexes to any significant extent. In general, complexation reactions are rapid and, therefore, assumption of the equilibrium condition is reasonable. Unlike major cations, many trace metals in water are predominantly in the form of complexes as opposed to free ions. Some of the most common species of common trace metals, based on equilibrium condition in natural waters, are listed in Table 6.10. The speciation of metal complexes varies with *pH* and *pE*, particularly when there are complexes involving hydroxides, carbonates, or other *pH* sensitive ligands.

6.1.11.1 Surface complexation of ions

In almost every natural setting, there is a large amount of mineral surface area in contact with water, particularly groundwater. Metals and ligands in mineral solids are incompletely coordinated when they are at the surface and, therefore, have a tendency to coordinate with ligands and metal ions present in the interstitial water. Coordination of atoms and functional groups that are part of these mineral surfaces is essentially similar to complexation in the aqueous phase.

Many mineral solids, particularly clay minerals, carry a net charge on their surfaces. This causes the water in contact with these minerals to become structured with respect to ion concentrations. The majority of clays have a negative charge on their surfaces. In the layer of water closest to the surface, water molecules are orientated with their hydrogen ions towards the surface and the concentration of cations is higher relative to that of anions. Cations in this zone are known as *counter ions* because they counter the charges

on the mineral surfaces. The negative mineral surface charge and the net positive charge on the surrounding water together form a charge distribution called a *double layer*. Metal cations in solution react with the coordinated hydroxyl ions present on mineral surfaces, displacing hydrogen and becoming sorbed onto the mineral surfaces. Similar to complexation in the aqueous phase discussed above, an inner sphere complex is formed when the metal ions bond covalently to atoms on the mineral surface and outer sphere complex is formed when the metal cation is loosely bound to the mineral surface with intervening water molecules. Since both inner- and outer- surface complexes are removed from the mobile aqueous phase, these are considered as sorbed.

Cations not only compete with H^+ ion for sorption sites, but also compete with each other. Ion exchange occurs when one type of ion displaces another from a coordination site at a mineral surface. The ion exchange process differs markedly from mineral to mineral. The *cation exchange capacity* (CEC) is a parameter that is used to quantify and compare the ion exchange capacity of different soils. CEC is defined as milli-equivalents of cation that can be exchanged per unit dry mass of the soil. The procedure for testing of CEC involves first rinsing the soil sample with ammonium acetate solution, a process that saturates all the exchangeable sorption sites with NH_4^+, after which the sample is flushed with an $NaCl$ solution resulting in Na^+ exchanging with NH_4^+. The degree of the exchange is quantified by measuring the amount of NH_4^+ flushed out from the sample. Because ion exchange processes vary from ion to ion, and also with *pH* as well as the concentration of the solution, the measured CEC may be of limited use in predicting cation sorption under a variety of conditions.

6.1.12 Nonpolar organic compounds

Many important contaminants, including hydrocarbon compounds and chlorinated solvents, have relatively nonpolar distribution of electric charge on molecules. Because of its polar nature and strong electrostatic attraction between its different molecules, water tends to exclude nonpolar molecules from aqueous solutions. Therefore, nonpolar molecules are also called *hydrophobic*. Hydrophobic molecules have finite but very low aqueous solubility, which tends to be lower for larger and more perfectly nonpolar molecules.

The tendency of water to exclude nonpolar solute molecules causes them to accumulate on the surfaces that repel nonpolar molecules to a lesser degree than water, giving a false impression of being sorbed onto these surfaces. Hydrophobic sorption is quite different from the chemical bonding and electrostatic attraction involved in ion sorption. But the net effect is the same: sorbed nonpolar molecules get immobilized and removed from flowing water.

6.2 Natural hydrochemical conditions in the subsurface

Changes in the natural groundwater quality start taking place immediately after its entry into top-soil, where infiltrating rainwater dissolves carbon dioxide produced by the biological activity taking place in the topsoil, which results in higher partial pressure of CO_2 in it, producing weak carbonic acid. This process assists removal of soluble minerals from the underlying such as calcite cements. Simultaneously, soil organisms consume some of the dissolved oxygen in the rainfall. In temperate and humid climates with significant recharge, groundwater moves relatively rapidly through the aquifer. Due to its short contact time with the rock matrix, groundwater in the outcrop areas of aquifers is likely to be low in overall chemical content, i.e. it has a low content of major ions and low TDS. Igneous rocks usually have less soluble constituents than sedimentary rocks (Hem 1985).

In the recharge area of the aquifer, oxidizing conditions occur and dissolution of calcium and bicarbonate dominates. As water continues to move down gradient, a sharp redox barrier is created beyond the edge of the confining layer, corresponding to complete consumption of the dissolved oxygen. The bicarbonate concentration increases and the *pH* rises until buffering occurs at about *pH* 8.3. Sulphate concentration remains stable in the oxidizing water but decreases abruptly just beyond the redox limit due to sulphate reduction. Groundwater becomes steadily more reducing while it flows in the aquifer down gradient, as demonstrated by the presence of sulphide, increase in the solubility of iron and manganese, and denitrification of nitrate present in water. After moving through the aquifer for a few kilometres, sodium concentration begins to increase at the expense of calcium, due to the ion exchange process, causing a natural softening of the water. Eventually, the available calcium in the water is exhausted but sodium continues to increase to a level higher than what could be accounted for purely by cation exchange. The point at which chloride begins to increase is indicative of recharging water moving slowly through the aquifer and mixing with the much older saline water present in the sediments (Fig. 6.4). The hydrochemical characteristics of groundwaters can thus be interpreted in terms of oxidation/reduction, ion exchange, and mixing processes.

In arid and semi-arid environments, evapotranspiration rates are much higher, recharge is lower, flow paths longer, and residence times much greater and hence much higher levels of natural mineralization, often dominated by sodium and chloride, can be encountered. Thus in arid and semi-arid regions, the levels of major ion concentration and TDS are often high. In some desert regions, even if groundwater can be found, it may be too salty (very high TDS) for it to be potable and the difficulty of meeting even the basic domestic requirements can have serious implications on health and livelihood.

In many tropical regions, weathered basement aquifers and alluvial sequences have low *pH*, and the reducing conditions that can promote mobilization of metals such as arsenic, and other constituents of concern to health. Thus the prevailing hydrochemical conditions of groundwater that are naturally present at a given point and get modified as it flows in the aquifer, need to be taken into account for: (i) developing schemes for groundwater abstraction for various uses and in protecting groundwater reservoir; and (ii) considering transport and attenuation of additional chemicals getting into groundwaters due to human activity.

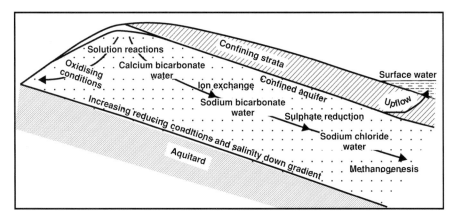

Fig. 6.4 Schematic representation of down-gradient change of hydrochemical facies in a typical regional groundwater systems. Redrawn from Rivett *et al.* (2006).

6.3 Presenting inorganic chemical data

Typically analytical chemistry data are reported in tables and numbers. However, when there are large numbers of analyses, these may be difficult to comprehend and interpret. Graphical methods of data presentation are, therefore, helpful for quick inspection and identification of general trends in the data.

Some of these graphical procedures are useful mainly for visual display purposes, for example, in audio-visual presentations or written reports on water quality to provide a basis for comparison of different analyses or to emphasize differences/similarities amongst them. Graphical procedures are more effective in this regard than the raw data presented in the form of tables. In addition, graphical procedures have been devised to help detect and identify mixing of waters of different origins/compositions and to identify some of the chemical processes that may take place as natural waters circulate through different environments. Some of the commonly employed graphical methods are presented here. Hem (1985) provides a more detailed account of these. Also one should not lose sight of the fact that the graphical method of presenting water quality data is a tool only to identify broad trends and not an end in itself.

Methods of graphical analyses are generally designed to simultaneously present the total solute concentration and the proportions assigned to each ionic species for a single analysis or a group of analyses. The units in which solute concentrations are generally expressed in these diagrams are milli-equivalents per litre ($meq\ l^{-1}$).

The chemical data of a limited number of samples can be depicted using bar charts or pie diagrams, which can be easily made by using database management and spreadsheet software. In bar charts (e.g. in Collins' ion-concentration graphical procedure; Collins 1923), each analysis is represented by drawing a vertical bar graph whose total height is proportional to the measured concentration of anions or cations, expressed in $meq\ l^{-1}$. The bar is divided into two parts by a vertical line with the left half representing the cations and the right half the anions. These segments are then divided by horizontal lines to show concentrations of major ions. These ions are identified by a single distinctive colours or patterns. Usually six divisions are made but more can be made if necessary. The concentrations of closely related ions are often added together and represented by a single pattern. A perfect charge balance results in columns of equal height.

The Collins' system, as described above, does not consider non-ionic constituents but they may be represented, if desired, by adding an extra bar or some other marker along with a supplementary scale. In Fig. 6.5a, the hardness of two waters is shown. On a pie chart, anions and cations are plotted in opposite hemispheres and subdivisions of

(a)   (b)

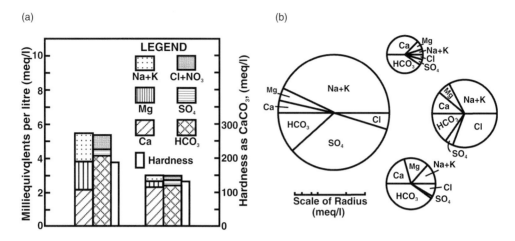

Fig. 6.5 (a) Bar charts and (b) Pie charts representing arbitrary water analyses. Units are milli-equivalents per litre (*meq l^{-1}*). The Collins' system of ion-concentration diagrams for bar charts does not consider non-ionic constituents, but if desired, they may be represented by adding extra bars with a supplementary scale as for hardness, as shown in (a). In pie diagrams, the scale of radii for the plotted graphs of the analysed data is so chosen as to make the area of the circle represent the total ionic concentration, as shown in (b). Sample codes are indicated above the bars or circles. Redrawn from Hem (1985). © U.S. Geological Survey.

the area represent proportions of different ions so that with perfect charge balance, anions and cations occupy half the circle each. A pie diagram can be plotted with the radii drawn to scale, proportional to the measured total concentration of different ions (Fig. 6.5b).

Another graphical method is the Stiff diagram (Stiff 1951), as shown in Fig. 6.6a. Polygons are drawn by plotting vertices at scaled distances to the left (cation) and the right (anions) of a central axis. Waters of different origins form polygons of different shapes in such plots. Normally only three par-

allel horizontal axes are used for plotting concentrations of three major cations and anions. A fourth axis for Fe^{2+} and CO_3^{2-} ions is often added at the bottom when these ions are present in significant concentrations. The ions are plotted in the same sequence. The width of the polygonal pattern is an approximate indicator of the total ionic content.

Bar charts, Pie charts, and *Stiff diagrams* are practical procedures for visual comparison of a small number of samples. A nomographic procedure proposed by Schoeller (1935), using vertically scaled axes for individual cations and anions, is a

(a)   (b)

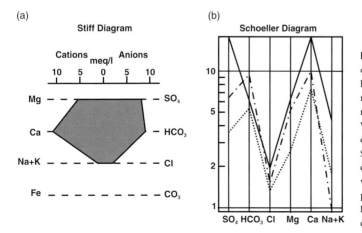

Fig. 6.6 (a) Stiff diagram for presenting data of major ion chemistry. Three or four parallel horizontal axes may be used. Each sample is represented by a polygon and sample numbers may be added below each polygon. Waters of differing origins reveal different-shaped polygons in such plots. (b) Schoeller diagram, which is a means of depicting groups of analyses. In this plot, waters of similar composition plot as near parallel lines. Different symbols and types of lines can be used to label and/or identify different samples.

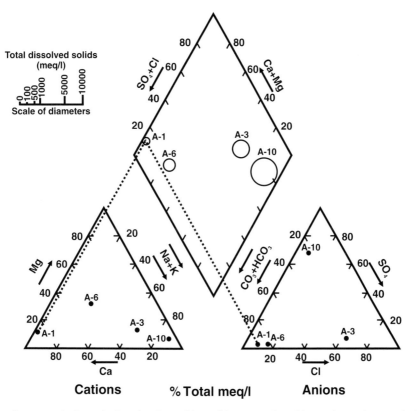

Fig. 6.7 Trilinear diagram to depict major ion chemistry of four arbitrary samples with sample numbers marked on the figure. The dotted lines show how the data for sample A-1 are projected from the two triangular plots to the diamond-shaped plot. The circles plotted in the diamond field have areas proportional to total concentration in $meq\ l^{-1}$ for each sample and the scale of diameters is shown in the upper left corner. Redrawn from Hem (1985). © U.S. Geological Survey.

means of depicting a group of analyses (Fig. 6.6b). As with the Stiff diagram, different *Schoeller plots* should have the same sequence of cations and anions for comparison. In this diagram, waters of similar composition plot as nearly parallel lines. This diagram, however, uses logarithmic scales and this may complicate interpretation of waters that differ considerably in composition.

A *trilinear* or *Piper diagram* (Fig. 6.7) is a convenient method of visualizing the results of a large number of analyses in a single plot (Piper 1944). The percentage values of total cations (in $meq\ l^{-1}$) is plotted in the lower left triangle, using Ca^{2+}, Mg^{2+}, and $(Na^+ + K^+)$ on the three axes of the triangle. For example, a sample in which Ca^{2+} is the only cation present would plot at the lower left vertex of the triangle. In a similar way, anions

plot in the lower right triangle using percentage $meq\ l^{-1}$ of $(HCO_3^- + CO_3^{2-})$, SO_4^{2-}, and Cl^-. The diamond-shaped part between the two triangles shows projections from the anion and cation triangles onto a field that shows percentage values of major anions and cations simultaneously. Fig. 6.7 shows four arbitrary samples plotted on a trilinear diagram. Waters are often classified on the basis of where they plot on a trilinear diagram; such a classification is called the *hydrochemical facies*. For example, Sample No. A-1 would be classified as a calcium–bicarbonate facies water. Other facies include sodium–chloride, sodium–sulphate, calcium–sulphate, etc. To enhance the visual impact of a Piper diagram, different ranges of TDS may be represented by symbols of different sizes or colours.

A simple application of the Piper diagram may be to show whether particular water may be a mixture of different waters for which analyses are available or whether it is affected by dissolution or precipitation of a single salt in the same water. Analysis of mixture of waters A and B should plot on the straight line joining A and B in the plotting field, provided the ions in A and B do not react chemically on mixing. If solutions A and C define a straight line pointing towards the $NaCl$ vertex, the more concentrated solution could form from a dilute solution spiked by addition of sodium chloride. Plotting of samples from wells successively located in the direction of the hydraulic gradient may show linear trends and other relationships that may have some geochemical implications.

6.4 Impact of human activities

All life forms interact with their environments in various ways. As a result of this, complex ecologic structures have evolved in which diverse life forms interact with one another in supportive as well as in predatory ways. One might view the present-day Earth-surface environment as having been shaped in many ways through the biological processes interacting with their surroundings over the long geological time span. Lovelock and Margulis (1974) suggested that metabolic processes of various organisms have resulted in maintaining the composition of the atmosphere at an optimum level for survival of various life forms on the Earth.

Many human activities have, however, adverse impacts on the environment. When these activities cause deterioration in the quality of natural waters, the result is 'water pollution'. Water pollution may be defined as the human-induced deterioration of water quality that is sufficiently severe to decrease the usefulness of the resource substantially, either for human beings or other life forms.

A serious consequence of human impact on water supply is the contamination and resulting pollution of the water source from diverse activities generating waste products such as: (i) industrial wastes, including mining, petrochemicals, etc.; (ii) agricultural non-point sources of pollution; (iii) wastewater treatment outfalls; (iv) accidental spills

and oil slicks; and (v) loading by sediments and nutrients. A *point source* is an identifiable source, such as a leaking septic tank, which may result in a well-defined plume of the pollutant emanating from it. On the contrary, non-point sources are more difficult to control and pose a greater risk to water quality. *Non-point sources* are larger in scale compared to point sources and produce relatively diffuse pollution plumes originating from either widespread application of polluting material in a given area or caused by a large number of smaller sources. The aggregate of point sources in a leaking sewerage system may be taken as an example of a non-point source of contamination to groundwater (usually described as multi-point pollution source).

Pollution of water sources ranges from small streams, rivers, lakes, and reservoirs to coastal waters and even groundwater. Pollution of streams, rivers, and groundwater can form plumes of a point source downstream. Common water pollutants and their sources are listed in Table 6.12.

Contamination of groundwater can occur from a number of potential sources such as: (i) leakage of liquid fuels and chemicals from underground storage tanks; (ii) septic tanks or cesspools used for disposal of domestic sewage or other wastewater; (iii) leachates from landfill refuse dumps or sewer pipes; (iv) injection wells disposing off liquid wastes in the subsurface; (v) pesticides, herbicides, and fertilizers sprayed on agricultural fields as part of farming operations; (vi) leachates from mine workings and tailings; (vii) spraying of salt for de-icing the snow/ice-covered roads in winter; and (viii) sea water intrusion in coastal areas due to excessive pumping of groundwater in the coastal aquifers. Groundwater contamination can also affect surface water and soil vapours.

Much of the subject of water-pollution control and problems relating to it are beyond the scope of this book. However, the basic principles of natural water chemistry can be applied to understand, predict, and remedy/mitigate pollution problems. It should be appreciated that environmental change attributable to anthropogenic factors is to a large extent unavoidable and that some deterioration in water quality may have to be accepted if the available alternatives entail unacceptable social costs.

Table 6.12 Common pollutants in water and their origin (adapted from http://www.epa.vic.gov.au/students/ water/pollutants.asp)

Pollutant	Origin
Sediments	• Land surface erosion • Pavement and vehicle wear and tear (tyres, brakes, etc.) • Atmospheric particulate matter • Spillage/illegal wastewater discharge • Organic matter (e.g. leaf litter, grass, bird and animal excreta) • Waste water generated from washing of cars and other vehicles • Weathering of buildings/structures
Nutrients	• Organic matter • Fertilizers • Sewer overflows/leaks from septic tanks • Animal/bird droppings • Detergents (household laundry/car washing) • Atmospheric particular matter • Accidental spillage/illegal wastewater discharges
Oxygen demanding substances	• Decaying organic matter • Atmospheric emissions of pollutants, e.g. CO_2, SO_2 • Sewer overflows/leaks from septic tanks • Animal/bird droppings • Spillage/illegal discharges
pH (acidity)	• Atmosphere • Spillage/illegal discharges • Decaying organic matter • Erosion of roofing material
Micro-organisms	• Animal/bird droppings • Sewer overflows/ improperly designed septic tanks/ cesspools • Decaying organic matter
Toxic organics	• Pesticides • Herbicides • Spillage/illegal discharges • Sewer overflows/leaks from septic tanks
Heavy metals	• Atmospheric particulate matter • Vehicle wear and tear • Sewer overflows/leaks from septic tanks • Weathering of buildings/structures • Spillage/illegal discharges
Oils and surfactants	• Asphalt pavements • Spillage/illegal discharges from automobiles/tankers • Leaks from vehicles • Washing of vehicles • Organic matter from various sources
High water temperatures	• Runoff from impervious surfaces • Removal of riparian vegetation • Industrial discharges (e.g. power production)

In recent years, as a result of improved technology and application of analytical chemistry, detection of progressively lower levels of organic or inorganic pollutants has become relatively easy. But the mere presence of a particular substance does not establish the existence of pollution. Various other aspects need to be considered. Major problems that need in-depth scientific study include determination of form, stability, transport rates, and mechanisms for generation of pollutant species; predicting probable effects of current or foreseeable uture practices in waste disposal or product consumption causing contamination; assessing relative impacts of artificial sources vis-à-vis natural sources; and providing methods for identifying the most significant existing as well as potential pollution sources.

Some common ionic species may be dispersed into the environment with no serious repercussions. Chloride, for example, may not be harmful if maintained at a low level of concentration. Furthermore, chloride gets easily conveyed to the ocean where it causes no significant ill effects. Some other solutes, however, may tend to accumulate and become concentrated in such places as sediments in streams or in biota and can be released from such sources in unforeseen ways to cause troublesome local concentrations.

Although some polluted surface waters can be rejuvenated to a reasonable quality fairly rapidly by decreasing the waste loads or concentrations, the process can be expensive. A polluted groundwater, on the other hand, may be so slow to recover that one may think of the pollution of aquifers as almost irreversible. For this reason, great care is needed to protect groundwater, particularly deeper aquifers. Incidents of contamination of groundwater from septic tanks, sewage, and industrial waste-disposal systems, solid-waste disposal practices, and natural gas and petroleum storage leaks have been a growing problem both in urban as well as rural areas.

In an industrial society with developed agriculture, a large number of organic and inorganic products are produced and used, which in the natural course do not enter the environment. Inevitably, some of these products, their residues, or by-products enter the water cycle. During the past few decades, attention has been drawn, for example, to the presence of lead, mercury, and various common and exotic organic substances in the aqueous environment. A matter of more recent concern worldwide is the occurrence of rainfall having a low pH – the so-called 'acid rain', which is harmful both for the plant as well as the animal kingdom.

6.4.1 Pathogenic micro-organisms

Waterborne diseases remain one of the major health concerns globally and waterborne pathogens from various sources, in particular agricultural and urban sources, are the main disease causing agents. Typically, the health risk from chemicals is lower than that from pathogens. This is because, unlike pathogens, health effects many of the hazardous chemicals manifest after a prolonged exposure and tend to be limited to specific geographical areas or particular types of water source. Following is a brief summary of the current knowledge on the distribution of pathogens in water in general and groundwater in particular, and the factors that control their transport and attenuation.

The ability of a pathogen – a group of disease-causing micro-organisms – to inflict damage upon the host is controlled by a combination of factors, in particular the nature of the organism (e.g. the degree of its virulence) and the susceptibility of the host. Water offers an easy carrier for transmission of most pathogenic micro-organisms, some being natural aquatic organisms and others introduced into water from an infected host. Overall, pathogens in water that are the main concern to public health originate in the faeces of humans and animals and cause infection when water contaminated with faecal matter is consumed by a susceptible host. Waterborne pathogenic micro-organisms of concern and the health effects caused by these are summarized in Table 6.13.

Furthermore, currently unknown and, therefore, undetectable pathogens may also be present in water. To overcome this difficulty, a separate group of micro-organisms is used as an indicator for the presence of potential pathogens. The commonly used term for this group of organisms is faecal indicator organisms. Gleeson and Gray (1997) published an exhaustive review of the application of faecal indicator organisms in water

Table 6.13 Waterborne pathogenic micro-organisms and their associated health effects. Adapted from Macler and Merkle (2000).

Organism	Associated health effects
Viruses	
Coxsackie virus	Fever, pharyngitis, rashes, respiratory diseases, diarrhoea, haemorrhagic conjunctivitis, myocarditis, pericarditis, aseptic meningitis, encephalitis, reactive insulin-dependent diabetes, diseases of hand, foot, and mouth
Echovirus	Respiratory diseases, aseptic meningitis, rash, fever
Norovirus (formerly Norwalk virus)	Gastroenteritis
Hepatitis A	Fever, nausea, jaundice, liver failure
Hepatitis E	Fever, nausea, jaundice, death
Rotavirus A and C	Gastroenteritis
Enteric adenovirus	Respiratory diseases, haemorrhagic conjunctivitis, gastroenteritis
Calicivirus	Gastroenteritis
Astrovirus	Gastroenteritis
Bacteria	
Escherichia coli	Gastroenteritis, Haemolytic Uraemic Syndrome (enterotoxic *E. coli*)
Salmonella spp.	Enterocolitis, endocarditis, meningitis, pericarditis, reactive arthritis, pneumonia
Shigella spp.	Gastroenteritis, dysentery, reactive arthritis
Campylobacter jejuni	Gastroenteritis, Guillain-Barré syndrome
Yersinia spp.	Diarrhoea, reactive arthritis
Legionella spp.	Legionnaire's disease, Pontiac fever
Vibrio cholerae	Cholera
Protozoa	
Cryptosporidium parvum	Diarrhoea
Giardia lamblia	Chronic diarrhoea

Note: The coliform group of bacteria comprises several genera belonging to the family Enterobacteriaceae (Gleeson and Gray 1997). The relatively limited number of biochemical and physiological attributes used to define the group means that its members include a heterogeneous mix of bacteria. Although the total coliform group includes bacteria of faecal origin, it also includes species that are found in unpolluted environments. Thermo-tolerant coliforms are those bacteria from within the coliform group that grow at $44°C$. *E. coli* is a thermo-tolerant coliform bacteria.

quality monitoring. This group comprises: (i) total coliform bacteria; (ii) thermo-tolerant coliform bacteria; (iii) *E. coli*; (iv) faecal streptococci; and (v) bacteriophages.

6.4.1.1 Transport and attenuation of micro-organisms in the subsurface

Many instances of groundwater contamination have occurred, possibly by rapid transport pathways accidentally introduced by human intervention and connecting the contamination source to the groundwater abstraction point. Such pathways could be provided by improper completion of springs, wells, and boreholes, and conduits left carelessly connecting the source of

contamination to the groundwater abstraction point/system. Existence of rapid transport pathways cannot, however, explain all occurrences of groundwater source contamination and it is now widely accepted that the transport of microbial pathogens through groundwater systems is an important mechanism for transmission of waterborne diseases.

6.4.1.1.1 Unsaturated zone

Hydrogeological processes in the unsaturated zone are complex and the behaviour of micro-organisms is often difficult to predict. Nevertheless, the unsaturated zone can play an important role in retarding (and in some cases eliminating) pathogens.

Attenuation of pathogens is generally more effective in the top soil layers where significant microbial activity occurs. Within this zone, presence of protozoa and other predatory organisms, rapid changes in moisture content and temperature of the soil, competition from the established microbial community, and the effect of sunlight at the surface combine together to reduce the level of pathogens.

Movement of pathogens from the surface into the subsurface requires presence of moisture. Even during relatively dry periods, soil particles on the surface retain sufficient moisture for pathogens to migrate. Within the thin film of moisture the organisms present are brought into close contact with the surface of particulate matter, thus increasing the opportunity for adsorption onto the surfaces of particulate matter, which further retards their movement. If the soil moisture content decreases, the strength of association between organisms and particle surfaces will increase to a point where organisms are attached irreversibly to particle surfaces. In laboratory experiments, soil moisture content between 10 and 15% has been shown to be optimal for survival of several strains of enteric viruses (Bagdasaryan 1964; Hurst *et al.* 1980a, b).

Increase in the moisture content of the unsaturated zone, on the other hand, may increase the vulnerability of the aquifer to pathogen contamination in two ways by: (i) providing rapid transport pathways; and (ii) remobilizing the adsorbed micro-organisms. During periods of high groundwater recharge, for example during prolonged heavy spells of rain, the inter-granular spaces in the unsaturated zone become waterlogged, providing hydraulic pathways for rapid transport of pathogens.

In the interval between individual recharge events, the chemistry of water in the unsaturated zone changes as it equilibrates with the soil matrix. In some soil types, these changes may favour adsorption of micro-organisms onto surfaces in the soil matrix. A lowering of the ionic strength or salt content of the surrounding medium, which can occur during a rainfall event may, however, be sufficient to cause desorption of the organisms allowing their further migration into the soil matrix. Variability in the size of different micro-organisms can, to some extent, control their mobility in the subsurface. Pore sizes of soil and rock particles are also variable and the size ranges of the two are known to overlap (Fig. 6.8). Thus in soils that are composed of fine-grained particles, typically clayey silts, the pore space is sufficiently small ($<4 \ \mu m$) to physically prevent the passage of pathogens, such as bacteria and protozoa, into the subsurface. This removal process is called physical filtration or straining.

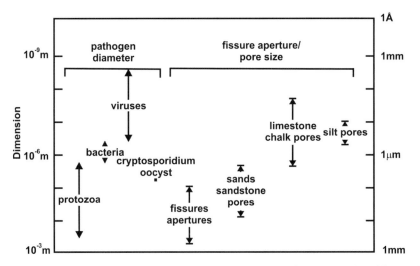

Fig. 6.8 Comparison of pathogen size diameter with pore size diameter in an aquifer matrix. Adapted from Pedley *et al.* (2006).

6.4.1.1.2 Saturated zone

Shallow groundwater (<5 m deep) has the highest probability of contamination, irrespective of the lithology of the unsaturated zone. As depth to the water table increases so does the capacity of the unsaturated zone to attenuate micro-organisms, although this also depends upon composition and structure of the unsaturated zone. On reaching the saturated zone, microbial contaminants are subject to the same processes of attenuation as in the unsaturated zone but under conditions of natural or artificially induced groundwater flow. Thus death, adsorption, filtration, predation, and dilution contribute to attenuation of pathogens in the saturated zone. However, there may be large variations in hydraulic conductivity of an aquifer depending on the nature of aquifer material, which can significantly influence mobility of micro-organisms in the aquifer. Micro-organisms are transported through groundwater by advection, dispersion, and diffusion, as defined in Section 7.6. It may be recalled that advection refers to transport of dissolved solute mass due to bulk flow of groundwater and to a non-reactive solute for which the advection velocity would equal the bulk groundwater velocity through pores. Diffusion and dispersion, arising from the tortuosity of flow within an aquifer, cause mixing of a solute and modify its concentration as the flow proceeds. The result of both advection and dispersion is migration and spreading of the contaminant in the aquifer, resulting in a decrease in its concentration both in space and time. This may result in contamination of increasingly large volumes of the aquifer as the pollutant moves down gradient with the groundwater. Although transport of pathogens in some aquifer types can be both rapid and extensive, there are several factors that may attenuate pathogens in groundwater (West et al. 1998).

Inactivation rates of bacteria and viruses in groundwater vary considerably, not only between different groups of bacteria and viruses but also between different strains within each group. There is also variation between the results of different measurements. Usually inactivation proceeds faster at higher temperatures, although highly dependent on type of micro-organism. Often, inactivation

of micro-organisms can be described reasonably well by a first-order rate process, especially under relatively mild conditions, such as temperatures between 5 and $20°C$ and pH values in the range 6 to 8. Thus:

$$C_t = C_0 e^{-\mu t} \quad \text{or} \quad ln\left(\frac{C_t}{C_0}\right) = -\mu t \quad \text{or}$$

$$log_{10}\left(\frac{C_t}{C_0}\right) = -\frac{\mu}{2.3}t \tag{6.45}$$

where C_t is the concentration of micro-organisms surviving after time t; C_0 is the initial concentration at $t = 0$; and μ is the inactivation rate coefficient [T^{-1}]. For ease of interpretation, μ is often divided by a factor of 2.3 (equal to the natural logarithm of 10). The inactivation rate coefficient then reflects the number of $log10$ units per unit time; for example, a decrease in virus count every 10 days by 2 units of $log10$ is equal to reduction by a factor of 100.

Under more extreme conditions, the rate of inactivation of a virus, e.g., is often found to proceed initially at a higher rate followed by a lower rate, as if two or more sub-populations differing in their stability exist simultaneously (Hurst et al. 1992).

6.4.1.2 Subsurface transport and attenuation of chemicals

In the classical conceptual model of contaminant from a point source to a receptor, the chemical contaminant is leached from a near-surface leachable source zone through the process of dissolution in the infiltrating water (Fig. 6.9a). Subsequently, a dissolved phase chemical solute plume emerges in water draining from the base of the contaminant source zone and moves vertically downwards through the unsaturated zone. This solute plume ultimately reaches the water table and subsequently migrates laterally in the aquifer along with groundwater flow. If the source (e.g. landfill, chemical waste lagoon, contaminated soils from industrial sites, pesticide residues in field soils, etc.) has sufficient pollutant mass, the contaminant plume can last for decades. In this manner, continuous dissolved phase plumes extending from the source through the groundwater pathway can grow with time and may eventually reach a distant location.

(a)   **(b)**

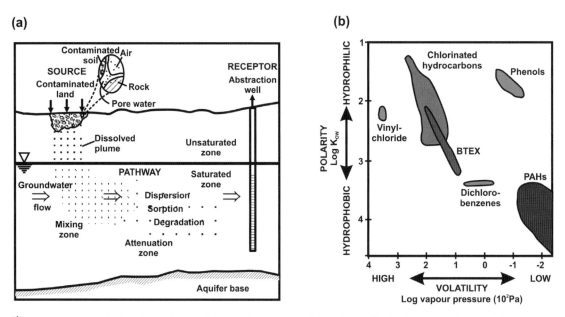

Fig. 6.9 (a) A conceptual model for transportation and attenuation of chemical and hydrophilic organic pollutants in groundwater starting from a point source up to the location of the observer. (b) Polarity-volatility diagram for select organic contaminants. Redrawn from Rivett *et al.* (2006).

The near surface leachable source–dissolved plume conceptual model as described above is frequently invoked for vulnerability and protection of groundwater and also for groundwater risk assessment. It is important to note, however, that the above conceptualization may be too simplistic and alternative conceptual models may be needed in some cases, notably for non-aqueous phase liquids (NAPLs).

The processes involved in retardation of a plume movement are: (i) sorption by which chemicals or organisms become attached to the aquifer solids; (ii) cation exchange by which interchanging of cations occurs in solution with those on the surfaces of clay particles or organic colloids; and (iii) filtration that affects relatively large-size particulate contaminants by preventing their movement by advection.

In addition, abiotic (i.e. not mediated by bacteria) reactions such as precipitation, hydrolysis, complexation, elimination, substitution, etc. that transform the chemicals present into some other chemicals and potentially alter their phase/state (solid, liquid, gas, dissolved) are also involved in trans-

portation and attenuation of contaminant plumes in the underground.

Biodegradation (biotic reactions) is a reaction process that is facilitated by microbial activity, for example, by bacteria present in the subsurface. Typically contaminant molecules are degraded (broken down) to molecules of simpler structure that often have lower toxicity.

6.4.2 Organic compounds

Natural sources invariably contribute some organic compounds to water, though at low levels. Natural organic matter comprises water-soluble compounds of a rather complex nature having a broad range of chemical and physical properties.

Typically, natural organic matter in surface- and groundwater is composed of humic substances (mostly fulvic acids) and non-humic materials (e.g. proteins, carbohydrates, and hydrocarbons) (Stevenson 1994; Thurman 1985). Natural organic matter is a complex, heterogeneous mixture varying in size, structure, functionality, and reactivity. It can originate from terrestrial sources (allochthonous natural organic matter) and/or algal and bacterial sources within the water body (autochthonous natural organic matter). *Dissolved organic carbon (DOC)* is considered to be a suitable parameter for quantifying the organic matter present in groundwater. However, *DOC* is a bulk organic quality parameter and does not provide specific identification data and may also incorporate organic compounds resulting from human activity. Natural organic matter, although considered innocuous, may still indirectly influence water quality. For example, contaminants may bind to organic-matter colloids, facilitating their transport with water flow.

Also, routine chlorination of water supplies containing natural organic matter may form disinfection by-products such as trihalomethanes that are known to be carcinogenic. However, natural organic substances are of little direct health concern and, therefore, are not addressed any further here.

Human activity has contributed to the release of a vast range of anthropogenic organic chemicals to the environment; some of these may adversely impact groundwater quality. In the following, the focus is specifically on the subsurface transport of non-aqueous phase liquids as part of the anthropogenic suite of organic chemicals.

6.4.2.1 Subsurface transport models for non-aqueous phase liquids

As already mentioned (see Section 6.1.12), due to the polar nature of water molecules, it causes nonpolar molecules to be excluded from aqueous solutions – called *hydrophobic* as opposed to polar molecules that are *hydrophilic*. Hydrophobic molecules do have finite but very low aqueous solubility, which tends to be lower for larger and more perfectly nonpolar molecules.

Hydrophobic non-ionic organic contaminants preferentially sorb onto the low polarity components of geo-solids, for example, any organic material present. Sorption is inversely related to the solubility of an organic compound; the more hydrophobic and less soluble an organic solute is, the greater its intrinsic potential for sorption onto any organic material present in the aquifer solids. Hence hydrophilic organics have negligible sorp-

tion and mild to moderately hydrophobic organics, such as the volatile organic compounds (VOCs), which show limited sorption. In contrast, hydrophobic, high molecular weight, large organics such as polynuclear aromatic hydrocarbons (PAHs) and polychlorinated biphenols (PCBs) of low solubility exhibit high sorption. Hydrophobicity and sorption retardation (indicative of solubility) trends of select organic contaminants against vapour pressure (a measure of volatilization tendency) are shown in Fig. 6.9b in the form of a polarity-volatility diagram. It is seen that PAHs are unlikely to volatilize and undergo a high degree of sorption and this also implies that in soil/unsaturated zone solids, concentrations of PAHs could often be high, which is frequently the case. There is relatively limited development of PAH plumes in groundwater; often in real-life situations relatively small plumes are encountered. The chlorinated hydrocarbons, in contrast, are volatile but with low sorption potential. It is likely that they get vaporized, and potentially become a vapour hazard to living organisms at the soil surface. Chlorinated hydrocarbons are also leached into the groundwater, resulting in their low concentrations in soils and unsaturated samples, which is quite often the case.

The classical near-surface leachable source zone dissolved plume model (presented in the Section 6.4.1.2; Fig. 6.9a) is not applicable to all organic substances. Of key importance is the realization that different organic chemicals have different affinities for water, ranging from hydrophilic to hydrophobic.

Most organic liquids, however, are so hydrophobic that they form a separate organic phase within the water (aqueous) phase. They are immiscible with water and a phase boundary exists between the organic phase and the aqueous phase. The organic phase is generally referred to as the non-aqueous phase liquid (NAPL). When a separate organic NAPL exists, it is important to consider the density of the NAPL relative to water, as this controls whether the NAPL will be above or below relative to the water phase. Most hydrocarbon-based organic liquids have a density $<1\ g\ cm^{-3}$, for example, the density of benzene is $0.88\ g\ cm^{-3}$ and that of pentane is $0.63\ g\ cm^{-3}$. When in contact with water, these liquids form the upper phase and 'float' on the water phase, which has density of 1 $g\ cm^{-3}$. Such 'light' organic compounds are generally referred to as LNAPLs.

In contrast, other hydrophobic organics have a relatively high density due to incorporation of dense chlorine (or other halogen) atoms in their structure. For example, chlorinated solvents such as trichloroethene (TCE), 1,1,1-trichloroethane (1,1,1-TCA), and polychlorinated biphenyl (PCB) mixtures have densities in the range 1.1–1.7 $g\ cm^{-3}$. Due to their higher density, such organic phases will form the lower phase and 'sink' below the water phase. Such 'dense' organic compounds are generally referred to as DNAPLs.

Although hydrophobic, LNAPL and DNAPL organics still have the potential for some of their organic molecules to dissolve into the adjacent aqueous phase. These organics are 'sparingly soluble' and have a finite solubility in water, leading to their finite dissolved concentrations in the water phase.

In contrast to hydrophilic miscible organics, hydrophobic immiscible organics, that is NAPLs, exhibit quite different behaviour. Conceptual models for LNAPL and DNAPL releases (Mackay and Cherry 1989) are shown in Fig. 6.10.

If sufficient pressure head exists to overcome the entry pressure to the pores or fractures, NAPLs may migrate as a separate phase and displace air and water from the pores into which they penetrate. NAPL migration is also controlled by its density and viscosity. For example, petrol and chlorinated solvents have viscosities lower than water and migrate more easily in the subsurface. In contrast, PCB oils or coal tar (PAH-based) hydrocarbons may be very viscous and perhaps take years for the NAPLs to come to a resting stable position in the subsurface. Chlorinated solvents, such as PCE, have high densities and may penetrate to significant depths through aquifer systems in very short time periods. Whereas dissolved pesticides may take a couple of years to decades to migrate through a 30 m unsaturated zone, DNAPLs may migrate through such a zone in the matter of a few hours or days (Pankow and Cherry 1996). At the water table, LNAPLs, being lighter than water, form a floating layer on the water table, often forming a slightly elongated plume in the direction of the hydraulic gradient of the water table. DNAPLs, in contrast, may penetrate as a separate immiscible layer below the water table. Its predominant

(a)

(b)

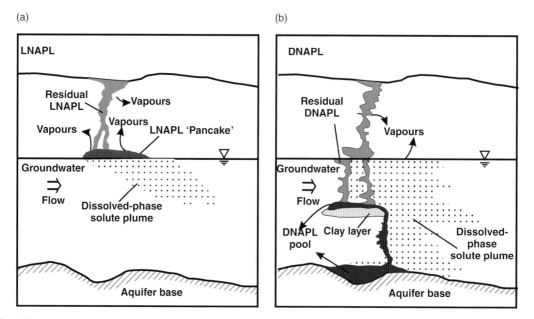

Fig. 6.10 Conceptual models of: (a) light non-aqueous phase liquid (LNAPL) release; and (b) dense non-aqueous phase liquid (DNAPL) release. Redrawn from Rivett *et al*. (2006).

movement in a vertical (downward) direction is due to its density, but some lateral spreading does occur as it encounters a lower permeability strata. If sufficient volume and pressure head exist, the DNAPLs may penetrate the entire aquifer depth up to the underlying aquitard/bedrock (Kueper *et al.* 1993). Migrating NAPLs leave behind a trail of immobile residual NAPL droplets along their migration pathways held by capillary forces causing NAPLs to spread across the entire aquifer thickness. DNAPLs accumulating on low permeability features, often referred to as pools, are potentially mobile. They may ultimately penetrate the formation due to changes in hydraulic pressure arising from their continued spillage, pumping, or remediation attempts or during drilling (for boreholes, piling, etc.) through the layer. Often NAPLs remain relatively localized at the site.

Risks posed to groundwater resources and supplies are generally concerned with migration of the dissolved-phase plume formed through contact of flowing groundwater with spilled NAPLs. Often the mass of NAPLs is so large and the dissolution of NAPLs into water so slow that the entire NAPL body subsequent to spilling should be regarded as a largely immobile source zone that can continuously generate a dissolved-phase solute plume of organics moving down-gradient for years to decades, or even centuries for low-solubility NAPLs. In general, DNAPLs tend to pose the greatest threat to groundwater as they reside deep in groundwater systems and many, being chlorinated, are less susceptible to attenuation. In contrast, LNAPLs are restricted to shallower groundwater table depths and are more susceptible to attenuation via biodegradation.

6.5 Geochemical modelling

An important issue that one is often confronted with relates to understanding the reactions that may occur along a flow path – given the parameters that are measured by aqueous geochemistry. This is essentially the process of inverse geochemical modelling vis-à-vis forward modelling, wherein one attempts to figure out the geochemical evolution of surface water when flowing over a given terrain or groundwater flowing through an aquifer, and how

the model-computed results compare with observations. Essentially, one has to look for tools to understand the evolution of chemical species in water at specific sites, for example: (i) to see if certain minerals would dissolve or precipitate in the water; (ii) to model water–rock interaction (dissolution, precipitation, ion exchange); and (iii) to estimate changes in chemical properties of water during mixing of different water masses.

6.5.1 Computer models

Many of the earlier computer models were designed for specific problems related to aqueous speciation (Allison *et al.* 1991; Nordstrom *et al.* 1979). Some important recent applications include modelling the disposal of high-activity radioactive waste, environmental issues associated with mining operations, landfill leachates, injection of hazardous chemical/biological wastes into deep wells, water resources-related issues, and artificial recharging of aquifers, particularly deep aquifers (Zhu and Anderson 2002).

Existing models use the same basic approach, calculating thermodynamic equilibrium state of a specified system that includes water, solutes, surfaces, and solid and gas phases. These models comprise four components:

1) *Input*: specific information that defines the system under consideration, such as concentration of solutes, temperature, partial pressure of gases, and composition of solid phases;
2) *Equations* that are to be solved by the model;
3) *Equilibrium* and *kinetic formulations* between solutes of interest;
4) *Output*: in tabular or graphic form.

The computer codes require initial input constraints that generally consist of water chemistry analyses along with, units used for chemical measurements, temperature, dissolved gas content, *pH*, and redox potential (*Eh*). The models convert the chemical concentrations, usually reported in wt./wt. or wt./volume units such as $mg\ kg^{-1}$ or $mg\ l^{-1}$ to *moles*, and then solving a series of simultaneous non-linear algebraic equations (involving chemical reactions, charge balance and mass balance equations) to determine the activity–concentration relationship for all the chemical species in a specified system. The models usually require electro-neutrality conditions and impose the charge balance condition with one of the designated components, as they solve a set of non-linear equations formulated to describe the problem. The capabilities of modern codes include calculation of pH and Eh, speciation of aqueous species, equilibration with gases and minerals, oxidation and reduction reactions (redox), kinetic reactions, and reactions with surfaces.

The non-linear algebraic equations are solved using an iterative approach by the Newton-Raphson method (Bethke 1996). The equations to be solved are drawn from a database that contains equations in the standard notations of chemical mass action. In principle, any reaction such as sorption of solutes onto surfaces that can be represented in this form can be incorporated into the model. Reactions are assumed to reach equilibrium state (the point of lowest free energy in the system) when there is no change in the concentration of reactants and products.

Kinetic reactions, that involve time, are included by assuming that chemical reactions proceed to reach the equilibrium at a specified rate. Examples of kinetic reactions include mineral dissolution and precipitation, redox reactions, microbial growth, and metabolism of solutes. The rate laws used in the codes vary but all codes with kinetic capabilities include simple first-order rate laws and may include more complex rate formulations such as cross-affinity, Michaelis-Menten, and Monod formulations (Bethke 1996).

Computer models are divided into two basic types: speciation models and reaction-path models. In both cases, the models are fundamentally static, that is, there is no explicit transport function. However, some forms of transport can be simulated by manipulation of the models. More complex reaction-transport models that explicitly incorporate transport are briefly described below.

All equilibrium models are speciation models in that they can calculate the speciation (distribution) of aqueous species for any element or compound included in the database.

Speciation models calculate activities (chemically reactive concentration), species distribution

for elements in the database, saturation indices and ion ratios for specified conditions of pH, and redox potential (ORP or Eh). Most of the models allow choice of the method employed for activity calculation (Davies, Debye-Hückle, extended Debye-Hückle, Pitzer). Some models incorporate surface reactions such as adsorption and multiple kinetic formulations. Only one model, PHREEQC, has provision for inverse modelling option. This feature uses mass balance constraints to calculate the mass transfer of minerals and gases that would produce the final water composition, given a specified starting water composition (Garrels and Mackenzie 1967). This method does not model mass transport but calculates and provides only statistical measures of fit to obtain possible solutions to the mass balance between starting and final water compositions.

The next step in the complexity is the reaction path (mass transfer) models. These models use speciation calculation as the starting point and then make forward predictions of changes along the specified reaction path (specified changes in T, P, pH, and addition of new reactants such as another fluid or solid). The program makes small incremental steps with step-wise addition or removal of mass (dissolution or precipitation) and can also include changes in temperature or pressure along the reaction path.

There are, however, limitations associated with any of the above models. The input field data may become corrupted due to bad analysis, missing parameters, or violation of electro-neutrality. Speciation models assume equilibrium conditions, which may not be the case in real-life situations. The databases are also a source of uncertainty. They do not always contain all the elements or species of interest, the data invariably has some uncertainty, and some data may be inaccurate (Drever 1997). Some of the available codes attempt to minimize this problem by including popular databases such as the MINTEQ database (EPA-approved database on metals), WATEQ (USGS database specifically on minerals), and the LLNL database (the most complete database available, which is compiled and maintained by the Lawrence Livermore National Laboratory). For environmental applications, lack

Table 6.14 A comparison of capabilities and features of select computer codes commonly used in geochemical modelling. Adapted from Thyne (2005).

Program	Source	Speciation	Reaction Path	Tabular output	Graphic output	Surface Rxns.	Kinetics	Inverse	Transport	Multiple databases
EQ3/6	LLNL	yes	yes	yes	no	no	yes	no	no	no
GWB	Rock-ware*	yes	yes	yes	yes	yes	yes	no	yes	yes
HYDRO-GEOCHEM2	SSG*	yes	yes	yes	no	yes	yes	no	yes	no
MINTEQ	EPA	yes	no	yes	no	yes	no	no	no	no
MINEQL+	ERS*	yes	no	yes	some	no	no	no	no	no
PHREEQC	USGS	yes	yes	yes	no	yes	yes	yes	yes	yes

* - Commercial programs, others are freeware.
EQ3/6 – http://www.llnl.gov/IPandC/technology/software/softwaretitles/eq36.php
GWB – Rockware – http://www.rockware.com
HYDROGEOCHEM – http://www.scisoftware.com/environmental_software/software.php
MINEQL+ – http://www.mineql.com/
MINTEQ – http://soils.stanford.edu/classes/GES166_266items%5Cminteq.htm
PHREEQC – http://wwwbrr.cr.usgs.gov/projects/GWC_coupled/phreeqc/

of adequate data for organic compounds remains a matter of concern.

Other limitations include the redox reactions that are of particular importance in metal transport. These reactions are difficult to model correctly since redox reactions may have different rates, producing natural systems that are not in equilibrium (Lindberg and Runnels 1984). This problem can be addressed by modelling redox reactions as rate-limited (kinetic) formulations if appropriate data are available.

Some of the commonly used programs in geochemical modelling, their sources, and some of their useful capabilities are listed in Table 6.14.

6.6 Chemical tracers

In many cases, water pollution is an indirect consequence of human activities in the sense that the polluting material is not being directly let out into the receiving water as a waste effluent. One such form of indirect pollution of water is the intrusion of seawater into the coastal aquifers. This may be considered as an example in which both hydrologic and chemical knowledge are essential in controlling the pollution, and chemical species often act as tracers.

6.6.1 Seawater intrusion

Along the sea coast there is a saltwater–freshwater contact zone, both in streams and in aquifers that extend under the seabed. The relation between sea water and fresh water in aquifers along the sea coast is generally described by the hydraulic relationship known as the Ghyben-Herzberg equation (see Section 3.6.4, in Chapter 3; Eqn. 3.51). The effect of excessive withdrawal of groundwater from the landward parts of these aquifers may have far-reaching effects on the position of the saltwater–freshwater interface.

Hydrologists generally consider that the boundary between fresh water and salt water in coastal aquifers depends on the balance of forces in a dynamic situation. Normally, there is a continuous seaward movement of fresh water at a rate that is related to the head above mean sea level in the freshwater aquifer. Cooper (1959) described the movement of fresh- and salt water along a contact zone, which tends to produce a diffuse zone of mixing rather than a sharp interface as predicted by the Ghyben-Herzberg relation. So long as a high head of fresh water is maintained inland in the aquifer, freshwater discharge maintains the zone of contact with saline water in the aquifer up to a considerable distance offshore. Pumping the inland aquifers reduces the head of the fresh water and as head changes are transmitted rapidly to maintain the hydrostatic equilibrium, the seaward flow of fresh water decreases. The head may decline accordingly to completely stop the seaward flow of fresh water past the interface. With reduced freshwater flow, the system becomes unstable and salt water invades the aquifer. The saltwater front moves inland to the point where the reduced freshwater head is sufficient to produce a balancing seaward movement of fresh water past the interface.

Overdevelopment of coastal aquifers can greatly decrease the freshwater head and can bring about conditions favourable for the intrusion of salt water inland. The migration of the saltwater front, however, is rather slow, as it represents actual movement of water in the system under low hydraulic gradients against high resistance offered by the aquifer material. The appearance of salty water in a well may, however, not occur until some years after the head decline in the coastal aquifers has reached serious proportions.

The rate of movement of some of the ions in the saltwater front is also influenced by ion exchange, and diffusion and head fluctuation will cause the interface to become a broad transition zone rather than a sharp front as given by the Ghyben-Herzberg relation. Saltwater intrusion into highly developed aquifers is a serious problem in many places along continental margins. Hydrologists are frequently confronted with the need to recognize incipient stages of saltwater intrusion so that steps can be taken to remedy the situation in time.

Chloride is the major anion in sea water, which moves through an aquifer at nearly the same rate as the intruding sea water. Increase in chloride concentrations with time may well be the first indication of the onset of a seawater intrusion front. In an area where no other known source of saline water contamination exists, high chloride concentrations in groundwater can be considered definite evidence for seawater contamination. However, if significant amounts of chloride are contributed by other sources, establishment of unambiguous proof of the seawater source may be difficult. Components of sea water other than chloride may also be used to identify contamination but

with some caution. Magnesium is present in sea water in much greater concentration than calcium. A low calcium/magnesium ratio may sometimes be indicative of seawater contamination. Presence of sulphate in anionic proportions similar to that of sea water may also be indicative. Because of possible cation exchange reactions and sulphate reduction in the aquifers that can be expected to occur when sea water is introduced, the proportions of anions and cations in the first contaminated water to reach the sampling point cannot be expected to be exactly the same as those in a simple mixture of sea water and fresh water. It is indeed likely that, even after moving only for a short distance through an aquifer, the water in the advancing saltwater front will have superficial resemblance to a simple mixture of sea water and groundwater. An example of cation exchange altering the Ca/Na balance in sea water intruded in the coastal karstic limestone aquifer from part of Saurashtra coast of Gujarat, India, was reported by Desai et al. (1979).

In some instances, minor constituents of sea water may aid in determining whether a particular aquifer has been contaminated by sea water or by some other saline source. Incipient stages of contamination are, however, difficult to detect by these constituents. Piper et al. (1953) reported some success in differentiating seawater contamination of an aquifer from contamination by connate brine by comparing iodide, boron, and barium concentrations in the suspected source of contamination. The constituents that might be useful in identifying the contributing sources should be selected by using knowledge of the composition of contaminating solution and by considering chemical- and exchange behaviour of the solutes.

6.6.2 Injected chemical tracers

An introduction to hydrologic tracers is given in Section 7.2. Various kinds of 'chemical' measuring techniques suitable for surface streams have been developed (Corbett et al. 1945; Rantz et al. 1982). A slug injection technique has been applied extensively to determine the rate of solute movement or time of travel of water through a given reach of a river. This estimate cannot be made accurately from stream-discharge data. The procedure involves adding a readily detectable solute to the stream in the form of a concentrated slug and the length of time required for the material to appear at a downstream point is measured. An index of time-of-travel studies by US Geological Survey was compiled by Boning (1973).

The salt-dilution method of measuring the flow rate of a stream consists of adding a known quantity of tracer, usually sodium chloride, at a constant rate and measuring its concentration in the water upstream from the point of addition, as well as far enough downstream so that mixing of the tracer in the stream is complete. The flow must remain reasonably constant during the interval while the measurements are made and enough time must be allowed so that the tracer concentration at the downstream point becomes stable (i.e. time invariant). This provides enough data for calculating discharge rate of the stream. The method can be used in streams in which other procedures are not possible because of inaccessibility or extreme conditions due to the presence of turbulence or during a flood. Amounts of salt added should not be excessively large, as it adversely affects water quality (see also Section 2.6.1.4).

In areas where interconnection between surface water and groundwater systems is of interest, detailed studies often include seepage estimation. These consist of a series of measurements of river flow and tributary inflow taken in a downstream direction, with gains/losses between measuring points being ascribed to groundwater inflows into the stream/losses of water from the stream to the groundwater reservoir. However, if water samples are taken at appropriate measuring sites and analysed, regions of inflow and outflow may be better identified to understand the hydrologic system. Data of this kind were used, for example, to help evaluate stream-aquifer interconnections along Sabarmati River in Gujarat, India (Bhandari *et al*. 1986).

Several investigations of direction and rates of groundwater movement have been made by injecting slugs of salt, dye, or radioactive material (Kaufman and Orlob 1956). Any tracer material that may be added must be similar in density and temperature to that of the groundwater. Large amounts of tracer might constitute unacceptable levels of pollution. In aquifers where groundwater movement is through large fissures or cavernous openings, a tracer technique becomes simple and easy to interpret. Organic dyes have been used to trace water movement through limestone aquifers and to identify pollution sources in such systems. Kaufman and Orlob (1956) observed that chloride ions seemed to move at effectively the same rate as water through porous material.

Materials that are naturally present in groundwater are also potentially useful for tracing groundwater flow and estimating rates of its movement. Use of environmental radioisotopes for studying groundwater movement is discussed in the Section 7.3.

6.7 Groundwater – numerical modelling of solute transport

As discussed in Chapter 3, the process of groundwater flow is governed by Darcy's law (Eqn. 3.24) and the law of conservation of mass (Eqn. 3.25). The purpose of a model that simulates solute transport in groundwater is to compute the concentration of a dissolved chemical species in an aquifer at any given time and place. Theoretical basis for the equation describing solute transport is well documented in the literature (Bear 1979; Domenico and Schwartz 1998). A conceptual framework for analysing and modelling physical solute-transport processes in groundwater has been provided by Reilly *et al*. (1987). Changes in chemical concentration occur in a dynamic groundwater system, primarily due to four distinct processes:

1) *Advective transport*, in which dissolved chemicals move with the flowing groundwater;
2) *Hydrodynamic dispersion*, in which molecular and ionic diffusion and small-scale variations in the flow velocity through the porous media cause the paths of dissolved molecules and ions to diverge or spread from the average direction of groundwater flow;
3) *Fluid sources*, where water of one composition is introduced and mixed with water of a different composition.

4) *Reactions*, in which some amount of a particular dissolved chemical species may be added to or removed from the groundwater as a result of chemical, biological, and physical reactions taking place in water or between the water and the solid aquifer materials or other separate liquid phases (e.g. NAPL).

6.7.1 Governing equations

A general form of the equation (Eqn. 3.36) describing the transient flow of a compressible fluid in a non-homogeneous anisotropic aquifer can be derived by combining Darcy's law with the continuity equation as:

$$S_s \frac{\partial h}{\partial t} + W = \frac{\partial}{\partial x}\left(K_x \frac{\partial h}{\partial x}\right) + \frac{\partial}{\partial y}\left(K_y \frac{\partial h}{\partial y}\right)$$
$$+ \frac{\partial}{\partial z}\left(K_z \frac{\partial h}{\partial z}\right) \qquad (6.46)$$

An equation describing the transport and dispersion of a dissolved chemical in flowing groundwater may also be derived from the principle of conservation of mass that requires that the net mass of solute entering or leaving a specified volume of aquifer during a given time interval must equal the accumulation or loss of mass stored in the volume during the interval. This relationship may then be expressed mathematically by considering all fluxes into and out of a representative elementary volume (REV). A generalized form of the solute-transport equation was presented by Grove (1976), in which appropriate terms are incorporated to represent chemical reactions and solute concentration both in the pore fluid and on the solid surfaces, as:

$$\frac{\partial (\varepsilon C)}{\partial t} = \frac{\partial}{\partial x_i}\left(\varepsilon D_{ij} \frac{\partial C}{\partial x_j}\right) - \frac{\partial}{\partial x_i}(\varepsilon C V_i)$$
$$- C'W^* + CHEM \qquad (6.47)$$

where CHEM equals:

$$-\rho_b \frac{\partial \bar{C}}{\partial t}$$

for linear equilibrium controlled sorption or ion-exchange reactions,

$$\sum_{k=1}^{s} R_k$$

for s chemical rate-controlled reactions, and (or)

$$-\lambda\left(\varepsilon C + \rho_b \bar{C}\right) \text{ for decay}$$

where ε is the effective porosity; D_{ij} is the coefficient of hydrodynamic dispersion (L^2T^{-1}); C' is the concentration of the solute in the source or sink fluid; $\bar{C}$ is the concentration of the species adsorbed on the solid (mass of solute/mass of solid); ρ_b is the bulk density of the sediment [ML^{-3}]; R_k is the rate of production of the solute in reaction k [$ML^{-3}T^{-1}$]; and λ is the decay constant (equal to $ln\, 2/T_{1/2}$) [T^{-1}] (Grove 1976).

The first term on the right-hand side of Eqn 6.47 represents change in the concentration due to hydrodynamic dispersion. This expression is analogous to Fick's Law describing the diffusive flux. This Fickian model assumes that the driving force is the concentration gradient and that the dispersive flux occurs in a direction from higher concentration towards lower concentration. The second term represents advective transport and describes movement of solutes at the average seepage velocity of groundwater. The third term represents effect of mixing with a source fluid that has a concentration different from that in the groundwater at the location of recharge or injection. The fourth term lumps together all of the chemical, geochemical, and biological reactions that cause transfer of mass between the liquid and solid phase or conversion of dissolved chemical species from one form to another. Chemical attenuation of inorganic chemicals can occur by sorption/desorption, precipitation/dissolution, or oxidation/reduction. Organic chemicals can adsorb or degrade by microbiological processes. There has been considerable progress in modelling these reactions. However, a comprehensive review of the reaction processes and their representation in transport models is beyond the scope of this book.

If reactions are limited to equilibrium-controlled sorption or exchange and first-order irreversible rate (decay) reactions, then the general governing equation (Eqn 6.47) can be written as:

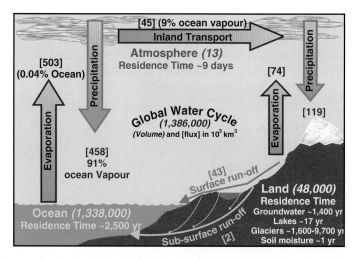

Plate 1.3 The global water cycle, involving water in all three of its phases – solid (as snow/ice), liquid, and gas – operates on a continuum of temporal and spatial scales, and exchanges large amounts of energy as water undergoes phase changes and moved dynamically from one part of the Earth system to another.

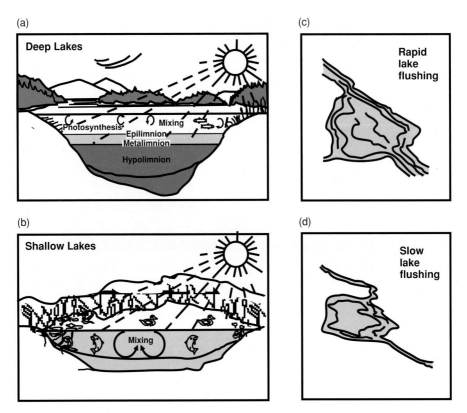

Plate 2.1 (a) Wind and solar radiation driven seasonal stratification of water in deep lakes. (b) Shallow lakes do not show such depth stratification. The ratio of lake volume to inflowing volume determines if the lake is (c) rapidly flushing or (d) slowly flushing. Redrawn after Michaud (1991). © Washington State Department of Ecology.

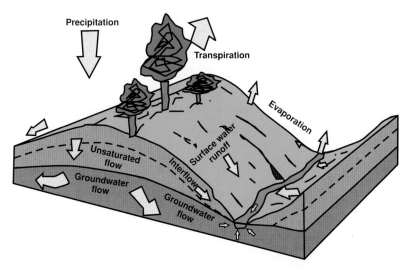

Plate 3.15 Partitioning of precipitation falling on land and the interconnection between the surface- and groundwater.

Plate 7.4 (a) Photograph of a specially designed foldable stand with a conical aluminium base, which holds a high density PVC bag filled with 100 litres of water sample. The supernatant water is decanted by piercing the bag after the carbonate precipitates settle in the conical base of the bag; (b) Carbonate precipitates are transferred from the PVC bag into a 1.2-litre soda-lime glass bottle without exposure to the atmosphere.

(a)

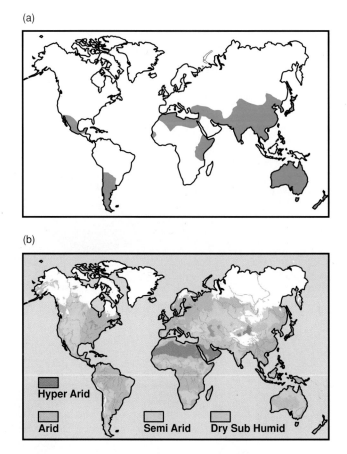

(b)

Hyper Arid

Arid   Semi Arid   Dry Sub Humid

Plate 7.11 (a) Global distribution of areas affected by endemic fluorosis. Redrawn from
http://www.unicef.org/programme/wes/info/fl_map.gif. (b) Geographical distribution of arid and semi-arid regions. Redrawn
from www.wateryear2003.org/en/ev.php-URL_ID=5137. It may be noted that most fluoride affected areas are in the arid and
semi-arid regions.

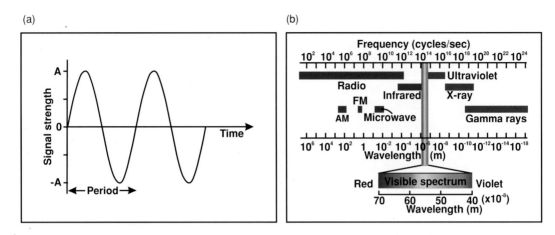

Plate 9.1 (a) Sinusoidal waveform of electromagnetic radiation (EMR). Frequency (v) is the number of oscillations per unit time, and wavelength (λ) is the distance travelled in time taken for one oscillation; (b) electromagnetic spectrum with increasing frequency is shown from left to right. Also included are the terms used for the main spectral regions. The wavelength scale shown at the bottom in this representation of EM spectrum increases from right to left. Redrawn from http://rst.gsfc.nasa.gov/Intro/Part2_2a.html accessed on 18th Nov., 2008.

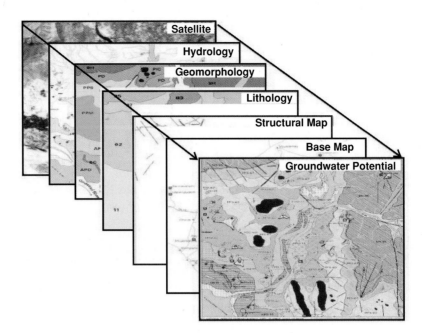

Plate 9.12 Schematic showing a typical sequence of remote sensing and GIS application to groundwater. Starting with the satellite images and various derived geo-spatial inputs, groundwater prospecting map is generated in the GIS environment. Redrawn from Navalgund *et al.* (2007) with permission from Current Science.

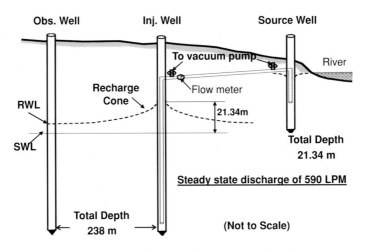

Plate 11.5 Schematic illustration of an injection groundwater recharge system at Ahmedabad, India. The Sabarmati riverbed sand was used to filter the river runoff water through a shallow depth source well in the river bed. The hydrostatic head difference between the supply well and the deep injection well on the river bank was used to transfer water between the two wells using a siphon.

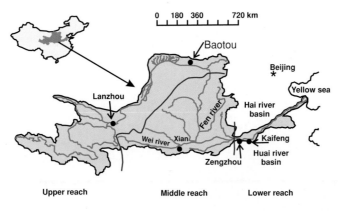

Plate 13.1 The Yellow River Basin, China. By Chinese convention, the Yellow River is divided into three reaches, namely, upper, middle, and lower, as demarcated by red lines. Redrawn from Giordano *et al.* (2004) © Comprehensive Assessment Secretariat.

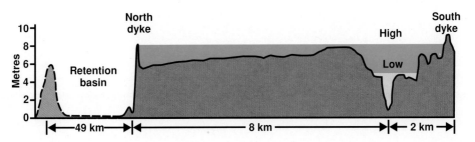

Plate 13.2 A representative cross-section of the 'suspended' Yellow River. The diagram is not drawn to scale. Redrawn from Ronan (1995) © Cambridge University Press.

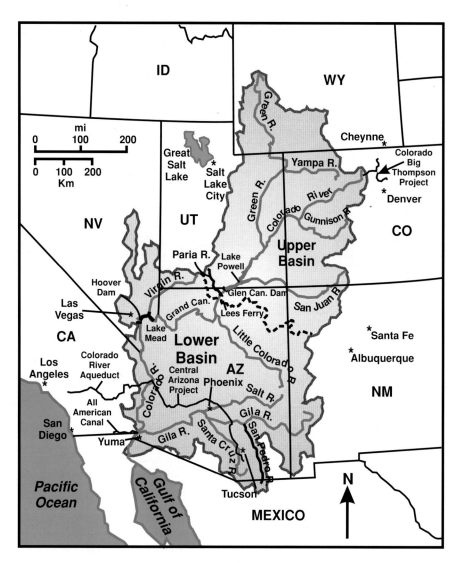

Plate 13.7 Colorado River Basin, Southwestern United States. Redrawn from NRC (2007)
© International Mapping Associates.

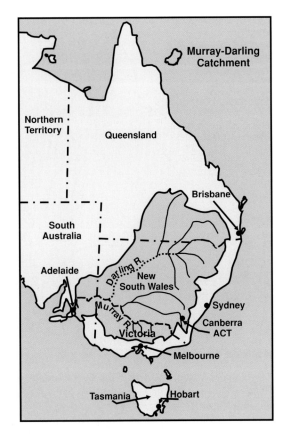

Plate 13.12 Location map of the Murray-Darling Basin.
Redrawn from http://en.wikipedia.org/wiki/Murray Darling_Basin.

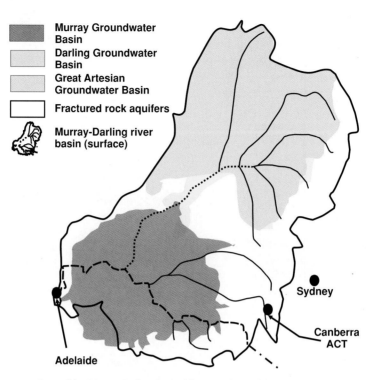

Murray Groundwater Basin

Darling Groundwater Basin

Great Artesian Groundwater Basin

Fractured rock aquifers

Murray-Darling river basin (surface)

Sydney

Canberra
ACT

Adelaide

Plate 13.14 Groundwater regions of the Murray-Darling Basin. The map shows the Murray Groundwater Basin, the portion of the Great Artesian Groundwater Basin within the MDB, the shallow aquifers of the Darling River Groundwater Basin, and the areas of fractured rock aquifers. *Source*: MDBC (1999).

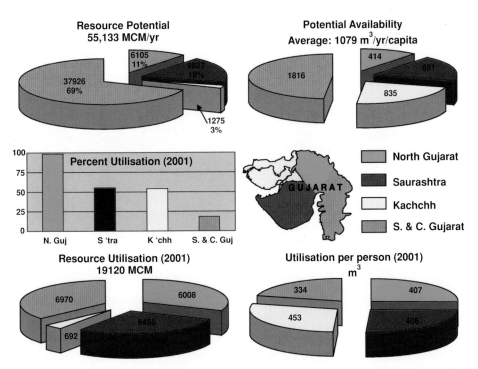

Plate 13.15 Water resource scenario in the State of Gujarat, India. In terms of per capita natural endowment of water, North Gujarat is the most stressed region, with nearly 100% of its resource already being utilized. The South and Central regions, with the Narmada, Mahi, and Tapi rivers passing through them, have surplus water that is being/will be transferred to water-deficient regions through a network of interlinked canals.

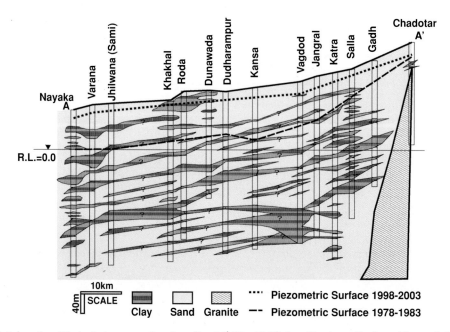

Plate 13.17 Sub-surface lithological cross-section along line AA′ (Fig. 13.16) from Nayaka to Chadotar. The sandy layers forming aquifer horizons are seen to be laterally continuous and vertically interspersed with thin semi-permeable clay/silt layers that may not have lateral continuity over a large area. The uncertainty in the continuity of horizons in view of their large separation is indicated by the symbol "?". Tube wells tap all the water-bearing horizons up to their maximum depth. Redrawn from Gupta *et al.* (2005a). Reproduced with kind permission of Springer Science+Business Media.

$$\frac{\partial C}{\partial t} + \frac{\rho_b}{\varepsilon}\frac{\partial \bar{C}}{\partial t} = \frac{\partial}{\partial x_i}\left(D_{ij}\frac{\partial C}{\partial x_j}\right) - \frac{\partial}{\partial x_i}(CV_i)$$

$$+ \frac{C'W^*}{\varepsilon} - \lambda C - \frac{\rho_b}{\varepsilon}\lambda\bar{C} \qquad (6.48)$$

Any temporal change in sorbed concentration in Eqn 6.48 can be represented in terms of the solute concentration using the chain rule of calculus, as follows:

$$\frac{d\bar{C}}{dt} = \frac{d\bar{C}}{dC}\frac{\partial C}{\partial t} \qquad (6.49)$$

The quantities $d\bar{C}/dC$ as well as $\bar{C}$ are functions of C alone for equilibrium sorption and exchange reactions. Therefore, the equilibrium relation for $\bar{C}$ and $d\bar{C}/dC$ can be substituted into the governing equation to reduce the partial differential equation in terms of only C. The resulting single transport equation is solved for solute concentration. Sorbed concentration can then be calculated using the equilibrium relation. The linear-sorption reaction considers that the concentration of solute sorbed onto the porous medium is directly proportional to the concentration of the solute in the pore fluid, according to the relation:

$$\bar{C} = K_d C \qquad (6.50)$$

where K_d is the distribution coefficient [L^3M^{-1}]. This reaction is assumed to be instantaneous and reversible. The curve relating sorbed concentration to dissolved concentration is known as an *isotherm*. If this relation is linear, the slope (given by the derivative) of the isotherm, $d\bar{C}/dC$, is known as the equilibrium distribution coefficient, K_d. Thus, in the case of a linear isotherm:

$$\frac{d\bar{C}}{dt} = \frac{d\bar{C}}{dC}\frac{\partial C}{\partial t} = K_d\frac{\partial C}{\partial t} \qquad (6.51)$$

After substituting and rewriting Eqn 6.48:

$$\frac{\partial C}{\partial t} + \frac{\rho_b K_d}{\varepsilon}\frac{\partial C}{\partial t} = \frac{\partial}{\partial x_i}\left(D_{ij}\frac{\partial C}{\partial x_j}\right) - \frac{\partial}{\partial x_i}(CV_i)$$

$$+ \frac{C'W^*}{\varepsilon} - \lambda C - \frac{\rho_b K_d}{\varepsilon}\lambda C \qquad (6.52)$$

Defining a dimensionless retardation factor, R_f, as:

$$R_f = 1 + \frac{\rho_b K_d}{\varepsilon} \qquad (6.53)$$

and substituting this relation into Eqn 6.52:

$$R_f\frac{\partial C}{\partial t} = \frac{\partial}{\partial x_i}\left(D_{ij}\frac{\partial C}{\partial x_j}\right) - \frac{\partial}{\partial x_i}(CV_i)$$

$$+ \frac{C'W^*}{\varepsilon} - R_f\lambda C \qquad (6.54)$$

As R_f is constant under these assumptions, solution to this governing equation is identical to the solution of the governing equation with no sorption, except that the velocity, dispersive flux, and source strength are reduced by a factor, R_f. The transport process thus appears to be 'retarded' because of instantaneous equilibrium sorption onto the particle surfaces in the porous medium.

In the conventional formulation of the solute-transport equation (Eqn 6.47), the coefficient of hydrodynamic dispersion is defined as the sum of mechanical dispersion and molecular diffusion. The mechanical dispersion is a function of both the intrinsic properties of the porous medium (such as heterogeneities in hydraulic conductivity and porosity) as well as of the fluid flow. Molecular diffusion in a porous medium differs from that in free water because of the effects of porosity and tortuosity. These relations are commonly expressed as:

$$D_{ij} = \alpha_{ijmn}\frac{V_m V_n}{|V|} + D_m \qquad i, j, m, n = 1, 2, 3$$

$$(6.55)$$

where α_{ijmn} is the dispersivity of the porous medium (a fourth-order tensor) [L]; V_m and V_n are components of the flow velocity of the fluid in the m and n directions, respectively [LT^{-1}]; D_m is the effective coefficient of molecular diffusion [L^2T^{-1}]; and $|V|$ is the magnitude of the velocity vector [LT^{-1}], defined as (Bear 1979; Domenico and Schwartz 1998; Scheidegger 1961):

$$V = \sqrt{V_x^2 + V_y^2 + V_z^2}$$

The dispersivity of an isotropic porous medium can be defined by two constants – the longitudinal dispersivity of the medium, α_L, and the transverse dispersivity of the medium, α_T. These are related to the longitudinal and transverse dispersion coefficients by $D_L = \alpha_L|V|$ and $D_T = \alpha_T|V|$. Most applications of transport models to groundwater problems are based on this conventional formulation.

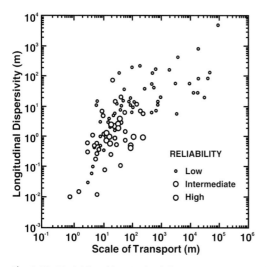

Fig. 6.11 Model-fitted longitudinal dispersivity in saturated media versus scale of modelled plume. Macro-dispersivities were determined by calibrating solute transport models with the observed solute plumes. Reliability classification follows original diagram of Gelhar *et al*. (1992). © American Geophysical Union.

Although conventional theory holds that α_L is generally an intrinsic property of the aquifer, it is found in practice to be proportional to the scale of the measurement (Fig. 6.11). But this trend is much less evident when reliability of data (Fig. 6.11) is considered (Gelhar *et al*. 1992). Most reported values of α_L fall in the range from 0.01 to 1.0 times the scale of the measurement, although the ratio of α_L to scale of measurement tends to decrease at larger spatial scales (Anderson 1984; Gelhar *et al*. 1992). Field-scale dispersion (commonly called macro-dispersion) results from large-scale spatial variations in hydraulic properties. Consequently, use of relatively large values of dispersivity together with uniform hydraulic properties (*Kij* and ε) is inappropriate for describing transport in geological systems (Smith and Schwartz 1980). If a model applied to a system having variable hydraulic conductivity uses mean values and thereby does not explicitly represent the variability, it is likely that the model calibration will yield values for the dispersivity coefficients that are larger than what would be measured locally in the field. Similarly, representing the transient flow field by a mean steady-state flow field, as is commonly done, inherently ignores some of the variability in velocity and must be compensated

for by using higher values of dispersivity (primarily transverse dispersivity) (Goode and Konikow 1990). Overall, the higher the accuracy with which a model can represent or simulate the true velocity distribution in space and time, the uncertainty concerning representation of dispersion processes will be correspondingly a smaller problem.

A special form of the solute-transport equation can be used for direct simulation of groundwater ages (Goode, 1996, 1999). This is accomplished by adding a zero-order growth term, which represents production of the solute [$ML^{-3}T^{-1}$] within the system itself. In developing an age transport equation, concentrations are replaced with corresponding ages representing a volume-averaged groundwater age in the aquifer and the zero-order growth rate having a value equal to unity. Decay and sorption reactions are assumed to be absent and, in general, the age of incoming water (analogous to C') is specified as zero. This type of analysis allows a direct comparison of groundwater modelling results with measured environmental tracer data, while accounting for effects of dispersion and other transport processes (see Section 7.6).

6.8 Relation between use and quality of water

An obvious purpose of water quality investigation is to determine if a given water supply is satisfactory for the intended use(s). Some discussion on water quality, in terms of physical and chemical parameters and pathogens, is given in Section 1.2. Standards for water meant to be used for drinking and other domestic purposes have been established in many countries. Published literature contains tolerance levels and related data for constituents of water to be used for agriculture, in industry, for development of fisheries, and for a number of other specific purposes. Water that is meant to be used for domestic supply may be employed for many purposes. Therefore, the standards used to evaluate the suitability of water for public supplies are generally more stringent than those applied to water for a small domestic or farm supply. Water from zones of mineralization and hot springs is used medicinally in many places and the

Table 6.15 Web links with details of drinking water quality standards and guidelines.

WHO	http://www.who.int/water_sanitation_health/dwq/gdwq3rev/en/index.html
USA	http://www.epa.gov/safewater/standards.html
India	http://www.chennaimetrowater.tn.nic.in/qualitymainpage.htm
WHO/EU	http://www.lenntech.com/WHO-EU-water-standards.htm

mystic qualities of natural hot springs have been of great interest to man since prehistoric times.

6.8.1 Domestic uses and public supplies

Besides being chemically safe for human consumption, water for domestic use should be free of undesirable physical properties such as colour or turbidity and should not have an unpleasant taste or odour. Harmful micro-organisms should be virtually absent, even though these are not usually considered in routine chemical analyses. Presence of harmful micro-organisms is considerably more difficult to ascertain than other properties of water, but they are of utmost concern. Additional risks arise from toxic chemicals and radiological hazards.

Mandatory standards for dissolved constituents believed to be harmful to humans were first established in the United States in 1914 by the US Public Health Service. In 1974, the federal Safe Drinking Water Act was legislated and standards for concentration of inorganic constituents in public water supplies became effective in 1977. Presently the United States has one of the safest water supplies in the world, but drinking water quality is still an issue of concern for human health in developing as well as many of the developed countries worldwide. Drinking water quality varies from place to place, depending on condition of the source from which it is drawn and the treatment it receives prior to supply. WHO has stipulated international norms on water quality and human health in the form of guidelines that are used worldwide as a basis for setting up regulatory guidelines in developing as well as in developed countries. For example, WHO guidelines for drinking water are used as a basis for the standards in the Drinking Water Directive in European Union, but with some differences. In India, the Bureau of Indian Standards (BIS) has notified standard drinking water specifications through BIS

10500: 1990. The web links in Table 6.15 can be accessed to obtain details of some of the drinking water quality standards and guidelines.

The limiting concentrations of radioactive substances in drinking water are viewed somewhat differently from those of non-radioactive solutes. It is generally agreed that the effects of radioactivity are harmful, and unnecessary exposure should be avoided. Strontium-90 is a fission product, but radium occurs naturally. Both nuclides are preferentially absorbed in bone structure and are, therefore, especially undesirable in drinking water.

6.8.2 Agricultural use

Water required for non-domestic purposes on farms includes that consumed by livestock and for irrigation. Water for livestock is subject to similar quality considerations as those related to drinking water for human consumption. Most animals, however, can tolerate water that has a considerably higher concentration of dissolved solids than that which is considered safe for humans.

The chemical quality of water is an important factor to be considered in evaluating its usefulness for irrigation. Features of the chemical composition that need to be considered include concentration of total dissolved matter, concentrations of certain potentially toxic constituents, and relative proportions of some specific constituents. Suitability of particular water for irrigation also depends on many factors not directly associated with composition of water. A brief discussion of some of these factors is given here to highlight the complexity of the problem for deciding whether or not given water is suitable for irrigation.

Part of the irrigation water that is actually consumed by plants or evaporated is virtually free of dissolved material. Growing plants selectively retain some nutrients and a part of the mineral matter originally dissolved in the water, but the amount of major cations and anions thus retained is only a small part of their total content in the irrigation water. Eaton (1954) showed that this consists mostly of calcium and magnesium salts. The bulk of the soluble material originally present in irrigation water stays behind in solution in residual water. Concentration of solutes in soil moisture cannot be allowed to rise too high, in order to avoid interference with the osmotic process by which plant root membranes assimilate water together with nutrients. Some compounds of low solubility, especially calcium carbonate, which is virtually harmless, may precipitate in the soil as solute concentrations increase, but the bulk of the residual solutes must be managed effectively to maintain productivity of irrigated soils.

The extent and severity of salt accumulation problems in irrigated areas depend on several factors. These include: (i) chemical composition of the water supply; (ii) nature and composition of the topsoil and subsoil; (iii) topography of the land; (iv) amount of water used; (v) method of irrigation employed (flooding the fields/sprinkler/drip); (vi) types of crops grown; (vii) climate of the region, especially the amount and distribution of rainfall; and (viii) groundwater conditions (depth to water table, quality) and nature of surface-water drainage system.

In most areas, excess of the soluble material left in the soil from previous irrigation is removed by leaching of the topsoil and percolation below the root zone of a part of the resulting solution into the groundwater reservoir. In areas where the water table beneath the irrigated land can be kept sufficiently below the surface, this process of drainage is reasonably effective. The leaching may be accomplished by rainfall in areas where precipitation is sufficient to saturate a large depth of soil. Leaching of soluble salts also occurs during irrigation when an excess amount of water is added, with the aim to store the extra supply of water in the soil or to use up the surplus amount of water that happens to be available at a particular time. The need for leaching of the soil with a view to remove excess salts is generally recognized by farmers who use highly mineralized irrigation water.

For long-term successful operation of an irrigation project, all the ions present in the irrigation water and those extracted by plants must be disposed off either by carrying them away from the area or by storing them safely within the area. The net ion load in an irrigated area can be expressed in terms of the salt balance, i.e. the difference between ion inflow and outflow. Besides the general increase in major solute ion concentrations that irrigation drainage may cause in the groundwater underlying the irrigated land, there may be additions of specific solutes that are undesirable. A major problem in some irrigated regions has been the increasing concentration of nitrate in groundwater received from the drainage of irrigated fields on which nitrogenous chemical fertilizers (particularly urea) has been applied. Some types of pesticides may also persist in the drainage water.

In addition to problems of excessive concentration of dissolved solids, certain constituents in irrigation water are especially undesirable, even when present in trace concentrations. Boron, for example, is an essential plant nutrient and is sometimes added to fertilizers in small amounts because some soils in humid regions are deficient in boron. But even a small excess of boron over the plant tolerance level is toxic to some types of plants, particularly citrus fruits and walnut trees. Lithium in water in small concentrations (0.06–0.10 mg l^{-1}) has been shown to cause damage to citrus plants (Bradford 1963). Soils of high salinity interfere with crop growth and a high pH may decrease the solubility of some essential elements.

Some minor constituents of irrigation water, notably molybdenum, selenium, and cadmium, may accumulate in plant tissues and cause toxicity when the plants or their seeds are consumed by humans/animals.

The process of cation exchange also occurs in irrigated soils and may influence soil properties, especially when concentrations of solutes are high. Irrigation water with a high ratio of sodium to total cations tends to put sodium ions in the exchange positions on the soil-mineral particles. In water having mostly divalent cations, this process is reversed. In soils, clay minerals have the highest exchange capacity per unit weight. Physical properties of soils are optimal for plant cultivation and growth when their exchange sites are occupied by divalent ions of calcium and magnesium. However, when exchange positions become saturated with sodium, soils tends to become deflocculated and, therefore, impermeable to water. A soil of this type is difficult to cultivate and may not support plant growth.

The cation-exchange process is reversible and can be controlled either by adjusting the composition of the water or by using soil amendments. The condition of a sodium-saturated soil can be improved by liberal application of gypsum, which releases calcium to occupy exchange positions. The soil also may be treated with sulphur, sulphuric acid, ferrous sulphate, or other chemicals that tend to lower the *pH* of the soil solution. The lower *pH* brings calcium into solution by dissolving carbonates or other calcium minerals. The tendency of water to replace adsorbed calcium and magnesium with sodium can be expressed by the sodium-adsorption ratio (SAR):

$$SAR = \frac{(Na^+)}{\sqrt{1/2\left[(Ca^{2+}) + (Mg^{2+})\right]}} \quad (6.56)$$

where ion concentrations (in parentheses) are expressed in *meq* l^{-1}.

Two other related indices are: (i) *Soluble Sodium Proportion (SSP)* indicating proportion of sodium ions in solution in relation to the total cation concentration in water, defined as:

$$SSP = \frac{Soluble\ Sodium\ Concentration\ (meq.l^{-1})}{Total\ Cation\ Concentration\ (meq.l^{-1})}$$
$$\times 100 \quad (6.57)$$

and (ii) *Exchangeable Sodium Percentage (ESP)* in soil defined as:

$$ESP = \frac{\begin{array}{c}Exchangeable\ Sodium\\(meq.(100g)^{-1}\ soil)\end{array}}{\begin{array}{c}Cation\ Exchange\ Capacity\\(meq.(100g)^{-1}\ soil)\end{array}} \times 100 \quad (6.58)$$

Whereas SSP is an indicator of the sodium hazard from the irrigation water, ESP, on the other hand, indicates the extent to which the adsorption complex of a soil is occupied by sodium. It has been observed that where irrigation water and drainage conditions are good, the ESP value of the soil varies only slightly from season to season or year to year. This implies that the cation exchange material of the soil has reached a steady state relative to the cations in the soil solution, which are derived from the irrigation water. Under such conditions, an empirical relation (Eqn 6.59) has been observed between ESP and SAR:

$$ESP = \frac{100\ (-0.0126 + 0.01475 \times SAR)}{1 + (-0.0126 + 0.0145 \times SAR)} \quad (6.59)$$

On the basis of this relationship (Eqn 6.59), SAR appears to be a useful index for designating sodium hazard of waters used for irrigation.

Eaton (1950) suggested that if much of the calcium and magnesium originally present were precipitated, the residual water would be considerably enriched in sodium relative to the other cations. Some waters, in which the bicarbonate content is higher than the amount equivalent to the total amount of calcium and magnesium, could thus evolve into solutions containing mostly sodium and bicarbonate and would have a high *pH* and potential for deposition of sodium carbonate (commonly known as black alkali). Residual sodium carbonate (RSC) is defined as an excess of carbonate or bicarbonate that water contains after subtracting an amount equivalent to the calcium plus the magnesium, i.e.:

$$RSC = \left(CO_3^{2+} + HCO_3^-\right)$$
$$- \left(Ca^{2+} + Mg^{2+}\right)\ in\ meq.l^{-1} \quad (6.60)$$

RSC is another alternative measure of the sodium content in relation to Mg and Ca. This value may appear in some water quality reports, although it is not frequently used. Water is considered safe for irrigation if RSC <1.25 and not appropriate if the RSC >2.5.

From this brief discussion it should be evident that the relationship between water quality and suitability of water for irrigation is not simple. Further complications can arise as salinity of water increases. A diagram widely used for evaluating suitability of waters for irrigation, published by the US Salinity Laboratory (US-SL 1954), is given in Fig. 6.12. In this diagram specific conductance or electrical conductivity (*EC*), as an index of dissolved solids concentration, is plotted on one axis and the sodium-adsorption ratio on the other. The

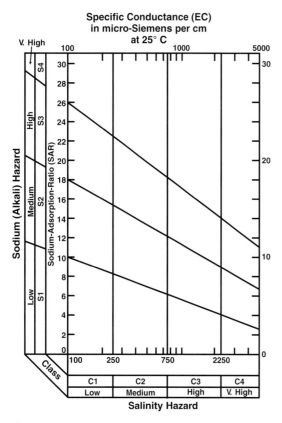

Fig. 6.12 Diagram used in interpreting the analysis of irrigation water. Redrawn from US Salinity Laboratory (US-SL 1954). © U.S. Geological Survey.

diagram is divided into 16 areas that are used to categorize/specify the degree to which a particular water source may contribute to salinity problems and undesirable ion-exchange effects in soil. Water having $EC > 5000\ \mu S\ cm^{-1}$ is also being used with some success in certain areas where proper soil conditions exist for growing suitable crops, and appropriate irrigation techniques are employed. A hydrologist needs to consider the local experience rather than arbitrarily deciding whether given water is suitable for irrigation at a given place. Salinity problems, however, may be slow to develop and may be observable only through indirect means, such as reduced crop yields or other indicators that are not easy to evaluate. A water of high salinity must always be viewed with caution until proof of it being safe for a specific use is established.

6.9 Industrial use

Quality requirements for industrial water supplies vary widely as almost every industrial application has its own norms. For some uses, such as single-pass condensing of steam or for cooling or concentrating ores, chemical quality is not particularly crucial and almost any water may be used. At the other extreme, water of nearly distilled water quality is required for processes such as manufacturing of high-grade paper or pharmaceuticals, as impurities in water can seriously impact the product quality. In nuclear reactors, water of very high purity is desirable to minimize the radioactivity induced by neutron activation of the dissolved constituents.

Technically it is possible to treat any water to make it suitable for a specific use. However, if extensive treatment is required and large volumes of water are involved, it may not be economically viable to use some of the supply sources. Industrial plants requiring large quantities of water, therefore, need to be suitably located by considering availability of water of the desired quality.

Although not a chemical property, the temperature of a water supply source and its seasonal fluctuations are major considerations in its use in industry for cooling purposes. In some areas, groundwater is used extensively for this purpose because its temperature is uniform and is below ambient air temperatures during warm weather and above ambient air temperatures during cold weather. Some industries recharge groundwater aquifers with cold water from surface streams during winter and withdraw it in the summer when the surface water is too warm for efficient cooling. In recent years the practice of discharging excessively warm water into streams is regulated to prevent ecological stress due to depletion of dissolved oxygen in water at elevated temperatures.

Much of industrial use, unlike agricultural use, is non-consumptive in the sense that the water is not evaporated or incorporated into the finished product but is discarded after a single use without significant change in its quantity, but generally with an increased load of dissolved or suspended material. As water supplies generally become more fully committed in the course of time, many industries find it necessary to conserve and reuse water which in the past would have been allowed to flow down a sewer or released into a surface stream. In several instances, reclaimed sewage is being used for certain non-critical industrial applications in terms of water quality.

6.9.1 Recreational and aesthetic uses

Considerable attention is now being paid to uses of rivers and lakes for such purposes as religious ceremonies, swimming, fishing, boating, and aesthetic and recreational purposes. Many of the surface water bodies in India are intimately linked to religious practices and periodic congregation of large numbers of devotees on the banks of streams and lakes and holy dips in rivers are a common practice. Restoring such water bodies entails enormous costs because of the high level of pollution but there is strong public support with the aim of creating or protecting them for religious purposes.

Water for swimming and other sports in which water is in contact with human skin must obviously conform to sanitary standards. Fish require clean water with a good supply of dissolved oxygen. Certain metal ions may be lethal to fish and other aquatic life forms when present at levels close to the limits specified for public water supplies. Copper, zinc, and aluminium, which are not amongst the metals for which limits are prescribed for public water supplies, are toxic to fish and many other aquatic life forms. Assimilation of dissolved metal ions by aquatic biota has a tendency of increasing concentrations in species higher up in the food chain. One of the more insidious effects of mercury-containing wastes that enter rivers and lakes is an increase in mercury content of fish, to the extent that they become dangerous for human consumption.

6.10 Tutorial

Ex 6.1 How many moles of hydrogen ions are contained in 1 litre solution of pH 10? How many grams and how many numbers of hydrogen ions are contained in the same solution?

[Ans. 10^{-10}, 10^{-10}, 6.022×10^{-13}]

Ex 6.2 What is the pH of pure water in equilibrium with the atmosphere ($p_{CO2} = 10^{-3.5}$ atm) at $25°C$?

Solution Using Eqn 6.30, $[H_2CO_3{}^*] = 10^{-1.47} \times 10^{-3.5} = 10^{-4.97}$. Now using Eqn 6.31, $[H^+][HCO_3{}^-] = 10^{-6.35} \times 10^{-4.97} = 10^{-11.32}$. Because the solution must have neutral charge, i.e. concentrations of cations and anions must balance so that $[H^+] = [HCO_3{}^-]$; $[H^+]^2 = 10^{-11.32}$. Or $[H^+] = 10^{-5.66}$; so that $pH = 5.66$. One thus expects the pH of rain to be around 5.66. In reality the pH of rain is quite variable, influenced by other solutes derived from the atmosphere.

Ex 6.3 Calculate the pH of a 0.0250 M solution of CO_2 in water.

[Hint. Use Eqn 1.31. Ans. $pH = 3:97$]

Ex 6.4 A 100-ml sample of natural water whose pH is 6.6 requires 12.2 ml of 0.10 M HCl for titration to the methyl orange end point and 5.85 ml of 0.10 M NaOH for titration to the phenolphthalein end point. Assuming that only carbonate species are present in significant quantities, find the total alkalinity, carbon dioxide acidity, and the total acidity of the water.

Solution Since the pH is below 8.3, bicarbonate is the major alkalinity species.

Total alkalinity: 12.2 mM l^{-1}.

CO_2 *acidity* (conversion of CO_2 to $HCO_3{}^-$): 5.85 mM l^{-1}

Total acidity (conversion of all carbonate species to $CO_3{}^{2-}$): Because it is impractical to carry out this titration, one can make use of the data already available. Conversion of the initial CO_2 to $HCO_3{}^-$ would require 5.85 mM l^{-1} of NaOH, and then an additional equal amount of $HCO_3{}^-$. Similarly, the 12.2 mM l^{-1} of $HCO_3{}^-$ initially present will require the same quantity of NaOH for conversion to $CO_3{}^{2-}$. The total acidity is thus (5.85 + 5.85 + 12.2) = 23.9 mM l^{-1}.

Ex 6.5 Calculate the amount of helium dissolved in air-saturated water under normal atmospheric conditions at $25°C$, given Henry's constant for oxygen at $25°C$, i.e. $H_i = 2.865 \times 10^3$ $atm/(mol\,l^{-1})$ can be calculated as follows. Under normal atmospheric conditions there is 5.24×10^{-4} *mole per cent* helium, which makes the partial pressure of helium 5.24×10^{-6} atm. Using Henry's law, Eqn 6.25, the

Table 6.16

S. No.	pH	EC μmhos. cm^{-1}	TDS mg.l^{-1}	Ca^{+2} + Mg^{+2}	Na$^+$	K$^+$	CO$_3^{-2}$	HCO$_3^{-1}$	Cl$^-$	F$^-$	SO$_4^{-2}$ ppm	SAR	RSC meq.l^{-1}	SSP %
							meq.l^{-1}							
1	8.62	1023	560	5.8	5.08	0.03	1.3	6.8	2.8	0.30	0.0			
2	9.20	3442	1910	3.1	32.87	0.01	1.1	14.2	17.5	0.34	2.6			
3	9.40	1945	1040	4.8	15.56	0.01	1.4	6.3	10.5	0.44	0.8			
4	9.00	2739	1456	4.2	26.30	0.00	1.6	10.7	14.0	0.24	1.0			
5	8.41	2006	1078	7.9	12.00	0.02	1.2	6.6	11.0	0.24	3.0			
6	8.60	3304	1715	6.7	24.87	0.02	1.4	11.9	19.5	0.26	0.0			
7	8.80	231	145	3.6	0.50	0.00	0.6	2.9	3.6	0.30	0.0			
8	8.65	1189	660	3.8	9.78	0.01	1.0	9.6	2.0	1.10	0.0			
9	8.68	614	350	4.1	4.50	0.03	1.6	4.9	1.0	0.22	0.0			
10	8.80	1620	940	12.3	7.50	0.03	0.9	4.8	8.5	0.20	0.5			
11	8.00	3281	1880	4.6	29.78	0.02	1.2	15.0	16.0	0.52	1.3			
12	8.45	1831	1070	3.5	15.30	0.02	1.3	6.3	11.0	0.39	0.8			
13	8.75	640	410	3.2	3.39	0.01	1.1	4.1	2.2	0.56	0.0			

concentration of helium is [5.24×10^{-6} atm /H_i], which is 1.83×10^{-9} mol l^{-1} or 1.83×10^{-6} mmol l^{-1}.

Ex 6.6 Calculate the amount of carbon dioxide dissolved in 1 litre of soda pop if the manufacturer uses a pressure of 2.4 atm of CO_2 to carbonate the soda pop. Given H_i for $CO_2 = 2.976 \times 10^1$ atm/(mol l^{-1}).

Ex 6.7 Estimate the amount of nitrogen that a diver must lose from his bloodstream (~ 5 l) in rising from a depth of 100 m to the surface, in order to avoid formation of nitrogen bubbles in his bloodstream. Given H_i for nitrogen $= 1.55 \times 10^8$ Pa/(mol l^{-1}).

Solution The density of water is about 1 kg l^{-1} or 1000 kg m^{-3}. A column of water 100 m thick would have a mass of 100,000 kg m^{-2} at its base, which would exert an additional force of 980,665 Nm^2 or 980,665 Pa. (The total pressure would be 980,665 kPa plus 101.325 kPa or 1081.990 kPa.)

The pressure change of 980,665 Pa would produce a concentration change of 980,665/(1.55 $\times$ 10^8) = 6.33 mmol l^{-1}. Therefore, the amount of nitrogen that must be lost is 5 × 6.33 = 31.65 mmol or >750 ml of nitrogen at room temperature and pressure. This is enough nitrogen gas to create massive bubbles in the bloodstream.

Ex 6.8 Is a diver is safer using a helium/oxygen mixture than a nitrogen/oxygen mixture when diving to longer depths or remaining submerged for longer periods of time? Given H_i for helium = 2.83×10^8 Pa/(mol l^{-1}).

Ex 6.9 Table 6.16 gives results of some chemical analyses of groundwater from Bhiloda Taluka (N. Gujarat), India. Source: Acharya et al. (2008).

Plot the data of these samples as bar charts, pie charts, Stiff diagram, Schoeller diagram, and trilinear diagram. Complete the table by computing values of SAR, RSC, and SSP. What conclusions can be drawn about irrigation water quality and other regional aspects from these analyses?

7

Hydrologic tracing

The most common measurement that hydrologists make pertains to the volumetric flow rate of water in its various physical forms at different locations and at different times – as liquid along streams, as rain or snow, as soil moisture in the soil profile, as groundwater in the subsurface, and as vapour in the atmosphere. The other important hydrologic parameter is the volume of the reservoir into which the water, in whatever form, flows in or flows out. As already mentioned in previous chapters, major objectives of various hydrologic measurements are to: (i) generate hydrographs; (ii) develop water budgets; (iii) characterize flow paths; (iv) estimate residence or storage time in the hydrologic reservoir; (v) provide insight into hydrologic processes operating at different spatial and temporal scales; and (vi) study interaction/transport across hydrologic boundaries.

While transforming hydrologic measurements to useful information for resource management, several simplifying assumptions and process models, both empirical and physical, are used. Flow measurements are limited in temporal and spatial domains by practical considerations. This has constrained the development of theories/models to small-scale catchments for short time scales. A major problem in hydrology, that still persists, is the scaling up of the inferences/results from small- to large-catchments and from short duration observations to longer periods. This is because hydrologic processes and parameters vary on wide space-time scales, leading to highly complex, strongly nonlinear catchment response due to strong interactions and feedbacks between the various processes.

Water systems are also getting ever larger and more complex and they are affected by complex patterns of land use and other anthropogenic factors. These developments require prediction, in space and time, of functioning of hydrologic systems and their impacts on future availability of water. Therefore, tools that track water movement within and across hydrologic reservoirs (i.e. hydrologic tracers) and focus on hydrologic processes over a wide range of temporal- and spatial scales are needed.

Water tracing techniques are useful, particularly due to the fact that tracing of water enables direct insight into the dynamics of surface- and subsurface water movement. As a result, tracer techniques provide useful tools to understand transport processes, phase changes (evaporation, condensation, sublimation), and genesis of water masses and their quality. Tracer techniques are particularly useful in arid and semi-arid regions for quantifying groundwater flow and water movement in the *vadose zone*. Tracer methods have become a major tool for calibration and validation in catchment modelling and for identification and quantification of runoff generation processes. Tracer techniques can be extremely useful in assessment of groundwater–surface water interactions, dating of groundwaters, quantifying water–rock interactions, and evaluating water resource vulnerability to various natural and anthropogenic factors.

Modern Hydrology and Sustainable Water Development, 1st edition. © S.K. Gupta
Published 2011 by Blackwell Publishing Ltd.

As shown subsequently, while advances are being made in all tracer techniques, isotopic tracers, in particular, comprise a large and growing family of hydrologic tracers.

7.1 Isotopes and radioactivity

Most elements found in nature consist of one or more isotopes, which are atoms of the same element having the same number of protons but different number of neutrons in their nucleus. As a result, isotopes of an element have different atomic masses. The *mass number* of an isotope is the sum of the number of protons and neutrons in its nucleus and is written as a superscript to the left of the element symbol. For example, amongst the hydrogen isotopes, deuterium (denoted as D or ^2H) has one neutron and one proton, whereas tritium (denoted as T or ^3H) has one proton and two neutrons. Isotope names are usually pronounced with the name of the element first followed by its mass number, as in 'oxygen-18' for ^{18}O instead of '18-oxygen'. In many texts, especially older ones, the mass number is shown to the right of the element abbreviation, as in C-13 or C^{13} for ^{13}C, pronounced as 'carbon-13'.

Radioactive isotopes are *nuclides* (isotope-specific atoms) that spontaneously disintegrate over time, with a characteristic *half-life*, to form other isotopes (radioactive or stable). During disintegration, radioactive isotopes emit *alpha* (α) *particles* or *beta* (β) *particles* and sometimes also *gamma* (γ) *rays*. An alpha particle is a helium nucleus that consists of two protons and two neutrons. Beta particles are indistinguishable from electrons but with the difference that they originate in the nucleus of a β emitting radioisotope. Gamma rays are electromagnetic radiation of very short wavelength. Typical gamma photon energies are several *MeV*. The *half-life* of a radioactive isotope (designated as $t_{1/2}$) is the duration of time it takes for half of the radioactive atoms in a sample to decay. The half-life of a given isotope remains invariant; it does not depend on the number of atoms or how long they have been around or on the external factors, such as temperature and other environmental parameters. Therefore, decays occur at a faster rate when there are more numbers of

radioactive atoms and the decays are fewer when there are less numbers of atoms. In fact, the number of disintegrating atoms of a radioactive substance at a given time, t, is directly proportional to the number present in the substance at that time, that is, $dN/dt = N\lambda$. The half-life is invariant but the number of atoms remaining after each successive half-life gets smaller by one-half. The decay equation (Eqn 7.1) expresses change in the concentration (activity) of a radioactive nuclide over time:

$$A_t = A_0 . e^{-\lambda t} \qquad (7.1)$$

where A_0 is the initial activity of the parent nuclide, A_t is its activity after time 't'; λ *is* the radioactive decay constant which equals $ln(2/t_{1/2})$; with $ln (x)$ defined as the natural logarithm (i.e. to the base 'e') of the variable 'x'.

The activity of a radioactive substance is given by the number of disintegrations per second of its atoms. The original unit for measuring the amount of radioactivity was the *curie (Ci)* – defined to correspond to the number of disintegrations per second of one gram of radium-226 and more recently defined as:

1 *curie* $= 3.7 \times 10^{10}$ *radioactive decays*

per second [exactly].

In the International System of Units (SI), the *curie* has been replaced by the *becquerel (Bq)*, where:

1 *becquerel* $= 1$ *radioactive decay per second*

$= 2.703 \times 10^{-11}$ *Ci.*

Many of the units used in radioactivity are expressed into smaller units or as multiples, using standard metric prefixes. Thus, a *kilobecquerel (kBq)* is 1000 *becquerels*, a *nanogram* is 10^{-9} *gram*, and a *picocurie* is 10^{-12} *curie*.

The type and energy of the ionizing radiation emitted are characteristic of the decaying radioisotope. Therefore, data on the emitted ionizing radiation, particularly its energy, yield qualitative as well as quantitative information on the radioisotope. A number of techniques are used to measure the energy of the emitted ionizing radiation, namely alpha *spectrometry* (for alpha particles), beta spectrometry (for beta particles), and gamma spectrometry (for gamma rays).

In *alpha spectrometry*, the test sample is deposited on a circular metal disk (~25 *mm* diameter) by electrolysis or by placing a drop of solution containing the radioactive substance to be analysed, drying it to give a uniform coating, and placing it in contact with an ionization detector (e.g. silicon surface barrier detector) in an evacuated chamber. This is because the range of α-particles in materials is small (1.12 cm for 2 *MeV* alpha particles in air at 15°*C* and 1 *atm* pressure). The electronics for counting of α-particles includes a pulse sorter (multi-channel analyser) and associated amplifiers and data readout devices. If the layer formed on the disk is too thick, the lines of the spectrum get broadened. This is because some of the energy of the α-particles is lost during their passage through the layer of the active material. An alternative method is to use the internal liquid scintillation technique, in which the sample is mixed with a scintillation cocktail and the emitted light is then counted. The detector records the amount of light energy per radioactive decay event. Due to imperfections of the liquid scintillation method, such as failure to detect all the emitted photons, it may be difficult to count cloudy or coloured samples. Furthermore, random quenching can reduce the number of photons generated per radioactive decay. Therefore, it is possible to get a broadening of the alpha energy spectra obtained through liquid scintillation. It is likely that the liquid scintillation spectra will be subject to a Gaussian broadening rather than the distortion exhibited when the layer of an active material on a disk is too thick.

Geiger-Müller (G-M) counters, gas proportional, or scintillation counters with heavy shielding and appropriate electronics for data collection and analysis are used for *beta* spectrometry, as in tritium and radiocarbon measurements.

Gamma (γ) spectrometry has been by far the most widely used method to measure the radioactivity. It is a powerful and useful measuring technique to analyse radioisotopes in various kinds of radioactive samples, because gamma rays exhibit discrete and unique energies that are intrinsic to each radionuclide. Normally, a thallium-activated sodium iodide [*NaI (Tl)*] scintillation detector or a solid-state germanium [*Ge*] detector is used to determine the energy(ies) of the emitted ionizing radiation. The *Ge* semiconductor detector has a far better energy resolution capability compared to the *NaI (Tl)* detector; hence most present-day gamma-spectrometry systems (used for radioactivity monitoring, activation analysis, and research purposes) incorporate a *Ge* detector.

Stable isotopes are nuclides that do not undergo radioactive decay, even on geological timescales, though they may themselves be produced by the decay of radioactive isotopes. Differences in isotopic composition of a given element in different reservoirs in exchange with each other (e.g. components of the hydrologic cycle) arise because during the process of exchange or phase change between different reservoirs (e.g. hydrologic processes), the heavier isotopes are somewhat sluggish compared to their lighter counterparts and tend to preferentially stay with the denser phase. Such differences manifest themselves as *isotope fractionation* effects between reservoirs. The degree of isotope fractionation thus is a measure of the extent to which the hydrologic processes proceed. Isotopic compositions are normally expressed in δ-*notation*, as deviations of heavy to light isotopic ratios relative to an international standard of known composition, expressed as parts per thousand and (denoted as ‰), or per million. The δ values are calculated as:

$$\delta(\text{in } ‰) = (R_x/R_s - 1) \times 1000 \qquad (7.2)$$

where R_x and R_s denote the ratio of heavy to light isotope (e.g. $^{13}C/^{12}C$, $^{18}O/^{16}O$, or D/H) in the sample and the standard, respectively.

7.2 Hydrologic tracers

Water is a universal solvent, which dissolves almost everything that comes into contact with it, including atmospheric gases. Any substance that can be used for tracking water movement through a given environment can be termed a *hydrologic tracer*. An *ideal tracer* behaves in the system exactly as the traced material, at least as far as the parameters of interest are concerned. However, it should have at least one characteristic property that distinguishes it from the traced material. It should be *conservative*, that is, it should not have sources or sinks (decay, sorption, or precipitation) in the system. In practice, a substance which has known

HYDROLOGIC TRACERS

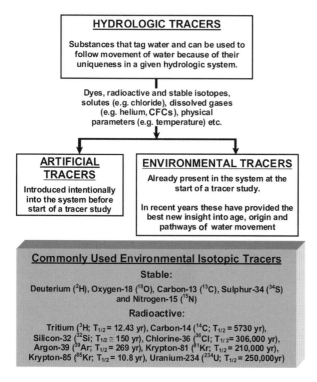

HYDROLOGIC TRACERS

Substances that tag water and can be used to follow movement of water because of their uniqueness in a given hydrologic system.

Dyes, radioactive and stable isotopes, solutes (e.g. chloride), dissolved gases (e.g. helium, CFCs), physical parameters (e.g. temperature) etc.

ARTIFICIAL TRACERS

Introduced intentionally into the system before start of a tracer study

ENVIRONMENTAL TRACERS

Already present in the system at the start of a tracer study.

In recent years these have provided the best new insight into age, origin and pathways of water movement

Commonly Used Environmental Isotopic Tracers

Stable:

Deuterium (^{2}H), Oxygen-18 (^{18}O), Carbon-13 (^{13}C), Sulphur-34 (^{34}S) and Nitrogen-15 (^{15}N)

Radioactive:

Tritium (^{3}H; $T_{1/2}$ = 12.43 yr), Carbon-14 (^{14}C; $T_{1/2}$ = 5730 yr), Silicon-32 (^{32}Si; $T_{1/2} \cong$ 150 yr), Chlorine-36 (^{36}Cl; $T_{1/2}$ = 306,000 yr), Argon-39 (^{39}Ar; $T_{1/2}$ = 269 yr), Krypton-81 (^{81}Kr; $T_{1/2}$ = 210,000 yr), Krypton-85 (^{85}Kr; $T_{1/2}$ = 10.8 yr), Uranium-234 (^{234}U; $T_{1/2}$ = 250,000yr)

Fig. 7.1 Commonly used hydrologic tracers and their characteristics.

sources or sinks can also be regarded as a suitable tracer, if these can be properly accounted for, or if their influence is negligible in terms of the desired accuracy.

Hydrologists make use of both *environmental tracers* (natural or anthropogenic compounds present in the environment) and injected tracers (e.g. dyes or chloride) that are introduced into a water body externally for studying its behaviour in a system. This basic grouping of commonly used hydrologic tracers, based on their mode of introduction into a hydrologic system along with their half-life (in case of radioactive isotopic tracers), is given in Fig. 7.1. Various hydrologic applications of commonly used elemental and isotopic tracers are listed in Table 7.1.

7.2.1 Artificial tracers

Injected or *artificially introduced tracers* are used to determine the direction and rate of groundwater movement, the rate of river flow, and mixing characteristics of various water bodies.

In the case of an artificially injected tracer, the input function is a sharp, well defined pulse with respect to time and location, mostly a delta-function. The output function is the measured concentration-time distribution of the tracer at the measuring or sampling location(s).

Some of the simple injected tracers are dyes and salts (e.g. Rhodamine or common salt). These can be used to measure the *residence time* of waters in a system. A known amount and concentration of tracer can be added to a system and changes in water- fluorescence, electrical conductivity, or water chemistry can be monitored at a given location by employing appropriate analytical methods. These methods may also be employed for groundwater tracing, but it should be noted that the subsurface movement of water can also introduce salts or the tracer can be adsorbed onto the aquifer matrix. Knowledge of the bedrock geology of the study site can, therefore, be helpful. While using ion chemistry to measure salt movement through a system, it is important to ensure the conservative

Table 7.1 Common hydrologic applications of various elemental and isotopic tracers.

Tracer	Common hydrologic application
^{2}H	Tracing precipitation sources and estimating evaporation rates
^{3}H, ^{3}He	Dating young groundwaters
^{4}He	Dating old groundwaters, engineering hydrology
^{14}C	Groundwater dating and tracing of a variety of hydrologic processes
^{15}N, ^{34}S	Tracing the nitrogen source for surface- and groundwater, identifying source(s) of pollution
^{18}O	Determining precipitation sources and evaporation rates. Can also record palaeoclimatic and palaeohydrologic information
He, Ne, Ar, Kr, Xe	Estimate palaeo-recharge temperature
^{32}Si	Tracing/estimating the groundwater recharge
^{36}Cl	Dating very old groundwaters (~ 1 Ma)
Fe, Se	Ascertaining redox condition of a given water body
^{86}Kr	Dating recent groundwater recharge
^{129}I	Constraining the groundwater age estimates and tracing groundwater flows
Os	Heavy metal pollution tracing in the environment
^{210}Pb	Identifying source(s) of lead pollution and determination of sedimentation rates
^{222}Rn	Estimating groundwater discharge into surface water bodies and groundwater dating
^{137}Cs	Estimating erosion and sedimentation rates

nature of the salt tracer, that is, the ion measured at the outlet of the system is not prone to biological uptake or reactions with water or sediment that would cause it to drop out of the dissolved state. Chloride represents a useful tracer since it is relatively inert and not used by biota to any significant degree. Nitrate is a poor tracer as it may be taken up by biota before it leaves the system.

In groundwater applications, hydrologists inject a tracer into a well (the injection well), and monitor nearby wells for the arrival of the tracer that indicates movement of the groundwater from the injection well to the monitoring/sampling well, thereby enabling determination of direction and rate of groundwater flow velocity. In rivers, measurement of the rate at which a river carries an introduced tracer enables estimation of river discharge. In practice, an injected tracer substance should possess good solubility in water, physical and chemical stability, and high resistance against adsorption onto the surrounding substratum or the suspended sediments. In addition, a tracer should be detectable at low levels of concentration and should be independent of temperature and pH of water. Furthermore, its natural background concentration should be as low as possible. In addition, questions such as whether the substance is

eco-friendly, available in the desired quantity, and is cost-effective, also play a role in its choice as an appropriate tracer.

Depending on the method of analysis, artificial tracers can be classified into four broad groups – chemical, radioactive, activable, and particulate tracers. *Chemical tracers* may be simple ionic compounds such as *NaCl* whose concentration can be determined by measuring conductivity or by ion selective electrodes, or metallic compounds such as *EDTA*, which can be measured by atomic absorption spectrometry. Although the number of different ions that may be used is large, cations are usually lost from water by exchange. Organic dyes are frequently used but their major disadvantage is that they are not fully conservative and tend to be lost from water by adsorption, particularly on clays. Analysis is generally done by filter fluorometers or colorimeters, which are robust enough for field use. *Radioactive tracers* are used extensively in water pollution studies because of high sensitivity of their detection that can be achieved and, with γ emitting tracers, the ability to be measured accurately *in situ* in the field. These tracers may be obtained as labelled compounds such as tritiated water (*HTO*) or from a soluble radioactive salt such as $K^{82}Br$, which has been shown to be a very good water tracer with negligible adsorptive loss. The short half-life of ^{82}Br (35.4 h) makes it highly attractive for tracer tests extending over only a few days. *Activable tracers* are elements that are stable when used in field tests but are made radioactive for analysis. A number of substances, notably the rare earths, may be used but their cumbersome analysis procedure has not made them particularly attractive for water tracing, although activation analysis is used for identifying pollutants. The same may be said for *particulate tracers*. These may be plant spores or micro-organisms such as bacteria or bacteriophages. An exception may be in the tracing of sewage effluents in coastal waters, where the strain *Serratia indica* is used. Samples are collected, cultivated, and the organisms counted by microscopic examination. *Radioactive particulate tracers*, on the other hand, have many advantages of tracing the movement of solids or salts through water, for example, in sewage. Treated sewage sludge may be coated with a gold or silver amine complex to provide a convenient tracer. However, surface labelled materials do not contain the tracer in proportion to their mass but to their surface area, so that quantitative errors may arise if the size distribution of particles in the sample is not taken into account. Artificial glass containing radioactive elements such as ^{46}Sc, powdered to match the size of the grains in the sediment or silt being traced, however, does not have this disadvantage.

In recent years, *gaseous tracer* methods are also being used for various applications. These include dissolved inert gases used both as geochemically conservative tracers in groundwater systems providing structural information, flow paths, transit times, and physical transport properties (Agarwal *et al.* 2006; Kipfer *et al.* 2002; McNeill *et al* 2001), as well as non-conservative, bulk partitioning agents to determine re-aeration coefficients in surface streams (Mackinnon *et al.* 2002; Murphy *et al.* 2001). Gaseous tracers are also used for characterizing the presence and extent of non-aqueous phases in subsurface groundwater systems, for example, NAPLs at contaminated land sites (Mohrlok *et al.* 2002). *Noble gases* as tracers are being particularly investigated because of their low natural background levels, high sensitivity of determination (by GC-MS), and lack of any taste, odour,

colour, and toxicity problems. These properties render noble gases as 'environmentally-friendly' type of tracers to protect potable water (e.g. public) supplies under investigation. In addition, they do not influence the intrinsic reactive processes occurring in the system being traced. For a detailed description of various applications and techniques of using artificial tracers in hydrologic studies, reference is made to Divine and McDonnell (2005) and Eilon *et al.* (1995).

7.2.2 Environmental tracers

Environmental tracers, of natural as well as anthropogenic origin, are present in the environment and there is no control in terms of their amount, location, and time of introduction into the environment but are nevertheless useful for hydrologic investigations. Variations, within and/or across hydrologic reservoirs, in the composition of a large number of substances, elements, or their isotopes (Fig. 7.1) that dissolve in water or constitute the water molecule itself, namely ^{2}H (or D), ^{3}H (or T) in the case of hydrogen, and ^{17}O and ^{18}O in case of oxygen, are also used for hydrologic tracing. The broad objective is to obtain information on the hydrologic system under investigation in terms of its component reservoirs and the degree and rate of exchange or mixing across the reservoir boundaries or within the system itself. The information on rates of hydrologic processes is obtained either through real-time monitoring or by making use of radioactive isotopes, which due to their characteristic radioactive decay rates provide a measure of time – just like a clock.

While using environmental tracers, the input function depends on the temporal and spatial distribution of the tracer concentration in precipitation and on the recharge rate into the unsaturated zone, as well as on the physico-chemical and microbiological conditions during the tracer transport to the water table. Therefore, determination of the input function is one of the main problems in ascertaining the validity of measurements of an environmental tracer. The output function is the concentration-time distribution of the tracer at the measuring or sampling point(s).

Table 7.2 Cosmogenic nuclide production rates and their inventories for radioisotopes of half-lives of more than 2 weeks. Based on Lal and Peters (1967).

Nuclide	Half- life (years) unless stated otherwise	Production rate $(cm^2\ sec)^{-1}$		Integrated inventory		Global inventory (gm)
		Troposphere	Total	(dpm/cm^2)	$(atoms/cm^2)$	
^{10}Be	1.5×10^6	1.5×10^{-2}	4.6×10^{-2}	2.70	3.07×10^{12}	2.6×10^8
^{26}Al	7.1×10^5	3.8×10^{-5}	1.4×10^{-4}	8.40×10^{-3}	4.52×10^9	1.0×10^6
^{36}Cl	3.1×10^5	4.0×10^{-4}	1.1×10^{-3}	6.60×10^{-2}	1.50×10^{10}	4.6×10^6
^{81}Kr	2.3×10^5	5.2×10^{-7}	1.18×10^{-2}	7.10×10^{-5}	1.24×10^7	8.5×10^3
^{14}C	5730	1.10	2.50	1.50×10^2	6.52×10^{11}	7.7×10^7
^{39}Ar	268	4.3×10^{-3}	1.29×10^{-2}	7.75×10^{-1}	1.58×10^8	5.2×10^4
^{32}Si	~150	5.4×10^{-5}	1.60×10^{-4}	9.60×10^{-3}	1.09×10^6	3.0×10^2
^3H	12.3	8.4×10^{-2}	2.50×10^{-1}	1.50×10^1	1.40×10^8	3.6×10^3
^{22}Na	2.6	2.4×10^{-5}	8.60×10^{-5}	5.16×10^{-3}	1.02×10^4	1.9
^{35}S	$87\ days$	4.9×10^{-4}	1.40×10^{-3}	8.40×10^{-2}	1.52×10^4	4.5
^7Be	$53\ days$	2.7×10^{-2}	8.10×10^{-2}	4.86	5.35×10^5	3.2×10^1
^{37}Ar	$35\ days$	2.8×10^{-4}	8.30×10^{-4}	4.98×10^{-2}	3.62×10^3	1.1
^{33}P	$25.3\ days$	2.2×10^{-4}	6.80×10^{-4}	4.08×10^{-2}	2.14×10^3	6.0×10^{-1}
^{32}P	$14.3\ days$	2.7×10^{-4}	8.10×10^{-4}	4.86×10^{-2}	1.44×10^3	3.9×10^{-1}

A special class of environmental radioisotopes is of *cosmogenic* origin. These isotopes are produced naturally by cosmic radiation in the Earth's atmosphere. Cosmic rays produce nine radio-nuclides of half-lives ranging between 10 years and 1.5 *Ma*, and 5 radio-nuclides, with half-lives ranging between 2 weeks and 1 year (Table 7.2). These have been used as tracers for measuring groundwater movement over time-scales ranging from a few weeks to millions of years. The various radio-nuclides listed in Table 7.2 have been extensively studied in the atmosphere, in wet precipitation, and in the hydrosphere, and in some cases, for example ^{10}Be, in sediments (Lal 1999; Lal and Peters 1967). Their dispersion in different terrestrial reservoirs is controlled principally by two factors: their chemical properties and half-lives (Lal and Peters 1967). In the case of long-lived isotopes, it is now possible to measure their concentration using the *Accelerator Mass Spectrometry* (*AMS*), usually at levels of more than 10^6 atoms in a sample. AMS has been successfully used to measure many isotopes for environment/hydrologic research. AMS detects charged ions that have been separated based on their mass. The main advantages of this technique are high precision, low detection limits,

and requirement of small quantities of sample. However, applicability of AMS is also limited by the relative quantities required for the isotopes of interest (e.g. ^{32}Si in silica).

Depending on the application, hydrologic tracers can be classified into two broad categories, namely studies involving: (i) movement of water, including its source identification, direction, and rate of movement and age; and (ii) hydrologic processes, including exchange across various reservoirs. Injected tracers are mostly used for studying movement of water, whereas environmental tracers are used for both types of investigations.

Use of isotopes for transport, retardation, and residence time distribution in groundwater hydrology is already well established (Mook 2000–2001). It is only during the last two decades that isotope techniques have also been used for studying the hydro-geochemical response in catchment hydrology (Kendall and McDonnell 1998). Typically, isotopes are used to: (i) determine the mechanisms and processes involved in generation of stream flow from various types of catchments; (ii) estimate *residence time* distribution of water within a catchment; (iii) determine the origin of stream flow components; (iv) calibrate or validate catchment stream flow models; (v) predict catchment response to a given storm; and (vi) estimate hydrologic parameters of a catchment.

In the subsequent sections, the focus is on some basic concepts related to environmental isotope tracer hydrology for groundwater studies. Appropriate theoretical background information on some individual tracer methods that have been applied in the North Gujarat regional aquifer system is also given.

7.3 Tracers and groundwater movement

7.3.1 Groundwater age

Groundwater age is generally considered as the average travel time for a water parcel from either the surface or from the water table (point of recharge) to a given point in an aquifer. A schematic illustration of how a groundwater parcel ages as it flows along a streamline is shown in Fig. 7.2a. Groundwater dating (i.e. age estimation) involves estimation of the groundwater age by one or more available techniques. The term 'residence time' is often used synonymously with 'age'. Hence, the residence time of groundwater is defined as the average travel time between the point of recharge and the point of discharge, for example, to a river or a lake or to any monitoring point in the groundwater zone. The tracer age estimate normally gives the average age of a water sample. This is a good approximation in cases where the flow system is simple, and can be approximated by a piston flow model (implying insignificant mixing and dispersion during the course of flow). However, where significant mixing and dispersion occur, for example, in long screens in water supply wells or in groundwater bodies with a significant volume of low permeability units ('stagnant zones') the estimated tracer model age may either underestimate or overestimate the actual mean age of the water parcel (Bethke and Johnson 2002; Weissmann *et al*. 2002). A sound knowledge of the geological setting and physical and chemical processes occurring in aquifers is, therefore, important for a proper interpretation and application of the environmental tracers and estimated groundwater ages.

7.3.1.1 Groundwater age estimation

There are basically three different ways of estimating groundwater age at a groundwater well or monitoring point: (1) by environmental tracers (Agarwal *et al*. 2006; Gupta 2001; Plummer *et al*. 1993); (2) by groundwater flow modelling (Engesgaard and Molson 1998; Gupta 2001); and (3) by a combination of both (Bauer *et al*. 2001; Troldborg 2004). Approximate range (in years) of dating employing commonly-used tracers is shown in Fig. 7.2b.

It is usual to distinguish between young (modern) groundwater and old groundwater. *Young groundwater* is considered to be groundwater recharged to the aquifers since 1950 AD, while that recharged before 1950 is considered to be *old*

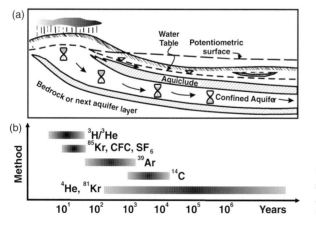

Fig. 7.2 (a) Aging of groundwater along a flow line in an aquifer; (b) most commonly used *environmental tracers* for groundwater age estimation and their useful dating ranges.

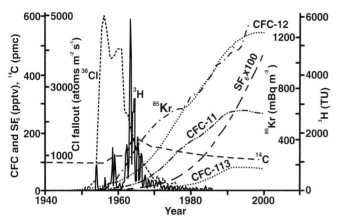

Fig. 7.3 Concentration variation of several environmental tracers in the atmosphere during the period 1940-2000. These tracers are applied either as relative or absolute groundwater dating tools. Redrawn from Clark and Fritz (1997) and Hinsby *et al.* (2001). © Geological Society of London.

groundwater (Cook and Böhlke 2000; Plummer *et al.* 1993).

7.3.1.1.1 Dating young groundwaters

During the past decade, several new environmental tracers for dating of young groundwaters have been introduced. Especially tracers such as *tritium* (^3H), *CFCs, SF*$_6$, and ^{85}Kr) are being increasing applied for identification of modern water components possibly containing contaminants, and for groundwater dating (Kipfer *et al.* 2002; Manning *et al.* 2005). Fig. 7.3 shows the varying atmospheric concentrations of the most common tracers for groundwater dating. The radionuclides ^{36}Cl, ^3H, and ^{14}C were introduced into the atmosphere by nuclear test explosions in the 1950s and early 1960s, while ^{85}Kr escapes into the atmosphere by the processing of the fuel rods from nuclear power plants. *CFCs* and *SF*$_6$ are gases used in the industry, although the use of CFCs is now banned due to their ozone depletion property. The tracers ^{85}Kr, *CFCs*, and *SF*$_6$ may be used for estimation of absolute ages, that is, under optimal conditions they can estimate the year of recharge of a given water sample or the average residence time (i.e. age) of groundwater. The tracers ^{36}Cl, ^{14}C, and ^3H are in this context considered as event markers (Cook and Böhlke 2000; Plummer *et al.* 1993), and may only be used for identification of human impacts or a relative age (e.g. recharged after 1960, 1990, etc.). However, ^3H can be used together with its daughter nuclide ^3He to estimate absolute ground-

water ages (Manning *et al.* 2005; Solomon and Cook 2000).

7.3.1.1.2 Dating old groundwaters

Tracers, such as ^{39}Ar, ^{14}C, ^4He, and ^{81}Kr, are used for dating groundwaters of pre-industrial age up to a million years old or more, in combination with groundwater flow modelling (Agarwal *et al.* 2006; Lehmann and Purtschert 1997). In small catchments, with short distances between recharge and discharge points, the groundwater used for drinking water supply is generally less than 1000 years old. However, in regional aquifer systems in large sedimentary basins, for example, Canada and Australia (Cresswell *et al.* 1999a; Fröhlich *et al.* 1991; Lehmann and Purtschert 1997), France (Marty *et al.* 1993), and India (Agarwal *et al.* 2006), groundwater may be very old (>100 000 years). These are often referred to as 'very old groundwaters' and defined as groundwater without measurable ^{14}C (Fröhlich *et al.* 1991).

In the following, some of the commonly used isotopic tracers used for groundwater dating are discussed in some detail in terms of their production, introduction into the hydrologic cycle, measurement, and data interpretation.

7.3.1.2 Tritium

Hydrogen has three isotopes, two stable (^1H and ^2H), and one radioactive (^3H). The stable isotopes of hydrogen are considered together with stable isotopes of oxygen in Section 7.4. The radioactive isotope tritium (^3H) can be used for dating very young groundwaters (younger than 50 *years*). Natural tritium is produced in the stratosphere by interaction of the cosmic ray produced neutrons with ^{14}N according to the nuclear reaction $^{14}_{7}N + ^{1}_{0}n \rightarrow ^{12}_{6}C + ^{3}_{1}H$. It combines with oxygen to produce tritiated water (H^3HO), which subsequently enters the hydrologic cycle. Tritium decays to a rare, stable isotope of helium (^3He) by beta (β) emission with a half-life of 12.32 *a* (*a* denotes for annum = year) (Lucas and Unterweger 2000). Natural production of tritium in the atmosphere is very low. Lithogenic tritium is produced by bombardment of the lithium present in rocks by neutrons produced during the spontaneous fission of uranium and thorium. This production is limited by the amount of lithium present in rocks. Tritium can also be produced from ^{10}B through neutron capture. In most cases, lithogenic production is negligible compared to other sources. The lithogenic tritium enters the groundwater directly. The natural production of tritium has been supplemented by anthropogenic production. Beginning in the 1950s, large amounts were produced from the atmospheric testing of thermonuclear bombs. Atmospheric concentrations of tritium decreased after 1963 (Fig. 7.3) as a result of the treaty banning above-ground testing. However, anthropogenic tritium continues to be released from nuclear power plants as a result of which present-day concentrations of tritium in the atmosphere have not returned to the pre-1950 natural concentrations. The tritium levels , however, continue to gradually decrease as it undergoes radioactive decay. All environmental tritium, whether of cosmogenic or anthropogenic origin, is rapidly incorporated into water molecules and becomes a part of the meteoric precipitation to enter the hydrologic cycle.

Tritium concentrations are represented as absolute concentrations in tritium units (*TU*) and so no reference standard is required. One tritium unit is equal to one molecule of ^3H per 10^{18} molecules of ^1H and has an activity of 0.118 *Bq kg^{-1}* (3.19 *pCi kg^{-1}*).

The large pulse of tritium that entered the hydrologic cycle in the 1960s (Fig. 7.3) can be used to establish the age of groundwater recharge recently. High levels of tritium (>30 *TU*) indicate water that was recharged during the late 1950s or early 1960s; moderate concentrations indicate modern recharge; levels close to detection limit (~ 1 *TU*) are likely to indicate sub-modern or palaeo-groundwaters that have mixed with shallow modern groundwaters (Clark and Fritz 1997). Bomb-produced tritium has also been used to study recharge of groundwater by tracing the movement of soil moisture through the *vadose zone* (Sukhija and Rama 1973) and in shallow water table aquifers (Gupta 1983). Artificially injected tritium has also been used for tracing soil moisture through the vadose zone (Datta *et al.* 1979). Environmental tritium can be used as a tracer in dating young groundwaters to help determine flow rates and directions, mean residence times, and hydraulic parameters such as conductivity, and can also be helpful in observing preferential flow paths and in investigating the mixing of waters from different sources. Its use, however, is somewhat limited by a number of factors, including uneven global distribution and local variations due to continued nuclear releases. Some approaches that may be used to counter the problems using ^3H in dating groundwaters include use of time series analysis to monitor the bomb spike for ^3H in an aquifer, to provide an indication of its mean *residence time*.

Groundwater can also be dated quantitatively using tritium and its daughter, ^3He. Age is determined by (Tolstikhin and Kamenskiy 1969):

$$\tau = \frac{1}{\lambda} . ln \left(1 + \frac{^3He_{tri}}{^3H} \right) \qquad (7.3)$$

where λ is the decay constant of tritium ($= ln\ 2/12.32 = 0.05626\ a^{-1}$); τ is the time since isolation of groundwater from the point of its contact with the atmosphere and is the ^3H-^3He age; and $^3He_{tri}/^3H$ is the concentration ratio of the two isotopes, expressed in *TU*.

Dating by the ^3H-^3He method also presents problems, since the total ^3He in groundwater comes from a variety of sources: the atmosphere, ^3H decay, subsurface nuclear reactions, and the Earth's mantle (Kipfer *et al.* 2002). The measured concentration of ^3He must be corrected for these contributing sources. ^3He is also not a routinely sampled or measured isotope. Other concerns are fractionation of ^3He if a gas phase is present and the fact that the solubility of He is temperature dependent (Weiss 1971).

Tritium is measured by counting β decay events in a liquid scintillation counter (LSC). A 10 ml sample aliquot is mixed with the scintillation compound that releases a photon when struck by a β particle. Photomultiplier tubes in the counter convert the photons to electrical pulses that are counted over a period of several hours (much smaller counting time is required when using artificial tritium due to its high concentration). Results are calculated by comparing the counts with those of calibrated standards and blanks. Increased precision is achieved through concentration by electrolytic enrichment of ^3H in water before counting, or by conversion to a suitable gas (CH_4 or C_3H_8) for gas proportional counting.

7.3.1.3 Radiocarbon

Carbon has three isotopes, two stable (^{12}C and ^{13}C) and one radioactive (^{14}C). Natural variation of the two stable isotopes of carbon can be useful for understanding carbon sources and the carbon cycle in ecosystems. W.F. Libby was awarded Nobel Prize in 1960 for his work concerning development of ^{14}C as a tool for archaeological dating. The half-life of ^{14}C (= 5730 a) makes it useful for Late Quaternary chronology. Even now it is used extensively to date groundwater (up to ~40 ka), as well as for tracing hydrologic processes, such as groundwater flow and ocean circulation.

Radiocarbon is formed in two different ways. Cosmogenically, ^{14}C is produced by interaction of cosmic ray produced neutrons with ^{14}N according to the nuclear reaction $^{14}_{7}N + ^{1}_{0}n \rightarrow ^{14}_{6}C + ^{1}_{1}p$. Similar to tritium, a considerable amount of ^{14}C was added to the atmosphere anthropogenically due to the nuclear bomb tests in the 1950s and the use of nuclear power. Whichever way formed, ^{14}C is rapidly oxidized to $^{14}CO_2$, which enters the Earth's plant and animal life cycle through photosynthesis and the food chain. The rapidity of the dispersal of ^{14}C into the atmosphere has been demonstrated by measurements of radioactive carbon produced from thermonuclear bomb testing. ^{14}C also enters the Earth's oceans through exchange across the

ocean surface–atmosphere boundary and as dissolved carbonate through terrestrial water influx. Plants and animals that utilize carbon in the biological food chain take up ^{14}C during their life times. They exist in equilibrium with ^{14}C production and its radioactive decay rates in the atmosphere. Thus, the ^{14}C concentration in active carbon pools in exchange with the atmosphere stays the same over a period of time. This is referred to as 'modern carbon' ^{14}C concentration and corresponds to A_0 in Eqn 7.1. The activity of modern carbon is defined as 95% of the ^{14}C activity in 1950 of the NBS oxalic acid standard. This is close to the activity of wood grown in 1890 in a fossil-CO_2-free environment and equals 13.56 $dpm\ g^{-1}$ carbon. All ^{14}C measurements are referred to as percent modern carbon (pmC). As soon as a plant or animal dies, it ceases the metabolic function of carbon uptake; there is no replenishment of radioactive carbon but radioactive decay continues.

The solid carbon method for ^{14}C isotope counting, originally developed by Libby and his collaborators, was replaced by gas counting methods in the 1950s. Liquid scintillation counting, utilizing benzene, acetylene, ethanol, methanol, etc., was developed around the same time. Today, the vast majority of radiocarbon laboratories utilize these two methods for radiocarbon dating. A recent development of major interest is the development of the *Accelerator Mass Spectrometry (AMS)* method, in which all the ^{14}C atoms can be counted directly, rather than only those decaying during the counting interval allotted for each analysis, as in gas or liquid scintillation counting methods. The main advantages of the AMS technique are high precision and low detection limits.

7.3.1.3.1 Radiocarbon dating of groundwater

Radiocarbon dating of groundwater is based on measuring the loss of amount of parent radionuclide (^{14}C) in a given sample. This assumes two key features of the system. The first is that the initial concentration of the parent is known and has remained constant in the past. The second is that the system is closed to subsequent gains or losses of the parent ^{14}C, except through radioactive decay. But, the very process of incorporating radiocarbon into groundwater by dissolution of plant root respiration and decay-derived carbon dioxide (with 100 pmC) in the presence of old/radioactively dead (~ 0 pmC) lime/dolomite in the soil zone, strongly dilute the initial ^{14}C activity in total dissolved inorganic carbon (TDIC), according to the principal chemical reactions:

$$CaCO_3 + H_2O + CO_2 \Leftrightarrow Ca^{2+} + 2HCO_3^- \quad (7.4)$$

$$CaMg(CO_3)_2 + 2H_2O + 2CO_2 \Leftrightarrow Ca^{2+}$$
$$+Mg^{2+} + 4HCO_3^- \quad (7.5)$$

This causes an artificial 'aging' of groundwater by dilution of ^{14}C when using Eqn 7.1. Subsequent evolution of the carbonate system can also lead to some changes in the ^{14}C activity of groundwater. Unravelling the relevant processes and distinguishing ^{14}C decay from ^{14}C dilution is an engaging geochemical problem.

There are several models that aim to estimate the contribution of soil carbonates to the TDIC and estimate the applicable value of A_0. This is done either through a stoichiometric approach for the various chemical reactions involving carbon or by estimating dilution of active carbon using an isotope mixing approach based on the ^{13}C content of each species involved, or a combination of the two approaches. Various methods for estimating A_0 can be found in Mook (1976) and Fontes and Garnier (1979). The error due to incorrect estimation of A_0, however, is $< t_{1/2}$ of ^{14}C, except in a special case of carbonate aquifers where continuous exchange between TDIC and the aquifer matrix may reduce A_0 to $<50\,pmC$. Since most of the chemical and isotope exchange occurs in the unsaturated soil zone during the process of groundwater recharge, and between TDIC and the soil CO_2, the A_0 in several groundwaters has been found to be $85 \pm 5\,pmC$ (Vogel 1967, 1970).

In the North Gujarat-Cambay study reported subsequently (Chapter 13), the theoretical value of A_0 (after equilibrium between soil CO_2, soil carbonate (at $^{14}C = 0\,pmC$; $\delta^{13}C = 0\,\%o$), and infiltrating water) is estimated using the following equation (Münnich 1957, 1968):

$$A_0 = \frac{\delta^{13}C_{TDIC}}{\delta^{13}C_{soil} - \varepsilon} 100 \quad (7.6)$$

where $\delta^{13}C_{TDIC}$ is the $\delta^{13}C$ value of the groundwater TDIC; $\delta^{13}C_{soil}$ is the $\delta^{13}C$ of soil CO_2 ($\sim -22\,\%o$); and ε is equilibrium fractionation between the soil CO_2 and the TDIC of groundwater ($\sim -9\,\%o$). The $\delta^{13}C$ values are computed using Eqn 7.2 with Pee Dee Belemnite (PDB) limestone as the standard reference material. The model of Eqn 7.6 is used taking into account that application of any other model would give radiocarbon ages differing by $< \pm 2$ ka. Also because in regional aquifers the difference in groundwater ages between any two locations, after the confinement of the groundwater in the aquifer becomes effective, is virtually independent of the applicable value of A_0.

For ^{14}C dating, about 100 litres of groundwater is piped directly into a collapsible high-density PVC bag through a narrow opening. The PVC bag is kept in the folded condition in a stand designed specifically for this purpose and assembled from its prefabricated parts at the site (Fig. 7.4a). The PVC bag unfolds only when the groundwater fills it. Before piping in the groundwater, a few pellets of $NaOH$ ($\sim 10g$) are added to the PVC bag to raise the solution pH to >10 for immobilizing the dissolved CO_2 in the form of CO_3^{2-} and its eventual precipitation as barium carbonate. At $pH > 10.3$ most of the dissolved CO_2 is in the form of CO_3^{2-}, since at this pH, activity of HCO_3^- drops and activity of CO_3^{2-} rises rapidly (Drever 1997).

Depending upon the alkalinity and sulphate concentration of groundwater samples (measured in the field), a pre-determined amount of barium chloride ($BaCl_2$) is then added to the 'groundwater-$NaOH$' solution to ensure complete precipitation of dissolved carbonates (Clark and Fritz 1997). Following vigorous stirring, the mixture is left undisturbed for precipitates to settle in the conical base of the PVC bag (Fig. 7.4a). It usually takes 4 to 5 hours for the precipitates to settle. After decanting the supernatant liquid, precipitates are transferred to glass bottles (Fig. 7.4b) and sealed. Care is taken to prevent/minimize sample exchange with atmospheric CO_2 during the entire extraction procedure.

On reacting with orthophosphoric acid, the barium carbonate precipitate liberates CO_2. The liberated CO_2 is first converted to acetylene and then trimerized into benzene (C_6H_6) and the ^{14}C

(a)

(b)

Fig. 7.4 (a) Photograph of a specially designed foldable stand with a conical aluminium base, which holds a high density PVC bag filled with 100 litres of water sample. The supernatant water is decanted by piercing the bag after the carbonate precipitates settle in the conical base of the bag; (b) Carbonate precipitates are transferred from the PVC bag into a 1.2-litre soda-lime glass bottle without exposure to the atmosphere. See also Plate 7.4.

activity in the benzene counted by liquid scintillation spectroscopy (Gupta and Polach 1985). A small aliquot of the sample CO_2 is sealed in glass ampoules for $\delta^{13}C$ measurement, using a stable isotope ratio mass-spectrometer (SIRM).

When using *AMS* for ^{14}C measurement, only about a litre of water sample sealed in a glass bottle is collected in the field and all subsequent processing of the sample is carried out in the laboratory.

7.3.1.4 Silicon-32

Silicon-32 (^{32}Si) is a cosmogenic isotope produced in the atmosphere by spallation of Argon-40. With a half-life of ~150 years, ^{32}Si is ideally suited to provide chronology in the range 50–1000 years. Because the nuclide is produced in the atmosphere, the highest activity per unit volume of water is found in fresh precipitation. The concentration in soil is several orders of magnitude higher. This is due to accumulation of ^{32}Si from exchange between dissolved silica in the infiltrating water and silica adsorbed onto the soil sediments.

Detection of natural ^{32}Si is, however, difficult due to the extremely low levels of concentration and isotopic ratios. When using AMS, for measurement of silicon, the ratio of $^{32}Si/^{28}Si$ must exceed 10^{-15} (Morgenstern *et al*. 2000). It is possible to satisfy this condition for precipitation samples. However, for groundwater and soil samples where ^{28}Si

is abundant, the ratio is $\ll 10^{-15}$ (Morgenstern *et al*. 2000). Therefore, scintillation counting is the preferred method for analysing ^{32}Si in groundwaters and soils.

In order to determine the activity of ^{32}Si by decay counting, the activity of the daughter product, ^{32}P, is measured. ^{32}P, rather than ^{32}Si, is selected for analysis because of its shorter half-life (14.3 *d* vs. ~150 *a* of ^{32}Si) and the higher energy of its beta decay ($E_{max} = 1.7\ MeV$ vs. $0.22\ MeV$ of ^{32}Si). To relate the activities of the parent and daughter, a *secular equilibrium* must be achieved (Fig. 7.5). This occurs when the ratio of the activity of the parent and the daughter remains constant with time. At this point, the ratio of the isotope abundances is equal to the ratio of their decay constants (λ) that are known. Therefore, if the activity of the daughter ($A^{32}P$) is measured when secular equilibrium has been established, the activity of the parent ($A^{32}Si$) can be calculated from:

$$A^{32}Si = A^{32}P^*(\lambda^{32}P/\lambda^{32}Si) \qquad (7.7)$$

Elaborate chemical processing of a water sample for ^{32}Si measurement involves several steps, such as: (i) addition of silica carrier; (ii) co-precipitation of ferric hydroxide and silica; (iii) purification of precipitated silica; (iv) allowing sufficient time for ^{32}Si to come into *secular equilibrium* with ^{32}P; (v) addition of carrier phosphate; (vi) precipitation of ammonium molybdophosphate (AMP);

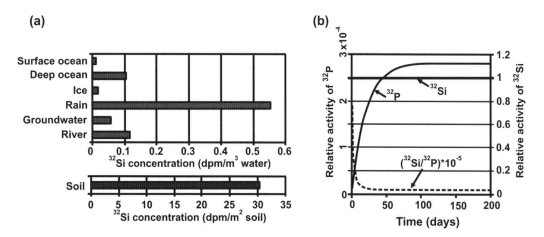

Fig. 7.5 (a) Activity of ^{32}Si in various waters and soil; (b) Activity of ^{32}Si in a sample is measured by counting decay events of the daughter product ^{32}P after about two *months* when secular equilibrium has been established between the activities of the parent and daughter isotopes.

(vii) precipitation of ammonium magnesium phosphate; (viii) ignition of the phosphate at $1000°C$; (ix) separation of phosphorus from silica; and finally (x) decay counting of ^{32}P.

^{32}Si is has been found as a useful tracer for the study of ocean circulation (Lal *et al*. 1976), study of atmospheric circulation involving exchange processes between stratosphere and troposphere (Lal 1999), understanding groundwater flow (Gupta *et al*. 1981; Nijampurkar *et al*. 1966), and dating of marine siliceous biota (Brzezinski *et al*. 2003).

7.4 Stable isotopes of oxygen and hydrogen

Amongst various isotopes used as tracers in hydrology, stable isotopes of oxygen (^{18}O) and hydrogen (^{2}H or D) are the most important. Being an integral part of the water molecule, these are ideally suited to trace the movement of water throughout the hydrologic cycle. Some basic information on these isotopes is given in Table 7.3.

In hydrologic parlance, the two isotopes are also referred to as *water isotopes*. Importance and applications of water isotopes to hydrologic studies have been discussed at length and demonstrated in several parts of the world (Araguas-Araguas *et al*. 1998; Clark and Fritz 1997; Dincer *et al*. 1974; Gat and Matsui 1991). Several Indian case studies have

also been reported over the last few decades (Bhattacharya *et al*. 2003, 1985; Das *et al*. 1988; Datta *et al*. 1994, 1996, 1991; Deshpande *et al*. 2003; Krishnamurthy and Bhattacharya 1991; Kumar *et al*. 1982; Navada *et al*. 1993; Navada and Rao 1991; Shivanna *et al*. 2004; Sukhija *et al*. 1998; Yadav 1997) and in recent years by Gupta and Deshpande (2003b, 2005a, b, c) and Gupta *et al*. (2005b).

Isotopic compositions are expressed in δ-*notation* calculated as in Eqn 7.2 with *SMOW* (*Standard Mean Ocean Water*; Craig, 1961b) or the equivalent *VSMOW* (Gonfiantini 1978) as a reference standard.

Isotopic fractionation occurs mainly by equilibrium isotopic exchange reactions and kinetic processes. Equilibrium exchange reactions involve redistribution of the isotopes between the products and reactants (or the two phases during phase

Table 7.3 Natural abundance of oxygen and hydrogen isotopes (compiled from the *CRC Handbook*).

Name (Symbol)	Oxygen (O)	Hydrogen (H)
Atomic number	8	1
Atomic mass	15.9994 *g/mol*	1.00794 *g/mol*
Stable isotopes (Natural abundance)	^{16}O (99.76%) ^{17}O (0.038%) ^{18}O (0.21%)	^{1}H (99.985%) ^{2}H (0.015%)

change) in contact with each other. As a 'rule of thumb', amongst different phases of the same compound (e.g. H_2O), the denser the material, the more it tends to be enriched in the heavier isotopes (D and ^{18}O).

In systems that are not in equilibrium, forward and backward reaction rates are not identical, and isotope reactions become unidirectional, for example, when reaction products are physically isolated from the reactants. Such reactions are called kinetic. A third type of reaction, where fractionation of isotopes occurs, is the diffusion of atoms or molecules across a concentration gradient. This can be diffusion in another medium or diffusion of a gas into a vacuum. In this case, fractionation arises from the difference in diffusive velocities of isotopic molecular species.

Isotope fractionation is expressed by a *fractionation factor*, α, which is the ratio of the isotope ratios for the reactant and the product ($\alpha = R_{reactant}/R_{product}$). For example, in the water→ vapour system, the fractionation factor is given by:

$$\alpha^{18}O_{(water \rightarrow vapour)} = \frac{\left(^{18}O/^{16}O\right)_{water}}{\left(^{18}O/^{16}O\right)_{vapour}} \qquad (7.8)$$

Craig (1961b) showed that despite great complexity in different components of the hydrologic cycle, $\delta^{18}O$ and δD in fresh waters correlate on a global scale. Craig's *Global Meteoric Water Line (GMWL)*

defines the relationship between $\delta^{18}O$ and δD in global precipitation as:

$$\delta D = 8 \times \delta^{18}O + 10 \quad (\text{‰SMOW}) \qquad (7.9)$$

This equation indicates that the isotopic composition of meteoric waters behaves in a fairly predictable manner. The *GMWL* is the average of many local or regional meteoric water lines, which may somewhat differ from each other due to varying climatic and geographic parameters. A *local meteoric water line (LMWL)* can differ from GMWL in both slope as well as the intercept on the deuterium axis. Nonetheless, GMWL provides a reference for interpreting hydrologic processes and provenance of different water masses at a given place.

To improve precision of the Craig's GMWL, Rozanski *et al.* (1993) compiled isotope data in precipitation from 219 stations of the IAEA/WMO operated Global Network for Isotopes in Precipitation (GNIP). This refined relationship between ^{18}O and D in global precipitation (Fig. 7.6) is given by:

$$\delta D = 8.17(\pm 0.07) \times \delta^{18}O$$

$$+11.27(\pm 0.65) \quad (\text{‰VSMOW}) \qquad (7.10)$$

The evolution of $\delta^{18}O$ and δD values of meteoric waters begins with evaporation from oceans, where the rate of evaporation controls the water–vapour exchange and hence the degree of isotopic equilibrium. Higher rates of evaporation impart a kinetic or non-equilibrium isotope effect

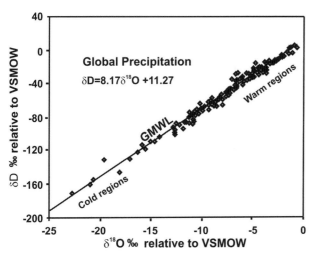

Fig. 7.6 The linear regression line between $\delta^{18}O$ and δD of global precipitation samples. Data are weighted average annual mean values for precipitation monitored at 219 stations of the IAEA/WMO global network. Redrawn from Rozanski *et al.* (1993). © American Geophysical Union.

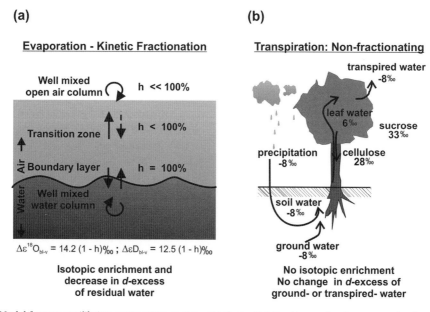

Fig. 7.7 (a) Model for non-equilibrium evaporation over a water body. Relative fluxes of water across mixed water column ↔ boundary layer ↔ mixed open air column through transition zone are shown by arrows. 'Kinetic' isotope fraction, namely, depletion of heavy isotopes in the overlying air column and enrichment in the water is caused by differences in the rate of diffusion of ^{18}O to ^{16}O and ^{2}H to ^{1}H. (b) Because soil water taken up by plant roots is quantitatively released by leaves to the atmosphere, transpiration, unlike evaporation, is a non-fractionating process. Both (a) and (b) redrawn from Clark and Fritz (1997). © Geological Society of London.

to the vapour. Kinetic effects are influenced by surface temperature, wind speed (shear at water surface), salinity, and most importantly humidity. For lower values of humidity, evaporation becomes an increasingly non-equilibrium process.

The most accepted model for non-equilibrium evaporation from a water body involves diffusion of water vapour across a hypothetical boundary layer (*bl*) with a thickness of a couple of microns above the air–liquid water interface (Fig. 7.7a). The boundary layer has virtually 100% water saturation. This layer is in isotopic equilibrium with the underlying water column. Between the boundary layer and the mixed atmosphere above is a transition zone through which water vapour is transported in both directions by molecular diffusion. It is within the transition zone that non-equilibrium fractionation arises as a result of diffusivity of $^{1}H_2{}^{16}O$ in air being greater than that of $^{2}H^{1}H^{16}O$ or $H_2{}^{18}O$. The additional isotopic enrichment ($\Delta\varepsilon$) of evaporating water due to kinetic fractionation at relative humidity (*h*; fractional) is approximated by the follow-

ing two empirical relationships (Gonfiantini 1986) that ignore all other controlling factors except humidity:

$$\Delta\varepsilon^{18}O_{bl-v} = 14.2 \times (1-h)\text{‰} \qquad (7.11)$$

$$\Delta\varepsilon^{2}H_{bl-v} = 12.5 \times (1-h)\text{‰} \qquad (7.12)$$

It is seen from these equations that the relative magnitude of kinetic fractionation for oxygen isotopes is significantly higher than that for hydrogen isotopes. The total enrichment between the water column and open air is the sum of the enrichment factor for equilibrium water vapour exchange ($\varepsilon_{l\to v}$) and the kinetic factor ($\Delta\varepsilon_{bl-v}$). For ^{18}O, this would be:

$$\delta^{18}O_l - \delta^{18}O_v = \varepsilon^{18}O_{l\to bl} + \Delta\varepsilon^{18}O_{bl\to v} \qquad (7.13)$$

Because the boundary layer is at 100% saturation, $\varepsilon^{18}O_{l\to bl}$ corresponds to equilibrium fractionation in water→vapour system. This calculation represents enrichment of water with respect to vapour. The depletion in vapour with respect to water results in negative isotopic enrichment. Under conditions of 100% relative humidity ($h = 1$), the vapour is in isotopic equilibrium with sea water ($\Delta\varepsilon^{18}O_{bl\text{-}v} = 0$). When relative humidity is low (e.g. $h = 0.5$), the vapour is strongly depleted in ^{18}O compared to D. The global atmospheric vapour is formed with an average relative humidity around 85% ($h = 0.85$). This is why the Craig's GMWL has a deuterium intercept of 10 ‰.

Thus, formation of atmospheric vapour masses is a non-equilibrium process due to humidity being lower than its saturation value. However, the reverse process, that is, condensation to form clouds and precipitation, takes place in an intimate mixture of vapour and water droplets with near saturation humidity so that equilibrium fractionation between vapour and water is easily achieved. As a result, isotopic evolution of precipitation during rainout is largely controlled by temperature. Because of this, the slope of GMWL (equal to 8) is largely in agreement with that given by the ratio of equilibrium fractionation factors for D and ^{18}O:

$$S \approx \frac{10^3 \ln\alpha\, D_{v \to l}}{10^3 \ln\alpha\, {}^{18}O_{v \to l}} = 8.2 \; at \; 25^{\circ}C \qquad (7.14)$$

However, slopes of various LMWLs vary from 9.2 to 8.0 for commonly encountered range of temperature between 0 and 30°C. It is thus seen that the relationship between ^{18}O and D in meteoric waters arises from a combination of the non-equilibrium fractionation from the ocean surface (at ~85% relative humidity) and equilibrium condensation from the vapour mass. However, during rainout, further partitioning of ^{18}O and D between different regions is governed by the *Rayleigh distillation* equation.

As an air mass moves from its vapour source area over continents along a trajectory, it progressively cools and loses its water content through the rainout process. During rainout, ^{18}O and D in vapour and the condensing phases (rain or snow) within the cloud get partitioned through equilibrium fractionation. But along the trajectory, as each episode of rainout removes some of the vapour mass, the heavier isotopes from the vapour are preferentially removed compared to the lighter ones, so that remaining vapour becomes progressively depleted in ^{18}O and D. Each rainout event gives isotopically enriched rain (or snow) with respect to its contributing parent vapour. It, however, gets depleted with respect to the preceding rainout spell because the vapour from which it was formed is isotopically depleted with respect to the vapour of the earlier rainout event (Fig. 7.8). One can model the isotope systematics during a rainout process according to the Rayleigh distillation equation as:

$$R_v = R_{v0} f^{(\alpha - 1)} \qquad (7.15)$$

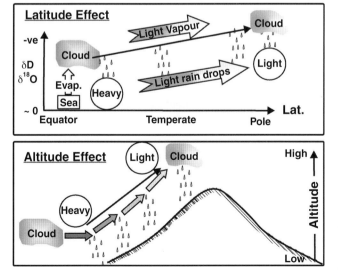

Fig. 7.8 Rainout of vapour leads to progressive heavy isotope depletion in successive rainout events away from the vapour source. This process, together with decrease in condensation temperature, explains the latitudinal and the altitudinal decrease in δ-values observed in global precipitation data.

Since $R_l / R_v = \alpha = R_{l0} / R_{v0}$, this equation can also be written in terms of liquid water, i.e.:

$$R_l = R_{l0} f^{(\alpha-1)} \qquad (7.16)$$

In the above equation, R_{v0} is the initial isotope ratio ($^{18}O/^{16}O$ or D/H) of vapour in the cloud and R_v the ratio after a given proportion of the vapour has rained out to yield the precipitation with the isotopic ratio, R_l, from liquid equilibrium value (R_{l0}) corresponding to R_{v0}. The residual fraction in the vapour reservoir is denoted by f. The factor α denotes the equilibrium water→vapour *fractionation factor* at the prevailing *in-situ* cloud temperature.

Along the trajectory, when part of the rained out vapour is returned to the atmosphere through evapotranspiration, the Rayleigh law, however, is not applicable. The downwind effect of evapotranspiration flux on the isotopic composition of atmospheric vapour and precipitation depends on details of the evapotranspiration process. Transpiration returns precipitated water essentially unfractionated back to the atmosphere, despite the complex fractionation in leaf water (Forstel 1982; Zimmermann *et al*. 1967). This is because of quantitative transfer of soil water taken up by the roots to the atmosphere (Fig. 7.7b).

Thus transpiration, by returning vapour mass with isotopic composition $R_V (= R_l/\alpha)$ in the downwind direction, in a way restores both the vapour mass and the heavy isotope depletion caused by the rainout in such a way that the subsequent rainout event is not depleted as much as it would have been without the transpiration flux. Under such circumstances, the change in isotopic composition along the air mass trajectory is only due to the fraction, f_{net}, representing the net loss of water from the air mass, rather than being a measure of the total amount of rainout. This causes apparent reduction in the downwind isotopic gradient. The evaporated water, on the other hand, is usually depleted in heavy species relative to that of transpired vapour (i.e. $<R_V$), thus restoring the vapour mass to the downwind cloud but depleting its heavy isotope composition. This causes apparent increase in the downwind isotopic gradient.

The isotopic imprints of evaporation are also recorded in the form of a parameter '*d*-excess'

in the evaporating water body, the evaporated vapour, and the precipitation originating from an admixture of atmospheric vapour and the evaporated flux. Since the kinetic fractionation for ^{18}O is more than that for D, as seen from Eqn 7.11 and Eqn 7.12, the relative enrichment of the residual water for an evaporating water body is higher for ^{18}O than for D. Correspondingly, for the resulting vapour, the depletion is more for ^{18}O than for D. The extent to which ^{18}O is fractionated more compared to D can be represented by a parameter '*d*-excess', defined by Dansgaard (1964) as:

$$d - \text{excess} = d = \delta D - 8 \times \delta^{18}O (\text{‰}) \qquad (7.17)$$

The *d*-excess, as defined above, represents the δD excess over 8 times $\delta^{18}O$ for any water body or vapour. Since magnitude of equilibrium fractionation for D is about 8 times that for ^{18}O, any value of δD in vapour in excess of 8 times $\delta^{18}O$ is indicative of the effect of kinetic fractionation due to evaporation. As already mentioned, the intercept (~ 10‰) of GMWL also signifies the kinetic fractionation during evaporation but the difference between intercept and *d*-excess is that the intercept of a meteoric or any other water line is valid for an entire dataset, whereas the *d*-excess parameter can be calculated even for a single water sample whose δD and $\delta^{18}O$ values are known.

As evaporation proceeds, because of the relatively higher enrichment of ^{18}O in the residual water, the *d*-excess of the evaporating water body decreases and that of the resulting vapour increases. Therefore, if the original water was meteoric in origin, the residual water not only becomes enriched in heavier isotopes but also shows progressively lower *d*-excess values as evaporation proceeds, that is, its position on the $\delta^{18}O$-δD plot will lie below the LMWL. The resulting vapour on the other hand shows the opposite effect. Furthermore, since condensation and consequently rainout is an equilibrium process (with slope ~ 8), it does not significantly alter the *d*-excess. Thus *d*-excess provides an additional handle to identify vapours of different histories and their mixing. Due to the effect of evaporation, most meteoric and subsurface processes shift the ($\delta^{18}O$-δD) signatures of water to a position below the LMWL. It is rare to find precipitation or groundwaters that plot above the LMWL, that is, show higher *d*-excess. However, in low-humidity regions, re-evaporation of precipitation from local surface waters and/or soil water/water table creates vapour masses with isotopic content that plots above the LMWL. If such vapours are re-condensed significantly before mixing with the larger tropospheric reservoir, the resulting rainwater will also plot above LMWL (Clark and Fritz 1997), along a condensation line with a slope of about 8. It is, however, important to recall that recycling of water back to the atmosphere in the form of vapour from soil moisture by plant transpiration is a non-fractionating process and, therefore, does not affect the *d*-excess either of soil water or of the atmospheric moisture.

A small fraction of rain percolates down through the soil layer and eventually becomes groundwater. For many groundwaters, their isotopic composition has been shown to equal the mean weighted annual composition of precipitation (Bhattacharya *et al.* 1985; Douglas 1997; Hamid *et al.* 1989; Krishnamurthy and Bhattacharya 1991; Rank *et al.* 1992). However, significant deviations from precipitation values are also found in several cases. Such deviations from local precipitation are more pronounced in arid regions due to extensive evaporation from the unsaturated zone or evaporative losses even from the water table (Allison *et al.* 1984; Dincer *et al.* 1974). Considering that only a small fraction of precipitation actually reaches the water table in most situations, the meteoric signal in groundwater can get significantly modified. Isotope variations in precipitation get attenuated and seasonal influences in recharge are reflected in the freshly recharged groundwater. This bifurcation of the hydrologic cycle between precipitation and surface water on the one hand and groundwater on the other ends where groundwater discharges and rejoins surface runoff in streams and rivers. Thus environmental isotopes, with their ability to characteristically label hydrologic components depending on the extent of the hydrologic process, play a significant role in quantifying the relative contribution of groundwater to stream flow and in understanding the hydrologic processes operating in a catchment.

7.5 Dissolved noble gases

Atmospheric gases enter the meteoric water cycle by gas partitioning during air–water exchange with the atmosphere in accordance with their solubility governed by respective *Henry coefficients*. Therefore, ubiquitous presence of atmospheric gases in the meteoric water cycle in solubility equilibrium defines a natural baseline. As a result, in most cases, the gas abundance in water can be understood as a binary mixture of two distinct gas components – a well-constrained atmospheric solubility equilibrium component and a residual component that may either be of atmospheric (e.g. trapped air bubbles) or non-atmospheric origin (e.g. radiogenic or terrigenic noble gases). Nitrogen (N_2), oxygen (O_2), and carbon dioxide (CO_2) are *reactive gases* and participate in many biogeochemical reactions which, depending on the particular reaction and the prevailing condition, also act as their sources and sinks, thereby causing additional variations in their concentration in waters of meteoric origin. On the other hand, noble gases, being inert, are conservative. Their low abundance also makes them an *ideal tracer* of many hydrologic processes.

As gas exchange phenomena proceed fairly rapidly, with controlling gas transfer velocities of the order of 10^{-5} m s^{-1} (Schwarzenbach *et al.* 2003), the surface waters of open water bodies are expected to have noble gases in equilibrium with the atmosphere at the prevailing physical conditions. This is supported by experimental evidence (Aeschbach-Hertig *et al.* 1999; Craig and Weiss 1971) and holds for all atmospheric gases that have no additional sources or sinks and hence can be considered as bio-geochemically conservative.

Gas partitioning at the free air–water interface can be reasonably well described by *Henry's law*, according to which concentrations in the two phases are directly proportional to each other:

$$c_{ia} / c_{iw} = k_i \left(T, c_{jw}, \ldots \ldots \right) \approx k_i \left(T, S \right) \qquad (7.18)$$

where c_{ia} and c_{iw} denote the concentrations of gas *i*, in 'air' and 'water' phases, respectively, under the atmospheric solubility equilibrium condition. In this form, Henry coefficient (k_i) is inverse of the 'solubility of a gas in water'. Poorly soluble gases

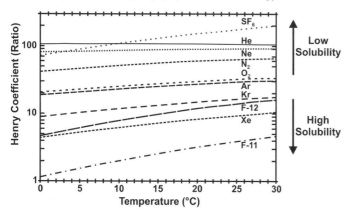

Fig. 7.9 Variation of gas solubility (in terms of Henry coefficient/ constant) in water with temperature of equilibration. Note the logarithmic scale for ordinate axis. The unit of Henry coefficient is the ratio l_{gas}/l_{water}. Data Source: Ar - (Weiss, 1970); He, Ne - (Weiss, 1971); Kr - (Weiss and Kyser, 1978), Xe - (Clever, 1979); CFCs - (Warner and Weiss, 1985); SF$_6$ - (Bullister *et al.*, 2002). Redrawn from Aeschbach-Hertig (personal Communication).

have large values of k_i, whereas highly soluble gases have low values of k_i.

Atmospherically derived noble gases make up the largest and most important contribution to the noble gas abundance in meteoric waters. The gas specific Henry coefficient can often be assumed to depend only on temperature and salinity of the water so that equilibrium concentrations of noble gases implicitly carry information on the physical properties of the water during gas exchange at the air–water interface, that is, air pressure, temperature, and salinity of the exchanging water mass.

The dependence of noble gas solubility equilibrium on the physical conditions during gas exchange, in particular the sensitivity of the Henry coefficient to temperature (Fig. 7.9), has been successfully used in groundwater studies to reconstruct the soil temperature prevailing at the time of groundwater recharge. If an aquifer contains groundwater that was recharged during different climatic conditions in the past, noble gas concentrations provide information about the past temperature (Andrews and Lee 1979; Mazor 1972). This approach has been applied to reconstruct the prevailing continental temperature of the Pleistocene–Holocene transition in the tropics (Stute *et al.* 1995a; Weyhenmeyer *et al.* 2000), as well as in mid-latitudes (Aeschbach-Hertig *et al.* 2002; Beyerle *et al.* 1998; Stute *et al.* 1995b; Stute and Schlosser 2000).

The experimental part of studying the dissolved conservative gases in water can be divided into four steps: (i) sample collection, storage and transporta-

tion to the laboratory; (ii) gas extraction from the water sample; (iii) purification and separation of the extracted gases; and (iv) quantitative chromatographic (for *CFCs* and *SF*$_6$) or mass spectrometric (for noble gases) analysis.

Depending on the dissolved gas species of interest, their physical properties vis-à-vis their permeation through and interaction with the material of the container in which the collected sample is stored for transportation and ultimate laboratory analyses, special field procedures and protocols have been devised. The main aim of each such procedure is to ensure that no loss of the dissolved gas and interaction with the atmosphere or with the material of the container occurs during collection and storage of the water sample.

For detailed discussion of methods for noble gas analysis in waters (and other terrestrial fluids), the reader is referred to Bayer *et al.* (1989), Beyerle *et al.* (2000), Clarke *et al.* (1976), Gröning (1994), Ludin *et al.* (1997), Rudolph (1981) and Stute (1989).

The various *CFC* species (*CFC*-12, *CFC*-11, and *CFC*-113) and *SF*$_6$ are measured in the laboratory using a purge-and-trap gas chromatography procedure with an electron capture detector (ECD). For a detailed description, reference is made to Busenberg and Plummer (2000) and Plummer and Busenberg (2000).

7.5.1 Dissolved helium

Helium (^4He) is the second most abundant element in the universe after hydrogen, constituting 23% of its elemental mass. It is the second lightest element in the Periodic Table after hydrogen. It is a colourless, odourless, non-toxic, and virtually inert monoatomic gas. Although there are eight known isotopes of helium, only two isotopes, namely ^4He (2 protons and 2 neutrons) and ^3He (2 protons and 1 neutron), are stable. The remaining six isotopes are radioactive and extremely short-lived. The isotopic abundance of helium, however, varies greatly, depending on its origin. On the Earth, helium is produced by radioactive decay of heavy elements such as uranium and thorium. The alpha particles produced during radioactive alpha decay are actually fully ionized ^4He nuclei. In the ^{238}U, ^{235}U, and ^{232}Th series, the decay chains yield 8, 7, and 6 atoms of ^4He, respectively (Fig. 7.10). Helium thus produced is released from grains by etching, dissolution, and fracturing and by alpha recoil and then expelled into the atmosphere by diffusion and temperature variations. Helium is produced in rocks and soils within the Earth and eventually escapes to outer space from the atmosphere due to its inertness and low mass. Its concentration, therefore, shows a gradient decreasing towards the ground surface–atmosphere interface. In the Earth's atmosphere, the concentration of helium is approximately 5.3 parts per million (ppm) by volume.

Water in solubility equilibrium with the atmosphere usually shows low concentrations of helium because of its low solubility ($\sim$1%) and low atmospheric concentration (Fig. 7.9). In groundwaters, however, dissolved helium can be high due to its radioactive production and release from earth materials. Since its diffusivity in water is low (7.78 $\times$ $10^{-5}\ cm^2s^{-1}$ at 25°C; CRC, 1980), it gets trapped in groundwater and collects additional radiogenic helium from the aquifer matrix while moving through it until the groundwater comes in contact with the atmosphere. This provides a basis for groundwater dating by the helium method, knowing the helium production rate in the aquifer matrix from its uranium and thorium content (Castro *et al.* 2000; Clark *et al.* 1998; Fröhlich and Gellermann 1987; Mazor and Bosch 1992; Torgersen 1980). Presence of fractures and fissures, particularly in hard rocks, provides preferential pathways for migration of radiogenic helium from the interconnected pore spaces between the subsurface grains. Therefore, depend-

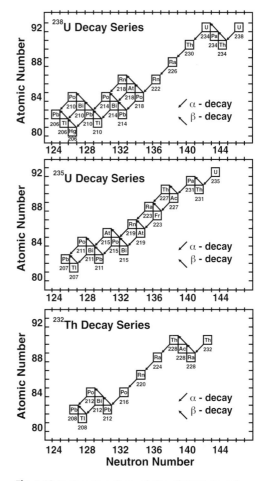

Fig. 7.10 Radioactive decay chains of Uranium and Thorium. The number below the box with element code is mass number (= Atomic number + Neutron number). Note that 8, 7, and 6 α-particles (the helium nuclei) are produced in the decay chains of ^{238}U, ^{235}U, and ^{232}Th, respectively. ^{222}Rn is produced in the decay chain of ^{238}U. Redrawn from Faure (1986). © John Wiley & Sons Inc.

ing on the collection volume of a particular fracture or fissure, helium concentration in the groundwater residing in the fracture zone can be significantly higher compared to its production from the rock volume in contact with groundwater. This forms the basis of delineating the deep subsurface structural zones (Filippo *et al.* 1999; Gulec *et al.* 2002; Gupta and Deshpande 2003a; Kulongoski *et al.* 2003, 2005; Minissale *et al.* 2000).

Furthermore, during and just before occurrence of an earthquake, enhancement of helium concentration in some groundwaters has been noticed (Barsukov *et al*. 1984; Rao *et al*. 1994; Reimer 1984). This is possibly due to rock dilation and fracturing during incipient fault movements that form a basis for using helium monitoring as a tool for earthquake prediction. The important point to note is that the entire radiogenic helium may not be released from the rock matrix in the normal course, giving a release factor of less than unity. When the groundwater has dissolved helium in excess of the atmospheric equilibration value, it is referred to as having a 'helium anomaly' or 'excess helium'. Groundwaters may have 'excess helium' due to other factors as well, such as: (i) occurrence of uranium mineralization within a particular aquifer zone or in its vicinity, forming a basis for radioactive mineral exploration (Reimer 1976); and (ii) occurrence of a natural gas reservoir below the aquifer system which can have high concentration of helium, forming a basis for petroleum exploration using helium surveys (Jones and Drozd 1983; Weismann 1980). In addition, depending on the physical nature of the capillary fringe zone overlying the saturated zone, air bubbles can be entrapped and transported into the saturated zone. Air bubbles can also be entrapped by rapid rise of the water table due to occasional heavy rainfall spells in semi-arid regions. Entrapment and subsequent dissolution of such air bubbles into groundwater introduces excess air and consequently all its components including helium. This contributes additional helium to groundwater over and above its atmospheric equilibration value (Andrews *et al*. 1985; Heaton 1981; Heaton and Vogel 1981).

Helium permeates readily through many materials and for this reason it is important to carefully choose the containers in which groundwater samples intended for a post-collection helium analysis at a later time are to be stored. The most appropriate material for such containers is oxygen-free high conductivity (OFHC) copper tubing in which samples for helium analyses are sealed by crimping at both the ends (Craig and Lupton 1976; Lupton *et al*. 1977). Gupta and Deshpande (2003a) showed that loss of dissolved helium from thick-walled soda-lime glass bottles, with bromobutyl synthetic rubber stoppers manufactured according to guidelines

of US Pharmacopoeia Standard II (USP Std-II) and secured by additional triple aluminium caps fixed by a hand-held crimping tool, was less than 0.15 % per day.

7.5.1.1 ^4He dating method

The helium-4 dating method for groundwater is based on estimating the amount and rate of accumulation of *in-situ* produced radiogenic ^4He in groundwater (Andrews and Lee 1979; Stute *et al*. 1992). If *secular equilibrium* and release of all the ^4He atoms produced in the interstitial water are assumed, groundwater ages can be calculated from the annual ^4He production rate estimated as below (Torgersen 1980):

$$J'_{He} = 0.2355 \times 10^{-12} U^* \tag{7.19}$$

where

$$U^* = [U]\{1 + 0.123([Th]/[U] - 4)\} \tag{7.20}$$

J'_{He} = production rate of ^4He (in $cm^3 STP\, g^{-1}$ rock a^{-1}); $[U]$ and $[Th]$ are concentrations (in *ppm*) of U and Th, respectively, in the rock/sediment.

Accumulation rate (AC'_{He}) of ^4He in $cm^3 STP\, cm^{-3}$water a^{-1} is, therefore, given by:

$$AC'_{He} = J'_{He}.\rho.\Lambda_{He}.(1 - n)/n \tag{7.21}$$

where Λ_{He} = helium release factor; ρ = rock density ($g\, cm^{-3}$); and n = rock porosity.

If helium measurements are made on equilibrated headspace air samples, the dissolved helium concentrations ($cm^3 STP\, cm^{-3}$water) would be in terms of Air Equilibration Units (*AEU*), which expresses the dissolved helium concentration in terms of the corresponding equilibrium dry gas phase mixing ratio at 1 atmosphere pressure at $25°C$. As a result, water in equilibrium with air containing 5.3 *ppmv* helium is assigned the dissolved concentration of 5.3 *ppmAEU*. Water of meteoric origin has a minimum helium concentration ($^4He_{eq}$) of 5.3 *ppmAEU*, acquired due to its equilibration with the atmosphere during the fall of raindrops. Excess helium ($^4He_{ex}$) represents the additional helium acquired by groundwater either from *in-situ* produced radiogenic ^4He and/or other subsurface sources.

Using a dimensionless Henry coefficient (H_x) of 105.7 for helium at $25°C$ (Weiss 1971), 5.3 *ppmAEU* corresponds to a moist air equilibrium concentration of 4.45×10^{-8} *cm³ STP He cm⁻³* water. Therefore:

$$AC_{He} = AC'_{He}.10^6.H_X.\left(T/T_0\right).P_0/(P_0 - e)$$

$$(7.22)$$

where AC_{He} = accumulation rate of ^4He in *ppmAEU* a^{-1}; $T_0 = 273.15°$ K; $P_0 = 1$ *atm*; and e = saturation water vapour pressure (0.031 *atm*) at $25°C$, $H_x = 105.7$.

For an average $[U] = 1$ *ppm* for alluvial sedimentary formations and $[Th]/[U] = 4$ (Ivanovich 1992), $n = 20$ %; $\Lambda_{He} = 1$; $\rho = 2.6$ *g cm⁻³*, an *in-situ* ^4He accumulation rate (AC_{He}) of 2.59×10^{-4} *ppmAEU* a^{-1} is obtained. Therefore, from the measured helium concentration of the sample (^4He_s), the age, t, of groundwater can be obtained by using:

$$t = \,^4He_{ex}/\,AC_{He} \qquad (7.23)$$

where $^4He_{ex}$ is obtained by subtracting $^4He_{eq}$ (= 5.3 *ppmAEU* at $25°C$ and 1 atmosphere pressure) from the measured concentration in groundwater sample (^4He_s).

The above formula, however, ignores 'excess air' helium ($^4He_{ea}$) due to super-saturation of atmospheric air as the groundwater infiltrates through the unsaturated zone. Various models have been proposed for estimating this component in groundwater (Aeschbach-Hertig *et al*. 2000; Kulongoski *et al*. 2003) based on measurement of other dissolved noble gases. However, from other studies (Holocher *et al*. 2002), it appears that $^4He_{ea}$ can be up to 10 to 30% of $^4He_{eq}$. Therefore, if it is ignored, there may be an apparent over-estimation of groundwater ages to the extent of about 6 *ka*.

Furthermore, (^4He_s) can contain other terrigenic helium components (Stute *et al*. 1992) that can cause overestimation of groundwater age. These terrigenic components are: (i) flux from an external source, for example, deep mantle or crust, adjacent aquifers, etc. (Torgersen and Clarke 1985); and (ii) release of geologically stored ^4He from young sediments (Solomon *et al*. 1996). Depending upon the geological setting, particularly in regions of active tectonism and/or hydrothermal circulation, contri-

bution of these sources may exceed the *in-situ* production by several orders of magnitude (Gupta and Deshpande 2003a). Additional measurements/data (e.g. $^3He/^4He$ ratio, and other noble gases) are required to resolve these components. For recent reviews on terrigenic helium, reference is made to Ballentine and Burnard (2002) and Castro *et al*. (2000). However, in many cases it seems possible to rule out major contributions from terrigenic He sources, since the helium flux may be shielded by underlying aquifers that flush the helium out of the system before it migrates across them (Castro *et al*. 2000; Torgersen and Clarke 1985). According to Andrews and Lee (1979), with the exception of a few localized sites and for very old groundwaters, 'excess He' in groundwater occurs due to *in-situ* production and is, therefore, often used for quantitative age estimations within the aquifer if the U and Th concentrations of the aquifer material are known. But in case there is evidence of deep crustal ^4He flux (J_0) entering the aquifer, the Eqn 7.23 becomes modified thus (Kulongoski *et al*. 2003):

$$t = \,^4He_{Ex}/[(J_0/nZ_0\rho_w)$$
$$+AC_{He}]/8.39 \bullet 10^{-9} \qquad (7.24)$$

where Z_0 is the depth (m) at which the ^4He flux enters the aquifer; and ρ_w is the density of water (1 *g cm⁻³*).

7.5.2 Dissolved radon

Radon (Rn) is a chemically inert radioactive noble gas formed by disintegration of *Ra* (radium) in the decay chains of uranium and thorium (Fig. 7.10). There are 20 known radioactive isotopes of Radon but it does not have any stable isotope. Amongst its isotopes, ^{222}Rn, formed by decay of ^{226}Ra (in the ^{238}U decay series; Fig. 7.10), has the longest half-life of 3.8235 days, decaying to ^{218}Po (polonium) by emitting an α–particle.

Natural radon concentration in the Earth's atmosphere is very low (1 in 10^{21} molecules of air) due to its radioactive decay, though it has been shown to accumulate in the air if there is a meteorological inversion. The natural waters in equilibrium with the atmosphere have low concentration of approximately $7 \times 10^{-4} dpm \; l^{-1}$ (Gupta *et al.* 2002).

Groundwaters can have higher concentration of ^{222}Rn due to its subsurface production. Due to its diffusion into the atmosphere, the radon content in the unsaturated zone, similar to other inert gases, is lower than that in the saturated zone. The mobility of radon in groundwaters is to some extent linked to its parent radium, which is particle- and salinity-sensitive (Krishnaswami *et al.* 1991).

As with helium, faults and fractures in the crust act as preferential pathways for migration of radon. Unlike helium that can go on accumulating in stagnant or confined groundwaters (Gupta *et al.* 2002), radon is expected to be in steady state between production, decay, and mobilization from the rock–soil matrix. The steady state conditions may not prevail during and for a short time following any disturbance, such as an earthquake. Similar to helium anomalies, radon anomalies have also been used as seismic precursors (Shapiro 1981; Virk *et al.* 2001), for delineation of active fracture zones (Reddy *et al.* 2006), radioactive mineral exploration (Jones and Drozd 1983; Weismann 1980), and for groundwater dating using helium/radon ratios (Agarwal *et al.* 2006; Gupta *et al.* 2002). In the last application mentioned above, due to the fact that both the radiogenic gases are produced from the decay of ^{238}U, their ratio is independent of uranium concentration in the rock matrix (Torgersen 1980).

Radon is one of the heaviest gases at room temperatures and can accumulate in buildings and in drinking water and may cause lung cancer (BEIR 1999). Therefore, it is considered to be a health hazard. It is probably the most significant contaminant of indoor air quality worldwide.

For ^{222}Rn measurements, in view of its short half-life, water samples need to be measured within a short period after collection. There are several methods, but γ-spectroscopy provides a convenient tool by counting 609 *keV* gamma rays produced by the decay of its short-lived daughter ^{214}Bi using a high purity germanium (HPGe) gamma ray spectrometer. A repeat counting after a period of more than three weeks gives a measure of ^{222}Rn supported by ^{226}Ra in the groundwater. Activity values are decay corrected to take into account the storage time since collection.

7.5.2.1 $^4He/^{222}Rn$ dating method

Since both ^4He and ^{222}Rn in groundwater, being produced by the α decay of U and/or Th in the aquifer material, have a common origin, their simultaneous measurement in groundwater can also be utilized for calculating its age (Torgersen 1980).

As in the case of ^4He, the ^{222}Rn accumulation rate (AC_{Rn}) in $cm^3 STP \; cm^{-3}$ water a^{-1} is given by:

$$AC_{Rn} = J'_{Rn} . \rho . \Lambda_{Rn} . (1 - n)/n \qquad (7.25)$$

where:

$$J'_{Rn} = 1.45 \times 10^{-14} [U] \qquad (7.26)$$

and J'_{Rn} = production rate of ^{222}Rn in $cm^3 STP \; g^{-1}$ rock a^{-1} and $[U]$ = concentration (in *ppm*) of U in the rock/sediment.

Thus, from the accumulation rate ratio of ^4He and ^{222}Rn $(= AC_{He}/AC_{Rn})$, the age of groundwater can be calculated as follows:

$$Age(t) = (\Lambda_{Rn} \Lambda_{He}) (AC_{Rn}/AC_{He}) (C_4/A_{222})$$

$$(7.27)$$

where $\Lambda_{Rn}/\Lambda_{He}$ is the release factor ratio for radon and helium from the aquifer material to groundwater; C_4 is concentration (*atom* l^{-1}) of ^{44}He; and A_{222} is activity (*disintegration* $l^{-1} \; a^{-1}$) of ^{222}Rn in groundwater. From Eqn 7.19 to Eqn. 7.21 and Eqn 7.25 to Eqn 7.27, it is seen that $^4He/^{222}Rn$ ages are independent of porosity, density, and U concentration, but do require measurement of $[Th]/[U]$ in the aquifer material. The ratio $\Lambda_{Rn}/\Lambda_{He}$ depends upon grain size and recoil path length of both ^{222}Rn ($\sim$0.05 μm) and ^4He (30–100 μm) (Andrews 1977). Release of ^{222}Rn by α-recoil from the outer surface ($\sim$0.05 μm) of a grain ($\sim$2–3 mm) has been estimated to be $\sim$0.005% (Krishnaswami and Seidemann 1988). In addition to α-recoil, both ^{222}Rn and ^4He can diffuse out of rocks/minerals through a network of 100–200$\mathring{A}$ nanopores present throughout the rock or grain body (Rama and Moore 1984). Radon release factors (Λ_{Rn}) ranging from 0.01–0.2 have been indicated from laboratory experiments for granites and common rock-forming minerals (Krishnaswami and Seidemann 1988). On the other hand, Torgersen and Clarke (1985), in agreement with numerous other authors, have shown that the most likely value of $\Lambda_{He} \approx 1$. Converting C_4 ($atm\ l^{-1}$) to C_{He} ($ppmAEU$) units and A_{222} ($disintegrations\ l^{-1}\ a^{-1}$) to A'_{222} ($dpm\ l^{-1}$) units, Eqn 7.27 can be rewritten as:

$$Age(t) = 4.3 \times 10^8 .(\Lambda_{Rn}/\Lambda_{He})$$

$$.(AC_{Rn}/AC_{He}).C_{He}/A'_{222} \qquad (7.28)$$

Here, 1 $ppmAEU$ ^4He concentration corresponds to 2.26×10^{14} atoms of $^4He\ l^{-1}$ of water. Another implicit assumption in the $^4He/^{222}Rn$ dating method is that both ^4He and ^{222}Rn originate from the same set of parent grains/rocks and their mobilization in groundwater is similarly affected.

Andrews $et\ al$. (1982) used the following one-dimensional equation for calculating diffusive transport of ^{222}Rn in granites:

$$C_x = C_0 \exp(-\sqrt{\lambda/D}.X) \qquad (7.29)$$

where C_0 and C_x are concentrations of ^{222}Rn from an arbitrary point taken at $x = 0$ and $x = x$, respectively; D = diffusion coefficient in water ($\sim 10^{-5}\ cm^2\ s^{-1}$); and λ = decay constant for ^{222}Rn. They calculated that $C_x/C_0 = 0.35$ at a distance equal to one diffusion length $X = (D/\lambda)^{1/2}$ (= 2.18 cm). Therefore, even at high ^{222}Rn activity, its diffusion beyond a few metres distance is not possible. The average radon diffusion co-efficient in soils with low moisture content and composed of silty and clayey sand is even lower, $\sim 2 \times 10^{-6}\ cm^2\ s^{-1}$ (Nazaroff $et\ al$. 1988). Thus, ^{222}Rn measurements of groundwater depend essentially on the U concentration in the pumped aquifer horizons in the vicinity of the sampled tube well.

Therefore, for groundwaters that may have a component external to the aquifer, the measured ^{222}Rn, because of its short half-life ($t_{1/2} = 3.825\ d$), is indicative of its local mobilization; whereas ^4He, being stable, may be mobilized from the entire flow-path. In such cases, the resulting $^4He/^{222}Rn$ ages for groundwater samples having high 'excess He' may be over estimated.

7.5.3 Chemical tracers

Application of injected chemical tracers to stream flow gauging (see Section 2.6.1.4) and of environmental chemical tracers to study sea water intrusion (see Section 6.6) has already been mentioned. A wide range of chemically conservative solutes can be found in different reservoirs and transfer of water taking place in recharge/discharge zones within the catchment system can be studied by measuring them. Solutes, such as chloride, sulphate, silica, or bromide, have been used for hydrograph separation (Hooper and Shoemaker 1986; Kobayashi $et\ al$. 1990; Pinder and Jones 1969; Robson and Neal 1990) and as tracers to infer water pathways and response times of the catchment to a given storm (Espeby 1990; Jardine $et\ al$. 1989; Kennedy $et\ al$. 1984). The basic premise of applicability of chemical tracers to evaluating hydrologic processes is that the tracers, artificial as well as environmental, imprint their signature on the source waters similar to isotopes. The interpretation of such chemical data requires care in that the catchment is a major source of some constituents, such as base cations and silica, and the atmosphere is a major source of anions, such as sulphate and chloride, and each may change within a catchment. For example, sulphate budgets may balance in highly acidic catchments, whereas chloride budgets may not balance in the short term due to interaction with vegetation (Harriman $et\ al$. 1990; Peters 1991).

7.5.3.1 Fluoride

In an aqueous solution, the element fluorine (F) commonly occurs as the fluoride ion (F^-). In groundwaters, fluoride is one of the most important environmental pollutants resulting from natural and/or anthropogenic factors. Pure fluorine (F_2) is a corrosive pale yellow gas that is a powerful oxidizing agent. It is the most reactive and electro-negative of all the elements and readily forms compounds with most other elements including noble gases.

The ill effects of excessive fluorine-fluoride ingestion on human health have been extensively studied (Gupta and Deshpande 1998; Luke 1997; WHO 1970, 1984; Zero $et\ al$. 1992). The occurrence and development of endemic fluorosis has its roots in the high fluoride content in water, air, and soil of which water is perhaps the major contributor. The Bureau of Indian Standards (BIS 1990) recommends the permissible limit of fluoride in drinking water as 1 part per million (*ppm*), which is lower then the WHO (1984) drinking water limit of 1.5 *ppm*. Gupta and Deshpande (1998) have described the hazardous effects on human and cattle health from excessive fluoride in the NGC region of Gujarat, where it was observed that ingestion of excessive fluoride has adverse effects on human teeth and bones (known as dental and skeletal fluorosis).

The normal fluoride content in the atmospheric air is reported as between 0.01 and 0.4 $\mu g\ m^{-3}$. However, in industrial areas, it is known to range from 5 to 111 $\mu g\ m^{-3}$ (Bowen 1966). Average fluoride content of precipitation varies from almost nil to 0.089 $mg\ l^{-1}$ with as high a value as 1 $mg\ l^{-1}$ (Gmelin 1959; Sugawara 1967) in industrial areas. The fluoride content of rivers also varies greatly, depending on the fluoride content of effluent discharges of groundwater feeding the stream and on the amount of precipitation and runoff. The fluoride content in surface waters is generally higher during dry periods due to its evaporative concentration (Deshmukh *et al*. 1995 a, b). The mean concentration of fluoride in ocean waters ranges from 0.03 to 1.35 $mg\ l^{-1}$ and is found to increase with depth in many cases (Bewers 1971; Riley 1965). The groundwaters, particularly in the arid and semi-arid regions throughout the globe, are known to have high fluoride concentration (Handa 1977).

The fluoride content of groundwater greatly depends on the type of soil/rocks/aquifer it comes into contact with. There are more than 150 fluorine-bearing minerals in the form of silicates, halides, and phosphates (Strunz 1970; Wedepohl 1974). Fluorite (CaF_2) is the most widely distributed fluorine bearing mineral in Nature, while fluorapatite ($Ca_5F(PO_4)_3$) is a common member of the immiscible phase generated during early differentiation of mafic and ultramafic magmas forming apatite-magnetite rocks. In advanced stages of differentiation, fluorine is enriched into the residuum and, therefore, rocks such as granitoids or pegmatites have a high content of apatite resulting in average fluorine content of up to 2.97% (by weight) in these rocks. In sedimentary rocks, in addition to fluorite and fluorapatite in the clastic component,

clay-sized minerals such as micas (muscovite, biotite, phlogopite, zinnwaldite, and lepidolite) and clay minerals (montmorillonite, illite, and kaolinite) have a high content of fluorine due to replacement of OH^- by F^- or due to admixture of skeletal debris in which hydroxyl bonds are replaced by fluorine in the hydroxy-apatite structure (Carpenter 1969). Out of the total fluorine content of the clay-sized particles, 80–90% is hosted in the micaceous minerals and the remainder is associated with clay minerals (Koritnig 1951).

The fluorine content of soils depends mainly on the rocks from which they are derived and the climatic regime in which the soil is formed. However, in warm and humid climates, decomposition of organic remains can be the main source of fluorine. The average fluorine content in the soil ranges between 90 and 980 *ppm* (Fleischer and Robinson 1963). The leaching of fluoride from soils and the aquifer matrix into groundwater involves adsorption-desorption and dissolution-precipitation processes. Since the solubility of fluorite and fluorapatite is very low in natural waters (Deshmukh *et al*. 1995b), it is dissolved slowly by the circulating water through leaching. Fluoride from mica is leached out rapidly. However, on account of the ionic strength of complex forming ions, solubility of fluorite can be drastically modified. Calcium and sulphate ions significantly lower the fluorite solubility in natural waters, often causing precipitation of fluorite (Deshmukh *et al*. 1993; Handa 1977; Perel'man 1977). Distribution of *Ca* and *F* in groundwater is, therefore, antipathic (Deshmukh *et al*. 1995b; Dev Burman *et al*. 1995; Handa 1977; Srivastava *et al*. 1995). Fluorine from the clay minerals is readily desorbed in an alkaline environment. Fluoride to hydroxyl ion exchange in clay minerals depends upon the concentration of fluoride ion and *pH* of the circulating water (Hubner 1969). Thus, solubility of fluorine-bearing minerals is governed by various parameters such as *pH*, alkalinity, ionic strength, calcium, and sulphate ion concentration, etc.

It is seen from the global map of endemic fluorosis affected areas (Fig. 7.11a) that most of the fluoride affected regions are in arid to semi-arid climatic zones (Fig. 7.11b). In India, fluoride-rich groundwaters are found in arid

(a)

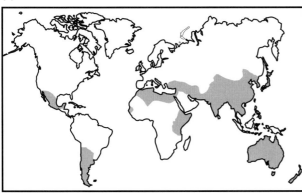

(b)

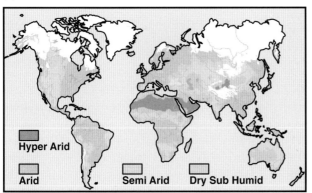

Hyper Arid

Arid   Semi Arid   Dry Sub Humid

Fig. 7.11 (a) Global distribution of areas affected by endemic fluorosis. Redrawn from http://www.unicef.org/ programme/wes/info/fl_map.gif. (b) Geographical distribution of arid and semi-arid regions. Redrawn from www.wateryear2003.org/en/ev.php-URL_ID=5137. It may be noted that most fluoride-affected areas are located in the arid and semi-arid regions. See also Plate 7.11.

regions, especially in Rajasthan, Gujarat, and interior parts of the southern peninsula characterized by episodic rainfall separated by extended dry periods (Agrawal *et al*. 1997; Jacks *et al*. 1993; Vasavada 1998). Fluoride excess is also reported from the arid climatic regions in North China (Yong and Hua 1991).

The fluoride content of thermal springs can increase with temperature due to increased rock–water interaction at elevated temperature (Chandrasekharam and Antu 1995) and/or scavenging from the large volume of rock during hydrothermal circulation (Gupta and Deshpande 2003a; Minissale *et al*. 2000), but this is not observed as a rule. There are thermal springs with a concentration of fluoride as low as <0.15 *mg l*$^{-1}$ and as high as 55.4 *mg l*$^{-1}$ (Gmelin 1959; Sugawara 1967).

7.6 Models for interpretation of groundwater age

In an aquifer system, an ensemble of water molecules arriving at a particular location within it comprises molecules that spend various time durations between their recharge and arrival. The concept of groundwater dating involves estimating the average time spent in the aquifer by the water molecules before reaching a given location. Thus, the age of groundwater at a given location is the average time spent by water molecules in the aquifer since the time of recharge until their arrival at the location. Depending on the conceptual mathematical model employed to describe the aquifer system, age can give additional information on the system. The common conceptual models in use for interpretation of the groundwater age estimates are:

(i) Piston Flow Model (PFM); (ii) Well-Mixed Reservoir Model (WMRM); and (iii) Dispersion-Advection Model (DAM).

Piston Flow Model (PFM) is the most commonly used groundwater flow model for estimating groundwater age. It assumes that as groundwater moves away from the recharge area, all the flow lines have the same velocity and that hydrodynamic dispersion as well as molecular diffusion of water molecules is negligible. Thus, water moves from the recharge area as if a parcel is being pushed by a piston with the mean velocity of groundwater. This implies that a radio tracer that appears at the sampling point at a given time, t, has entered the system at the time instant $(t - T)$, and from that moment its concentration decreases only due to radioactive decay during the time span, T, spent in the aquifer. Therefore:

$$C_{out}(t) = C_{in}(t - T).exp(-\lambda T) \qquad (7.30)$$

The term C represents the tracer concentration (or activity) of water with the suffix representing the appropriate location. Eqn 7.30 describes a dynamic system and is mathematically equivalent to describing the concentration of a radioisotope in a static water parcel isolated since its recharge, whereby:

$$C_t = C_0.exp(-\lambda t) \qquad (7.31)$$

Here, t is the radiometric age of the water that corresponds to the age, T, of the dynamic system. If x is the distance from the recharge boundary, $T = x/u$ can be used to estimate the flow rate (u) of groundwater in the aquifer:

$$C_t = C_0 exp(-\lambda x/u) \qquad (7.32)$$

Unlike the *PFM*, if it is assumed that the recharge flux completely mixes with the entire volume of the reservoir before its outflow, one gets another extreme situation and the model is known as the *Well Mixed Reservoir (WMR) model*. In applying this model to an aquifer system, it is assumed that the well-mixed reservoir comprises the entire volume between the recharge area and the sampling point. Under this condition for a radiotracer:

$$C_t = C_0/(1 + \lambda t) \qquad (7.33)$$

In Eqn 7.33, λ is the radioactive decay constant and τ is the ratio of the reservoir volume to the input

flux to the reservoir and represents the estimated mixing time (or the mean *residence time*) of water between the recharge area and the sampling location. It is seen that τ, as estimated from the tracer data, actually represents a dynamic parameter – the mixing time:

$$\tau = \frac{1}{\lambda}\left(\frac{C_0}{C_t} - 1\right) \qquad (7.34)$$

The phenomenon of mixing accompanying the movement of water molecules through porous media can also be described by a diffusion-advection equation in which the diffusion coefficient is replaced by a dispersion coefficient (Gupta 2001; Scheidegger 1961). For a radiotracer, the one-dimensional continuity equation in the groundwater flow system may be written as:

$$\frac{\partial C}{\partial t} = \frac{\partial \left(D^{\partial C}/\partial t - uC\right)}{\partial x} + W_1 - W_2 \qquad (7.35)$$

In the above equation, D is the diffusion coefficient of the tracer, and as in the case of *PMF*, x is the distance from the recharge boundary, u is the bulk flow velocity, and W_1 and W_2 are the rates of the introduction and removal of the tracer, respectively. With further assumptions of u and D not being a function of x and in the case of steady state (i.e. $\partial C/\partial t = 0$), the above equation reduces to:

$$D\frac{\partial^2 C}{\partial x^2} - u\frac{\partial C}{\partial x} + W_1 - W_2 = 0 \qquad (7.36)$$

In the case of radioactive tracers, the term W_2 includes, in addition to radioactive decay, loss of tracer due to non-radioactive processes (e.g. adsorption on the aquifer matrix). Considering the loss of tracer by radioactive decay alone and for $W_1 = 0$, the above equation can be rewritten as:

$$D\frac{\partial^2 C}{\partial x^2} - u\frac{\partial C}{\partial x} - C = 0 \qquad (7.37)$$

This equation for $D = 0$, and the boundary condition $C = C_0$ at $x = 0$, give the solution for the piston flow (Eqn 7.32).

In the case of finite dispersion, the solution of (Eqn 7.37) for the boundary conditions $C = C_0$ at $x = 0$ and $C = 0$ at ∞, is given by Gupta *et al.* (1981):

$$C/_{C_0} = \exp\left[\frac{xu}{2D}\left\{1 - \left(1 + 4\lambda D/_{u^2}\right)^{1/2}\right\}\right]$$

(7.38)

The tracer concentration decreases exponentially with distance, similar to *PFM* (Eqn 7.32). Therefore, a simplistic application of the *PFM* would give an apparent velocity:

$$u_a = \frac{u}{2}\left\{1 - \left(1 + 4\lambda D/_{u^2}\right)^{1/2}\right\}$$

(7.39)

There are several other mathematical models in use that conceptualize the flow of groundwater in aquifer systems differently, depending on specific aquifer geometry and flow paths. But for a confined aquifer with a well-defined recharge area, the above models are commonly used to interpret the environmental isotopic data and to determine the groundwater age evolution from recharge to discharge regions of the aquifer. This, however, is a challenging task for hydro-geochemists, because sampling locations are often randomly distributed over the area where water from the aquifer is pumped at various depths or where springs bring water to the surface. Several *environmental tracers* (including radionuclides) find application in determining magnitude and direction of groundwater flow, hydro-geological parameters of the aquifer, and age of groundwater (Andrews *et al.* 1989; Cserepes and Lenkey 1999).

7.6.1 Dual radiotracer dating

It is readily seen from Eqn 7.38 and Eqn 7.39 that a simplistic application of PFM would lead to considerable overestimation of flow velocity in highly dispersive groundwater flow systems ($D/(xu) \gg 1$). For a given value of the dispersion coefficient, the difference between the real and the apparent velocities becomes more pronounced for the smaller half-life tracer. For real velocity approaching zero, a minimum apparent flow velocity equal to $(\lambda D)^{1/2}$ is obtained.

For a confined aquifer, assuming $W_1 = W_2 = 0$, Eqn 7.38 shows that a plot of $\ln(C/C_0)$ versus xu/D

is a straight line, whose slope is a function of the parameter D/u^2. However, the value of abscissa (i.e. xu/D) in a given field situation will not be known *a priori*. Therefore, it is not possible to estimate both parameters D and u when using only one radiotracer. However, in the case of two radiotracers '1' and '2', the ratio of their relative concentrations is obtained from Eqn 7.38 as:

$$\ln(C/C_0)_1 = \frac{1 - \left(1 + 4\lambda_1 D/u^2\right)^{1/2}}{1 - \left(1 + 4\lambda_2 D/u^2\right)^{1/2}}\ln(C/C_0)_2$$

(7.40)

which is a straight line on a *log-log* plot, the slope of which is a function of parameter D/u^2 (Fig. 7.12). The distance from the recharge boundary (characterized by $C/C_0 = 1$ for both tracers) to any point in the plotting field is a linear measure of the dimensionless parameter, xu/D. Thus knowing x (i.e. distance of the sampling well from the recharge boundary), both u and D can be estimated.

7.6.1.1 General case: semi-confined aquifer

The steady state continuity equation for the general case of a semi-confined aquifer, taking into account the variation of u (due to leakage influx and outflux) along the aquifer and with other simplifying assumptions as for Eqn 7.35, can be written as:

$$D\frac{d^2C}{dx^2} = \frac{d(uC)}{dx} + pC_0 - EC - \lambda C - qC = 0$$

(7.41)

The term p represents the rate of leakage influx of relatively young water (activity $\approx C_0$) from the overlying unconfined aquifers and the term, q, represents the leakage out flux (activity $= C$) from the aquifer. With reference to Eqn 7.35, $pC_0 = W_1$ and $(E + \lambda + q)C = W_2$. If p is assumed constant, then $du/dx =$ constant, or zero implying $q =$ constant or zero. If we further assume that the aquifer is extensive enough so that tracer activity at discharge boundary is not controlled by the leakage flux (i.e. $dC/dx = 0$), the solution (Gupta *et al.* 1981) of Eqn 7.41 is:

$$C = \beta_1 \exp(m_1 x) + \beta_2 \exp(m_2 x) + pC_0/K$$

(7.42)

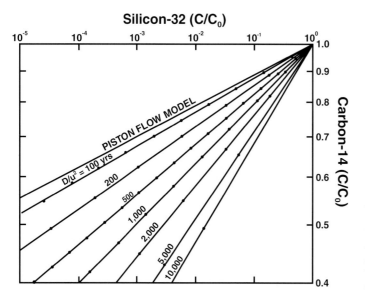

Fig. 7.12 Expected variation of concentrations of ^{32}Si and ^{14}C in a confined aquifer. The distance of 'dots' from $C/C_0 = 1$ for both tracers (corresponding to recharge boundary in the field) along each line corresponds to integer value of xu/D at intervals of 5 for $D/u^2 = 100$ and 200 *years*, and 1 for other values of D/u^2. Redrawn from Gupta *et al.* (1981). © American Geophysical Union.

with

$$K = \lambda + E + p$$

$$m_1 = u \left\{ 1 + \left(1 + 4KD/u^2\right)^{1/2} \right\} / 2D$$

$$m_2 = u \left\{ 1 - \left(1 + 4KD/u^2\right)^{1/2} \right\} / 2D$$

$$\beta_1 = \frac{(m_2/m_1)\, C_0 \,(1 - p/K)\, exp\,\{(m_2 - m_1)\, L\}}{\left[1 - (m_2/m_1)\, exp\,\{(m_2 - m_1)\, L\}\right]}$$

$$\beta_2 = C_0 \,(1 - p/K)\, [1 - (m_2/m_1)$$
$$exp\,\{(m_2 - m_1)\, L\}]$$

where L is the lateral extent of the aquifer. In the limit $L \to \infty$ and since $m_1 > m_2$, $\beta_1 \approx 0$ and $\beta_2 \approx C_0\,(1 - p/K)$. In this case, Eqn 7.42 becomes:

$$^C/_{C_0} = exp\,(mx) + \{p/(p+\lambda)\,\{1 - exp\,(mx)\}$$
(7.43)

where

$$m =$$

$$\frac{u_0 + (p-q)x - [\{u_0 + (p-q)\,x\}^2 + 4\,(\lambda + p)\, D]^{1/2}}{2\,D}$$

(7.44)

For the particular case $p = q$; $du/dx = 0$, the u is constant and m becomes independent of x and is given by:

$$m = \frac{u\,[1 - \{1 + 4\,(\lambda + p)\, D/u^2\}^{1/2}]}{2\,D}$$
(7.45)

It is easy to see that for $p \neq 0$, the tracer concentration given by Eqn 7.43 approaches an asymptotic value:

$$(C/C_0)_{asym} = p/(p + \lambda)$$
(7.46)

independent of eddy diffusivity, D. From Eqn 7.43 and Eqn 7.46, it is easy to see that the asymptotic value is reached first for the smaller half-life tracer as one moves away from the recharge boundary along the length of the aquifer. In a real field situation, unless the tracer half-life is compatible with aquifer dimensions, it may not be possible to observe the asymptotic value of the tracer concentration to enable estimation of the value of p.

Gupta *et al.* (1981) outlined a graphical procedure using Eqn 7.43 employing simultaneous measurements of two radiotracers to estimate the aquifer parameters of interest. It is seen (Fig. 7.13) from the log-log plot of activity ratios, C/C_0, of two tracers ^{32}Si and ^{14}C for $p \neq 0$, that the straight line behaviour observed in $p = 0$ is significantly altered and the lines bend towards the larger half-life tracer axis. Concentration of the shorter-lived tracer reaches an asymptotic value first; eventually terminal values are reached for both the tracers. In principle, if the sampling is extensive, one should first estimate D and u from the straight line part of the curve near the origin,

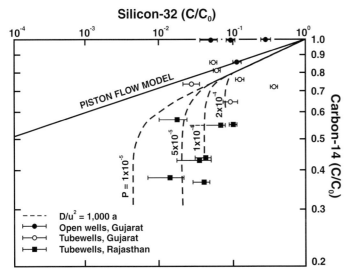

Silicon-32 (C/C_0)

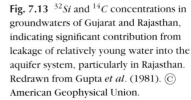

Fig. 7.13 ^{32}Si and ^{14}C concentrations in groundwaters of Gujarat and Rajasthan, indicating significant contribution from leakage of relatively young water into the aquifer system, particularly in Rajasthan. Redrawn from Gupta *et al.* (1981). © American Geophysical Union.

as can be done from Fig. 7.12 and using Eqn 7.40. Having estimated both *D* and *u*, the value of *p* can be estimated from an appropriate curve, as in Fig. 7.13. This diagram has some data points from the available ^{32}Si and ^{14}C measurements on groundwaters in Gujarat and Rajasthan taken from Nijampurkar (1974). Because of relatively few data points and large scatter, it is not possible to quantitatively estimate the various groundwater parameters from this data. However, significant leakage of relatively young groundwater is indicated, particularly in the case of samples from Rajasthan. Gupta (2001) used these concepts of dual radiotracer dating with data on ^{14}C and ^{36}Cl generated by Cresswell *et al.* (1999a, b) from Central Australia.

7.7 Tracers for estimation of groundwater recharge

Groundwater recharge may be defined as 'the downward flow of water reaching the water table, contributing to the groundwater reservoir' (Lerner *et al.* 1990). There are two main types of recharge: direct (vertical infiltration of precipitation where it falls on the ground) and indirect (infiltration following runoff). It is generally believed that in temperate climates most recharge is direct, whereas in arid regions most recharge occurs from surface runoff. However, this distinction does not always hold true: there are some situations in temperate

regions where indirect recharge dominates, most notably in karst areas, where recharge occurs from losing rivers and via swallow holes and other solution features. Without underestimating the significance of indirect recharge, the emphasis here is on direct recharge.

Groundwater recharge can be estimated using both environmental and injected tracers. Lerner *et al.* (1990) separated the methods into *signature* methods and *throughput* methods. In the signature method, a parcel of water containing the tracer is tracked and dated. Throughput methods, on the other hand, involve a mass balance of tracer, comparing the concentration in precipitation with the concentration in soil water or sometimes with the concentrations below the water table. Piston flow is generally assumed in most tracer studies. However, tracers can be used to investigate flow processes, including the occurrence of preferential pathways.

Throughput tracer studies may involve both the saturated and the unsaturated zones. Recharge estimates based on the use of environmental tracers (i.e. Cl^- and ^3H) in the saturated zone, give long-term (>1 *year*) averages and an areally-integrated value, which is difficult to obtain using soil physics methods. Regional recharge rates are required for resource planning. Studies based on monitoring of tracers in unsaturated profiles usually yield point (area $<0.1\ m^2$) recharge values. Thus the technique could be developed to measure spatial variability and to ascertain the effects of various factors on recharge.

7.7.1 Environmental chloride method

The most successful non-isotopic environmental tracer in hydrologic studies is chloride, which is deposited from the atmosphere both as dry and wet precipitation transported by winds. Chloride has been used for estimating areal recharge based on its concentration in the saturated zone (Eriksson and Kunakasen 1969; Kitching *et al.* 1980), and localized recharge by considering its depth distribution in soil water (Allison and Hughes 1978; Kitching *et al.* 1980; Peck *et al.* 1981; Sharma and Hughes 1985). From the depth distribution of chloride concentration of soil water (C) and its volumetric water content (θ), time-averaged vertical water flux (q_w) can be computed using a steady state water and solute flow model, i.e.:

$$q_w = \frac{1}{C}\left(J_s + D_s\,\theta\,\frac{\partial C}{\partial z}\right) \tag{7.47}$$

where J_s is the average chloride input to the system ($= P\,C_p$, with C_p as the time-averaged chloride concentration of the long-term average annual rainfall P); D_s is the diffusion-dispersion coefficient of the solute; and $\partial C/\partial z$ is the slope of the observed chloride concentration with respect to depth. The computed vertical water flux below the root zone depth of the local vegetation is interpreted as the average recharge rate to the groundwater.

Chloride concentrations of groundwater from saturated zone have also been used to estimate the ratio of annual recharge to total groundwater storage. In this case, the groundwater recharge is given by the ratio of chloride deposition ($mg\ m^{-2}\ a^{-1}$) and its concentration ($mg\ m^{-3}$) in groundwater.

Complications may arise in interpreting chloride data, for example, when water movement through the profile deviates significantly from vertical, or when water flow through the profile is not steady and uniform. Even in relatively uniform, sandy profiles, about 50% of annual recharge may occur via movement of water through preferred pathways, bypassing the soil matrix (Sharma and Hughes 1985; Sukhija *et al.* 2003). In some forested lateritic profiles with clayey subsoil, recharge through

macro-pores amounts to more than 95% (Peck *et al.* 1981) of the total infiltration at the given site. Several aspects of interpretation of the observed chloride profiles still remain unresolved.

7.7.2 Environmental tritium method

Tritium, generated by atmospheric nuclear tests, is the most common environmental isotope used in recharge studies (Smith *et al.* 1970). Localized recharge rates have been estimated from distribution of tritium in the unsaturated part of the profile (Sukhija and Rama 1973) and found to agree with other estimates, such as from the chloride method and injected tritium method.

When using the environmental tritium method, to begin with the tritium input function from precipitation is estimated, as shown in Fig. 7.14a. In this case study (from Ahmedabad, India) tritium concentration measurement data for the period 1962–1970 was available. For the period 1952–1962, the input function was extrapolated from correlation with Ottawa (Canada) data (Sukhija and Rama 1973). The bomb tritium peak for 1953–1964 is clearly seen. The total tritium fallout ($TU \times cm$) is given by:

$$T = \sum A_i \cdot P_i \tag{7.48}$$

where
   T = total tritium fallout ($TU \times cm$);
   A_i = tritium concentration (TU) of precipitation in month (or year), i;
   P_i = precipitation amount (cm) in month (or year), i.

The summation is carried out for the period 1952 (onset of thermonuclear era) to the time of carrying out the investigation (November 1967 or November 1969 in the two cases shown in Fig. 7.14a).

Similarly for percolation function:

$$t = \sum a_j \cdot m_j \tag{7.49}$$

where
   t = total amount of tritium ($TU \times cm$) present in the soil column;
   a_j = tritium content (TU) of the soil segment j;
   m_j = Moisture content (cm) of the soil segment j.

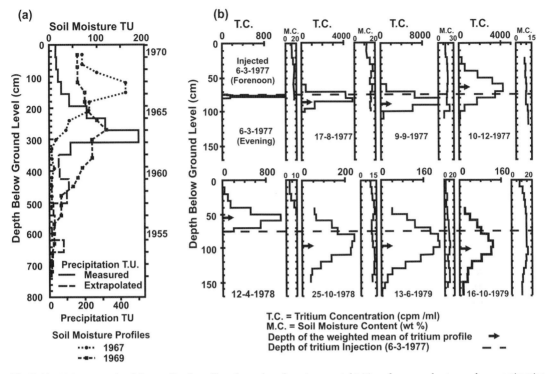

Fig. 7.14 (a) An example of the application of bomb-produced environmental tritium for groundwater recharge estimation using soil moisture tracing in North Gujarat, India. Redrawn from Sukhija *et al.* (2003). (b) An example of the application of injected tritium as tracer for soil moisture movement and groundwater recharge estimation at a site near Ahmedabad, Gujarat, India. Redrawn from Bhandari *et al.* (1986). The peak of various tritium profiles is seen to move seasonally upward/downward in tandem with soil moisture movement, but from one rainy season to the next a net downward movement in response to groundwater recharge can be clearly seen. Both types of measurements also indicate that rainfall of a given year takes a couple of years of transport through the unsaturated soil zone before reaching the water table. Redrawn from Sukhija *et al.* (2003). © Springer.

Since 'T' equals total fallout of tritium at the site and 't' equals the amount percolated (net amount after evapotranspiration and runoff losses are taken into account), t/T represents the fraction of rainfall that goes to recharge the groundwater. Therefore, recharge, as a percentage of precipitation by the integral method, is given by:

$$r\,(\%) = t/T \bullet 100 \quad Tritium\,Integral\,Method$$

$$(7.50)$$

The tritium peak method aims to locate the position of 1963 precipitation in the soil profile, assuming a layered movement of soil moisture (piston flow). The recharge is computed by estimating the total soil moisture above the peak position and assigning the same to precipitation since 1963 up to the time of investigation (November 1967 or November 1969 in Fig. 7.14a). Thus:

$$r\,(\%) = S/P \bullet 100 \quad Tritium\,Peak\,Method \quad (7.51)$$

where

$S =$ Soil moisture content (in *cm*) in the column from surface to the depth of tritium peak;

$P =$ Precipitation since 1963 up to the time of investigation.

Overall, recharge rates computed by the tritium peak method represent an average of relatively small duration (5–7 years), while the tritium integral method gives recharge averaged over longer duration (15–20 years), though the results are likely to be biased towards years with higher tritium content in precipitation.

The environmental tritium peak methods have been successfully applied throughout the world, and in many different climates (Scanlon *et al.* 2006). In low-rainfall, arid conditions, infiltrating rainwater moves intermittently and remains in the unsaturated zone for decades. In moister climates where infiltration is high, artificial tritium can be injected as a tracer to determine the rate of recharge. This method is applicable for estimating piston flow recharge anywhere, provided layered movement of soil moisture can be reasonably assumed.

7.7.3 Tritium injection method

While estimating recharge using tritium injection, tritiated water is generally injected as a slug involving a set of 5 to 10 injections at a properly selected site to form a layer below the root zone of the local vegetation (depth: 60–80 *cm*) before commencement of the rainy season, and soil profiles are sampled during the post-rainfall dry period up to depths of 2 to 6 *m*. At each site, several such sets of close-by injections are made for repeat sampling for a period of 1 to 3 *years* after the injections are done. An example of one such study is shown in Fig. 7.14b. Interesting observations regarding soil moisture movement in arid and semi-arid regions can be made from this diagram: (i) the tracer peak in the soil moisture gradually broadens with time due to diffusion/dispersion and movement; and (ii) there is a net downward layered movement of tracer over the full year cycle. The upward layered movement during the dry period is also clearly discerniable. Therefore, recharge estimates from tritium injection methods are likely to show higher variability depending on climatic conditions prevailing at the site during the time period between injection and sampling. To obviate some of these problems, tritium injection stations were sampled over a period of two to three consecutive years during 1977–1979.

Datta *et al.* (1980b) conceptualised the unsaturated transport of moisture as pulses of infiltrating sheet through a series of interconnected expandable mixing cells sub-dividing the soil profile. In response to infiltration, each hypothetical soil cell expands to hold soil moisture up to its field capacity and drains any excess water downwards

under the influence of gravity to the soil cell directly below, making the recharge pulse move downwards towards the water table. By applying the mass balance equation successively to each hypothetical mixing cell subdividing the soil profile at discrete time intervals, the process of dispersion and mixing leading to broadening of the tracer peak (Fig. 7.14b) is mathematically approximated and closely reproduces the field data. A close correlation was found to exist between the rainfall and estimated number of recharge pulses from the field data (Datta *et al.* 1980b).

Tritium concentrations in the saturated zone have been used to estimate the ratio of annual recharge to total groundwater storage in the aquifer. From the fact that evaporation is accompanied by isotope enrichment, depth distributions of ^{18}O and ^{2}H have been used to estimate recharge, particularly in areas where direct evaporation is a major component of total vapour-transpiration (Allison and Hughes 1978; Sharma and Hughes 1985).

7.8 Tutorial

Ex 7.1 The figure below (Fig. 7.15) shows a salt wave recorded during a slug-injection measurement experiment in a mountainous stream. The volume of injected solution was 6.35 *l*. This volume resulted from mixing 1 *kg* of salt with 6 *l* of water (total solution volume 6.36 *l*), followed by extracting 10 *ml* of injection solution for use in the calibration procedure. The stream EC data were

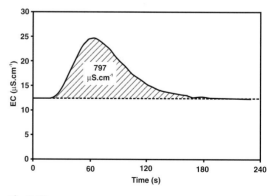

Fig. 7.15

logged at 1 second intervals, and the calibration constant was 2.99×10^{-6} cm/μS.

Solution:

$$Q = \frac{V}{k.\Delta t.\left[EC(t) - EC_{bg}\right]}$$

$$= \frac{6.35 \times 10^{-3} m^3}{(2.99 \times 10^{-6} cm/\mu S)(1s)(797 \mu S/cm)}$$

$$= 2.66 m^3.s^{-1}$$

Ex 7.2 Protactinium-231 is a radioactive isotope, decaying into the daughter isotope Actinium-227 with a half-life of 32 760 years. What type of decay does protactinium-231 undergo?

Ex 7.3 If a particular confined groundwater contains 3% of the original activity of carbon-14 in the recharge area, what would be its approximate age?

Ex 7.4 In a 1-day-old sample of radioactive zinc ($_{30}^{71m}Zn$) with a half-life of 3.97 *hours*, how much of the original zinc has decayed?

Ex 7.5 What is the total number of radioactive atoms present in 1 *gram* of each of the following substances:

Ex 7.6 Using the following chart identify the isotope best suited to examine date an artefact from the early Bronze Age in Egypt, around the year 3000 BC?

Ex 7.7 Abundance ratio of $^2H/^1H$ and $^{18}O/^{16}O$ in VSMOW are 1.5575×10^{-4} and 2.0672×10^{-2}, respectively. Determine abundance ratios of $^2H/^1H$ and $^{18}O/^{16}O$ in water samples with $\delta^{18}O$ and δD values as: (–10, –70), (0, 11), (5, 62), (–22, –135). Calculate *d*-excess for these water samples.

[Hint: Use Eqn 7.2 and Eqn 7.17]

Ex 7. 8 The average $\delta^{18}O$ of the base flow water of a high mountain stream was estimated as –10 ‰, and $\delta^{18}O$ of the snowpack meltwater was estimated as –16 ‰. During a particular melt season, the $\delta^{18}O$ of the stream water monitored at four different times was –11, –13, –13.5, and –12 ‰. Calculate the proportion of the snowmelt component in the total stream flow water at the time instants of the four measurements.

[Hint: Use two-component mixing model with base flow as one and snowpack as the other component]

Table 7.4

Cosmogenic Nuclides				
Nuclide	Symbol	Half-life	Source	Natural Activity
Carbon-14	^{14}C	5730 a	Cosmic-ray interactions, $^{14}N(n, p)^{14}C$	6 pCi g^{-1} (0.22 Bq g^{-1}) in organic material
Hydrogen-3 (Tritium)	^3H	12.3 a	Cosmic-ray interactions with N and O, spallation from cosmic-rays, $^6Li(n, alpha)^3H$	0.032 pCi kg^{-1} (1.2 $\times$ 10^{-3} Bq kg^{-1})
Beryllium-7	^7Be	53.28 days	Cosmic-ray interactions with N and O	0.27 pCi kg^{-1} (0.01 Bq kg^{-1})

Table 7.5

Isotope		Half-life of parent (a)	Effective dating range (a)
Parent	Daughter		
Uranium-235	Lead-207	710 million	>10 million
Potassium-10	Argon-49	1.3 billion	10 000 to 3 billion
Carbon-14	Nitrogen-14	5730	Up to 50 000

Ex 7.9 The groundwater samples up gradient from a lake clearly showed meteoric origin, as their isotopic composition was similar to that of rainfall (Fig. 7.16). Water samples from two in-lake wells had isotopic compositions similar to meteoric water and plotted along the meteoric water line, indicating that groundwater inflow occurs at these sites. The isotopic compositions of groundwater down-gradient from the lake had enriched values relative to meteoric water and plotted along a mixing line described by the expression, $\delta D = 4.6 \times$ $\delta^{18}O - 1.3$. The mixing line connects the isotopic composition of the surface water and groundwater end members, for example, evaporated lake water and groundwater up-gradient from the lake. Intersection of the GMWL and the mixing line is at (-3.5, -17.5) and represents the average isotopic composition of groundwater. Use isotope mass-balance calculations, using $\delta^{18}O$ and δD, to estimate the fraction of lake water contribution at each well location.

Well No.	$\delta^{18}O$ (‰)	δD (‰)	F_{sw} ($\delta^{18}O$)	F_{sw} (δD)
W-80	0.95	2.50	0.63	0.59
W-20	−2.25	−11.0	0.18	0.19
W-40	−1.00	−5.55	0.35	0.35
W-60	−0.45	−2.00	0.43	0.46
Lake	3.60	16.5	1.0	1.0

F_{sw} denotes fraction of lake water mixing with ground water

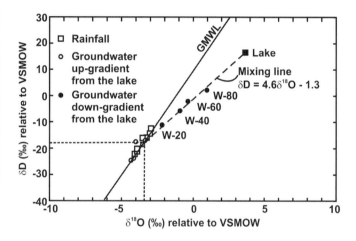

Fig. 7.16

8

Statistical analyses in hydrology

Without realizing it we inadvertently use *statistics* in our day-to-day life. For example, we make an estimate in our mind of the time required to cover a given distance using a particular mode of transport or how much money we need to carry when leaving home each morning. Similarly while investing money; we weigh risk versus gain when choosing a particular investment option – all based on previous realizations. Professional analysts use statistics for collection, analysis, interpretation, and presentation of numerical data with a view to understanding a variety of phenomena. Many hydrologic data represent random phenomena – atmospheric temperature, rainfall, wind, etc. – measured using various instruments or derived from other measurements.

All measurements can be placed into one of the two overriding categories, namely continuous and discontinuous or discrete. *Continuous measurements* are those where fractions can be included and invariably contain some error in the recorded value, because one can always measure them more accurately by using a device capable of measuring in smaller units – say in millimetres rather than centimetres for length measurement. Flow rate, rainfall intensity, water-surface elevation, etc. are some other examples of continuous measurements in hydrology. *Discontinuous* or *discrete measurements* can only take certain values, usually integers. Numbers of storm events occurring in a specified time period, and the number of overtopping flood events per year for a levee on a stream, are some examples of discrete measurements in hydrology.

Values derived from such data should be rounded off to the nearest whole number.

In addition to the above, there are other categories in which data may fall: (i) *Qualitative measurements* describing quality rather than a quantity, for example, type of species (male/female) or colour that describe quality, and normally the data that follow are numbers/counts for each of these categories; (ii) *Quantitative measurements* normally indicating a variable that can be measured quantitatively, for example, length and weight are quantitative variables; and (iii) *Derived variables* are data not measured directly but calculated from other measurements. The commonly derived variables are proportions (ratios and percentages).

In statistics, any set of people or objects with some attribute in common are defined as a *population*. One can have a population of students, monthly rainfall at a certain station, storm events in a given region during a given time interval – anything one may be interested in studying can be defined as a population. The objective of statistical analysis is to characterize a given population, and draw inferences or identify trends inherent in it. For example, one may want to test whether a new drug is effective for a specific group of people. Most of the time, it is not possible to administer every one with a new drug. Instead the drug is administered to a sample group of people from the population to see if it is effective. This subset of the population is called a *sample*. In this sense all types of hydrological data, limited in number

Modern Hydrology and Sustainable Water Development, 1st edition. © S.K. Gupta.
Published 2011 by Blackwell Publishing Ltd.

of realizations/measurements, are samples of their respective parent populations. When something is measured or inferred about a population, it is called a *parameter*. When the same thing is measured in a sample, it is called a *statistic*. For example, if one could measure the average age of the entire Indian population, it would be a parameter. When the same is measured in a sample of Indians, it would be called a statistic. Thus, a population is to a parameter as a sample is to a statistic. This distinction between the sample and the population is important because the aim is to draw inferences about populations from different samples. Similarly symbols to denote populations and samples differ too. For denoting population, one uses Greek letters for parameters and Roman letters for denoting a measure (statistic) from a sample.

8.1 Descriptive statistics

Before analysing data, one needs to know some details about the datasets (samples) obtained. First, one needs to know about the frequency pattern (distribution) inherent in the data. Second, one also needs to know how close the sample fits the population. Descriptive statistical methods are used mainly to characterize the sample. These are used, in the first instance, to get a feel for the data and then to ascertain its suitability for carrying out relevant statistical tests on them. One also needs to indicate the errors associated with the results derived from the statistical tests that were applied.

8.1.1 Measures of central tendency

8.1.1.1 Mean

The most common description of the central tendency is the *mean* or *average* value of a number of measurements and is defined as the sum of all the individual values divided by the number of measurements carried out. Thus:

$$\bar{x} \equiv \frac{1}{N} \sum_{i=1}^{N} x_i \equiv (x_1 + x_2 + x_3 + \cdots + x_N)/N \quad (8.1)$$

N is the number of measurements; i denotes the serial number of an arbitrary measurement; and x

is the parameter measured. The summation sign is usually written as $\sum x_i$, omitting $i = 1$ and N. As an example, the mean value of a set of 7 measurements of a variable having values 28.5, 18.75, 22.9, 25.4, 24.55, 23.7, and 23.9, using Eqn 8.1, is 23.96. Since, in practice, the number of measurements is limited, a better estimate of the population mean can be obtained if the sample size is large. However, if we could increase this number to infinity, we would end up with the mean parameter of the population, defined as:

$$\mu \equiv \lim_{N \to \infty} \left(\frac{1}{N} \sum x_i \right) \quad (8.2)$$

If parameter values x_i occur f_i times, respectively (i.e. occur with frequencies f_i), the mean is:

$$\bar{x} = \frac{\sum f_i x_i}{\sum f_i} = \frac{\sum f_i x_i}{N} \quad (8.3)$$

where $N = \sum f_i$ is the total frequency (i.e. total number of cases/number of measurements).

While summarizing large volumes of raw data, it is often useful to classify the data into *classes*, or *categories*, and to determine the number of individual data belonging to each class, called the *class frequency*. A tabular arrangement of data by classes together with corresponding frequencies is called the *frequency distribution*, or *frequency table*. Data organized and summarized as a frequency distribution are often called *grouped data*. Although in the grouping process much of the original detail is generally lost, a clear 'overall' picture and some vital relationships may emerge. The size or width of a class interval is the difference between the lower and upper end members of the class and is referred to as *class width*, *class size*, or *class length*. If all class intervals of a frequency distribution have equal widths, this common width is denoted by Δx. The *class mark* or *class midpoint* is the midpoint of the class interval and is obtained by adding the lower and upper class limits and dividing by 2 or adding $\Delta x/2$ to the lower class limit. In subsequent mathematical analysis, all observations belonging to a given class are assumed to coincide with the class mark. Thus, for example, in Fig. 8.1, the frequency in the class interval 19.5–20.5 is considered to correspond to 20. A graphical representation of frequency distribution with a set of rectangles

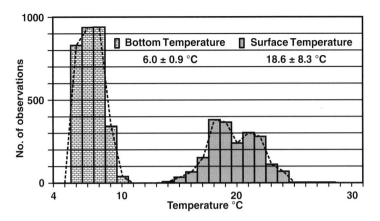

Fig. 8.1 Frequency histogram of epilimnetic- and hypolimnetic temperature measured in a lake. Dotted line connecting the midpoints of the top of the rectangles of the histogram is the frequency polygon. The values of mean (6.0 and 18.6) and standard deviation (0.9 and 8.3) show, what is also seen from the histogram, that it is significantly cooler at the bottom with much less scatter than for the hypolimnetic temperatures. Based on http://waterontheweb.org/curricula/ws/unit_05/Mod_17/mod17-1.ppt.

having (a) base on the horizontal axis (the *x*-axis) with centre as the class mark and width equal to class width, and (b) area proportional to class frequency is called a histogram or *frequency histogram*. Line connecting the midpoints of the top of the rectangles of histogram is the frequency polygon (Fig. 8.1).

If in constructing the histogram, Δx, or the class width chosen is too large, nearly all data may lie within a single column, suggesting a good statistical certainty, but a bad resolution. If, on the other hand, Δx is made too small, that is, the resolution is increased, few data values fall within any given column and the reliability appears to be less (scattering histogram).

8.1.1.2 Median

The mean, as defined above, can distort the picture if there are a few extreme yet genuine values. The *median* parameter can help in this case and is found by locating the 'middle' value. The *median* is defined as the value such that half the measurements of the dataset are below and the other half are above it. In the set of seven values given above (18.75, 22.9, 23.7, 23.9, 25.4, 24.55, 28.5), the median is 23.9. If N is an even number, the median is the mean of the two middle values. For a symmet-

rical distribution, the mean and the median values are the same.

8.1.1.3 Mode

Mode is the value that occurs in a dataset most often and may not exist in many datasets, including the one above. Unlike mean or median, many datasets may not have a unique value of mode. For example, the mode of the sample (1, 3, 6, 6, 6, 6, 7, 7, 12, 12, and 17) is 6 (unique). Given the list of data (1, 1, 2, 4, 4), the mode is not unique.

8.1.1.4 Geometric mean

The *geometric mean*, G, of a set of N positive numbers, x_i, is the Nth root of the product of all the numbers:

$$G = \sqrt[N]{x_1 x_2 x_3 \cdots x_N} \qquad (8.4)$$

8.1.1.5 Harmonic mean

The *harmonic mean*, H, of a set of N numbers, x_i, is the reciprocal of the arithmetic mean of the reciprocals of numbers:

$$H = \frac{1}{\frac{1}{N}\sum_{i=1}^{N}\frac{1}{x_i}} = \frac{N}{\sum \frac{1}{x_i}} \qquad (8.5)$$

8.1.1.6 The root mean square (RMS)

The *root mean square* (*RMS*) or the *quadratic mean* of a set of numbers, x_i, is defined as:

$$RMS = \sqrt{\frac{\sum_{i=1}^{N} x_i^2}{N}} = \sqrt{\frac{\sum x_i^2}{N}} \qquad (8.6)$$

8.1.1.7 Quartiles, deciles, and percentiles

Extending the idea of median, one can think of values that divide the given dataset into four equal parts. These values, denoted by Q_1, Q_2, and Q_3, are called the first, second, and third *quartiles*, respectively; the value Q_2 equals the median.

Similarly, the values that divide the data into 10 equal parts are called *deciles* and are denoted by D_1, D_2, ..., D_9, while the values dividing the data into 100 equal parts are called *percentiles* and are denoted by P_1, P_2, ..., P_{99}. The 5th decile and the 50th percentile correspond to the median. The 25th and 75th percentiles correspond to the 1st and the 3rd quartiles, respectively.

8.1.2 Measures of dispersion and variability

One can characterize data in a better way using some other parameters, such as *Range* and *Standard Deviation*. *Range* is given by the highest and lowest values in a dataset. In the above dataset, the range is 18.75-28.5. *Standard Deviation* (*SD*) is useful to measure the variation of a sample around the mean value. To appreciate the significance of mean and standard deviation, let us consider measurements of epilimnetic and hypolimnetic temperatures of a lake (Fig. 8.1). The mean values indicate that the lake is significantly cooler at the bottom. But it is also seen that the variation at the bottom is much less and so is the standard deviation – the value being 0.9 for the epilimnetic temperatures compared to 8.3 for the hypolimnetic temperatures.

Since by definition, the average deviation of the measured value from the mean equals zero, the sum of the squares of the deviations of individual values from the mean is taken to characterize the

distribution. The resulting value is referred to as *variance of population* and designated by σ^2:

$$\sigma^2 \equiv \lim_{N \to \infty} \left[\frac{1}{N} \sum (x_i - \mu)^2 \right]$$
$$= \lim_{N \to \infty} \left(\frac{1}{N} \sum x_i^2 \right) - \mu^2 \qquad (8.7)$$

If we now consider a real set of measurements (sample), the *standard deviation* of the set (sample) is:

$$s = \sqrt{\left(\frac{1}{N-1} \sum (x_i - \bar{x})^2 \right)} \qquad (8.8)$$

The fact that $(N - 1)$ instead of N appears in the denominator is due to loss of one degree of freedom while estimating $\bar{x}$. The need to do this can be appreciated by considering the extreme case of only one measurement. Because a single measurement cannot give any idea on the precision of the measurement, the value of s obtained using $(N - 1)$ in the denominator of Eqn 8.8 will not be a real number in accordance with the situation.

If parameter values, x_i, occur in the dataset with corresponding frequencies, f_i, the standard deviation can be written as:

$$s = \sqrt{\left(\frac{1}{N-1} \sum f_i(x_i - \bar{x})^2 \right)} \qquad (8.9)$$

The mean and SD are often combined to characterize a given dataset as $\bar{x} \pm s$. For a large dataset, $s \approx \sigma$ and the distinction between the two is often ignored.

Coefficient of Variation is yet another parameter to show how much variation occurs in a given dataset. It is calculated using:

$$Coefficient\ of\ variation = \frac{s}{\bar{x}} \times 100 \qquad (8.10)$$

The higher its value, the more the number of data points one needs to collect to be confident that the sample is representative of the population. It can also be used to compare variation between different datasets.

Example 8.1. A set of 8 annual rainfall values recorded from a mountainous catchment are: 15, 20, 21, 20, 36, 15, 25, and 15. Estimate various measures of central tendency described above.

Solution. :

i	x_i	x_i (arranged)	$(x_i)^2$	$x_i - \bar{x}$	$(x_i - \bar{x})^2$	$1/x_i$
1	15	15	225	−5.875	34.51563	0.066667
2	20	15	400	−0.875	0.765625	0.05
3	21	15	441	0.125	0.015625	0.047619
4	20	20	400	−0.875	0.765625	0.05
5	36	20	1296	15.125	228.7656	0.027778
6	15	21	225	−5.875	34.51563	0.066667
7	25	25	625	4.125	17.01563	0.04
8	15	36	225	−5.875	34.51563	0.066667
Σ	167		3837	0	350.875	0.415397

With reference to calculations shown in the table above:

Mean $(\bar{x}) = 167/8 = 20.875$; Median $= 20$;
Mode $= 15$
Geometric mean (G)
$\quad = \sqrt[8]{15 \times 20 \times 21 \times 20 \times 36 \times 15 \times 25 \times 15}$
$\quad = 19.992$
Harmonic mean $= 8/0.415397 = 19.259$
RMS $= \sqrt{(3837/8)} = 21.900$
Range $= 36$ to $15 = 21$
Std. Deviation $(\sigma) = \sqrt{(350.875/7)} = 7.080$; Variance $= \sigma^2 = 50.125$

Although one can calculate these univariate statistical parameters by hand, it gets quite tedious when we have more than a few values and a number of variables. However, most statistical programs are capable of calculating them easily.

8.1.3 *Measures of shape and distribution*

8.1.3.1 *Moments*

If x_i are the N values assumed by a variable, x, in a distribution, the quantity obtained by raising all values of x_i to power r and adding them and then dividing the sum with total number of x_i, i.e.:

$$\bar{x^r} = \frac{x_1^r + x_2^r + \cdots x_N^r}{N} = \frac{\sum\limits_{i=1}^{N} x_i^r}{N} = \frac{\sum x_i^r}{N} \quad (8.11)$$

is called the rth moment of the distribution. The first *moment* with $r = 1$ gives the arithmetic mean $\bar{x}$.

The rth moment about the mean $\bar{x}$ is defined as:

$$m_r = \frac{\sum\limits_{i=1}^{N}(x_i - \bar{x})^r}{N} = \frac{\sum(x_i - \bar{x})^r}{N} = \overline{(x_i - \bar{x})^r} \quad (8.12)$$

If $r = 1$, then $m_1 = 0$. If $r = 2$, then $m_2 = s^2$, the variance.

For grouped data with parameters, x_i, occurring with frequencies, f_i, the above moments are given by:

$$\bar{x^r} = \frac{f_1 x_1^r + f_2 x_2^r + \cdots f_K x_K^r}{N} = \frac{\sum\limits_{i=1}^{K} f_i x_i^r}{N} = \frac{\sum x_i^r}{N} \quad (8.13)$$

$$m_r = \frac{\sum\limits_{i=1}^{K} f_i(x_i - \bar{x})^r}{N} = \frac{\sum f_i(x_i - \bar{x})^r}{N} = \overline{(x_i - \bar{x})^r} \quad (8.14)$$

where $N = \sum_{i=1}^{K} f_i = \sum f_i$.

To generalize, one can define dimensionless moments about the mean that are independent of any specific system of units as:

$$a_r = \frac{m_r}{s_r} = \frac{m_r}{(\sqrt{m_2})^r} = \frac{m_r}{\sqrt{m_2^r}} \quad (8.15)$$

where $s = \sqrt{m_2}$ is the standard deviation. Since $m_1 = 0$ and $m_2 = s^2$, we have $a_1 = 0$ and $a_2 = 1$.

8.1.3.2 *Skewness*

Skewness is a measure of the degree of asymmetry, or departure from symmetry, of a distribution around the mean/central maximum. If smoothed frequency curve of a distribution has a longer tail to the right of the central maximum than to the left, the distribution is said to be *skewed to the right* or to have *positive skewness*. If the reverse is true, it is said to be *skewed to the left* or to have *negative skewness* (Fig. 8.2a).

For skewed distributions the mean tends to lie on the same side of the mode as the longer tail, that is, to the right of the mode for right skewed

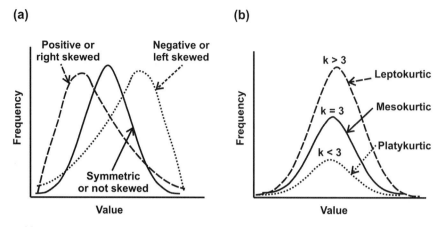

Fig. 8.2 Types of frequency curves based on (a) Skewness and (b) Kurtosis.

and to the left of the mode for left skewed. Thus a measure of asymmetry is provided by the difference (mean – mode). This can be made dimensionless if one divides it by a measure of dispersion, such as standard deviation, leading to the definition:

$$Skewness = \frac{mean - mode}{standard\ deviation} = \frac{\bar{x} - mode}{s} \tag{8.16}$$

The following empirical relation is found to hold for unimodal frequency distribution curves that are moderately skewed:

$$mean - mode = 3(mean - median) \tag{8.17}$$

Using Eqn 8.17, the coefficient of skewness can be defined as:

$$Skewness = \frac{3(mean - median)}{standard\ deviation}$$
$$= \frac{3(\bar{x} - median)}{s} \tag{8.18}$$

Eqn 8.16 and Eqn 8.18 are called *Pearson's first and second coefficients of skewness.*

8.1.3.3 Kurtosis

Kurtosis indicates the peakness of a distribution, usually taken relative to the *normal distribution* (also called *Gaussian distribution*) – a symmetrical bell-shaped curve. A distribution having relatively high peak, such as in Fig. 8.2b, is called *leptokur-*

tic. The curve that is less peaked or flat topped is called *platykurtic*. The normal distribution that is neither very peaked nor very flat topped is called *mesokurtic.*

One measure of the kurtosis employs the *4th* moment about the mean expressed in a dimensionless form and is given by:

$$Moment\ coefficient\ of\ kurtosis = k = a_4 = \frac{m_4}{s^4}$$
$$= \frac{m_4}{m_2^2} \tag{8.19}$$

which is often denoted by k or b_2. For a normal distribution, $k = b_2 = a_4 = 3$. For this reason, $k > 3$ defines a leptokurtic distribution; $k < 3$ defines a platykurtic distribution.

8.2 Probability theory

Due to the vast complexities of nature, hydrologic events such as rainfall and stream flow that affect our day-to-day life and are closely linked to our climate are highly variable. As a result, probability theory and statistics have to be employed to predict occurrences of these events, or many other natural phenomena. For example, water resources managers want to know if the probability of the spring and summer stream flow of major rivers in the year 2010 will be, say 30% lower than their respective long-term average values? Or what is the probability that there will be a warm December

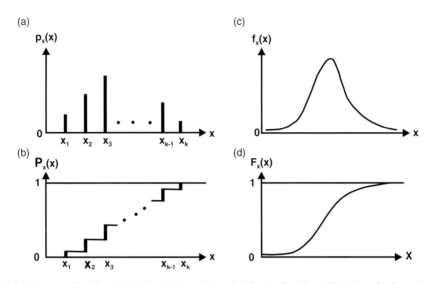

Fig. 8.3 (a) Probability mass function (*PMF*), p_x. (b) Cumulative distribution function (*CDF*), P_x, of a discrete random variable. (c) Probability density function (*PDF*), f_x, and (d) Cumulative distribution function (*CDF*), F_x, of a continuous random variable. Based on Tung *et al.* (2006).

month in 2010? *Probability* is defined as the number of times a specific event occurs out of a total number of observed events (that defines the sample size):

$$P(X) = n_x/n, \quad lim\ n \to \infty \qquad (8.20)$$

where

> $P(X) =$ Probability of occurrence of event $X(t)$;
>
> $n_x =$ Number of occurrences of $X(t)$ event;
>
> $n =$ Sample size.

A hydrologic variable is called a random variable because its occurrence is associated with a certain degree of uncertainty or probability. A probability distribution represents the relative frequency of occurrence of a population. Often, if a sample is large enough, one can plot histograms of the data with small class intervals. By fitting the histograms with a curve, an approximate probability distribution of the data X occurring in each class interval is obtained. This curve is often called the *probability density function* (*PDF*). The shape of this curve gives an idea about the nature of *PDF*. For example, if it is bell-shaped and symmetric, it is likely to be a Gaussian (normal) distribution. Probability distributions commonly encountered in hydrology are shown in Fig. 8.5, which also shows interrelation-

ships amongst them. Some of these distributions are discussed in Sections 8.2.2 and 8.2.3.

The *cumulative distribution function* (*CDF*, or simply distribution function (*DF*), of a random variable X is defined as:

$$F_x = P(X \leq x) \qquad (8.21)$$

The *CDF* of a function $F_x(x)$ gives the non-exceedance probability, which is a non-decreasing function of the argument x, that is $F_x(a) \leq F_x(b)$ for $a < b$. For a discrete random variable X, the *probability mass function* (*PMF*), is defined as:

$$p_x(x) = P(X = x) \qquad (8.22)$$

The *PMF* of a discrete random variable and its associated *CDF* are sketched schematically in Fig. 8.3(a, b). For a continuous random variable, the *PDF*, $f_x(x)$, is defined as:

$$f_x(x) = \frac{dF_x(x)}{dx} \qquad (8.23)$$

Similar to the discrete case, the *PDF* of a continuous random variable must satisfy two conditions: (i) $f_x(x) \geq 0$; and (ii) $\int f_x(x)\ dx = 1$. Thus the *PDF* of a continuous random variable, $f_x(x)$, is the slope of its corresponding *CDF*. Graphical representations of PDF and CDF are shown in Fig. 8.3(c, d). Given the *PDF* of a random variable X, its *CDF* can be obtained as:

$$F_x(x) = \int_{-\infty}^{x} f_x(u)du \qquad (8.24)$$

in which u is a dummy variable of integration. It should be noted that the *PDF* has meaning only when it is integrated between two points. The probability of a continuous random variable taking on a particular value is zero, whereas this may not be the case for discrete random variables.

The rth-order moment of a random variable X about a reference point $X = x_0$ is defined, for the continuous case, as:

$$E[(X - x_0)^r] = \int_{-\infty}^{\infty} (x - x_0)^r f_x(x)dx$$

$$= \int_{-\infty}^{\infty} (x - x_0)^r dF_x(x) \qquad (8.25)$$

whereas for the discrete case:

$$E[(X - x_0)^r] = \sum_{k=1}^{N} (x_k - x_0)^r p_x(x_k) \qquad (8.26)$$

where $E[\cdot]$ is a *statistical expectation operator*. The first three moments ($r = 1, 2, 3$) as above are used to describe the central tendency, variability, and asymmetry of the distribution respectively. It may be noted that Eqn 8.26 is identically equal to Eqn 8.13 and Eqn 8.14 with the values of x_0 taken as zero and $\bar{x}$, respectively. As in the case of a discrete variable, two types of moments are commonly used for continuous variables also, that is, moments about the origin where $x_0 = 0$, and the central moments where $x_0 = \mu_x$, with $\mu_x = E[X]$. The rth-order central moment is denoted as $\mu_r = E[(X - \mu_x)^r]$, whereas the rth-order moment about the origin is denoted as $\mu'_r = E(X^r)$.

From Eqn 8.25, it is easy to see that the central tendency of a continuous random variable X can be represented by its *expectation*, which is the first-order moment (or *mean*) about the origin:

$$E(X) = \mu_x = \int_{-\infty}^{\infty} xf_x(x)dx = \int_{-\infty}^{\infty} xdF_x(x)$$

$$= \int_{-\infty}^{\infty} [1 - F_x(x)]dx \qquad (8.27)$$

The following two operational properties of the expectation are useful:

1. The expectation of the sum of several random variables (regardless of their dependence) equals the sum of the expectation of the individual random variables, i.e.:

$$E\left(\sum_{k=1}^{K} a_k X_k\right) = \sum_{k=1}^{K} a_k \mu_k \qquad (8.28)$$

in which $\mu_k = E(X_k)$, for $k = 1, 2, \ldots, K$.

2. The expectation value of multiplication of several independent random variables equals the product of the expectation of the individual random variables, i.e.:

$$E\left(\prod_{k=1}^{K} X_k\right) = \prod_{k=1}^{K} \mu_k \qquad (8.29)$$

The *median*, x_{md}, of a continuous random variable divides the distribution into two equal halves and is also the 50th percentile satisfying the condition:

$$F_x(x_{md}) = \int_{-\infty}^{x_{md}} f_x(x)dx = 0.5 \qquad (8.30)$$

The *mode* x_{mo} at which the value of a *PDF* peaks can be obtained by solving the following equation:

$$\left[\frac{\partial f_x(x)}{\partial x}\right] = 0 \qquad (8.31)$$

The variance is the second-order central moment (i.e. about the mean) and for the continuous case is defined as:

$$Var(x) = \mu_2 = \sigma_x^2 = E[(X - \mu_x)^2]$$

$$= \int_{-\infty}^{\infty} (x - \mu_x)^2 f_x(x)dx \qquad (8.32)$$

Three important properties of the variance are:

1. $Var(a) = 0$ when a is constant;
2. $Var(X) = E(X^2) - E^2(X) = \mu'_2 - \mu_x^2$;
3. The variance of the sum of several independent random variables equals the sum of variances of the individual random variables, that is:

$$Var\left(\sum_{k=1}^{K} a_k X_k\right) = \sum_{k=1}^{K} a_k^2 \sigma_k^2 \qquad (8.33)$$

where a_k is constant, and σ_k is the standard deviation of random variable X_k, $k = 1, 2, \ldots, K$.

Two measures of the *skewness coefficient* that indicate the asymmetry of a distribution are defined by Eqn 8.16 and Eqn 8.18. Another measure of the asymmetry of the *PDF* of a random variable is related to the third-order central moment as:

$$\gamma_x = \frac{\mu_3}{\mu_2^{1.5}} = \frac{E[(X - \mu_x)]}{\sigma_x^3} \tag{8.34}$$

The sign of the skewness coefficient indicates the degree of symmetry of the probability distribution function. If $\gamma_x = 0$, the distribution is symmetric about its mean. The distribution has a long tail to the right when $\gamma_x > 0$; whereas $\gamma_x < 0$ indicates that the distribution has a long tail to the left (Fig. 8.2a).

Kurtosis, κ_x, is a measure of the peakness of a distribution. As in Eqn 8.19, it is related to the fourth-order central moment of a random variable and may also be written as:

$$\kappa_x = \frac{\mu_4}{\mu_2^2} = \frac{E[(X - \mu_x)^4]}{\sigma_x^4} \tag{8.35}$$

with $\kappa_x > 0$. For a random variable having a normal distribution, its kurtosis equals 3. For all possible distribution functions, the skewness coefficient and kurtosis must satisfy the following inequality:

$$\gamma_x^2 + 1 \le \kappa_x \tag{8.36}$$

Example 8.2. The time to failure, T, of a pump in a water distribution system, is a continuous random variable having PDF as:

$$f_t(t) = exp(-t/1250)/\beta \quad for \ t = 0, \beta > 0$$

in which t is the time elapsed (in hours) before the pump fails; and β is the parameter of the distribution function. Determine the constant β and the probability that the operating life of the pump is longer than 200 h. Determine the first two moments about the origin for the time to failure of the pump. Then calculate the first two central moments. Also find values of mean, mode, median, and 10 percentile for the random time to failure, T.

Solution. The shape of the *PDF* is shown in Fig. 8.9a (for $\alpha = 1$). The *PDF* must satisfy two conditions, that is: (i) $f_t(t) \ge 0$, for all t; and (ii) the area under $f_t(t)$ must equal unity. Compliance with

condition (i) can be proved easily. Condition (ii) is used to determine the value of the constant β as:

$$\int_0^\infty f_t(t)dt = 1 = \int_0^\infty \frac{e^{-t/1250}}{\beta}dt$$

$$= \left[\frac{-1250e^{-t/1250}}{\beta}\right]_0^\infty = \frac{1250}{\beta}$$

Therefore, the constant $\beta = 1250$ h/failure. This particular *PDF* follows exponential distribution. To determine the probability that the operational life of the pump would exceed 200 h, one has to calculate $P(T \ge 200)$:

$$P(T \ge 200) = \int_{200}^\infty \frac{e^{-t/1250}}{1250}dt = [-e^{-t/1250}]_{200}^\infty$$

$$= e^{-200/1250} = 0.852$$

According to Eqn 8.25, moments about the origin are:

$$E(T^r) = \mu_r' = \int_0^\infty t^r \left(\frac{e^{-t/\beta}}{\beta}\right)dt$$

Performing integration by parts, one obtains

for $r = 1$, $\mu_1' = E(T) = \mu_t = \beta = 1250 \ h$
for $r = 2$, $\mu_2' = E(T^2) = 2\beta^2 = 3,125,000 \ h^2$

The central moments can be determined using Eqn 8.25, with $x_0 = \mu_t$:

for $r = 1$, $\mu_1 = E(T - \mu_t) = 0$
for $r = 2$, $\mu_2 = E[(T - \mu_t)^2] = \beta^2 = 1,562,500 \ h^2$

The mean value of the time to failure, called the mean time to failure (MTTF), is the first-order moment about the origin, which is $\mu_t = 1250 \ h$. From the shape of the *PDF* for the exponential distribution, as shown in Fig. 8.9a (for $\alpha = 1$), one can readily see that the mode, representing the most likely time of pump failure, is at the beginning of pump operation, that is $t_{mo} = 0 \ h$.

To determine the median time to failure of the pump, one can first derive an expression for the *CDF* from the given exponential *PDF* as:

$$F_t(t) = P(T \le t) = \int_0^t \frac{e^{-u/1250}}{1250}du = 1 - e^{-t/1250}$$

$$for \ t \ge 0$$

in which u is the dummy variable of integration. The median time to failure, t_{md}, can be obtained

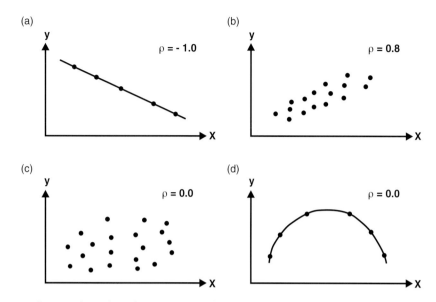

Fig. 8.4 Some sample cases of correlation between two random variables X and Y: (a) linear negative correlation; (b) strong linear correlation; (c) no correlation; (d) correlated nonlinearly but uncorrelated linearly. Redrawn from Tung et al. (2006).

from Eqn 8.30:

$$F_t(t_{md}) = 1 - e^{-t_{md}/1250} = 0.5$$

which yields t_{md} =866.43 h.

Similarly, the 10 percentile value, $t_{0.1}$, that gives the time elapsed over which the pump would fail with a probability of 0.1, can be found in the same way as the median, except that the value of the CDF here is 0.1, that is:

$$F_t(t_{0.1}) = 1 - e^{-t_{0.1}/1250} = 0.1$$

which yields $t_{0.1}$ =131.7 h.

8.2.1 Covariance and correlation coefficient

When a problem involves two dependent random variables, the degree of linear dependence between the two is obtained from the *correlation coefficient*, $\rho_{x,y}$, which is defined as:

$$Corr(X, Y) = \rho_{x,y} = Cov(X, Y)/\sigma_x\sigma_y \qquad (8.37)$$

where $Cov(X, Y)$ is the covariance between random variables X and Y, defined as:

$$Cov(X, Y) = E[(X - \mu_x)(Y - \mu_y)]$$

$$= E(XY) - \mu_x\mu_y \qquad (8.38)$$

The correlation coefficient defined by Eqn 8.37 is called the *Pearson product-moment correlation coefficient* or simply *correlation coefficient* in common usage. It is easily seen that $Cov(X'_1, X'_2) = Corr(X_1, X_2)$, with X'_1 and X'_2 being the normalized random variables. In the realm of statistics, a random variable can be normalized as:

$$X' = (X - \mu_x)/\mu_x \qquad (8.39)$$

Hence a normalized random variable has zero mean and unit variance. Normalization does not affect the skewness coefficient and kurtosis of a random variable, because these parameters are dimensionless. Some commonly encountered cases of correlation between two random variables are shown schematically in Fig. 8.4.

8.2.2 Some discrete univariate probability distributions

As mentioned above, *probability distributions* are classified into two types – discrete and continuous, based on the nature of the random variable. In this section, two discrete distributions, namely the Binomial distribution and the Poisson distribution that are commonly encountered in hydrosystems, are described. Several frequently used

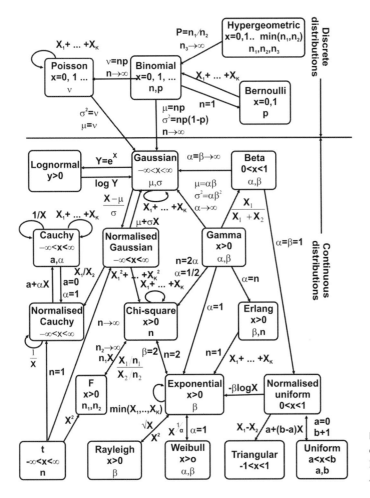

Fig. 8.5 Inter-relationships amongst different univariate distributions. After Leemis (1986). © American Statistical Association.

univariate continuous distributions are described below. The inter-relationships between the various distributions discussed in this chapter, as well as some other distributions not discussed here, are shown in Fig. 8.5.

8.2.2.1 Binomial distribution

The *binomial distribution* is applicable to random processes with only two types of outcomes, for example, a system is either functioning (i.e. successful) or non-functioning (i.e. failure). Consider an experiment involving a total of n independent trials with each trial having two possible outcomes, say success or failure. In each trial, if the probability

of having a successful outcome is p, the probability of having k successes in n trials can be computed as:

$$p_x(k) = C_{n,k} p^k q^{n-k} \qquad (8.40)$$

where $C_{n,k}$ is the binomial coefficient; and $q = (1 - p)$, the probability of failure in each trial. Computationally, it is more convenient to use the following recursive formula to evaluate the binomial *PMF* (Drane *et al.* 1993):

$$p_x(k|n, p) = \left(\frac{n+1-k}{k} \right) \left(\frac{p}{q} \right) p_x(k-1|n, p) \qquad (8.41)$$

for $k = 0, 1, 2, \ldots, n$, with the initial probability $p_x(k=0|n, p) = q^n$. A simple recursive scheme

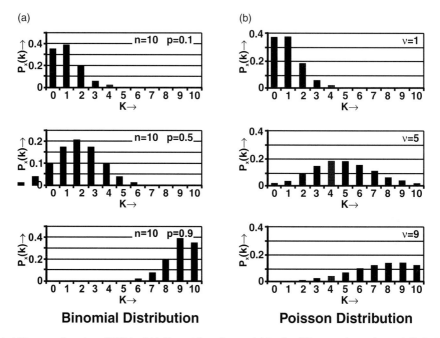

Binomial Distribution   **Poisson Distribution**

Fig. 8.6 Probability mass functions (PMFs) of (a) Binomial random variables for different values of p; and (b) Poisson random variables for different parameter values. The parameter k represents the number of realizations of the event. Based on Tung *et al.* (2006) with permission from Mcgraw-Hill.

for computing the binomial cumulative probability was given by Tietjen (1994).

A random variable X having a binomial distribution with parameters n and p has the expectation $E(X) = np$ and variance $Var(X) = npq$. Shape of the *PMF* of a binomial random variable depends on the values of p and q. The skewness coefficient of a binomial random variable is $(q - p)/\sqrt{(npq)}$. Hence the *PMF* is positively skewed if $p < q$, symmetric if $p = q = 0.5$, and negatively skewed if $p > q$. Plots of binomial *PMF*s for different values of p with a fixed value of n are shown in Fig. 8.6a.

8.2.2.2 Poisson distribution

The *PMF* of a *Poisson distribution* is:

$$p_x(k|v) = \frac{e^{-v}v^k}{k!} \quad for\ k = 0, 1, 2, \ldots \quad (8.42)$$

where the parameter, $v > 0$, represents the mean of a Poisson random variable. Unlike the binomial random variables, Poisson random variables have no upper bound. A recursive formula (Drane *et al.*

1993) for calculating the Poisson *PMF* is:

$$p_x(k|v) = \left(\frac{v}{k}\right) p_x(k - 1|v) \quad for\ k = 1, 2, \ldots$$

$$(8.43)$$

with $p_x(k=0 \mid v) = e^{-v}$.

For a Poisson random variable, the mean and the variance are identical to v. Plots of Poisson *PMF*s corresponding to different values of v are shown in Fig. 8.6b. The skewness coefficient of a Poisson random variable is $1/\sqrt{v}$, indicating that the shape of the distribution tends to become symmetrical as v becomes large.

The Poisson distribution has been applied widely in modelling the number of occurrences of a random event within a specified time or space interval. Eqn 8.42 can be modified as:

$$p_x(k|\lambda, t) = \frac{e^{-\lambda t}(\lambda t)^k}{k!} \quad for\ k = 0, 1, 2, \ldots$$

$$(8.44)$$

in which the parameter λ can be interpreted as the average rate of occurrence of the random event in the time interval $(0, t)$.

Example 8.3. A bridge is designed to transmit a flood with a return period of 50 years. In other words, the annual probability of the bridge being overtopped by the flood is 1-in-50 or $1/50 = 0.02$. What is the probability that the bridge would be overtopped during its expected useful life of 100 years? Also, use Poisson distribution to compute the same result.

Solution. The random variable X gives the number of times the bridge will be overtopped over a 100-year period. Each year is an independent trial in which the bridge can be either overtopped or allow the flood to pass through normally without overtopping. Because the outcome of each 'trial' is binary, the binomial distribution without overtopping applicable.

The probability of the event of overtopping of the bridge in each trial (i.e. each year), is 0.02. The period of 100 years represents 100 trials. Hence, in the binomial distribution model, the parameters are $p = 0.02$ and $n = 100$. The probability that overtopping occurs in a period of 100 years can be calculated, using Eqn 8.40, as

P (overtopping in a 100-year period) = P (overtopping occurs at least once in 100-year period) = $P(X \geq 1 \mid n = 100, p = 0.02)$:

$$= \sum_{k=1}^{100} p_x(k) = \sum_{k=1}^{100} C_{100,k}(0.02)^k (0.98)^{100-k}$$

This requires evaluation of 100 binomial terms, which is quite cumbersome. But if the problem is viewed differently, that is, the non-occurrence of overtopping, which is $= 1 - P$ (no overtopping in a 100-year period) $= 1 - p(X{=}0 \mid n{=}100, p{=}0.98)$ $= 1 - (0.98)^{100} = 1 - 0.1326 = 0.8674$.

Using the Poisson distribution, one has to determine the average number of overtopping events in a given 100-year period. For a 50-year event, the average rate of overtopping is $\lambda = 0.02/year$. The average number of overtopping events in the 100-year period can be obtained as $v = (0.02)(100) = 2$. Therefore, probability of overtopping in a 100-year period, using the Poisson distribution, is obtained as with binomial distribution, by estimating

the probability of no overtopping and subtracting it from unity, i.e.:

$= 1 - P$(no overtopping occurs in a 100-year period);
$= 1 - p(X = 0 \mid v = 2) = 1 - e^{-2}$;
$= 1 - 0.1353 = 0.8647$.

Comparing it with the result from that obtained using binomial distribution, it is seen that use of the Poisson distribution yields a slightly smaller risk of overtopping.

8.2.3 Some continuous univariate probability distributions

Several continuous *PDF*s are used frequently in hydrological reliability analysis. These include normal, lognormal, gamma, Weibull, and exponential distributions. Other less commonly used include beta and extremal distributions.

Many of the probability distributions are not a single distribution but comprise, in fact, a family of distributions. This is due to the distribution having one or more location, scale, and shape parameters. A location parameter shifts the graph to the left or right on the horizontal axis. The effect of a scale parameter greater than unity is to stretch the PDF. Shape parameters allow a distribution to take on a variety of shapes, depending on its value. These distributions are particularly useful in modelling various applications, since they are flexible enough to model a variety of datasets.

8.2.3.1 The Gaussian or normal distribution

The *Gaussian* or *normal* distribution is the most well-known probability distribution involving two parameters, namely, the mean and the variance. A normal random variable having the mean, μ_x, and variance, σ^2_x, is denoted here as $X \sim N(\mu_x, \sigma_x)$ with the PDF as:

$$f_x\left(x|\mu_x, \sigma_x^2\right) = \frac{1}{\sqrt{2\pi}\sigma_x} exp\left[-\frac{1}{2}\left(\frac{x-\mu_x}{\sigma_x}\right)^2\right]$$

$$for -\infty < x < \infty \qquad (8.45)$$

The normal distribution is bell-shaped and symmetric with respect to the mean, μ_x (Fig. 8.7).

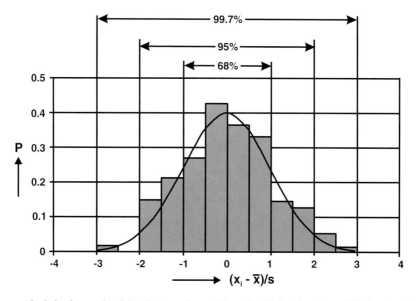

Fig. 8.7 Histogram (shaded) of normalized distribution of a variable x classified with a class width Δx, that is, between x_i and $x_i + \Delta x$. The normalization of the variable is achieved by dividing the deviations of actual observations from the mean value ($\bar{x}$) by the standard deviation (s) of the dataset. The normalized variable, therefore, has mean equal to zero and standard deviation equal to unity. The normalized frequency or the number of observations of values within the specific class interval divided by the total number of observations (i.e. n_i/N) is plotted on the y-axis. The smooth curve gives the Gaussian distribution, which is a hypothetical result corresponding to an infinite number of measurements. It also represents the probability distribution (P) of data around the mean value. On top of the graph, the integral or summed probabilities are shown, giving the probability of observing values between ($\bar{x} + s$) and ($\bar{x} - s$) as 68%, between ($\bar{x} + 2s$) and ($\bar{x} - 2s$) as 95%, and between ($\bar{x} + 3s$) and ($\bar{x} - 3s$) as 99.7%. Redrawn from Mook (2000) © UNESCO/Springer.

Therefore, the skewness coefficient of a normal random variable is zero. Owing to the symmetry of the PDF, all odd-ordered central moments are zero. The kurtosis of a normal random variable is $\kappa_x = 3.0$. A linear function of several normal random variables is also normal, that is, the linear combination of K normal random variables $W = a_1X_1 + a_2X_2 + \ldots + a_KX_K$, with $X_k \sim N(\mu_k, \sigma_k)$, for $k = 1, 2, \ldots K$ is also a normal random variable with the mean, μ_w, and the variance, σ^2_w, respectively, as given by:

$$\mu_w = \sum_{k=1}^{K} a_k\mu_k$$

$$\sigma^2_w = \sum_{k=1}^{K} a_k^2\sigma_k^2 + 2\sum_{k=1}^{K-1}\sum_{k'=k+1}^{K} a_k a_{k'} Cov(X_k, X_{k'})$$

$$(8.46)$$

The normal distribution often provides a practical alternative to approximate the probability of a non-normal random variable. The accuracy of such an approximation, of course, depends on how close is the given distribution to the normal distribution.

An important theorem relating to the sum of independent random variables is the *central limit theorem*, which roughly states that distribution of the sum of a number of independent random variables, regardless of their individual distributions, can be approximated by a normal distribution as long as none of the variables has a dominant effect on the sum. The larger the number of random variables involved in the summation, the better the approximation. Because many natural processes can be thought of as a combination of a large number of independent component processes, none dominating the others, the normal distribution is, therefore, a reasonable approximation to model these processes. Dowson and Wragg (1973) showed that when only the mean and the variance are specified, the maximum entropy on the interval $(-\infty, +\infty)$ follows the normal distribution. Thus, when only the first two moments are specified, use of a normal distribution implies closer approximation to the true nature of the underlying process than by any other distribution.

For a normal distribution, the location and scale parameters correspond to the mean and the standard deviation. Therefore, probability computations for normal random variables are made by first transforming the original variable to a standardized normal variable, Z, which removes the effect of these parameters by using Eqn 8.47, as:

$$Z = (X - \mu)/\sigma_x \qquad (8.47)$$

in which Z has a zero mean and a variance of unity. Since Z is a linear function of the normal random variable X, it is, therefore, normally distributed, that is, $Z \sim N(\mu_z = 0, \sigma_z = 1)$. The *PDF* of Z, called the standard normal distribution, can be obtained readily from Eqn 8.45:

$$\phi(z) = \frac{1}{\sqrt{2\pi}} exp\left(-\frac{z^2}{2}\right) \quad for \; -\infty < z < \infty$$

$$(8.48)$$

General expressions for the product-moments of the standard normal random variable, using Eqn 8.25, are:

$$E(Z^{2r}) = \frac{(2r)!}{2^r \times r!} \quad and \quad E(Z^{2r+1}) = 0 \qquad (8.49)$$

Computations of probability for X $(N(\mu x, \sigma x))$ can be made as:

$$P(X \le x) = P(Z \le z) = \Phi(z) \qquad (8.50)$$

where $z = (x - \mu_x)/\sigma_x$ and $\Phi(z)$ is the standard normal CDF defined as:

$$\Phi(z) = \int_{-\infty}^{z} \phi(z)dz \qquad (8.51)$$

The shape of *PDF* of the standard normal random variable is shown in Fig. 8.7. The integral in Eqn 8.51 is not analytically solvable. A table of the standard normal CDF can be found in many textbooks on statistics (Abramowitz and Stegun 1972; Blank 1980; Devore 1999; Haan 1974). For numerical computation, several highly accurate approximations are available for determining $\Phi(z)$. One

such approximation is the polynomial approximation (Abramowitz and Stegun 1972):

$$\Phi(z) = 1 - \phi(z)\left(b_1t + b_2t^2 + b_3t^3 + b_4t^4 + b_5t^5\right)$$

$$for \; z \ge 0 \qquad (8.52)$$

in which $t = 1/(1 + 0.2316419z)$; $b_1 = 0.31938153$; $b_2 = -0.356563782$; $b_3 = 1.781477937$; $b_4 = -1.821255978$; and $b_5 = 1.33027443$. The maximum absolute error of the approximation is 7.5×10^{-8}, which is reasonably accurate for most practical applications. Note that Eqn 8.52 is applicable only to the non-negative values of z. For $z < 0$, the value of a standard normal CDF can be computed as $\Phi(z) = 1 - \Phi(|z|)$, using the symmetry of the function $\phi(z)$.

The inverse operation of finding the standard normal percentile, z_p with the specified probability level, p, can be easily done using the table of standard normal CDF and carrying out some interpolation. However, for practical algebraic computations using a computer, the following rational approximation can be used (Abramowitz and Stegun, 1972):

$$z_p = t - \frac{c_0 + c_1t + c_2t^2}{1 + d_1t + d_2t^2 + d_3t^3}$$

$$for \; 0.5 < p \le 1 \qquad (8.53)$$

in which $p = \Phi(z_p)$; $t = \sqrt{-2 \, ln(1 - p)}$; $c_0 = 2.515517$, $c_1 = 0.802853$, $c_2 = 0.010328$, $d_1 = 1.432788$, $d_2 = 0.189269$; and $d_3 = 0.001308$. The corresponding maximum absolute error introduced by this rational approximation is 4.5×10^{-4}. Eqn 8.53 is valid for the values of $\Phi(z)$ that lie between [0.5, 1]. When $p < 0.5$, Eqn 8.53 can still be used by letting $t = \sqrt{-2 \, ln(p)}$ and putting a minus sign to the computed percentile value.

Example 8.4. Referring to Example 8.3, determine the probability of more than five overtopping events over a 100-year period using a normal approximation.

Solution. For this problem, the random variable X of interest is the number of overtopping events in a given 100-year period. The exact distribution of X is binomial with parameters $n = 100$ and $p = 0.02$ or alternatively the Poisson distribution with a parameter $\nu = 2$ can also be used. The exact probability of having more than five occurrences of overtopping in 100 years can be computed as:

$$P(X > 5) = P(X \geq 6)$$

$$= \sum_{x=6}^{100} \binom{100}{x} (0.02)^x (0.98)^{100-x}$$

$$= 1 - P(X \leq 5)$$

$$= 1 - \sum_{x=0}^{5} \binom{100}{x} (0.02)^x (0.98)^{100-x}$$

$$= 1 - 0.9845 = 0.0155$$

There are a total of six terms under the summation sign on the right-hand side. Although the computation of probability is possible manually, the following approximation gives a reasonably accurate value. Using a normal probability approximation, the mean and variance of X are:

$$\mu_x = np = (100)(0.2) = 2.0$$

$$\sigma_x^2 = npq = (100)(0.2)(0.98) = 1.96$$

The above binomial probability can be approximated as:

$$P(X \geq 6) \approx P(X \geq 5.5) = 1 - P(X < 5.5)$$

$$= 1 - [Z < (5.5 - 2.0)/\sqrt{1.96}]$$

$$= 1 - \Phi(2.5) = 1 - 0.9938 = 0.062$$

DeGroot (1975) showed that when $(np)^{1.5} > 1.07$, the error of using the normal distribution to approximate the binomial probability does not exceed 0.05. The error gets progressively reduced as the value of $(np)^{1.5}$ becomes larger. For this example, $(np)^{1.5} = 0.283 < 1.07$, and the accuracy of approximation is not satisfactory, as seen above by the difference between the computation using binomial and normal distributions.

Example 8.5. The magnitude of annual maximum flood in a river has a normal distribution with a mean of 600 $m^3 \ s^{-1}$ and standard deviation of 400 $m^3 \ s^{-1}$. (a) What is the annual probability that the flood magnitude would exceed 1000 $m^3 \ s^{-1}$? (b)

Determine the magnitude of the flood with a return period of 100 years.

Solution. (a) Let X be the random annual maximum flood magnitude. Since X has a normal distribution with a mean $\mu_x = 600 \ m^3 \ s^{-1}$ and standard deviation $\sigma_x = 400 \ m^3 \ s^{-1}$, the probability of the annual maximum flood magnitude exceeding 1000 $m^3 \ s^{-1}$ is obtained by using the Table of Standard Normal Probability, $\Phi(z) = P(Z \leq z)$ or Eqn 8.52:

$$P(X > 1000)$$

$$= 1 - P[Z \leq (1000 - 600)/400]$$

$$= 1 - \Phi(1.00) = 1 - 0.8413 = 0.1587$$

(b) A flood event with a 100-year return period represents the event; the annual probability of its magnitude being exceeded is 0.01. Thus $P(X \geq q_{100}) = 0.01$, in which q_{100} is the magnitude of the 100-year flood. Therefore, the problem is to determine q_{100} from:

$$P(X \leq q_{100}) = 1 - P(X \geq q_{100})$$

because

$$P(X \leq q_{100}) = P\{Z \leq [(q_{100} - \mu_x)/\sigma_x]\}$$

$$= P[Z \leq (q_{100} - 600)/400]$$

$$= \Phi[(q_{100} - 600)/400] = 0.99$$

From the Table of Standard Normal Probability, $\Phi(z) = P(Z \leq z)$ or from Eqn 8.53, $\Phi(2.33) = 0.99$. Therefore:

$$(q_{100} - 600)/400 = 2.33$$

which gives the magnitude of the 100-year flood event as $q_{100} = 1532 \ m^3 \ s^{-1}$.

8.2.3.2 Lognormal distribution

The *lognormal* distribution is a commonly used continuous distribution for positively valued random variables that are closely related to normal random variables. A random variable X is said to have a lognormal distribution if its logarithmic transform $Y = ln(X)$ has a normal distribution with mean, μ_{lnx}, and variance, σ^2_{lnx}. From the central limit theorem, if a natural process can be thought of as

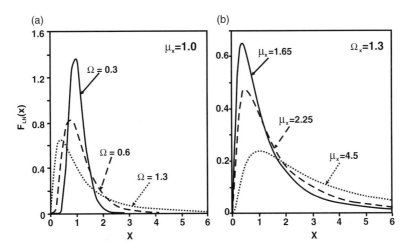

Fig. 8.8 Shapes of lognormal probability density functions for different values of mean (μ_x) and normalized standard deviation ($\Omega_x = \sigma_x/\mu_x$): (a) $\mu_x = 1.0$; (b) $\Omega_x = 1.30$. Based on Tung *et al.* (2006) with permission from Mcgraw-Hill.

a product of a large number of independent component processes, none dominating the others, the lognormal distribution is a reasonable approximation for these natural processes. With reference to Eqn 8.45, the PDF of a lognormal random variable can be written as:

$$f_{LN}\left(X|\mu_{\ln x}, \sigma_{\ln x}^2\right)$$

$$= \frac{1}{\sqrt{2\pi}\,\sigma_{\ln x}x}\,exp\left\{-\frac{1}{2}\left[\frac{\ln(x) - \mu_{\ln x}}{\sigma_{\ln x}}\right]^2\right\}$$

$$\textit{for } x > 0 \tag{8.54}$$

Statistical properties of a lognormal random variable in the original scale can be computed from those of log-transformed variables as:

$$\mu_x = exp\left(\mu_{\ln x} + \frac{\sigma_{\ln x}^2}{2}\right) \tag{8.55}$$

$$\sigma_x^2 = \mu_x^2\left[exp\left(\sigma_{\ln x}^2\right) - 1\right] \tag{8.56}$$

$$\textit{or} \quad \Omega_x^2 = \sigma_x^2/\mu_x^2 = exp(\sigma_{\ln x}^2) - 1 \tag{8.57}$$

and skewness coefficient:

$$\gamma_x = \Omega_x^3 + 3\Omega \tag{8.58}$$

Conversely, the statistical moments of $ln(X)$ can be computed from those of X by using Eqn 8.55 and Eqn 8.56:

$$\mu_{\ln x} = ln\,(\mu_x) - \frac{1}{2}\sigma_{\ln x}^2 = \frac{1}{2}ln\left[\frac{\mu_x^2}{1 + \Omega_x^2}\right] \tag{8.59}$$

$$\sigma_{\ln x}^2 = ln(1 + \Omega_x^2) \tag{8.60}$$

One can see from Eqn 8.55, that the shape of a lognormal PDF is always positively skewed (Fig. 8.8). Eqn 8.55 and Eqn 8.56 can be derived easily from the moment-generating function (Tung and Yen 2005).

It is interesting to note from Eqn 8.57 that the variance of a log-transformed variable ($\sigma_{\ln x}^2$) is dimensionless. Since the sum of normal random variables is normally distributed, the product of lognormal random variables is also lognormally distributed (Fig. 8.5). This useful property of lognormal random variables can be stated as: if $X_1, X_2, \ldots, X_K$ are independent lognormal random variables, then $W = b_0 \prod_{k=1}^{K} X_k^{b_k}$ has a lognormal distribution with mean and variance as:

$$\mu_{\ln w} = ln(b_0) + \sum_{k=1}^{K} b_k\mu_{\ln x_k} \quad and \quad \sigma_{\ln w}^2$$

$$= \sum_{k=1}^{K} b_k^2\sigma_{\ln x_k}^2 \tag{8.61}$$

Example 8.6. Redo the Example 8.5 for when the annual maximum flood magnitude in the river flow follows a lognormal distribution.

Solution. (a) Since Q has a lognormal distribution, $ln(Q)$ is also normally distributed with a mean and variance that can be computed from Eqn 8.57, Eqn 8.60, and Eqn 8.59, respectively, as:

$$\Omega_x^2 = \sigma_x^2/\mu_x^2 \quad or \quad \Omega_x = 400/600 = 0.667$$

$$\sigma_{ln\,x}^2 = ln(1 + 0.667^2) = 0.368$$

$$\mu_{ln\,x} = ln(600) - 0.368/2 = 6.213$$

The probability of the magnitude of annual maximum flood exceeding 1000 m³ s⁻¹ is, therefore, given by:

$$P(X > 1000) = P[ln\ X > ln(1000)]$$

$$= 1 - P(Z \leq 6.908 - 6.213)/\sqrt{0.368}$$

$$= 1 - \Phi(1.145) = 1 - 0.8736 = 0.1261$$

(b) A 100-year flood q_{100} represents an event, the magnitude of which corresponds to $P(X \geq q_{100}) = 0.01$, which can be estimated from:

$$P(X \leq q_{100}) = 1 - P(X \geq q_{100}) = 0.99$$

$$P(X \leq q_{100}) = P(lnX \leq ln\ q_{100})$$

$$= P[Z \leq (ln\ q_{100} - \mu_{ln\,x})/\sigma_{ln\,x}]$$

$$= P[Z \leq (ln\ q_{100} - 6.213)/\sqrt{0.368}]$$

$$= \Phi[ln\ q_{100} - 6.213/\sqrt{0.368}] = 0.99$$

From the Table of Standard Normal Probability, $\Phi(z) = P(Z \leq z)$ or from Eqn 8.53, one finds that $\Phi(2.33) = 0.99$. Therefore, $[ln\ q_{100} - 6.213]/\sqrt{0.368} = 2.33$, which yields $ln(q_{100}) = 7.626$. The magnitude of the 100-year flood event is, therefore, $q_{100} = exp(7.626) = 2051\ m^3\ s^{-1}$.

8.2.3.3 Gamma distribution and its variation

The *gamma* distribution is a versatile continuous distribution associated with a positive-valued random variable. The two-parameter gamma distribution has a PDF defined as:

$$f_G(X|\alpha, \beta) = \frac{1}{\beta\Gamma(\alpha)}(x/\beta)^{\alpha-1}exp(x/\beta)$$

$$for\ x > 0 \tag{8.62}$$

in which $\beta > 0$ and $\alpha > 0$ are the scale and shape parameters, respectively, and $\Gamma(\bullet)$ denotes a *gamma function* defined as:

$$\Gamma(\alpha) = \int_0^\infty t^{\alpha-1}e^{-t}dt \tag{8.63}$$

The mean, variance, and skewness coefficient of a gamma random variable having the PDF as given by Eqn 8.62, are:

$$\mu_x = \alpha\beta \quad \sigma_x^2 = \alpha\beta^2 \quad \gamma_x = 2/\sqrt{\alpha} \tag{8.64}$$

When the lower bound of a gamma random variable is a positive quantity, the above two-parameter gamma PDF can be modified to a three-parameter gamma PDF with the additional location parameter (ξ) as:

$$f_G(X|\xi, \alpha, \beta)$$

$$= \frac{1}{\beta\Gamma(\alpha)}\left[\frac{x-\xi}{\beta}\right]^{\alpha-1}exp[-(x-\xi)/\beta]$$

$$for\ x > \xi \tag{8.65}$$

where ξ is the lower bound. The two-parameter gamma distribution can be reduced to a simpler form by letting $Y = X/\beta$, and the resulting one-parameter gamma PDF (called the standard gamma distribution) is:

$$f_G(X|\alpha) = \frac{1}{\Gamma(\alpha)}(y)^{\alpha-1}exp(y) \quad for\ y > 0 \tag{8.66}$$

Tables of the cumulative probability of the standard gamma distribution can be found in Dudewicz (1976). The versatility of some gamma distributions can be seen from the shapes illustrated in Fig. 8.9a. If α is a positive integer in Eqn 8.66, the distribution is called an *Erlang distribution*.

When $\alpha = 1$, the two-parameter gamma distribution reduces to an *exponential distribution* with the PDF given by:

$$f_{EXP}(x|\beta) = e^{-x/\beta}/\beta \quad for\ x > 0 \tag{8.67}$$

An exponential random variable with a PDF, as given by Eqn 8.67, has the mean and standard deviation equal to β (Example 8.2). Therefore, the coefficient of variation of an exponential random variable is equal to unity. The exponential distribution is used commonly for describing the life span of various electronic and mechanical components.

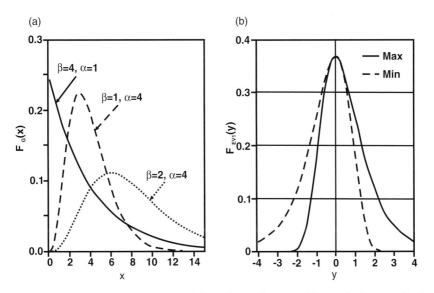

Fig. 8.9 (a) Various shapes of two-parameter gamma probability density functions. The standard gamma distribution, also called one-parameter gamma distribution, is obtained by letting $Y = X/\beta$. When $\alpha = 1$, the two-parameter gamma distribution reduces to an exponential distribution. (b) Probability density function of extreme-value Type-1 random variables. The random variable is transformed using $y = (x - \xi)/\beta$. Based on Tung *et al.* (2006) with permission from Mcgraw-Hill.

Two variations of the gamma distribution used frequently in hydrologic frequency analysis are the Pearson and log-Pearson Type-3 distributions. In particular, the log-Pearson Type-3 distribution is recommended by the US Water Resources Council (USWRC 1982) as the standard distribution for flood frequency analysis. A *Pearson Type-3* random variable has the PDF of the form:

$$f_{P3}(x|\xi, \alpha, \beta) = \frac{1}{\beta|\Gamma(\alpha)} \left(\frac{x - \xi}{\beta}\right)^{\alpha-1}$$
$$\times exp[-(x - \xi)/\beta] \quad (8.68)$$

with $\alpha > 0$; $x \geq \xi$ when $\beta > 0$ and with $\alpha > 0$; $x \leq \xi$ when $\beta < 0$. When $\beta > 0$, the Pearson Type-3 distribution is identical to the three-parameter gamma distribution. However, the Pearson Type-3 distribution has the flexibility to model negatively skewed random variables corresponding to $\beta < 0$.

Similar to the relationship between normal and lognormal distributions, the PDF of a *log-Pearson Type-3* random variable is given by:

$$f_{LP3}(x|\xi, \alpha, \beta) = \frac{1}{x|\beta|\Gamma(\alpha)} \left[\frac{ln\,(x) - \xi}{\beta}\right]^{\alpha-1}$$
$$\times exp\{-[ln\,(x) - \xi]/\beta\} \quad (8.69)$$

with $\alpha > 0$; $x \geq e^{\xi}$ when $\beta > 0$ and with $\alpha > 0$; and $x \leq e^{\xi}$ when $\beta < 0$. Kite (1988), Stedinger *et al.* (1993) and Rao and Hamed (2000) provide good accounts of Pearson Type-3 and log-Pearson Type-3 distributions. Evaluation of the probability of gamma random variables involves computation of the gamma function, which can be made by using the following recursive formula:

$$\Gamma(\alpha) = (\alpha - 1)\Gamma(\alpha - 1) \quad (8.70)$$

When the argument α is an integer number, then $\Gamma(\alpha) = (\alpha - 1)! = (\alpha - 1)(\alpha - 2)\ldots1$. However, when α is a real number, the recursive relation leads to $\Gamma(\alpha')$ as the smallest term, with $1 < \alpha' < 2$. The value of $\Gamma(\alpha')$ can be determined from the table of the gamma function or by numerical integration of Eqn 8.63. Alternatively, the following approximation could be applied (Abramowitz and Stegun, 1972):

$$\Gamma(\alpha') = \Gamma(x + 1)$$
$$= 1 + \sum_{i=1}^{5} a_i x^i \quad for\ 0 < x < 1 \quad (8.71)$$

in which $a_1 = -0.577191652$; $a_2 = 0.988205891$; $a_3 = -0.897056937$; $a_4 = 0.4245549$; and $a_5 = -0.1010678$. The maximum absolute error associated with Eqn 8.71 is 5×10^{-5}.

8.2.3.4 Extreme-value distributions

In many hydrosystems, the focus is on extreme events such as floods and droughts. Statistics of extremes is concerned with the statistical characteristics of $X_{max,n} = max(X_1, X_2, \ldots, X_n)$ and/or $X_{min,n} = min(X_1, X_2, \ldots, X_n)$, in which $X_1, X_2, \ldots, X_n$ are observed values of random processes. The exact distributions of extremes are, in fact, functions of the parent distribution that generate the random observations $X_1, X_2, \ldots, X_n$ and the number (n) of observations.

Of practical interest are the asymptotic distributions of extremes that assume the resulting distribution as a limiting form of $F_{max,n}(y)$ or $F_{min,n}(y)$, when the number of observations, n, approaches infinity. As a result, asymptotic distributions of extremes turn out to be independent of the sample size, n, and the parent distribution for random observations, that is:

$$lim_{n \to \infty} F_{max,n}(y) = F_{max}(y) \text{ and}$$

$$lim_{n \to 8} F_{min,n}(y) = F_{min}(y)$$

These asymptotic distributions of the extremes largely depend on the behaviour of the tail of the parent distribution in either direction towards both the extremes. The central portion of the parent distribution has little significance for defining the asymptotic distribution of extremes.

Three types of asymptotic distribution of extremes are derived, based on different character-istics of the underlying distribution (Haan 1977) as under:

1. *Type-1.* Parent distributions are unbounded in the direction of both extremes and all statistical moments exist. Examples of this type of parent distribution are normal (for both the largest and the smallest extremes), lognormal, and gamma distributions (for the largest extreme).
2. *Type-2.* Parent distributions are unbounded in the direction of both extremes but all moments do not exist. One such distribution is the Cauchy distribution (Table 8.2).
3. *Type-3.* Parent distributions are bounded in the direction of the desired extreme. Examples of this type of underlying distribution are the beta distribution (for both the largest and the smallest extremes) and the lognormal and gamma distributions (for the smallest extreme).

Owing to the fact that $X_{min,n} = -max(-X_1, -X_2, \ldots, -X_n)$, the asymptotic distribution functions of $X_{max,n}$ and $X_{min,n}$ satisfy the following relation (Leadbetter et al. 1983):

$$F_{min}(y) = 1 - F_{max}(-y) \qquad (8.72)$$

Consequently, the asymptotic distribution of X_{min} can be obtained directly from that of X_{max}. Three types of asymptotic distributions of the extremes are listed in Table 8.1.

8.2.3.4.1 Extreme-value Type-1 distribution

Other names given to this distribution are: *Gumbel distribution, Fisher-Tippett distribution,* and the *Double exponential distribution.* The CDF and PDF of the *extreme-value Type-1 (EV1)* distribution

Table 8.1 Three types of asymptotic Cumulative Distribution Functions (*CDF*s) of extremes. Source: Tung et al. (2006).

Type	Maxima	Range	Minima	Range
1	$exp(-e^{-y})$	$-\infty < y < \infty$	$1 - exp(-e^y)$	$-\infty < y < \infty$
2	$exp(-y^\alpha)$	$\alpha < 0, y > 0$	$1 - exp[-(-y)^\alpha]$	$\alpha < 0, y < 0$
2	$exp[-(-y)^\alpha]$	$\alpha > 0, y < 0$	$1 - exp(-y^\alpha)$	$\alpha > 0, y > 0$

Note: $y = (x - \xi)/\beta$.

have, respectively, the following forms:

$$F_{EV1}(x|\xi, \beta) = exp\left\{-exp\left[-\left(\frac{x-\xi}{\beta}\right)\right]\right\}$$

for maxima

$$= 1 - exp\left\{-exp\left[+\left(\frac{x-\xi}{\beta}\right)\right]\right\}$$

for minima (8.73)

$$f_{EV1}(x|\xi, \beta)$$

$$= \frac{1}{\beta}exp\left\{-\left(\frac{x-\xi}{\beta}\right) - exp\left[-\left(\frac{x-\xi}{\beta}\right)\right]\right\}$$

for maxima

$$= \frac{1}{\beta}exp\left\{+\left(\frac{x-\xi}{\beta}\right) - exp\left[+\left(\frac{x-\xi}{\beta}\right)\right]\right\}$$

for minima (8.74)

for $-\infty < x$; the location parameter $\xi < \infty$; and the scale parameter $\beta \geq 0$. The shapes of the *EV1* distribution are shown in Fig. 8.9b, in which the transformed random variable $Y = (X -\xi)/\beta$ is used. As can be seen, the *PDF* associated with the largest extreme is a mirror image of the smallest extreme with respect to the vertical line passing through the common mode, which happens to be the parameter ξ. The first three product-moments of an *EV1* random variable are:

$$\mu_x = \xi + 0.5772\beta \quad \textit{for the largest extreme}$$

$$= \xi - 0.5772\beta \quad \textit{for the smallest extreme}$$

$$\sigma_x^2 = 1.645\beta^2 \quad \textit{for both types}$$

$$\gamma_x = 1.13955 \quad \textit{for the largest extreme}$$

$$= -1.3955 \quad \textit{for the smallest extreme} \quad (8.75)$$

Shen and Bryson (1979) showed that if a random variable has an *EV1* distribution, the following relationship is satisfied when ξ is small:

$$x_{T_1} \approx \left[\frac{ln\,(T_1)}{ln\,(T_2)}\right]x_{T_2} \quad (8.76)$$

where x_T is the percentile corresponding to the exceedance probability of $1/T$.

Example 8.7. Redo Example 8.5 by assuming that the annual maximum flood follows the *EV1* distribution.

Solution. Based on the values of mean ($600\ m^3\ s^{-1}$) and standard deviation ($400\ m^3\ s^{-1}$), the values of the distribution parameters ξ and β can be determined as follows. For obtaining the maxima, β is computed from Eqn 8.75 as:

$$\beta = \frac{\sigma_x}{\sqrt{1.645}} = \frac{400}{1.2826} = 311.872\ m^3.s^{-1}$$

$$\xi = \mu_x - 0.577\beta = 600 - 0.577(311.872)$$

$$= 420.05\ m^3.s^{-1}$$

1. The probability of the flood discharge exceeding $1000\ m^3\ s^{-1}$, according to Eqn 8.73, is:

$$P(X > 1000)$$

$$= 1 - F_{EV1}(1000)$$

$$= 1 - exp\left[-exp\left(-\frac{1000 - 420.05}{311.872}\right)\right]$$

$$= 1 - exp[-exp(-1.860)]$$

$$= 1 - 0.8558 = 0.1442$$

2. On the other hand, the magnitude of a 100-year flood event can be calculated as:

$$y_{100} = \frac{q_{100} - \xi}{\beta} = -ln\,[-ln\,(1 - 0.01)] = 4.60$$

Hence:

$$q_{100} = 420.05 + 4.60 \times (311.87) = 1855\ m^3\ s^{-1}$$

8.2.3.4.2 Extreme-value Type-3 distribution

As seen from Table 8.1, the parent distributions are bounded in the direction of the desired extreme in the *extreme-value Type-3* (*EV3*) distribution. But for many hydrologic variables, the lower bound is zero, and the upper bound is infinity. Therefore, the *EV3* distribution for the maxima has limited applications. On the other hand, the *EV3* distribution of the minima is used widely for modelling the

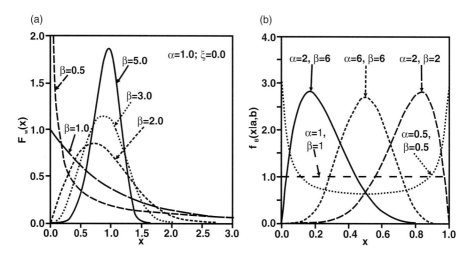

Fig. 8.10 (a) Probability density functions of Weibull random variables for different parameters. (b) Shapes of standard beta probability density functions for different parameter values. Based on Tung *et al.* (2006) with permission from Mcgraw-Hill.

smallest extreme such as drought or low-flow conditions. The *EV3* distribution for the minima is also known as the *Weibull distribution*, having a PDF defined as:

$$f_W(x|\xi, \alpha, \beta)$$
$$= \frac{\alpha}{\beta}\left(\frac{x-\xi}{\beta}\right)^{\alpha-1} exp\left[-\left(\frac{x-\xi}{\beta}\right)^{\alpha}\right]$$
$$for\ x \geq \xi\ and\ \alpha, \beta > 0 \qquad (8.77)$$

Various shapes of the Weibull distribution are generated using the shape parameter α. When $\xi = 0$ and $\alpha = 1$, the Weibull distribution reduces to the exponential distribution.

Fig. 8.10a shows that versatility of the Weibull distribution function depends on the parameter values. The CDF of Weibull random variables is:

$$F_w(x|\xi, \alpha, \beta) = 1 - exp\left[-\left(\frac{x-\xi}{\beta}\right)^{\alpha}\right] \qquad (8.78)$$

The mean and variance of a Weibull random variable can be derived as:

$$\mu_x = \xi + \beta\Gamma\left(1 + \frac{1}{\beta}\right) \quad and$$

$$\sigma_x^2 = \beta^2\left[\Gamma\left(1 + \frac{2}{\alpha}\right) - \Gamma\left(1 + \frac{1}{\alpha}\right)\right] \qquad (8.79)$$

8.2.3.4.3 Generalized extreme-value distribution

The *Generalized Extreme Value (GEV)* distribution combines the *Type-1*, *Type-2*, and *Type-3* extreme value distributions into a single family, to allow a continuous range of possible shapes. The CDF of a random variable corresponding to the maximum with a *GEV* distribution is:

$$F_{GEV}(x|\xi, \alpha, \beta) = exp\left\{-\left[1 - \frac{\alpha(x-\xi)}{\beta}\right]^{1/\alpha}\right\}$$
$$for\ \neq 0 \qquad (8.80)$$

When $\alpha = 0$, Eqn 8.80 reduces to Eqn 8.73 for the Gumbel or *EV1* distribution. For $\alpha < 0$, it corresponds to the *EV2* distribution having a lower bound $x > \xi + \beta/\alpha$. On the other hand, for $\alpha > 0$, it corresponds to the Weibul or *EV3* distribution having an upper bound $x < \xi + \beta/\alpha$. For $|\alpha| < 0.3$, the shape of the *GEV* distribution is similar to the Gumbel distribution, except that the right-hand tail is thicker for $\alpha < 0$ and thinner for $\alpha > 0$ (Stedinger *et al.* 1993). The first three moments of the *GEV* distribution, respectively, are:

$$\mu_x = \xi + \left(\frac{\beta}{\alpha}\right)[1 - \Gamma(1 + \alpha)] \qquad (8.81)$$

$$\sigma_x^2 = \left(\frac{\beta}{\alpha}\right)^2[\Gamma(1 + 2\alpha) - \Gamma^2(1 + \alpha)] \qquad (8.82)$$

$$\gamma_x = sign(\alpha) \times$$

$$\left(\frac{-\Gamma(1 + 3\alpha) + 3\Gamma(1 + 2\alpha)\Gamma(1 + \alpha) - 2\Gamma^3(1 + \alpha)}{[\Gamma(1 + 2\alpha) - \Gamma^2(1 + \alpha)]^{1.5}} \right)$$

$$(8.83)$$

where $sign(\alpha)$ is $+1$ or -1, depending on the sign of α. From Eqn 8.82 and Eqn 8.83, one notes that the variance of the *GEV* distribution exists when $\alpha > -0.5$, and the skewness coefficient exists when $\alpha > -0.33$. Recently the GEV distribution has been frequently used in modelling the random hydrologic extremes, such as precipitation and floods.

8.2.3.4.4 Beta distributions

In hydrosystems, such as reservoir storage and groundwater tables for unconfined aquifers, random variables that are bounded on both limits are often best described by the *beta distribution*. The general form of the *beta* PDF is:

$$f_{NB}(x|a, b, \alpha, \beta)$$

$$= \frac{1}{B(\alpha, \beta)(b - a)^{\alpha + \beta - 1}}(x - a)^{\alpha - 1}(b - x)^{\beta - 1}$$

$$for\ a \leq x \leq b \qquad (8.84)$$

in which a and b are the lower and upper bounds of the *beta* random variable, respectively; the two shape parameters are $\alpha > 0$, $\beta > 0$, and $B(\alpha, \beta)$ is a *beta function* defined as:

$$B(\alpha, \beta) = \int_0^1 t^{\alpha - 1}(1 - t)^{\beta - 1} dt$$

$$= \frac{\Gamma(\alpha)\Gamma(\beta)}{\Gamma(\alpha + \beta)} \qquad (8.85)$$

The *general beta PDF* can be transformed to the *standard beta PDF* using a new variable, $Y = (X - a)/(b - a)$, as:

$$f_B(y|\alpha, \beta) = \frac{1}{B(\alpha, \beta)}y^{\alpha - 1}(1 - y)^{\beta - 1}$$

$$for\ 0 < y < 1 \qquad (8.86)$$

The *CDF* of the *standard beta* function is given by:

$$F(y|\alpha, \beta) = \frac{B_y(\alpha, \beta)}{B(\alpha, \beta)} = I_y(\alpha, \beta) \qquad (8.87)$$

where $B_y(\alpha,\beta)$ is the *incomplete beta function*, which is a generalization of the beta function that replaces the definite integral of the beta function (Eqn 8.85) with an indefinite integral, i.e.:

$$B_y(\alpha, \beta) = \int_0^y t^{\alpha - 1}(1 - t)^{\beta - 1} dt \qquad (8.88)$$

and $I_y(\alpha,\beta)$ is the *regularized incomplete beta function*.

The beta distribution is also versatile and can have many shapes, as shown in Fig. 8.10b. The mean and variance of the standard beta random variable, Y, respectively, are:

$$\mu_y = \frac{\alpha}{\alpha + \beta} \quad and$$

$$\sigma_y^2 = \frac{\alpha\beta}{(\alpha + \beta + 1)(\alpha + \beta)^2} \qquad (8.89)$$

When $\alpha = \beta = 1$, the beta distribution reduces to a uniform distribution as:

$$f_U(x) = \frac{1}{b - a} \quad for\ a \leq x \leq b \qquad (8.90)$$

8.2.3.5 Distributions related to normal random variables

The Gaussian or normal distribution has played an important role in the development of statistical theories. In the following, two distributions related to the functions of normal random variables are briefly described.

8.2.3.5.1 χ^2 (chi-square) distribution

The sum of the squares of K independent standard normal random variables results in a χ^2 (chi-square) random variable with K degrees of freedom, denoted as χ_K^2. In other words:

$$\sum_{k=1}^{K} Z_k^2 \approx \chi_K^2 \qquad (8.91)$$

in which the Z_ks are independent standard normal random variables. The PDF of a χ^2 random variable with K degrees of freedom is:

$$f_{\chi^2}(x|K) = \frac{1}{2^{K/2}\Gamma(K/2)}x^{(K/2 - 1)}e^{-x/2} \quad for\ x > 0$$

$$(8.92)$$

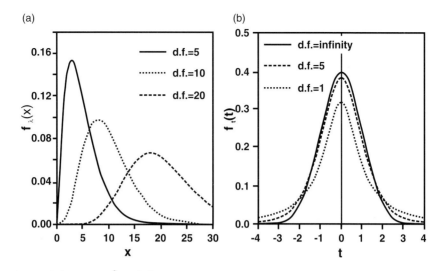

Fig. 8.11 (a) Shapes of chi-square (χ^2) probability density functions (*PDF*s). As the number of degrees of freedom (*d.f.*) increases, so do mean and variance of the distribution. (b) Shapes of *t*-distributions for different values of *d.f.* Based on Tung *et al.* (2006) with permission from Mcgraw-Hill.

Comparing Eqn 8.92 with Eqn 8.62, one notes that the χ^2 distribution is a special case of the two-parameter gamma distribution with $\alpha = K/2$ and $\beta = 2$. The mean, variance, and skewness coefficient of a χ_K^2 random variable, respectively, are:

$$\mu_x = K \quad \sigma_x^2 = 2K \quad \gamma_x = 2/\sqrt{K/2} \qquad (8.93)$$

Thus, as the value of K increases, the χ^2 distribution approaches a symmetric distribution. Fig. 8.11a shows some of the χ^2 distributions with various degrees of freedom. If $X_1, X_2, \ldots, X_K$ are independent normal random variables with the common mean μ_x and variance σ^2_x, the χ^2 distribution is related to the sample of normal random variables as follows:

The sum of K squared standardized normal variables $Z_k = (X_k - X)/\sigma_x \ k = 1, 2, \ldots, K$, has a χ^2 distribution with $(K - 1)$ degrees of freedom.

The quantity $(K - 1)s^2/\sigma^2_x$ has a χ^2 distribution with $(K - 1)$ degrees of freedom, in which s^2 is the unbiased sample variance.

8.2.3.5.2 *t*-distribution

A random variable having a *t*-distribution results from the ratio of the standard normal random vari-

able to the square root of the χ^2 random variable divided by its degrees of freedom, that is:

$$T_K = \frac{Z}{\sqrt{\chi_K^2/K}} \qquad (8.94)$$

in which T_K is a *t*-distributed random variable with K degrees of freedom. The *PDF* of T_K can be expressed as:

$$f_T(x|K) = \frac{\Gamma[(K+1)/2]}{\sqrt{\pi K}\,\Gamma(K/2)} \left(1 + \frac{x^2}{K}\right)^{-(K+1)/2}$$
$$for \ -\infty < x < \infty \qquad (8.95)$$

A *t-distribution* is symmetric with respect to the mean $\mu_x = 0$ when $K \geq 1$. Its shape is similar to the standard normal distribution (z), except that the tail portions of its PDF are thicker than $\phi(z)$. However, as $K \to \infty$, the PDF of a *t*-distributed random variable approaches the standard normal distribution. Fig. 8.11b shows some PDFs for *t*-random variables of different degrees of freedom. It may be noted that when $K = 1$, the *t*-distribution reduces to the Cauchy distribution for which all the product-moments do not exist. The mean and variance of a *t*-distributed random variable with K degrees of freedom are:

$$\mu_x = 0 \quad \sigma_x^2 = K/(K-2) \quad for \ K \geq 3 \qquad (8.96)$$

When the population variance of normal random variables is known, the sample mean $\bar{X}$ of K normal random samples from $N(\mu_x, \sigma^2_x)$ has a normal distribution with mean μ_x and variance σ^2_x/K. However, when the population variance is unknown but is estimated by s^2, the quantity $\sqrt{K}\,(\bar{X} - \mu_x)/s$, which is the standardized sample mean using the sample variance, has a t-distribution with $(K - 1)$ degrees of freedom.

8.3 Hydrologic frequency analysis

The aim of hydrosystems analysis is not to altogether eliminate all hydro-hazards but to reduce the frequency of their occurrences and thereby minimize the resulting damage. Therefore, the probabilities of the investigated event must be evaluated correctly. This is inherently complex because in many cases the 'input' is controlled by nature and therefore has a limited predictability. For example, variations in the amount, timing, and spatial distribution of precipitation and its resulting effects such as runoff are far from being perfectly understood.

Therefore, a statistical or probabilistic approach is often used as this does not require a complete understanding of the hydrologic phenomenon involved but examines the relationship between event and frequency of occurrence with the aim of finding some statistical pattern between these variables. In effect, the past trends are extrapolated into the future, assuming that whatever complex physical interactions and processes govern nature do not change with time. Therefore, the historical record can be used as a basis for predicting future events. In other words, the data are assumed to satisfy statistical *stationarity* by which the underlying properties of the distribution remain invariant with time. The hydrologic data commonly analysed in this manner are rainfall and stream flow records. An obvious example that violates the assumed statistical stationarity is the progressive urbanization within a watershed that could result in a tendency of increasing the magnitude of peak flow over time. Global change due to a variety of human activities, including greenhouse gas emissions, may also be contributing to violation of statistical stationarity.

Basic probability concepts and theories useful for frequency analysis have already been described in Section 8.2. In general, there is no *a priori* physical reasoning that stipulates the use of a particular distribution in the frequency analysis of geophysical data. However, since the maximum or the minimum values of hydrological events are usually of interest, extreme-value-related distributions have been found useful.

There are three basic types of data series extractable from geophysical events:

1. *A complete series*, which includes all the available data concerning the magnitude of a phenomenon. Such a data series is usually very large and since in some instances the interest is only in the extremes of the distribution (e.g. floods, droughts, wind speeds, and wave heights), following other data series are often more relevant.

2. *An extreme-value series* is one that contains the largest (or smallest) value of the data for individual equal time intervals. If, for example, the largest value of the data in the record of each year is used, the extreme-value series is called an annual maximum series. If the smallest value is used, the series is called an annual minimum series.

3. *A partial-duration series* consists of all data above or below the base value. For example, one might consider only floods in a river with a magnitude greater than $1000~m^3~s^{-1}$. When the base value is selected such that the number of events included in the data series equals the number of years of record, the resulting series is called an *annual exceedance series*. This series contains the n largest or n smallest values in n years of record.

Another issue related to selection of data series for frequency analysis is adequacy of the record length. The US Water Resources Council (USWRC 1967) recommended that at least 10 years of data should be available before a frequency analysis can be done. It has, however, been shown that if a frequency analysis is done using 10 years of record, a high degree of uncertainty can be expected in the estimate of high-return-period events.

The third issue related to the data series used for frequency analysis concerns the problem of data homogeneity. This refers to different accuracies with which various events of different magnitude are recorded, primarily due to difficulties in measuring high magnitude events. As an example, the probability distribution of measured floods can be greatly distorted with respect to the parent population. This further contributes to the uncertainty in flood frequency analysis.

8.3.1 Return period

The concept of return period (or *recurrence interval*) is commonly used in hydro-systems engineering. This is a substitute for probability and lends it some physical interpretation. The return period for a given event is defined as the period of time on the long-term average value at which a given event is equalled or exceeded. Hence, on average, an event with a two-year return period will be equalled or exceeded once every two years. The relationship between the probability and return period is given by:

$$T = \frac{1}{P(X \geq x_T)} = \frac{1}{1 - P(X < x_T)} \quad (8.97)$$

in which x_T is the value of the variate corresponding to a T-year return period. For example, if the probability that a flood will be equalled or exceeded in a single year is 0.1, that is $P(X \geq x_T) = 0.1$, the corresponding return period is $1/P(X \geq x_T) = 1/0.1 = 10$ years. Note that $P(X \geq x_T)$ is the probability that the event is equalled or exceeded in any one year and is the same for each year, regardless of the magnitudes that occurred in prior years. This is because the events are independent and the long-term probabilities are used without regard to the order in which they may occur.

A common error or misconception is to assume, for example, that if the 100-year magnitude event occurs this year, it will not recur for the next 100 years. In fact, it may occur again the very next year and then may not be repeated for several hundred years. This misconception resulted in considerable public outcry in the United States when the Phoenix area experienced two 50-year

and one 100-year floods in a span of 18 months in 1978–1979 and the Milwaukee area experienced 100-year floods in two consecutive years, June 1997 and June 1998.

The common unit used for the return period is year, although semi-annual, monthly, or any other time period may be used. The unit used for the time series is also the unit assigned to the return period. Thus an annual series has a unit for the return-period as year and a monthly series has the unit of return-period as month(s).

8.3.2 Probability estimates for data series: plotting positions (rank-order probability)

In order to fit a probability distribution to a data series, estimates of probability must be assigned to each term in the data series. Consider a data series consisting of the entire population of N values for a particular variable. If this series were ranked according to decreasing magnitude, probability of the largest variate being equalled or exceeded is $1/N$, where N is the total number of variates. Similarly, the exceedance probability of the second largest variate is $2/N$, and so on. In general:

$$P(X \geq x_{(m)}) = \frac{1}{T_m} = \frac{m}{N} \quad (8.98)$$

in which m is the rank of the data in descending order; $x(m)$ is the mth largest variate in a data series of size N; and T_m is the return period associated with $x(m)$. As the entire population is either not used or is not available, only an estimate of the exceedance probability based on a sample can be made. Eqn 8.98 giving the rank-order probability is called a plotting position formula because it provides an estimate of probability so that the data series can be plotted (as magnitude versus probability). Some modifications are done to avoid theoretical inconsistency when it is applied to the sample data series. For example, an exceedance probability of 1.0 for the smallest variate implies that all values must be equal to or larger than it. A number of plotting-position formulas have been introduced. Perhaps the most popular plotting-position formula is the Weibull formula:

$$P(X \geq x_{(m)}) = \frac{1}{T_m} = \frac{1}{n+1} \tag{8.99}$$

n being the sample size.

8.3.3 Graphical approach

After identifying the data series, ranking the events and calculating the plotting position, a graph of magnitude of x versus its probability [$P(X \geq x)$, $P(X < x)$, or T] can be plotted and a distribution fitted graphically. To facilitate this procedure, one often uses specially designed probability graph papers that have the probability scale chosen such that the resulting plot is a straight line if the chosen distribution fits the data perfectly. Probability graph papers are available for Gaussian (normal), lognormal, Gumbel, and some other distributions. By plotting the data using a particular probability scale and constructing the best-fit straight line through the data, a graphical fit is done to the distribution used in constructing the probability scale. This is a graphical approach to estimate the statistical parameters of the distribution.

8.3.4 Analytical approaches

An alternative to the graphical technique is to estimate the statistical parameters of a distribution from the sample data. The distribution model can then be used to solve for the variate value corresponding to any desired return period or probability as:

$$x_T = F_x^{-1}\left(1 - \frac{1}{T} \middle| \theta\right) \tag{8.100}$$

in which $F_x^{-1}(\theta)$ is the inverse cumulative distribution function with the model parameter vector θ. Eqn 8.100 is applicable when the inverse distribution functional forms are analytically tractable, such as the Gumbel, generalized extreme value, etc.

Example 8.8. Consider that the annual maximum floods follow a lognormal distribution with a mean of 490 $m^3 \ s^{-1}$ and a standard deviation of 80 m^3

s^{-1}. Determine the flood magnitude with a 1-in-100 chance of being exceeded in any given year.

Solution. From Eqn 8.57 and the Table of Standard Normal Probability $\Phi(z) = P(Z \leq z)$, the parameters of a lognormal distribution, for annual maximum flood Q, can be obtained as:

$$\sigma_{ln\ Q} = \sqrt{ln(\Omega_Q^2 + 1)} = \sqrt{ln\left[\left(\frac{\sigma_Q}{\mu_Q}\right)^2 + 1\right]}$$

$$= \sqrt{ln\left[\left(\frac{80}{490}\right)^2 + 1\right]} = 0.1622$$

$$\mu_{ln\ Q} = ln(\mu_Q) - \frac{1}{2}\sigma_{ln\ Q}^2 = ln(490) - \frac{1}{2}(0.1622)^2$$

$$= 6.1812$$

Since $ln(Q)$ follows a normal distribution with a mean of $\mu_{ln\ Q} = 6.1812$ and a standard deviation of $\sigma_{ln\ Q} = 0.1622$, the magnitude of the log-transformed 100-year flood can be calculated as:

$$\frac{ln\ (q_{100}) - \mu_{ln\ Q}}{\sigma_{ln\ Q}} = \Phi^{-1}\left(1 - \frac{1}{100}\right) = \Phi^{-1}(0.99)$$

$$= 2.34$$

Hence

$$ln(q_{100}) = 2.34 \times \sigma_{ln\ Q} + \mu_{ln\ Q} = 6.5607$$

and the corresponding 100-year flood magnitude is:

$$exp\ (6.5607) = 706.8 \ m^3 s^{-1}$$

For some distributions, such as Pearson Type-3 or log-Pearson Type-3, the appropriate probability paper or CDF inverse form is unavailable. In such a case, an analytical approach using the *frequency factor* K_T is applied:

$$x_T = \mu_x + K_T \times \sigma_x \tag{8.101}$$

in which x_T is the variate corresponding to a return period of T, μ_x, and σ_x are the mean and standard deviation of the random variable, respectively; and K_T is the frequency factor, which is a function of the return period T or $P(X \geq xT)$ and higher moments, if required. The x_T versus K_T plot of Eqn 8.101 is a straight line with slope σ_x and intercept μ_x.

In order for Eqn 8.101 to be applicable, the functional relationship between K_T and exceedance probability or return period must be determined for the distribution to be used. In fact, the frequency factor $K_T = (x_T - \mu_x)/\sigma_x$ is identical to a standardized variate corresponding to the exceedance probability of $1/T$ for a particular distribution model under consideration. For example, if the normal distribution is considered, $K_T = z_T = \Phi^{-1}(1 - T^{-1})$. The same applies to the lognormal distribution when the mean and standard deviation of log-transformed random variables are used. Hence the standard normal probability table provides values of the frequency factor for the sample data from normal and log-normal distributions. Once this relationship is known, a nonlinear probability or return-period scale can be constructed to replace the linear K_T scale, and thus a special graph paper can be constructed for any distribution so that plot of x_T versus P (or T) is linear.

The procedure for using the frequency-factor method is outlined as follows:

- Compute the sample mean $\bar{x}$, standard deviation s_x, and skewness coefficient γ_x (if needed) for the sample.
- For the desired return period, determine the corresponding value of K_T for the distribution.
- Compute the desired percentile value using Eqn 8.101 with $\bar{x}$ replacing μ_x and s_x replacing σ_x, that is:

$$\hat{x} = \bar{x} + K_T \times s_x \qquad (8.102)$$

It should be noted that the basic difference between the graphical and analytical approaches lies in estimating the statistical parameters of the distribution being used. By employing the graphical approach, the best-fit line is constructed that determines the statistical parameters. In the analytical approach, statistical parameters are first computed from the sample and the straight-line fit thus obtained is used. The straight line obtained in the analytical approach is in general a poorer fit to the observed data than that obtained from the graphical approach, especially if curve-fitting procedures are adopted. However, the US Water Resources Council (USWRC 1967) recommended use of the analytical approach because:

1. Graphical least-squares methods are avoided to reduce incorporation of random characteristics of the particular dataset (especially in the light of difficulty in selecting an appropriate plotting-position formula).
2. The generally larger variance of the analytical approach is believed to help compensate for the errors that arise in the case of typically small sized data sets.

8.3.5 *Limitations of hydrologic frequency analysis*

Several probability distributions have been proposed for application to hydrologic data. Some of them were proposed because the underlying concept of the distribution coincided with the goal of hydrologic frequency analysis. For example, the extremal distributions discussed below have favourable properties for hydrologic frequency analysis.

In the 1960s, a working group of hydrology experts was formed by the US Water Resources Council to evaluate the best/preferred approach to flood frequency analysis. The following key results emerged from the study conducted by this group (Benson 1968):

1. From physical considerations there is no *a priori* requirement that dictates use of a specific distribution in the analysis of hydrologic data.
2. Intuitively, there is no reason to expect that a single distribution will apply globally to all streams.
3. No single method of testing the computed results against the original data was acceptable to the working group. The statistical experts also could not offer a mathematically rigorous procedure.

Subsequent to this study, the USWRC (1967) recommended use of the log-Pearson Type-3 distribution for flood frequency analyses in the United States and this has become the officially recommended distribution for all flood frequency studies in that country. There is, however, no physical basis for application of this distribution to hydrologic data. It, however, has added flexibility over other two-parameter distributions (Gumbel, lognormal) because the skewness coefficient is the third independent parameter and the use of three parameters generally results in a better fit to the data.

Often frequency analysis is applied for the purpose of estimating magnitude of rare events, for example, a 100-year flood, on the basis of a short data series. Viessman *et al.* (1977), however, noted 'as a general rule, frequency analysis should be avoided in estimating frequencies of expected hydrologic events greater than twice the record length.' This general rule is followed rarely in practice because of the regulatory requirement to estimate the 100-year flood; for example, the USWRC (1967) gave its approval for frequency analyses using as little as 10 years of peak flow data. In order to estimate the 100-year flood on the basis of a short record, the analyst must rely on extrapolation. Klemes (1986) noted that there are many known causes for non-stationarity, ranging from the dynamics of the Earth's motion to anthropogenic changes in land use. In this context, Klemes (1986) reasoned that the notion of a 100-year flood has no meaning in terms of average return period, and thus the 100-year flood really provides a reference for design considerations rather than a true reflection of the frequency of occurrence of an event.

8.4 Nonparametric density estimation methods

In flood frequency analysis, one of the more important issues is the choice of the best probability distribution. The true distribution is always unknown in practice and often an arbitrary choice of or preference for a given distribution increases the estimation uncertainty. Some countries have tried to find standard distributions for flood frequency analysis in order to avoid arbitrariness in selection of the type of distribution adopted. Many distributions have been used to estimate flood flow frequencies from observed annual flood series. The general extreme value (GEV) distribution is recommended as a base model in the United Kingdom. After appraising many distributions, the USWRC issued a series of bulletins recommending the Log-Pearson Type-3 distribution as a base method for use by all US Federal Agencies. The most commonly used in Canada are the three parameter log-normal, the Generalized Extreme Value and the log-Pearson 3.

Recently, nonparametric density estimation methods have gained popularity in many fields of science, including hydrology. This model has several advantages. The shapes of nonparametric density functions are directly determined from the data (Faucher *et al.* 2001). It does not require any apriory assumptions about the distribution of the population of interest (Adamowski 1989) and the estimation of parameters (i.e. mean, variance, and skewness) are not also needed. The parametric distributions are limited to certain shapes. However, nonparametric densities can adapt to the often-irregular empirical distribution of random variables found in nature (Faucher *et al.* 2001).

The probability density function *f(x)* estimated by a nonparametric method is given by (Adamowksi 2000):

$$f(x) = \frac{1}{nb} \sum_{j=1}^{n} K\left(\frac{x - x_j}{b}\right) \tag{8.103}$$

where x_1 to x_n are the observations; $K(\ )$ is a kernel function, itself a probability density function; and b is a bandwidth or smoothing factor to be estimated from the data.

The following conditions are imposed on the kernel (Adamowksi 2000):

$$\int K(z)dz = 1 \tag{8.104}$$

$$\int zK(z)dz = 0 \tag{8.105}$$

$$\int z^2 K(z)dz = C \neq 0 \tag{8.106}$$

where C is the kernel variance.

The kernel distribution function is the integration of the density Eqn 8.103 from -1 to x (Adamowksi 2000):

$$F(x) = \frac{1}{n} \sum_{j=1}^{n} K_l\left(\frac{x - x_j}{b}\right) \tag{8.107}$$

where:

$$K_l(u) = \int_{-\infty}^{u} K(w)dw.$$

The kernel distribution function may serve to estimate percentiles corresponding to a given probability of exceedance. The flood quantile x_T with a return period of T years, of the kernel distribution function is (Adamowksi 2000):

$$x_T = F^{-1}\left(1 - \frac{1}{T}\right) \tag{8.108}$$

where $F^{-1}(\ )$ represents the inverse of $F(\ )$. The value of x_T can be determined by solving Eqn 8.107 numerically.

In nonparametric frequency analysis, the choice of kernel function is not critical since various kernels lead to comparable estimates (Adamowksi 2000). However, the calculation and the choice of the bandwidth, h, in Eqn 8.103 is critical. While the Gaussian kernel is often used, Table 8.1 shows some of the other commonly used kernel functions.

Several bandwidth estimation techniques are based on minimization of an estimate of the mean square error (MSE) or of the integrated mean square error ($IMSE$) function. $IMSE$ is obtained by integrating the MSE function over the entire domain of x.

The criterion of optimality is based on minimizing the IMSE given by Guo et al. (1996):

$$IMSE = \int_{-\infty}^{\infty} [\hat{f}(x) - f(x)]^2 dx \tag{8.109}$$

where $\hat{f}(x)$ is an estimate of the known density function $f(x)$. An optimal value of h can be obtained by minimizing the $IMSE$ for a given density $f(x)$ and sample. The value of h has to be derived empirically from the observed data. The optimal value of h, can be determined numerically by differentiating the objective function (Eqn 8.109) with respect to h and then equating it to zero, as given by Guo et al. (1996):

$$h = \frac{\sum\limits_{i=2}^{n}\sum\limits_{j=1}^{i-1}(x_i - x_j)}{\sqrt{5}n(n - 10/3)} \tag{8.110}$$

Table 8.2 Some commonly used kernel functions where $t = h^{-1}(x - x_i)$.

Kernel	$K(t)$
Epanechnikov	$K(t) = \frac{3}{4}(1 - t^2), \quad |t| < 1$
Rectangular	$K(t) = \frac{1}{2}, \quad |t| < 1$
Biweight	$K(t) = \frac{15}{16}(1 - t^2)^2, \quad |t| < 1$
Gaussian	$K(t) = \frac{1}{\sqrt{2\pi}}e^{-\frac{1}{2}t^2}$
Cauchy	$K(t) = \frac{1}{\pi(1+t^2)}$
EV1	$K(t) = e^{-t-e^{-t}}$

8.5 Error analysis

In the foregoing, it was implicitly assumed that the experimental observations have no uncertainty and every observation of a variable (e.g. daily stream flow) under consideration represents its true value. However, in real life most observations have a certain degree of uncertainty due to a variety of factors. As already mentioned in the Section 1.4, the term accuracy is defined to indicate closeness to the 'true' value, whereas the term precision is defined to indicate reproducibility of measurements irrespective of how close the result is to the 'true' value. In a way, accuracy and precision are related, as described below.

Precision is a measure of the magnitude of random errors. If one is able to reduce random errors, for instance by employing better equipment or procedures of data collection/analysis, results are more precise and reproducible. *Random errors* are the most common type of errors. These arise from limitations of the quality of instrumentation employed. These can only be partly overcome by refining the instrumentation/analytical method(s) employed, by repeating the measurements (reading temperature or pH several times), or extending the observation time (e.g. when using radio-isotopic measurements).

Accuracy of a measurement, on the other hand, is directly affected by systematic errors; avoiding or eliminating systematic errors makes the result more accurate and reliable. *Systematic errors* are reproducible discrepancies, often resulting from failure/fault in the instrumentation or a consistent mathematical inadequacy in the data analysis.

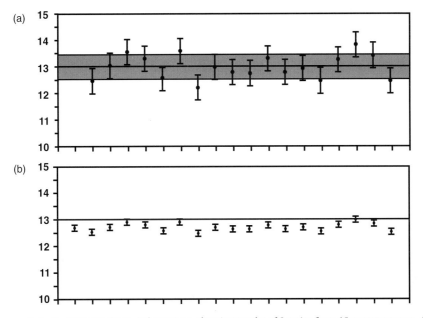

Fig. 8.12 An example to illustrate precision and accuracy, showing results of 2 series from 19 measurements. (a) The data are imprecise but accurate, giving the correct average value of 13 units. The grey area refers to the 1σ confidence level; that is, 68% of the data lie within this range. (b) The data are precise, but inaccurate probably because of a systematic error in the measurement, as the average value for this case is less than the correct value of 13 units. Redrawn from Mook (2000) © UNESCO/IAEA/IHPV.

These are more difficult to detect but may be found by repeating the analysis with different equipment or by recalculation using an alternative statistical procedure.

For studying and eventually reducing systematic errors it is important to collect data with small random errors, that is, with relatively high precision. On the other hand, it is no use increasing the precision if systematic errors are large. The difference between precision and accuracy is graphically depicted in Fig. 8.12.

8.5.1 Propagation of errors

8.5.1.1 Standard deviation

It is often required to determine a quantity, say A, which is a function of one or more variables, each with its own uncertainty. The uncertainty in each of these variables contributes to the overall uncertainty of A. If the uncertainty is due to statistical or random fluctuations and $A = f$

$(x, y, z, \ldots)$:

$$\sigma_A^2 = \sigma_x^2 \left(\frac{\partial A}{\partial x}\right)^2 + \sigma_y^2 \left(\frac{\partial A}{\partial y}\right)^2 + \sigma_{xz}^2 \left(\frac{\partial A}{\partial z}\right)^2 + \cdots \tag{8.111}$$

If the uncertainties are estimated on the basis of instrumental uncertainties, similar equations are to be used for calculating the uncertainty in the final result. For the general relation $A = f(x, y, z, \ldots)$ with the estimated instrumental uncertainties Δx, Δy, Δz, $\ldots$, the uncertainty in A is:

$$\Delta A^2 = \Delta x^2 \left(\frac{\partial A}{\partial x}\right)^2 + \Delta y^2 \left(\frac{\partial A}{\partial y}\right)^2$$

$$+ \Delta z^2 \left(\frac{\partial A}{\partial z}\right)^2 + \cdots \tag{8.112}$$

Equations for propagation of uncertainty to the dependent variable A for some common functions of two independent variables x and y, derived by using Eqn 8.112, are given in Table 8.3.

Table 8.3 Propagation of uncertainty in some commonly used functions for the dependent variable A with known uncertainty in independent variables x and y.

Function $A = f(x, y)$	Equation for ΔA
$A = ax + by$, and $A = ax - by$	$\Delta A^2 = a^2 \, \Delta x^2 + b^2 \, \Delta y^2$
$A = \pm a\,xy$, and $A = \pm a\,x/y$	$\Delta A^2 / A^2 = \Delta x^2 / x^2$ $+ \Delta y^2 / y^2$
$A = a\,e^{\pm bx}$	$\Delta A / A = \pm b\Delta x$
$A = a\,\ln(\pm bx)$	$\Delta A = a\Delta x / x$

8.5.1.2 Weighted mean

While calculating average values using Eqn 8.1, all the numbers are implicitly assumed to have the same precision and thus the same weight. If each number is assigned its own uncertainty corresponding to standard deviation (s), the weight of each result is inversely proportional to the square of the standard deviation ($1/s^2$), which is referred to as the weighting factor. Accordingly the mean is calculated as:

$$\bar{x} = \sum \frac{x_i}{\sigma_i^2} \Big/ \sum \frac{1}{\sigma_i^2} \qquad (8.113)$$

while the standard deviation of the mean is obtained from:

$$\sigma_x = \frac{1}{\sqrt{\sum(1/\sigma_i)^2}} \qquad (8.114)$$

If all the standard deviation values, s_i, are equal, the expression for s_x reduces to:

$$\sigma_x^2 = 1 \Big/ \sum (1/\sigma_i)^2 = 1/(N/\sigma_i^2) = \sigma_i^2/N$$

$$\text{or} \quad \sigma_x = \sigma_i/\sqrt{N} \qquad (8.115)$$

8.5.2 Least squares fit to a straight line

One is often interested in obtaining the best description of data in terms of some theory, which involves parameters whose values are initially unknown. Often the interest is to see if the relation between the dependent and the independent variables in a dataset is linear, that is, data lie on a straight line $y = a + bx$, and if so what are the values of the gradient (b) and the intercept (a) of the best fit line? We proceed by assuming that the data

consists of a series of points (x_i, $y_i \pm \sigma_i$) with no uncertainty in x. The σ_i is, in principle, the theoretical error that one would have expected. In practice, the observed experimental error for a given data point is used. The deviation from any value of y (i.e. y_i) from the straight line is given by:

$$\Delta y_i = y_i - f(x) = y_i - a - bx_1 \qquad (8.116)$$

Minimizing the sum of these deviations yields zero.

Thus, adding the absolute values of Δy_i does not result into a useful mathematical procedure to estimate the coefficients a and b, characterizing the straight line. But minimizing the weighted sum of the squares of the deviations [i.e. $S = \sum(\Delta y_i/\sigma_i)^2$], we obtain:

$$\frac{\partial}{\partial a} \sum \left(\frac{\Delta y_i}{\sigma_i}\right)^2 = \frac{\partial}{\partial a} \sum \left[\frac{(y_i - a - bx_i)}{\sigma_i}\right]^2 = 0 \qquad (8.117)$$

and

$$\frac{\partial}{\partial b} \sum \left(\frac{\Delta y_i}{\sigma_i}\right)^2 = \frac{\partial}{\partial b} \sum \left[\frac{(y_i - a - bx_i)}{\sigma_i}\right]^2 = 0 \qquad (8.118)$$

Resulting values of a and b, therefore, are:

$$a = \frac{1}{\Delta} \left(\sum \frac{x_i^2}{\sigma_i^2} \sum \frac{y_i^2}{\sigma_i^2} - \sum \frac{x_i^2}{\sigma_i^2} \sum \frac{x_i y_i}{\sigma_i^2}\right) \qquad (8.119)$$

and

$$b = \frac{1}{\Delta} \left(\sum \frac{1}{\sigma_i^2} \sum \frac{x_1 y_i}{\sigma_i^2} - \sum \frac{x_i^2}{\sigma_i^2} \sum \frac{y_i^2}{\sigma_i^2}\right) \qquad (8.120)$$

with

$$\Delta = \sum \frac{1}{\sigma_i^2} \sum \frac{x_i^2}{\sigma_i^2} - \left(\sum \frac{x_i}{\sigma_i^2}\right)^2 \qquad (8.121)$$

If the standard deviations of y are all equal, the values of a and b are:

$$a = \frac{1}{\Delta} \left(\sum x_i^2 \sum y_i - \sum x_i \sum x_i y_i\right) \qquad (8.122)$$

$$b = \frac{1}{\Delta} \left(N \sum x_i y_i - \sum x_i \sum y_i\right) \qquad (8.123)$$

$$\Delta = N \sum x_i^2 - \left(\sum x_i\right)^2 \qquad (8.124)$$

The standard deviations of the coefficients a and b are:

$$\sigma_a = \sqrt{\frac{1}{\Delta}\sum \frac{x_i^2}{\sigma_i^2}} \qquad (8.125)$$

and

$$\sigma_b = \sqrt{\frac{1}{\Delta}\sum \frac{1}{\sigma_i^2}} \qquad (8.126)$$

As above, the least-squares fit to any arbitrary curve, for instance a quadratic or second-degree polynomial, harmonic, or an exponential, can also be obtained by minimizing the weighted sum of squares (S) of deviation between observed values of y_i and the computed value for the particular x_i for the chosen curve and the arbitrary value of constants (e.g. a and b above) to be optimized:

$$S = \sum_i \left(\frac{\Delta y_i}{\sigma_i}\right)^2 = \sum_i \left(\frac{y_i^{obs} - y_i^{calc}}{\sigma_i}\right)^2 \qquad (8.127)$$

Several computer programs are available to obtain the optimized values of the parameters for fitting any arbitrarily chosen curve.

8.5.3 Chi-square test for testing of distributions

Choosing an equation to fit a set of observations implies hypothesizing a particular form for the dataset of the parent distribution. One would like to know if this hypothesis concerning the form of parent distribution is correct. In fact, it is not possible to give an unambiguous 'yes' or 'no' answer to this question, but one would still like to state how confidently the hypothesis can either be accepted or rejected.

Distributions are tested by the χ^2 method. When the experimentally observed y_i^{obs} for each experimental point is normally distributed with mean y_i^{th} and the variance σ_i^2, the S defined in Eqn 8.127 after replacing y_i^{calc} with y_i^{th} is distributed as χ^2 (defined in Section 8.2.3.5.1). Therefore, in order to test a hypothesis that the observed distribution is consistent with the expected distribution one needs to:

1. Construct S using Eqn 8.127 and minimize it with respect to the free parameters:
2. Determine the number of degrees of freedom (K) from:

$$K = n - p \qquad (8.128)$$

where n is the number of data points included in the summation for S, and p is the number of free parameters that are made to vary to arrive at the value of S_{min};
3. Refer to the relevant probability tables for the given degree of freedom, K, to see if χ^2 is greater than or equal to S_{min}. In such a case, the assumed distribution is likely to be consistent with the dataset.

As the term σ_i^2 appears in the denominator of Eqn 8.127, the magnitude of errors on individual points also determines if the hypothesis that the data are consistent with the expected distribution is reasonable.

More useful than the χ^2 distribution itself, is:

$$P_y(c) = P_K(\chi^2 > c) \qquad (8.129)$$

which gives the probability that, for the given degrees of freedom, the value of χ^2 will exceed a specified value of c. In deciding whether or not to reject a hypothesis, one can make the following two kinds of incorrect decisions.

4. *Error of the first kind* – In this case the hypothesis gets rejected even though it is correct. Thus it is erroneously concluded that the data are inconsistent with the distribution whereas, in fact, they are consistent. This happens even in a known fraction of tests by the maximum accepted value, S_{min}. The number of errors of the first kind can be reduced simply by increasing the limit on S_{min}, the cut-off limit beyond which the hypothesis is rejected. But this is liable to increase the number of errors of the second kind and, therefore, some compromise value of the limit must be chosen.
5. *Error of the second kind* – In this case one fails to reject the hypothesis even though it is false and some other hypothesis may be applicable. This means that one fails to detect that the data

are inconsistent with the distribution. This happens because the value of S_{min} accidentally turns out to be small even though the data are actually inconsistent with the distribution. In general, it is quite difficult to estimate how frequent this effect is likely to occur. It, however, depends on the magnitude of the cut-off used for S_{min} and on the size of errors on the individual points.

Though arbitrary, as a matter of practice, if the probability that S_{min} is exceeded for the given degree of freedom by less than 5% or 1%, the hypothesis can be accepted at 95% or 99% confidence levels, respectively.

Example 8.9. Consider a situation where one may be testing the linear distribution of a set of (say 12) data points and when one fits the expression $y = a + bx$ to the data, a value of 20.0 is obtained for S_{min} using Eqn 8.127. In this case there are 10 degrees of freedom (12 points minus 2, corresponding to the parameters a and b). From the table of percentage area in the tails of χ^2 distributions for various degrees of freedom, it may be seen that the probability of getting a value of 20 or more is about 3%. The linear distribution can be accepted at the 95% but not at the 99% confidence level.

8.5.4 Student's t-test

It is a statistical test of the null hypothesis that the mean values of two normally distributed populations are equal. Given two datasets, each characterized by its own mean, standard deviation, and number of data points, one can use the t-test to determine whether the two mean values are distinct, provided that the underlying distributions can be assumed to be normal.

It has been shown that a t-distribution has the mean equal to 0 and the variance is equal to $K/(K-2)$, where K is the degrees of freedom (Eqn 8.96). The variance is always greater than 1, although it is close to unity when there are many degrees of freedom. For infinite degrees of freedom, the t-distribution is the same as the standard normal distribution.

As with χ^2 distribution, one can define 95%, 99%, or other confidence intervals by using the table of t-distribution. As mentioned above, the probability of one being wrong in rejecting the hypothesis (i.e. making Type-1 error) is given by the area under the tail of the distribution exceeding the significance level of 0.05 or 0.01, or any other pre-assigned value obtained by computing the t-test statistics. If two random samples of sizes N_1 and N_2 are drawn from normal populations whose standard deviations s_1 and s_2 are equal ($\sigma_1 = \sigma_2$) and the two sample means are $\bar{X}_1$ and $\bar{X}_2$, respectively, the test statistics t is given by:

$$t = \frac{\bar{X}_1 - \bar{X}_2}{\sigma\sqrt{1/N_1 + 1/N_2}} \quad where \quad = \sqrt{\frac{N_1 s_1^2 + N_2 s_2^2}{N_1 + N_2 - 2}}$$

$$(8.130)$$

The degrees of freedom of the t-distribution are $K = N_1 + N_2 - 2$.

8.5.5 The F-test

Just as the sampling distribution of the difference in the mean ($\bar{X}_1 - \bar{X}_2$) of the two samples is important in some applications, one may also need the sampling distribution of the difference in variance ($s_1^2 - s_2^2$). It, however, turns out that this distribution is rather complicated. One may, therefore, consider the statistics s_1^2/s_2^2, since either a large or small ratio would indicate a large difference between the two, while a ratio close to unity would correspondingly indicate a small difference. The sampling distribution in this case is called the *F-distribution*.

More precisely, suppose that one has two samples of sizes N_1 and N_2 drawn from two normal (or approximately normal) populations having variance σ_1^2 and σ_2^2. The statistics F can be defined as:

$$F = \frac{\hat{s}_1^2/\sigma_1^2}{\hat{s}_2^2/\sigma_2^2} = \frac{N_1 s_1^2/(N_1 - 1)\sigma_1^2}{N_2 s_2^2/(N_2 - 1)\sigma_2^2}$$

$$where \quad \hat{s}_1^2 = \frac{N_1 s_1^2}{(N_1 - 1)} \quad and \quad \hat{s}_2^2 = \frac{N_2 s_2^2}{(N_2 - 1)}$$

$$(8.131)$$

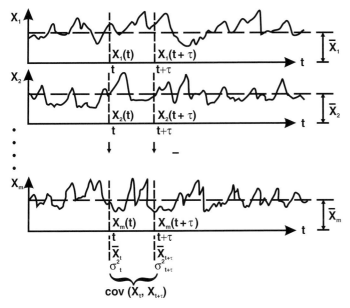

Fig. 8.13 An ensemble of m time series (X_1, X_2, ..., X_m). The values of X at t and ($t + \tau$) for any series k are designated by $x_k(t)$ and $x_k(t+\tau)$, respectively. The mean and variance across the m series at t and ($t+\tau$) are given by $\bar{x}(t)$, σ_t^2 and $\bar{x}(t + \tau)$, $\sigma_{t+\tau}^2$, respectively. The covariance of $x(t)$ and $x(t + \tau)$ across the m series is designated by $\text{cov}(x_t, x_{t+\tau})$. Redrawn from Yevjevich (1972). © Water Resources Publications,.

The degrees of freedom of the F distribution are $K_1 = (N_1 - 1)$ and $K_2 = (N_2 - 1)$. The PDF of the F distribution is given by:

$$f_F = \frac{CF^{(K_1/2)-1}}{(K_1 F + K_2)^{(K_1+K_2)/2}} \qquad (8.132)$$

where C is a constant depending on K_1 and K_2 such that the total area under the curve is unity. The percentile values for which the areas in the right-hand tail are 0.05 and 0.01, denoted by $F_{.95}$ and $F_{.99}$ respectively, can be read from the tables corresponding to appropriate degrees of freedom K_1 and K_2. These represent the 5% and 1% significance levels respectively, and can be used to determine whether or not the variance s_1^2 is significantly larger than s_2^2. In practice, the sample with the larger variance is chosen as the sample 1.

8.6 Time series analysis

A time series (TS) is an ordered sequence of data, measured/derived typically at discrete time intervals. Unlike analyses of random samples of observations in relation to various statistics, the analysis of TS is based on the assumption that successive values in the data represent consecutive measurements taken at discrete time steps/intervals. If for any reason some time steps are either skipped or no data is available for a certain time step (or number of time steps), it may sometimes be possible to interpolate the values of the parameter for the missing time steps. Hydrologic models can also generate TS data.

TS data are usually plotted as a line or bar graph. An ensemble of TS, each of size N in the discrete case, or length T in the continuous case, is a set of the TS. An example is shown in Fig. 8.13. This ensemble may be thought of as the magnitude of a parameter measured at the same place and under the same conditions, but for successive time steps T.

Time series are used to describe several aspects of the hydrologic cycle. To a hydrologist or a hydraulic engineer, the TS approach is useful to characterize the resources and/or conditions of a water basin. Hydrologists use TS methodology for displaying the amount of rainfall that has taken place within a catchment on various time scales – during the previous day, past year, or 10 years. As an example, Fig. 8.14 is a time series of annual rainfall over India since 1872. This information, combined with additional TS data (e.g. on El Nino and La Nina years, as given in Fig. 8.14), is used by meteorologists to find cause and effect relationships between the monsoon and the El Nino/La Nina, whereas hydrologists are interested in calculating

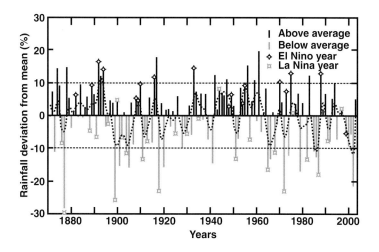

Fig. 8.14 A time series of annual rainfall over India since 1872, with El Nino and La Nina years marked. Redrawn from Kumar (2007).

the amount of storm water runoff and estimation of the total discharge from a given catchment. Hydraulic engineers use TS of measured discharge and water level of a river for designing new water control structures/better management of the existing river development projects. TS data are also used for planning any input or withdrawal of water from the river. For example, TS data are used to describe the amount of water that is being released from a sewage treatment plant into a river system or the amount of water being taken out of a river system for supplying an irrigation canal network.

There are two main objectives of TS analysis: (i) identifying nature of the phenomenon represented by the sequence of observations; (ii) forecasting (predicting future values of the TS variable). Both these objectives require that the pattern in the observed TS data is identified and more or less formally described. Once the trend is identified, one can extrapolate it to predict future events. Time series analysis is also used in many applications, such as economic- and sales- forecasting, budgetary analysis, stock market analysis, agricultural yield projections, process and quality control, inventory monitoring, and workload projections in industry, utility studies, census analysis, and many more.

As with most other hydrologic data analyses, the TS data are assumed to comprise: (i) a systematic

pattern (usually a set of identifiable frequency components); and (ii) random noise (error) components. Most TS analysis techniques involve some form of filtering out of noise in order to make the pattern more conspicuous/prominent. Methods for TS analyses can be divided into two broad classes: (i) Time-domain methods; and (ii) frequency-domain methods. A time domain analysis aims to describe the pattern of the series over time. A frequency domain analysis, on the other hand, aims to determine the strength/power of the periodicity inherent in the series within each given frequency band over a range of frequencies. In the following we first consider the time-domain methods of TS analysis.

8.6.1 General aspects of time series patterns

Time series patterns, in general, can be described in terms of two basic classes of components: trend and periodicity. The former represents a general systematic linear, or more often the nonlinear component, that varies over time and does not repeat at least within the time range encompassed by most parts of the data. The latter component (i.e. periodicity) may formally have a similar nature, but it repeats itself at periodic time intervals. The terms *cyclic* or *seasonal* are often used for components

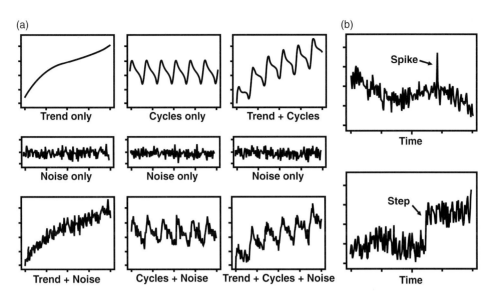

Fig. 8.15 (a) Combining trend, cycles, and noise in a time series. (b) Some important features to look out for during preliminary analysis of a time series. Redrawn from Wild and Seber (2000). © John Wiley & Sons Inc.

when the series is influenced by periodic factors that recur on a regular basis. Fig. 8.15a shows the various components schematically.

Sometimes the term cyclic is used only when the data exhibit rises and falls for purely periodic components that are not of any fixed period or season. The various TS components may co-exist in real-life data. Thus, in terms of its component parts, a TS comprises trend, cycle, seasonal, and irregular (or random error) components and may be written as:

$$x_t = T_t + S_t + C_t + E_t \tag{8.133}$$

for an additive model and as:

$$x_t = T_t \times S_t \times C_t \times E_t \tag{8.134}$$

for a multiplicative model. The symbols are: T_t = trend component; S_t = seasonality component; C_t = cyclic component; and E_t = random error component. If the magnitude of the TS varies with the level of the series, one should opt for a multiplicative model in preference to an additive model. The multiplicative model can be processed as an additive model by taking logarithms of the various terms in the series. This decomposition may enable one to study the TS components separately;

by de-trending; or to make seasonal adjustments if needed for further analysis.

8.6.1.1 Stationarity and ergodicity

If properties of the time series do not change with time' they are called *stationary* (time-invariant or non-evolutive). If properties of the sequence do not change with position on the line, a series is called homogenous (line-invariant, non-evolutive).

A sample of discrete series of size N, or a sample of continuous series of length T, is a realization of the underlying stochastic process. It is also called a *sample function*. Consider an ensemble of the m series, $I = 1, 2, \ldots, m$, each being T-time long (or N-size) for a continuous (or discrete) TS, as shown in Fig. 8.13. The mean of $x(t)$, at the position t from the origin of each series, determined over the entire ensemble where both $x(t)$ and $x(t+\tau)$ designate the value of x at t and $(t + \tau)$, respectively, is:

$$\bar{x}_t = \frac{1}{m} \sum_{i=1}^{m} x_i(t) \tag{8.135}$$

and of $x(t + \tau)$ at the position $(t + \tau)$ is:

$$\bar{x}_{t+} = \frac{1}{m} \sum_{i=1}^{m} x_i(t + \tau) \tag{8.136}$$

The covariance of $x(t)$ and $x(t + \tau)$ is:

$$\text{cov}(x_t, x_{t+\tau}) = \frac{1}{m} \sum_{i=1}^{m} x_i(t)x_i(t + \tau) - \bar{x}_t\bar{x}_{t+\tau}$$

(8.137)

If the mean of $x_{(t)}$ and $x_{(t+\tau)}$, given in the form $\bar{x}_t$ and $\bar{x}_{t+\tau}$ of Eqn 8.135 and Eqn 8.136, converges to the same population mean, μ, with the probability equal to one, for the case when $m \to \infty$, regardless of the position of t between zero and T, the time series $x(t)$ is said to be *stationary* in the mean, or *first-order stationary*.

If the covariance of Eqn 8.137 is independent of position, t, but dependent on the *lag* τ, and is a constant for a given *lag*, i.e.:

$$cov(x_t, x_{t+\tau}) = \rho_\tau \sigma_x^2, \quad for \quad m \to \infty \quad (8.138)$$

in which σ_x^2 is the population variance of x_t and ρ_τ is the population autocorrelation coefficient for *lag* τ, the TS is said to be *stationary in the covariance*. For $\tau = 0$, $\rho_\tau = 1$, which implies that the series is stationary in the variance at the same time. When the series is stationary both in mean and in the covariance, it is said to be *second-order stationary*. If all higher moments (3rd, 4th, 5th, ...) of $x(t)$, $x(t+_1)$, $x(t+_2)$, ... are independent of t, but dependent on t_1, t_2, ..., and they all converge with probability equal to 1 to corresponding higher-order population moments as $m \to \infty$, the series is *higher-order stationary*. This implies that the series is also stationary in mean and covariance. Stationarity of this kind is called strong stationarity, or stationarity in the strict sense as opposed to the weak stationarity when the series is only second-order stationary. In the field of hydrology, in actual practice one is often satisfied with second-order stationarity of TS.

Statistical parameters of each particular series of the ensemble of series, given here as *i*-series, $i = 1, 2, ..., m$, are

$$\bar{x}_i = \frac{1}{T} \int_0^T x_t dt$$

(8.139)

$$cov(x_t, x_{t+\tau}) = \frac{1}{T - \tau} \int_0^{T-\tau} (x_t - \bar{x}_t)(x_{t+\tau} - \bar{x}_{t+\tau}) \, dt$$

(8.140)

in which x_t and $x_{t+\tau}$ refer to the *i*th series only, and similarly for higher-order moments. If for any series $\bar{x}_i \to \mu$ (population mean with the probability unity for $T \to \infty$), and $cov(x_t, x_{t+\tau}) = f(\tau) = \rho_\tau \sigma_\tau^2$ for any τ, even as $m \to \infty$ and for higher-order moments, the series in the ensemble is said to be ergodic. In the opposite case, the series is non-ergodic.

8.6.1.1.1 Tests of stationarity and ergodicity

In practice, both the number of terms in a series and the length of a series are always finite. Statistical tests, therefore, must be performed to ascertain whether the stochastic process is stationary as well as ergodic. The following scheme of ensemble of a discrete series is assumed for discussion of these tests.:

$$
\begin{array}{cccc|c}
x_{1,1} & x_{1,j} & \cdots & x_{1,N} & \hat{\alpha}_{1,j} \\
x_{i,1} & x_{i,j} & \cdots & x_{i,N} & \hat{\alpha}_{i,j} \\
\cdots & \cdots & \cdots & \cdots & \cdots \\
x_{m,1} & x_{m,j} & \cdots & x_{m,N} & \hat{\alpha}_{m,j} \\
\hline
\hat{\alpha}_{i,1} & \hat{\alpha}_{i,j} & \cdots & \hat{\alpha}_{i,N} &
\end{array}
$$

Any parameter α is estimated either across the ensemble of the series as $\hat{\alpha}_{i,1}$, $\hat{\alpha}_{i,2}$, ..., $\hat{\alpha}_{i,N}$, or along the discrete series as $\hat{\alpha}_{1,j}$, $\hat{\alpha}_{2,j}$, ..., $\hat{\alpha}_{m,j}$, . Because of sampling fluctuations, neither $\hat{\alpha}_{i,1}$, $\hat{\alpha}_{i,2}$, ..., $\hat{\alpha}_{i,N}$, nor $\hat{\alpha}_{1,j}$, $\hat{\alpha}_{2,j}$, ..., $\hat{\alpha}_{m,j}$, are likely to be identical but may be shown by a proper test (say χ^2 test) as statistically indistinguishable from each other. In the first case, the stochastic process is inferred to be stationary in α, and if that is accepted, in the second case the process is inferred to be ergodic in α as well. When hydrologic processes are inherently stationary and ergodic, or made stationary and ergodic by eliminating non-stationarity and non-ergodicity where possible, the mathematical analysis and description become much simpler.

8.6.1.1.2 Self-stationarity

Hydrologic stochastic processes are usually described by a single series. The only test to be performed would be whether the properties of the series do not change as one considers the higher-order terms. In other words, the test is to ascertain whether parts of the series have the same properties as the total population, in the limits of sampling fluctuation. For example, if the series is divided into two parts and it is shown that the parameters of both parts are not significantly different from each other, both parts are considered as originating from the same process. If this is repeated for still smaller parts and the same results are obtained, the stochastic process is said to be *self-stationary*. In general, if the sub-series parameters are confined to within 95% tolerance limits around the corresponding value of the parameter for the entire series, the process is inferred to be self-stationary.

Example 8.10. A given year with 12 monthly precipitation values is considered a sub-series making n series of ensemble for n years of record. If the values of mean, variance, covariance, and parameters related to other moments are obtained for each month, it can be easily shown that the hydrologic series of monthly precipitation does not follow a stationary stochastic process. The reason for this is the presence of a periodic annual component in such series.

8.6.1.2 Trend analysis

If a time series has a monotonous trend (either consistently increasing or decreasing), that part of data analysis is generally not difficult and often the first step in the process of trend identification is smoothing.

8.6.1.2.1 Smoothing

Smoothing involves some form of local averaging of data such that the non-systematic components of individual observations cancel each other. Thus smoothing removes random variations from the data and brings out trends and cyclic components, if present. The most common technique is *moving average* smoothing, which replaces each element of the series by either the simple or weighted average of the surrounding n elements, where n is the width of the smoothing 'window' (Box and Jenkins 1976):

$$\hat{x}_t = M_t^{[1]} = \frac{x_t + x_{t-1} + x_{t-2} \cdots \cdots x_{t-n+1}}{n} \quad or$$

$$M_t^{[1]} = M_{t-1}^{[1]} + (x_t - x_{t-n})/n \qquad (8.141)$$

where $M_t^{[1]}$ *is* the moving average at time t and n is the width or the number of terms in the moving average. At each successive time interval, the most recent observation is included and the earliest observation is excluded for computing the average. Simple moving average is intended for data involving constants or without having any trend. For linear or quadratic trends, double moving average can be calculated. To calculate this moving average, $M_t^{[1]}$, is treated over time simply as individual data points and a moving average of these average values is obtained. As an example, simple decadal moving average values of annual rainfall series of India is shown in Fig. 8.14. Median values can be used instead of the mean values. The main advantage of using median smoothing, as compared to moving average smoothing, is that its results are less biased by outliers (within the smoothing window). Thus, if there are outliers in the data (e.g. due to measurement errors), median smoothing typically produces smoother or at least more 'reliable' curves than moving average based on the same window width. The main disadvantage of median smoothing is that in the absence of clear outliers, it may produce more 'jagged' curves than moving average and it does not allow for weighting. There are several other moving average models in use.

8.6.1.2.2 Fitting a function

Many monotonous TS data can be adequately approximated by a linear function; if there is a clear monotonous nonlinear component, the data first need to be transformed to remove the nonlinearity. Usually a logarithmic, exponential, or sometimes a polynomial function can also be used.

8.6.1.3 Analysis of seasonality

Cyclic dependence (also called seasonality, analogous to seasons) is another general component of the TS pattern. The concept is mathematically defined as co-relational dependence of order τ between each tth element of the series and the $(t + \tau)^{th}$ element, as measured by autocorrelation (i.e. a correlation between the two terms); τ is usually called the *lag*. If the measurement error is small, seasonality can be visually identified in the series as a pattern that repeats after every k elements (Fig. 8.15a).

The correlation between the pair $(x_t, x_{t+\tau})$ is given by the autocorrelation function (ACF) obtained as:

$$r_\tau = \frac{\sum_{t=1}^{t-\tau}(x_t - \bar{x})(x_{t+\tau} - \bar{x})}{\sum_{t=1}^{n}(x_t - \bar{x})^2} \quad (8.142)$$

It ranges from -1 to $+1$. The correlogram (autocorrelogram) displays graphically (as bar charts) and numerically, the r_τ for consecutive *lags* in a specified range of *lags*, as shown in Fig. 8.16. The degree of autocorrelation is of primary interest as one is usually interested only in very strong (and thus highly significant) autocorrelations.

A partial autocorrelation function (PACF) is used to denote the degree of association between x_t and $x_{t+\tau}$ when the cyclic effects of other time *lags* 1, 2, …$\tau-1$ are already removed. In other words, the partial autocorrelation is similar to autocorrelation except that when calculating it the (auto)correlations with all the elements within the *lag* period are removed.

Cyclic dependence for a particular *lag* τ can be removed by differencing the series, that is, converting each $(t + \tau)^{th}$ element of the series into its difference from the tth element. There are two major reasons for effecting such a transformation.

First, one can identify the nature of hidden seasonal dependence in the series. Because autocorrelations for consecutive *lags* are interdependent, removing some of the autocorrelations modifies other autocorrelation values, that is, it may eliminate them or make some other seasonality more apparent.

The other reason for removing seasonal dependence is to make the series stationary, which is necessary for ARIMA and other techniques, as discussed subsequently. (Section 8.6.2)

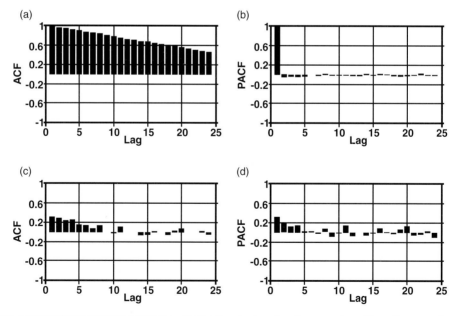

Fig. 8.16 (a) Autocorrelation Function (ACF) showing significant values for a large number of *lags*. However, the autocorrelations at *lag* 2 and above are possibly due to propagation of the autocorrelation at *lag* 1. This is confirmed by the Partial Autocorrelation Function (PACF) plot in (b) that shows a significant spike only at *lag* 1, implying that all the higher-order autocorrelations are effectively explained by the *lag* 1 autocorrelation, suggesting one order of non-seasonal differencing, that is, ARIMA (0, 1, 0) for making the series stationary; (c) ACF obtained for the series of (a) by applying ARIMA (0, 1, 0) model; (d) PACF of (c). Note from (c) that the correlation at *lag* 1 is significant and positive and from (d) that the PACF shows a sharper 'cut-off' than the ACF. In particular, the PACF has only two significant spikes, while the ACF has four. Redrawn from http://www.duke.edu/~rnau/411arim3.htm.

8.6.1.3.1 Hypothesis test on r_1

The first-order autocorrelation coefficient is especially important because dependence of physical systems on past values is likely to be strongest for the most recent past. The first-order autocorrelation coefficient, r_1, can be tested against the null hypothesis that the corresponding population value $\rho_1 = 0$. The critical value of r_1 for a given significance level (e.g. 95%) depends on whether the test is one-tailed or two-tailed. For the one-tailed hypothesis, the alternative hypothesis is usually that the true first-order autocorrelation is greater than zero:

$$H : \rho_1 > 0 \qquad (8.143)$$

For the two-tailed test, alternative hypothesis is that the true first-order autocorrelation is different from 0, with no specification as to positive or negative:

$$H : \rho_1 \neq 0 \qquad (8.144)$$

Which alternative hypothesis to use depends on the problem. If there is some reason to expect positive autocorrelation (e.g. with tree rings, from carryover food storage in trees), the one-sided test is the best. Otherwise, the two-sided test is the best.

For the one-sided test, the World Meteorological Organization (WMO 1966) recommends that the 95% significance level for r_1 be computed by using the equation:

$$r_1(95\%) = \frac{-1 + 1.645\sqrt{n-2}}{n-1} \qquad (8.145)$$

where n is the sample size. More generally, following Salas *et al.* (1980), the probability limits on the correlogram of an independent series are:

$$r_k(95\%) = \frac{-1 + 1.645\sqrt{n-\tau-1}}{n-\tau} \quad for\ one\ sided$$
$$(8.146)$$

$$r_k(95\%) = \frac{-1 \pm 1.96\sqrt{n-\tau-1}}{n-\tau} \quad for\ two\ sided$$
$$(8.147)$$

where n is the sample size and τ is the *lag*.

8.6.1.3.2 Effective sample size

If a time series of length N is autocorrelated, the number of *independent* observations is fewer than N. Essentially, the series is not random in time domain, and the information in each observation is not totally independent of the information in other observations. The reduction in the number of independent observations has implications for hypothesis testing.

Some standard statistical tests that depend on the assumption of random samples can still be applied to a time series, despite the autocorrelation in the series. The way of circumventing the problem of autocorrelation is to adjust the sample size for autocorrelation. The number of independent samples after adjustment is fewer than the number of observations of the series.

Below is an equation for computing so-called 'effective' sample size, or sample size adjusted for autocorrelation. More on the adjustment can be found elsewhere (Dawdy and Matalas 1964; WMO 1966). The equation was derived based on the assumption that the autocorrelation in the series represents *first-order* autocorrelation (dependence on lag-1 only). In other words, the governing process is *first-order autoregressive*, or *Markovian*. Computation of the effective sample size requires only the sample size and first-order sample autocorrelation coefficient. The 'effective' sample size is given by:

$$n' = n\frac{(1 - r_1)}{(1 + r_1)} \qquad (8.148)$$

where n is the sample size; n' is the effective sample size; and r_1 is the first-order autocorrelation coefficient. For example, a series with a sample size of 100 years and a first-order autocorrelation of 0.50 has an adjusted sample size of:

$$n' = 100\frac{(1 - 0.5)}{(1 + 0.5)} = 100\frac{0.5}{1.5} \approx 33\ years \qquad (8.149)$$

8.6.2 ARIMA modelling

The modelling and forecasting procedures discussed above necessitate knowledge of the mathematical model of the process. However, in real-life situations, as patterns of the data are not clear, and individual observations may involve considerable errors. One still needs not only to uncover the hidden patterns in the data but also to generate forecasts. The ARIMA methodology developed by Box and Jenkins (1976) enables one to do just this.

The following section introduces basic ideas of this methodology.

8.6.2.1 Two commonly employed processes

8.6.2.1.1 Autoregressive process

Most TS often consist of elements that are serially dependent in the sense that one can estimate a coefficient or a set of coefficients that describe elements of the series from specific, time-lagged (previous) elements. This can be summarized by the equation:

$$x_t = \xi + \phi_1 \times x_{(t-1)} + \phi_2 \times x_{(t-2)}$$
$$+ \phi_3 \times x_{(t-3)} + \cdots \phi_p \times x_{(t-p)} + \varepsilon_t \quad (8.150)$$

where ξ is a constant, usually equal to the mean; and $\phi_1, \phi_2, \phi_3, \ldots \phi_p$ are the autoregressive model parameters. Expressed in words, each observation consists of a random error (ε_t) component (also called random shock) and a linear combination of prior observations. The value of p is called the order of the AR model. AR models can be analysed by one of various methods, including standard linear least squares techniques, and have a straightforward interpretation.

Some formulations transform the series by subtracting the mean of the series from each data point. This yields a series with a mean value of zero. With this transformation; Eqn 8.150 can be rewritten as:

$$x_t = \phi_1 \times x_{(t-1)} + \phi_2 \times x_{(t-2)}$$
$$+ \phi_3 \times x_{(t-3)} + \cdots \phi_p \times x_{(t-p)} + \varepsilon_t \quad (8.151)$$

An autoregressive process is stable if the parameters lie within a certain range; for example, if there is only one autoregressive parameter; then it must be within the interval $-1 < \phi_1 < 1$. Otherwise, past effects would accumulate and the values of successive x_t would move towards infinity, that is, the series would not be stationary. If there is more than one autoregressive parameter, similar (general) restrictions on the parameter values can be specified.

8.6.2.1.2 Moving average process

Independent of the autoregressive process, each element in the series can also be affected by the past error (or random shock) that cannot be accounted for by the autoregressive component, i.e.:

$$x_t = \mu + \bar{\varepsilon}_t - \theta_1 \times \bar{\varepsilon}_{(t-1)} - \theta_2 \times \bar{\varepsilon}_{(t-2)}$$
$$- \theta_3 \times \bar{\varepsilon}_{(t-3)} - \cdots - \theta_q \times \bar{\varepsilon}_{(t-q)} \quad (8.152)$$

where μ is the series mean and $\theta_1, \theta_2, \theta_3, \ldots \theta_q$ are the moving average model parameters. This means that each observation is made up of a random error component (random shock, $\bar{\varepsilon}$) and a linear combination of prior random shocks. Thus, a moving average model is conceptually a linear regression of the current value of the series against the white noise or random shocks of one or more prior values of the series. The value of q is called the order of the MA model.

If the series is transformed by subtracting the mean, Eqn 8.152 can be rewritten as:

$$x_t = \bar{\varepsilon}_t - \theta_1 \times \bar{\varepsilon}_{(t-1)} - \theta_2 \times \bar{\varepsilon}_{(t-2)}$$
$$- \theta_3 \times \bar{\varepsilon}_{(t-3)} - \cdots - \theta_q \times \bar{\varepsilon}_{(t-q)} \quad (8.153)$$

Without going into detail, it is obvious that there is a 'duality' between the moving average process and the autoregressive process, that is, the moving average equation given above can be rewritten (inverted) into an autoregressive form (of infinite order). However, analogous to the stationarity condition described for the autoregressive model, this can only be done if the moving average parameters follow certain conditions, i.e. if the model is invertible. Otherwise, the series is not stationary.

8.6.2.1.3 Lag operator

In TS analysis, the *lag operator* or *backshift operator* operates on an element of TS to produce the previous element. For example, given a TS:

$$L x_t = x_{(t-1)} \quad for\ all\ t > 1 \quad (8.154)$$

where L is the *lag* operator. Sometimes the symbol B for backshift is used. Note that the *lag* operator can be raised to arbitrary integer powers so that:

$$L^{-1} x_t = x_{(t+1)} \quad and \quad L^k x_t = x_{(t-k)} \quad (8.155)$$

Also polynomials of the *lag* operator can be used, and this is a common notation for ARMA models.

For example:

$$\varepsilon_t = x_t - \sum_{i=1}^{p} \phi_i \times x_{(t-i)}$$

$$= \left(1 - \sum_{i=1}^{p} \phi_i L^i\right) x_t \qquad (8.156)$$

specifies an AR(p) model.

A polynomial of *lag* operator is called a *lag polynomial* so that, for example, the ARMA model can be mathematically specified as:

$$\phi x_t = \theta \varepsilon_t \qquad (8.157)$$

where ϕ and θ, respectively, represent the *lag* polynomials:

$$\phi = 1 - \sum_{i=1}^{p} \phi_i L^i \quad and$$

$$\theta = 1 + \sum_{i=1}^{q} \theta_i L^i \qquad (8.158)$$

In TS analysis, the first difference operator Δ is a special case of *lag* polynomial:

$$\Delta x_t = x_t - x_{(t-1)} = (1 - L)x_t \qquad (8.159)$$

Similarly, the second difference operator:

$$\Delta(\Delta x_t) = \Delta x_t - \Delta x_{(t-1)}$$

$$\Delta^2 x_t = (1 - L)\Delta x_t$$

$$= (1 - L)(1 - L)x_t$$

$$= (1 - L^2)x_t \qquad (8.160)$$

The above approach can be generalized to give the ith difference operator as:

$$\Delta^i x_t = (1 - L)^i x_t \qquad (8.161)$$

8.6.2.2 Autoregressive integrated moving average (ARIMA) model

The acronym ARIMA stands for 'Auto-Regressive Integrated Moving Average.' *Lags* of the differenced series appearing in the forecasting equation are called 'auto-regressive' terms, *lags* of the forecast errors are called 'moving average' terms, and a time series, which needs to be differenced to make it stationary, is said to be an 'integrated' version of a stationary series.

The non-seasonal ARIMA model introduced by Box and Jenkins (1976) includes three types of parameters in the model – the autoregressive parameters (p), the number of differencing passes (d), and moving average parameters (q). In the notation introduced by Box and Jenkins, models are summarized as ARIMA (p, d, q); thus for example, a model described as (0, 1, 2) means that it contains 0 (zero) autoregressive parameters and 2 moving average parameters that are computed for the series after it is differenced once. For instance, given a TS process X_t, a first-order autoregressive process is denoted by ARIMA (1, 0, 0) or simply AR (1) and is given by:

$$x_t = \phi_1 \times x_{(t-1)} + \varepsilon_t \qquad (8.162)$$

and the first-order moving average process is denoted by ARIMA (0, 0, 1) or simply MA (1) and is given by:

$$x_t = \varepsilon_t - \theta_1 \times \varepsilon_{(t-1)} \qquad (8.163)$$

The ultimate model may be a combination of these processes and of higher orders as well. Thus, an ARMA (p, q) process defined by Eqn 8.157 can be expanded as:

$$\left(1 - \sum_{i=1}^{p} \phi_i L^i\right) x_t = \left(1 + \sum_{i=1}^{q} \theta_i L^i\right) \varepsilon_t \qquad (8.164)$$

where L is the *lag* operator, the ϕ_i are the parameters of the autoregressive part of the model. The θ_i are the parameters of the moving average part and the ε_t are the error terms. The error terms ε_t are generally assumed to be independent, identically distributed variables sampled from a normal distribution with zero mean.

An ARIMA (p, d, q) process is obtained by integrating an ARMA (p, q) process, i.e.:

$$\left(1 - \sum_{i=1}^{p} \phi_i L^i\right)(1 - L)^d x_t = \left(1 + \sum_{i=1}^{q} \theta_i L^i\right) \varepsilon_t$$

$$(8.165)$$

where d is a positive integer that controls the level of differencing (or, if $d = 0$, this model is equivalent to an ARMA model). Conversely, applying term-by-term differencing d times to an ARMA (p, q) process gives an ARIMA (p, d, q) process. Note that it is only necessary to difference the AR side of the

ARMA representation, because the MA component is always I (0).

It should be noted that not all choices of parameters produce well-behaved models. In particular, if the model is required to be stationary, then conditions on these parameters must be satisfied.

Some well-known special cases arise naturally. For example, an ARIMA (0, 1, 0) model is given by:

$$x_t = x_{(t-1)} + \varepsilon_t \qquad (8.166)$$

which is simply a random walk model. A number of variations of the ARIMA model are commonly used.

ARIMA models can be extended to include seasonal autoregressive and seasonal moving average terms. Although this complicates the notation and mathematics involved in the model, the underlying concepts for seasonal autoregressive and seasonal moving average terms are similar to the nonseasonal autoregressive and moving average terms.

The most general form of the ARIMA model includes difference operators, autoregressive terms, moving average terms, seasonal difference operators, seasonal autoregressive terms, and seasonal moving average terms. In general, as with any modelling approach, only necessary terms should be included in the model.

8.6.2.2.1 Identification

As already mentioned, the input series for ARIMA needs to be stationary, that is, it should have a constant mean, variance, and autocorrelation over time. Therefore, usually the series first needs to be differenced until it is stationary (this also often requires log-transforming the data to stabilize the variance). The number of times the series needs to be differenced to achieve stationarity is reflected in the d parameter. In order to determine the necessary level of differencing, one should examine the plot of the data and autocorrelogram function (ACF). Significant changes in level (large, upward or downward changes) usually require first-order differencing (lag 1); large changes of slope usually require second-order differencing. Seasonal patterns require respective seasonal differencing. If the estimated autocorrelation coefficients decline slowly at longer lags, first-order differencing usually suffices. However, one should keep in mind that some TS may require little or no

differencing, and that an over- differenced series produces less stable coefficient estimates.

At this stage (which is usually called Identification phase) one also needs to decide how many autoregressive (p) and moving average (q) parameters are necessary to yield an effective but still parsimonious model (meaning a model with the least number of parameters and the largest number of degrees of freedom) of the process represented by the TS data.

Major tools used in the identification phase are plots of the series, correlograms of ACF, and partial autocorrelation function (PACF). The decision is not straightforward and in some cases may require not only experience but also a good deal of experimentation with alternative models (as well as with the technical parameters of ARIMA). However, majority of empirical time series patterns can be sufficiently well approximated using one of the five basic models that can be identified based on the shape of the ACF and the PACF. The following brief summary is based on practical recommendations of Pankratz (1983). Also, it is to be noted that since the number of parameters (to be estimated) of each kind does not normally exceed two, it is often practical to try alternative models on the same data:

- One autoregressive (p) parameter: ACF – exponential decay; PACF – spike at lag 1; no correlation for other lags.
- Two autoregressive (p) parameters: ACF – a sine-wave shape pattern or a set of exponential decays; PACF – spikes at lags 1 and 2, and no correlation for other lags.
- One moving average (q) parameter: ACF – spike at lag 1, and no correlation for other lags; PACF – damps out exponentially.
- Two moving average (q) parameters: ACF – spikes at lags 1 and 2, and no correlation for other lags; PACF – a sinusoidal pattern or a set of exponential decays.
- One autoregressive (p) and one moving average (q) parameter: ACF – exponential decay starting at lag 1; PACF – exponential decay starting at lag 1.

8.6.2.2.2 Estimation and forecasting

In the next step, the parameters are estimated (using function minimization procedures), and the sum of the squares of residuals is minimized. The estimates of the parameters are used in the last stage (i.e. forecasting) for calculation of new values of the series (beyond those included in the input dataset) together with the confidence intervals for the predicted values. The estimation process is performed on the transformed (differenced) data. Before the forecasts are generated, the series needs to be integrated (integration is the inverse of differencing) so that the values generated by forecasts are compatible with the input data. This automatic integration feature is represented by including the letter 'I' in the name of the methodology (ARIMA = Auto-Regressive Integrated Moving Average).

In addition to the standard autoregressive and moving average parameters, ARIMA models may also include a constant, as described above (Eqn 8.150 and Eqn 8.152). The interpretation of the (statistically significant) constant depends on the model that fits given data well. Specifically: (i) if there are no autoregressive parameters in the model, then the expected value of the constant is μ, the mean of the series; (ii) if there are autoregressive parameters in the series, then the constant represents the intercept. If the series is differenced, then the constant represents the mean or intercept of the differenced series. For example, if the series is differenced once, and there are no autoregressive parameters in the model, then the constant represents the mean of the differenced series and, therefore, the slope of the straight line represents the linear trend of the un-differenced series.

8.6.3 Single spectrum (Fourier) analysis

Spectrum analysis is a frequency-domain method of TS analysis and is concerned with the exploration of cyclical patterns of data. The purpose of the analysis is to decompose a complex time series with cyclical components into a few underlying sinusoidal (sine and cosine) functions of particular frequencies. In essence, performing spectrum analysis on TS is like putting white light through a prism in order to identify the wavelengths and the underlying cyclic components. As a result of successful analysis, one might uncover just a few recurring cycles of different lengths in the time series of interest, which at first sight would have looked more or less like random noise. A frequently

cited example of spectrum analysis is the cyclic nature of sunspot activity. It turns out that sunspot activity has a dominant 11-year cycle. Other examples to demonstrate this technique include analysis of celestial phenomena, weather patterns, fluctuations in commodity prices, economic activities, etc. To contrast this technique with ARIMA, the purpose of spectrum analysis is to identify the seasonal fluctuations of different lengths, while in the ARIMA model, the length of the seasonal component is usually known (or guessed) *a priori* and then included in some theoretical model of moving averages or autocorrelations.

A sine or cosine function is typically expressed in terms of the number of cycles per unit time (frequency), often denoted by the Greek letter v(*nu*) (some text books use f). Thus, if unit of time is 1 year, there are n observed cycles (for monthly cycles $n = 12$, i.e. $v = 12$ cycles per year). Of course, there are other likely cycles with different frequencies. For example, there might be annual cycles ($v = 1$), and weekly cycles ($v = 52$ cycles per year).

The period T of a sine or cosine function is defined as the length of time required for one full cycle. Thus, it is the reciprocal of the frequency, that is, $T = 1/v$. The period of monthly cycle, expressed in terms of year, is equal to $1/12 = 0.0833$. In other words, there is a period in the series of length of time as 0.0833 years.

One way to decompose the original series into its component sine and cosine functions of different frequencies is to formulate the problem as a linear Multiple Regression model, in which the dependent variable is the observed time series, and the independent variables are the sine functions of all possible (discrete) frequencies. Such a model may be written as:

$$x_t = a_0 + \sum \left[a_k \cdot \cos\left(\frac{\pi k}{L} t\right) + b_k \cdot \sin\left(\frac{\pi k}{L} t\right) \right]$$

$$for\ k = 1\ to\ \infty \qquad (8.167)$$

where the basic approach is to represent any function x_t of period $2L$ as a combination of periodic components, namely sines and cosines.

A term-by-term integration between negative and positive L values allows one to determine the coefficients Eqn 8.167:

$$a_0 = \frac{1}{2L} \int_{-L}^{L} x_t \, dt$$

$$a_k = \frac{1}{L} \int_{-L}^{L} x_t \cos\left(\frac{kt}{L}\right) t \, dt$$

$$b_k = \frac{1}{L} \int_{-L}^{L} x_t \sin\left(\frac{kt}{L}\right) t \, dt \qquad (8.168)$$

Eqn 8.167 is known as the Fourier series of x_t, with Eqn 8.168 as its Fourier coefficients. Series with $b_k = 0$ for all k are known as Fourier cosine series, whereas series with $a_k = 0$ for all k are known as Fourier sine series. One can thus estimate any periodic function x_t as a combination of sine and cosine functions, with the coefficients given by equation Eqn 8.168.

Example 8.11. Assuming that we have monthly means, and we average all years to obtain f_n, monthly averages corresponding to each month of the year. Δt is then 1 month, and $N = 12$. Typically, we can represent the annual cycle with at least two harmonics, to allow for a lack of symmetry between winter and summer:

$$f_{aCn} = a_0 + a_1 \cos\left(\frac{2\pi}{12/1}n\right) + b_1 \sin\left(\frac{2\pi}{12/1}n\right)$$

$$+ a_2 \cos\left(\frac{2\pi}{12/2}n\right) + b_2 \sin\left(\frac{2\pi}{12/2}n\right)$$

The a_0 term represents the annual average (with zero frequency); the a_1 and b_1 terms represent the periodic component with period of 12 months (fundamental frequency); and the a_2 and b_2 terms represent a periodic component with period of 6 months (first harmonic). The coefficients a_0, a_1, a_2, b_1, b_2 can be obtained as before, e.g.:

$$b_2 = \frac{1}{12} \sum_{n=1}^{12} \bar{f}_n \sin\left(\frac{2\pi}{12/2}n\right)$$

where the bar represents the monthly average over several years. Once the coefficients are obtained, the annual cycle can be subtracted from the time series in order to deal with *anomalies*.

8.6.3.1 Non-periodic functions: Fourier integrals and transforms

In a non-periodic function, the approach is to assume that the function x_t is periodic in the limit as $t \to \infty$. Instead of writing x_t as a summation of sines and cosines, the following integral representation is introduced:

$$x_t = \int_0^{\infty} [A(\omega) \cos \omega t + B(\omega) \sin \omega t] \, d\omega \qquad (8.169)$$

with

$$A(\omega) = \frac{1}{\pi} \int_{-\infty}^{\infty} y(v) \cos \omega v \, dv$$

$$B(\omega) = \frac{1}{\pi} \int_{-\infty}^{\infty} y(v) \sin \omega v \, dv \qquad (8.170)$$

Equation Eqn 8.169 is equivalent to Eqn 8.167 and Eqn 8.170 is equivalent to Eqn 8.168. So far no assumption is made about the periodicity of the function x_t. It is sometimes useful to represent Eqn 8.169 in the complex form using the Euler equation:

$$exp(i\,t) = \cos t + i \sin t \qquad (8.171)$$

Substituting Eqn 8.170 into Eqn 8.169), one can derive the complex Fourier integral as:

$$x_t = \frac{1}{2\pi} \int_{-\infty}^{\infty} \int_{-\infty}^{\infty} x(v) \exp(i\omega(t-v)) \, dv \, d\omega \qquad (8.172)$$

Rearranging the double integral in Eqn 8.172 gives:

$$x_t = \frac{1}{\sqrt{2\pi}} \int_{-\infty}^{\infty} \left[\frac{1}{\sqrt{2\pi}} \int_{-\infty}^{\infty} x(v) \exp(-i\omega t) \, dv \right]$$

$$\times \exp(i\omega t) \, d\omega \qquad (8.173)$$

In Eqn 8.173, the expression in the square brackets is the Fourier transform of x_t, as given by:

$$X(\omega) = \frac{1}{\sqrt{2\pi}} \int_{-\infty}^{\infty} x(t) \exp(-i\omega t) \, dt \qquad (8.174)$$

and thus Eqn 8.173 becomes the inverse Fourier transform of x:

$$x_t = \frac{1}{\sqrt{2\pi}} \int_{-\infty}^{\infty} Y(\omega) \, exp \, (i\omega t) \, d\omega \qquad (8.175)$$

8.6.3.2 Discrete sampling effects

The usual method of describing a physical process is in the time domain, that is, the value of some function, x_t, is described as a function of time, that is as $x(t)$. An alternate method is to describe processes in the frequency domain, such that a function, X, is described as a function of frequency, ω, i.e. as $X(\omega)$. The two functions, $x(t)$ and $X(\omega)$, are related through the Fourier transform Eqn 8.174 and Eqn 8.175.

These, however, refer to continuous functions. In reality, the experimental measurements are generally discrete, made only at certain time intervals. An important consequence of the discrete nature of measurements is the introduction of a minimum resolvable frequency. If a time series is formed with a sampling interval of 10 seconds, the minimum resolvable period is 20 seconds, as at least 2 points are required to determine a sine wave. This means that a time series collected at 0.1 Hz (1/10 s) can only resolve frequencies less than 0.05 Hz (1/20 s). In this example, 0.05 Hz is known as the Nyquist or cut-off frequency, defined as:

$$f_c = \frac{1}{2\Delta t} \qquad (8.176)$$

Sampling at discrete intervals for a certain period of time gives N measurements. The frequencies at which one can calculate the Fourier transform of f is limited to a set of N frequencies, determined by:

$$f_n = \frac{n}{N\Delta t}, \qquad n = -\frac{n}{2}, \cdots, \frac{N}{2} \qquad (8.177)$$

Therefore, one needs to replace continuous representation of the Fourier transform Eqn 8.174 with a discrete representation, in keeping with the discrete nature of sampling:

$$X(k) = \Delta t \sum_{n=0}^{N-1} x(n) exp(-i 2\pi n k/N) \qquad (8.178)$$

The angular frequency ω in Eqn 8.174 is replaced by $2\pi f$, with t in Eqn 8.174 replaced by $n\Delta t$. A discretized form of f is also substituted, giving:

$$f_k = \frac{k}{N\Delta t}, \qquad k = 0, \cdots, \frac{N}{2} \qquad (8.179)$$

8.6.3.3 Calculation of the power spectrum and estimation of the periodogram

To determine the 'dominant' frequencies in a time series, the power spectral density is defined as:

$$G(\omega) = \frac{2}{T}|X(\omega)|^2 \qquad (8.180)$$

where T is the length of the time series, and $X(\omega)$ is defined in Eqn 8.174. This is the continuous representation of the power spectral density that gives an estimate of the 'power' in the signal $x(t)$ at the frequency ω. The discrete representation of the power spectral density is:

$$G(f) = \frac{2\Delta t}{N}|X(f)|^2 \qquad (8.181)$$

Analysis of the power spectral density $G(f)$ enables one to investigate the dominant frequencies in a signal, as the dominant frequencies are likely to be important in characterizing a physical process.

A periodogram estimate of the power spectrum is used, which is defined at $(N/2 + 1)$ frequency as:

$$P(f_0) = \frac{1}{N^2}|F_0|^2$$

$$P(f_k) = \frac{1}{N^2} \left(|F_k|^2 + |F_{N-k}|^2 \right)$$

$$k = 1, 2, \cdots, \left(\frac{N}{2} - 1 \right)$$

$$P(f_c) = \frac{1}{N^2}|F_{N/2}|^2 \qquad (8.182)$$

The above estimate of the power at each frequency, f_k, is representative of some kind of an average value midway between the preceding frequency 'bin' to a value midway to the next one. This is not always particularly accurate, due to the phenomenon known as 'leakage', that is, there is a leakage of energy from one frequency to another in the periodogram estimate. One can reduce the

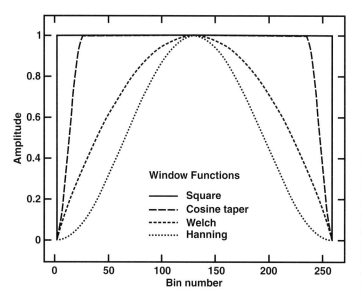

Fig. 8.17 Windowing functions that preserve all the power at the centre of the bin and allow little power at the extremities of the bin are used to reduce frequency leakage during spectral analysis of a time series. Some common windows and their leakage functions are shown. Redrawn from http://www.cwr.uwa.edu.au/~antenucc/spectral_analysis/spectral_analysis.html# Coherence_and_phase.html.

effects of frequency leakage through a technique known as 'data windowing' that aims to preserve the entire power at the centre of the bin and virtually no the power at its extremities, with a slow transition between the two. The introduction of a data window results in transformation of Eqn 8.178 to:

$$W(k) = \Delta t \sum_{n=0}^{N-1} x(n)w(j)\, exp\,(-i2\pi nk/N) \quad (8.183)$$

where $w(j)$ is the window function. Several types of window function are in common use. Some of these are:

Square window $w_j = 1$
Hanning or cosine squared window
$$w_j = \frac{1}{2}\left[1 - cos\frac{2\pi j}{N+1}\right]$$
Welch window
$$w_j = -1\left(\frac{j-0.5N}{0.5}\right)^2$$
The cosine taper window
$$w_j = \frac{1}{2}\left[1 - cos\frac{\pi j}{m}\right] \qquad j = 0,\cdots,m-1$$
$$w_j = 1 \qquad j = m,\cdots,N-m-1$$
$$w_j = \frac{1}{2}\left[1 - cos\frac{(N-j)}{m}\right] \qquad j = 0,\cdots,m-1$$
where m is usually equal to 0.1 N.

The leakage functions of these windows are shown in Fig. 8.17.

8.6.3.4 Coherence and phase relationships between signals

It is sometimes of interest to investigate the joint structure of two series, that is, their mutual dependence or coherence of either series with the other. Since one can observe relationships only at the same frequency in both series, the cross-correlation spectrum may be defined as:

$$G_{ij}(\omega) = \frac{2}{T}\left[X_i(\omega)X_j^*(\omega)\right] \quad (8.184)$$

The '*' superscript refers to the complex conjugate of the function, which for a complex function, y, is simply:

$$conj(y) = real(y) - i\,\{imag(y)\} \quad (8.185)$$

Unlike the power spectrum, the cross-spectrum is complex as it contains both amplitude and phase information. The real part of the cross-spectrum is known as the coincident spectrum (or co-spectrum), and the complex part of the cross-spectrum is known as the quadrature spectrum (or quad-spectrum). From the cross-spectrum, one can estimate the amplitude and phase relationships between the signals. The amplitude relationship is quantified through the *coherency squared*, calculated as:

$$S_{ij}^2(\omega) = \frac{|G_{ij}(\omega)|^2}{G_i(\omega)G_j(w)} \qquad (8.186)$$

A coherency squared value of unity indicates total dependence of one signal on another, whereas a coherency squared value of zero refers to non-dependence of one signal on another. Two signals can only be coherent at the same frequency.

Phase is calculated directly from the co-spectrum and quadrature spectrum as follows:

$$\phi_{ij}(\omega) = \arctan\left[\frac{\text{imag}(G_{ij})}{\text{real}(G_{ij})}\right] \qquad (8.187)$$

The phase shift estimates are measures of the extent to which each frequency component of one series leads the other.

8.7 Tutorial

Ex 8.1 What is the chance that exactly three 50-year floods will be equalled or exceeded in a given 100-year period? What is the chance that three or more floods will occur?

Table 8.4

exclusive events (if one occurs, the others cannot), or cut short the process by substituting 2, 1, and 0 in the formula and subtracting the sum of probabilities from 1.00 (certainty), since it is certain that either (a) 3 or more or (b) 2 or less events must occur. Substituting 2, 1, and 0 in the formula, we obtain 0.274, 0.271, and 0.133 in turn for P; 1 minus their sum is 0.322 or about 1 chance in 3.

Ex 8.2 What is the chance that a 1000-year flood will be exceeded in the estimated 50-year operational lifetime of a project?

Solution One can interpret this to include the chance that the flood will be exceeded exactly once or more than once. One should, therefore, solve for the chance that it will not occur ($k = 0$) and subtract that probability from 1.00, as explained in Ex 8.1. Note that for $k = 0$, the Eqn 8.40 simplifies to:

$$P = (1 - p)^n$$

Substituting 0.001 for p and 50 for n, we obtain 0.952 for P and, therefore, 0.048 for the answer, that is, about 1 chance in 20.

Ex 8.3 In Table 8.4 below, 20 years of data on average global temperature deviations with respect to the average are given. Plot the time series and comment on any trends or other features of the data.

Deviation (y_t) from the Average Global Temperature, in °C

Year	y_t	Year	y_t	Year	y_t	Year	y_t
1975	−0.09	1980	0.19	1985	0.09	1990	0.39
1976	−0.22	1981	0.27	1986	0.17	1991	0.39
1977	0.11	1982	0.09	1987	0.30	1992	0.17
1978	0.04	1983	0.30	1988	0.34	1993	0.21
1979	0.10	1984	0.12	1989	0.25	1994	0.31

Solution The general formula for the exact number of chance events, k, out of n trials is given by Eqn 8.40. Substituting 100 for N, 3 for k, and 0.02 for p, we obtain $P = 0.183$, or about 1 chance in 5 or 6. To obtain the answer to the second question, one could substitute, 3, 4, 5, etc. up to 100 in the formula and add the probabilities of these mutually

Ex 8.4 Table 8.5 gives a hypothetical data on reforestation of a catchment and the runoff of a stream flowing out of it. Draw time series plots for each of the variables 'Average runoff' and 'Fractional forested area'. Do these plots show that 'Fractional forested area' has affected the runoff, under the assumption of near constant rainfall?

Table 8.5

Year	Avg. Runoff $m^3 s^{-1}$	Forested area (%)	Year	Avg. Runoff $m^3 s^{-1}$	Forested area (%)	Year	Avg. Runoff $m^3 s^{-1}$	Forested area (%)
1973	0.26	30.5	1980	0.26	24.2	1987	0.21	34.1
1974	0.27	27.8	1981	0.26	23.9	1988	0.21	34.8
1975	0.28	25.6	1982	0.26	24.8	1989	0.16	41.2
1976	0.28	25.1	1983	0.25	26.3	1990	0.17	43
1977	0.28	25.4	1984	0.26	25.8	1991	0.15	47.5
1978	0.27	24.1	1985	0.23	26.7	1992	0.16	46.5
1979	0.27	24.1	1986	0.21	30.4	1993	0.15	45.8

Ex 8.5 One of the methods for removing a linear trend is by differencing, that is by plotting the series with terms $z_t = y_{t+1} - y_t$. Verify that if this differencing is applied to the straight line $y_t = a + bt$, then z_t does not contain a trend.

Ex 8.6 Table 8.6 gives quarterly rainfall at a particular meteorological station. Determine if the data shows any seasonality.

been largely smoothed out? Is it possible conclude that the above method can be used for smoothing out any cycle that repeats itself every c months or years? We simply take a c-point moving average. The key step is to determine the period c of the cycle.

Ex 8.7 De-seasonalize the values of (y_t) from Ex 8.6.

Table 8.6

Year	Quarter	(y_t)	4-point smoothing ($y'_{t+.5}$)	2-point smoothing z_t of $y'_{t+.5}$
2002	Jan–Mar	9.9		
	Apr–Jun	9.5	9.1	
	Jul–Sep	8.3	9.1	9.1
	Sep–Dec	8.7	8.925	9.0125
2003	Jan–Mar	9.9	8.6	8.7625
	Apr–Jun	8.8	8.4	8.5
	Jul–Sep	7.0	8.25	8.325
	Sep–Dec	7.9	7.925	8.0875
2004	Jan–Mar	9.3	7.9	7.9125
	Apr–Jun	7.5	7.65	7.775
	Jul–Sep	6.9		
	Sep–Dec	6.9		

   A 4-point moving average results in the column labelled ($y'_{t+.5}$) for the first four points. This corresponds to a time midway between the second and third points. A 2-point average brings this smoothed series back in step with the original series, giving the column labelled (z_t). Plot the smoothed series z_t along with individual points of the series y_t. You should check some of the values of z_t for yourself. Does this exercise result in z_t being approximately linear? Have the seasonal and random variation

Solution Removal of seasonal effect from a series and leaving any trend and the random ups and downs back in the data may give us a clearer picture of the situation. The resulting series gives us a de-seasonalized data. To achieve this in a reasonable manner, we need a suitable model of the process producing the original time series. Our model takes the form:

data = trend + cycle + error

To remove a quarterly cycle, for example, we begin by averaging all the first quarters, namely $[(y_1+y_5+y_9+....)/\text{no. of years}]$, then averaging all the second quarters, all the third quarters and finally all the fourth quarters, giving us just four numbers s_1, s_2, s_3 and s_4. We then subtract the mean $\bar{s}$ of these four numbers (which is the same as $\bar{y}$ of the original series) to get $s_t - \bar{s}$. The deseasonalized series is then given by $z_t = y_t - (s_t - \bar{s})$, where the definition of s_t is extended beyond the first year by simply repeating the same four numbers.

Table 8.7 demonstrates the calculations, with reference to y_t in Ex 8.6, using:

$$s_1 = s_5 = s_9 = \frac{1}{3}(y_1 + y_5 + y_9)$$

$$= \frac{1}{3}(9.9 + 9.9 + 9.3) = 9.7$$

$$\bar{s} = 8.38; \quad s_1 - \bar{s} = 9.7 - 8.38 = 1.32$$

$$\text{and} \quad z_1 = y_1 - (s_1 - \bar{s}) = 8.58$$

Plot the series z_t along with the points for y_t and note that without the seasonal cycle there appears to be an almost linear trend.

Table 8.7

Year	Quarter	(y_t)	Quarter s_t	Adjustment $(s_t - \bar{s})$	Deseasoned (z_t)
2002	Jan–Mar	9.9	9.70	1.32	8.58
	Apr–Jun	9.5	8.60	0.22	9.28
	Jul–Sep	8.3	7.40	−0.98	9.28
	Sep–Dec	8.7	7.83	−0.55	9.25
2003	Jan–Mar	9.9	9.70	1.32	8.58
	Apr–Jun	8.8	8.60	0.22	8.58
	Jul–Sep	7.0	7.40	−0.98	7.98
	Sep–Dec	7.9	7.83	−0.55	8.45
2004	Jan–Mar	9.3	9.70	1.32	7.98
	Apr–Jun	7.5	8.60	0.22	7.28
	Jul–Sep	6.9	7.40	−0.98	7.88
	Sep–Dec	6.9	7.83	−0.55	7.45
	Mean	8.38			

9 Remote sensing and GIS in hydrology

Remote sensing involves the use of instruments or sensors to 'capture' the spectral and spatial characteristics of objects and materials observable at a distance – usually from a certain height above the Earth's surface. In its simplest form, it is a rather familiar activity that we all do in our daily lives. Several of the human sensory organs perceive the external world almost entirely through a variety of signals, either emitted by or reflected from objects as electromagnetic waves or pulses. However, in practice we do not usually think of our sense organs as performing remote sensing in the way that this term is employed technically. The term 'remote sensing' is a relatively new addition to the technical lexicon. It seems to have been devised to take into account the view from space obtained by the early meteorological satellites, which were obviously more 'remote' from their targets than airplanes, which until then had provided aerial photos as a means for recording images of the Earth's surface. In its common or normal usage, remote sensing is a technology for sampling electromagnetic radiation to acquire and interpret non-contiguous data in a geographically defined co-ordinate system from which information can be extracted about features, objects, and classes on the Earth's land surface, oceans, atmosphere and, where applicable, on celestial bodies.

Many different Earth-sensing satellites, with diverse sensors mounted on sophisticated platforms, are orbiting the Earth and newer ones will continue to be launched. Thus remote sensing, with its ability to provide a synoptic view, repetitive coverage of a given location with calibrated sensors to detect temporal changes, and observations at different resolutions, provides a unique alternative for applications as diverse as monitoring of forest fires, grassland inventory, coastal zones, early cyclone warning, flood assessment, crop health, and assessment of crop yield around the world. Hydrologists have also realized the potential of satellite remote sensing technology since the 1970s. Remote sensing offers a method to avoid the logistic and economic limitations associated with obtaining continuous *in-situ* measurements of various hydrologic variables, particularly in view of difficulties encountered in remote regions. Further, microwave instruments can potentially provide all-weather, areally averaged estimates of certain parameters (e.g. precipitation, soil moisture, and snow water content), something that has not been possible in the past.

The hydrologic cycle integrates atmospheric, hydrospheric, cryospheric, and biospheric processes over a wide range of spatial and temporal scales and thus lies at the heart of the Earth's climate system. Studies of the integrated, global nature of the hydrologic cycle are crucial for a proper understanding of natural climate variability and for prediction of climatic response to anthropogenic forcing. Only in recent years, particularly with the advent of satellite remote sensing, the required data on a global scale seems to be within reach.

Modern Hydrology and Sustainable Water Development, 1st edition. © S.K. Gupta
Published 2011 by Blackwell Publishing Ltd.

(a)   (b)

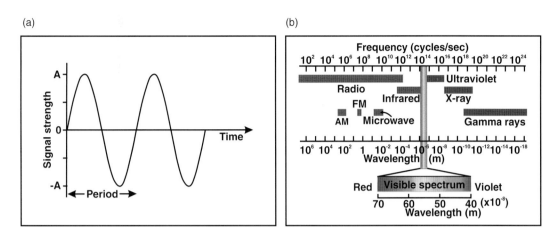

Fig. 9.1 (a) Sinusoidal waveform of electromagnetic radiation (EMR). Frequency (*v*) is the number of oscillations per unit time, and wavelength (λ) is the distance travelled in time taken for one oscillation; (b) electromagnetic spectrum with increasing frequency is shown from left to right. Also included are the terms used for the main spectral regions. The wavelength scale shown at the bottom in this representation of EM spectrum increases from right to left. Redrawn from http://rst.gsfc.nasa.gov/Intro/Part2_2a.html accessed on 18th Nov., 2008. See also Plate 9.1.

Although most hydrologists believe that remotely sensed data is valuable for global hydrologic studies and even for regional hydrologic modelling and field operations, these data are rarely used, possibly due to: (i) lack of necessary technical expertise in processing/interpreting the data; and (ii) the form of emitted and reflected radiances not being the type of data traditionally used to run and calibrate models. Remotely sensed data also represent averages over finite areas, or pixels, and thus obliterate much of the detail at individual points to which most hydrologists are accustomed. In addition, current remotely sensed observations are not optimized to provide the temporal resolution needed to measure certain changes in hydrologic processes. Furthermore, algorithms for converting these reflectances into physical quantities are often empirical in nature and are subject to noise present in the calibration data.

Nevertheless, remote sensing is beginning to prove useful in providing hydrologic information. Hydrologic remote sensing can reveal complex spatial variations that cannot be readily obtained through traditional *in-situ* approaches. Development of such datasets and models in which these data can be used requires field experiments that combine appropriate remote sensing measurements with traditional *in-situ* measurements in

regions that are hydrologically well understood. Once hydrologic models are developed for use with remote sensing data in well monitored basins, they can possibly be extended to regions having little or no *in-situ* measurements.

9.1 Principle of remote sensing

Measurement of the varying energy levels of the fundamental unit in the electromagnetic (EM) field, known as the *photon*, forms the basis of most remote sensing methods and systems. Variations in photon energies (expressed in *joules* or *ergs*) are related to the *wavelength* of the electromagnetic radiation or its inverse, *frequency*. EM radiation distributed through a continuum of photon energies from high to low energy levels comprises the *Electromagnetic Spectrum* (EMS). EM waves oscillate in a harmonic pattern, of which one common form is the sinusoidal type (Fig. 9.1a;). Radiation from specific parts of the EMS contains photons of different wavelengths whose energy levels fall within a discrete range. When any target material is excited either by internal processes or by interaction with incoming EM radiation, it either emits or reflects photons of varying wavelengths whose properties differ for different wavelengths that are characteristic of the material. Photon energy received at the detector is commonly expressed in power units such as watt per square metre per unit wavelength. A plot of variation of power with wavelength gives rise to a specific pattern or curve that constitutes the *spectral signature* of the object or feature being sensed (see Section 9.1.3).

9.1.1 The photon and the associated radiometric quantities

EM waves are transverse waves because particles within a physical medium are set into vibration normal (at right angles) to the direction of wave propagation. EM radiation can propagate even through empty space (vacuum) completely devoid of particles in the medium. Each *photon* is associated with an electric field (E) and a magnetic field (H); both being vectors oriented at right angles to each other. The two fields oscillate simultaneously as described by co-varying sine waves having the same wavelength λ (being the distance between two adjacent crests [or troughs] on the waveform) and the same frequency, ν (being the number of oscillations per unit time). The equation relating λ and ν is $c = \lambda\nu$, where the constant c is the speed of light and equals 299 792.46 *km sec^{-1}* (commonly rounded off to 300 000 *km sec^{-1}*). Since the speed of light is constant, as λ increases $\nu(= c/\lambda)$, therefore, decreases and vice versa, to maintain its constancy. An EM spectrum chart is shown in Fig. 9.1b, wherein frequency is seen to increase from left to right and, therefore, wavelength (shown at the bottom) increases from right to left. The terms corresponding to the principal spectral regions are also included in the diagram. When the electric field is made to be confined in one direction, the radiation is said to be plane polarized. The wave amplitudes of the two fields are also coincident in time and are a measure of radiation intensity (brightness). Units for λ are usually specified in the metric system and are dependent on the particular point or region of the EM spectrum being considered. Familiar wavelength units include the *nanometre*; the *micrometer* (or *micron*); the metre; and the *Angstrom* ($= 10^{-10}$ *m*). A fixed quantum of energy E (expressed in units of *ergs/joules/electron volts*) is characteristic of a photon emitted at a discrete frequency,

according to Planck's quantum equation: $E = h\nu = hc/\lambda$, where h is Planck's constant (6.6260... $\times 10^{-34}$ *Joule-sec*). From the Planck equation, it is evident that waves representing different photon energies oscillate at different frequencies. It also follows that photons of higher frequencies (i.e. shorter wavelengths) have more energy and those with lower frequency (i.e. longer wavelength) have less energy.

Example 9.1. Find the wavelength of a quantum of radiation with photon energy of 2.10 $\times$ 10^{-19} *Joule*.

Using $E = hc/\lambda$ or $\lambda = hc/E$, the wavelength = (6.626 $\times$ 10^{-34}) (3.00 $\times$ 10^8)/(2.10 $\times$ 10^{-19}) = 9.4657 $\times$ 10^{-7} *m* = 946.6 *nm*; = 9466 *Angstroms*; = 0.9466 μm (which lies in the near-infrared region of the spectrum).

Example 9.2. A radio station broadcasts at 120 *MHz* (megahertz or million *cycles sec^{-1}*) frequency. Find the corresponding wavelength of radio waves in *metres*.

From the equation: wavelength = c/frequency. Thus, wavelength = (3.00 $\times$ 10^8) (120 $\times$ 10^6) = 2.5 *m*.

The interaction of EMR with matter can be treated from two perspectives. The first, the *macroscopic* view, is governed by the laws of optics. More fundamental is the *microscopic* approach that works at the atomic or molecular level.

In remote sensing, the sensors used employ detectors that produce currents (and voltages, $V = IR$) whose magnitude for any given frequency depends on the photoelectric effect. This means that there is an emission of negative particles (electrons) when a negatively charged plate of an appropriate light-sensitive material is impinged upon by a beam of photons. Electrons flowing as a current from the plate, are collected, and counted as a signal. As the magnitude of the electric current produced (given by the number of photoelectrons flowing per unit time) is directly proportional to the light intensity, its changes can be used to measure changes in the photons (their numbers and intensity) that strike the plate (detector) during a given time interval. The kinetic energy of the emitted photoelectrons varies with frequency (or wavelength) of the incident radiation. However, different materials exhibit the photoelectric effect, releasing electrons over different wavelength regions. Each material has a threshold wavelength at which the photoelectric effect begins and a longer cut-off wavelength at which it ceases.

Furthermore, at a microscopic level a beam of radiation, such as from the Sun, is usually polychromatic, that is, it has photons of different wavelengths (energies). If an excited atom experiences a change in energy level from a higher level E_2 to a lower level E_1, Planck's equation gives:

$$\Delta E = E_2 - E_1 = h\nu \qquad (9.1)$$

where ν has a discrete value given by $(\nu_2 - \nu_1)$; corresponding to E_1 and E_2). In other words, a particular change in the energy of an atom is characterized by emission of radiation (photons) at a specific frequency, ν (and the corresponding wavelength λ) dependent on the magnitude of change in the energy.

When energies involved in processes from the molecular to subatomic level are involved (as in the case of photoelectric effect), these energies are measured in *electron volt* units. One *electron volt (eV)* is equal to the amount of energy gained by a single unbound electron when it accelerates through an electrostatic potential difference of one volt. In SI units, electron volt is expressed in *Joule*; $1\ eV = 1.602 \times 10^{-19}$ *Joule*. There are about 6.24 $\times 10^{18}$ electrons in one *Coulomb* of change.

9.1.2 Solar radiation: transmittance, absorptance, and reflectance

The primary source of energy that illuminates all natural objects on the Earth is the Sun. The wavelength region of the solar radiation emitted by the Sun is determined by the temperature of the Sun's photosphere (peaking around $5600°C$). The main wavelength region of the solar radiation reaching the Earth (called the solar insolation) is between 200 and 3400 *nm* (0.2 and 3.4 μm), with the maximum power around 480 *nm* (0.48 μm), which is in the visible (green) region. As solar radiation reaches the Earth, the atmosphere absorbs or backscatters a fraction of it and transmits the remainder to the sur-

face. Upon striking the land and ocean surface (and the objects present there) and the atmospheric constituents such as air, moisture, and clouds, the incoming radiation (irradiance) is partitioned into three components:

1. *Transmittance* (τ) – Some fraction (up to 100%) of the incoming radiation penetrates certain surface materials such as water and if the material is transparent and thin, it normally passes through, generally with some attenuation in intensity (Fig. 9.2).
2. *Absorptance* (α) – Some radiation is absorbed through reaction with electrons or molecules within the medium. A part of this energy is then re-emitted, usually at longer wavelengths, and some of it remains in the medium, which heats it up (Fig. 9.2).
3. *Reflectance* (ρ) – some radiation (commonly 100%) gets reflected (moves away from the target) at specific angles and/or is scattered away from the target at various angles, depending on the surface roughness and the angle of incidence of the rays (Fig. 9.2).

As these partitions involve ratios (with respect to irradiance), these three parameters are dimensionless numbers (between 0 and 1), but are commonly expressed as percentages. In accordance with the Law of Conservation of Energy, $\tau + \alpha + \rho = 1$.

A fourth situation, when the emitted radiation arises from internal atomic/molecular excitation, usually related to the temperature state of a body, is a thermal process.

When a remote sensing instrument is in the line-of-sight with an object that is reflecting solar radiation, the instrument collects the reflected energy and records it. Most remote sensing systems are designed to collect the reflected radiation from various objects.

Generally the remote sensing survey is conducted above the Earth's surface, either within or above the atmosphere. The gases present in the Earth's atmosphere interact with solar radiation and also with radiation emitted by the Earth's surface. The atmospheric constituents in turn get excited by these EM radiations and become another source of photons. A generalized diagram showing relative atmospheric radiation opacity for different

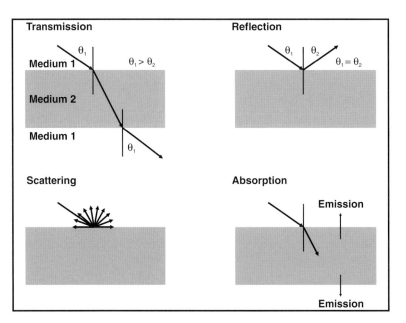

Fig. 9.2 A beam of photons from a source passing through medium 1 (usually air) that is incident upon an object or target (medium 2) will undergo one or more processes as illustrated. Redrawn from http://rst.gsfc.nasa.gov/Intro/Part2_3.html accessed on 18th Nov., 2008.

wavelengths is shown in Fig. 9.3. Grey zones (absorption bands) mark minimal passage of incoming and/or outgoing radiation, whereas white areas (transmission peaks) denote atmospheric windows in which the radiation does not interact much with air molecules and hence it is not absorbed.

Backscattering (scattering of photons in all directions above the target in the hemisphere that lies on the source side) is an important phenomenon in the atmosphere. *Mie scattering* refers to reflection and refraction of radiation by atmospheric constituents (e.g. smoke) whose dimensions are of the

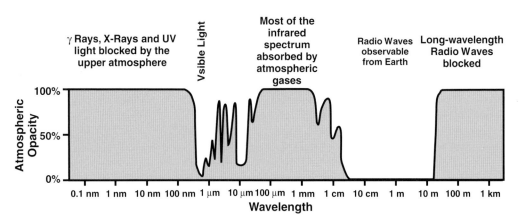

Fig. 9.3 Generalized diagram showing relative atmospheric radiation opacity for different wavelengths. Grey zones (absorption bands) mark minimal passage of incoming and/or outgoing radiation, whereas white areas (transmission peaks) denote atmospheric windows in which the radiation does not interact much with air molecules and hence it is not absorbed significantly. Redrawn from http://en.wikipedia.org/wiki/Electromagnetic_spectrum accessed on 7th Nov. 2008.

order of the wavelengths of the incident radiation. *Rayleigh scattering* results from constituents (e.g. molecular gases: O_2, N_2 and other nitrogen compounds, CO_2 and water vapour) such that the sizes of their molecules are much smaller than the wavelengths of radiation. Rayleigh scattering increases with decreasing wavelength, causing preferential scattering of blue light (the "blue sky effect"). However, the crimson red colour of the sky at sunset and sunrise results from significant absorption of the visible light (shorter wavelengths) owing to greater 'optical depth' of the atmospheric path when the Sun is near the horizon. Particles much larger than the wavelength of the radiation give rise to nonselective (wavelength-independent) scattering. Atmospheric backscattering can, under certain conditions, account for 80 to 90% of the radiant flux observed by a spacecraft sensor.

Remote sensing of the Earth traditionally involves reflected energy in the visible and infrared and emitted energy in the thermal infrared and microwave regions to gather radiation. The collected radiation can be analysed numerically or alternatively used to generate images whose tonal variations represent different intensities of photons associated with a range of wavelengths that are received by the sensor. Sampling of a (continuous or discontinuous) range(s) of wavelengths is the essence of what is usually termed *multispectral remote sensing*.

Images formed by varying wavelength/intensity signals coming from different parts of an object show variations in grey tones in black and white versions or in colours (in terms of hue, saturation, and intensity in coloured versions). Pictorial (image) representation of target objects and features in different spectral regions, usually employ different sensors (commonly in band pass filters) each tuned to accept and process the wavelengths that characterize a given region. The images normally show significant differences in the distribution (patterns) of colour or grey tones. It is this variation which produces an image or picture. Each spectral band produces an image that has a range of tones or colours characteristic of the spectral responses of the various objects in the scene. Images made from different spectral bands show different tones or colours.

Thus from a quantum phenomenological approach, remote sensing involves detection of photons of varying energies coming from the target after they are generated by selective reflectance or emitted from the target material(s) by passing them through frequency (wavelength)-dependent dispersion devices (filters, prisms) onto metals or metallic compounds/alloys, which through photoelectric effect produce electric currents. These currents are analysed and interpreted in terms of energy-dependent parameters (frequency, intensity, etc.) whose variations are controlled by the atomic composition of the targets. The spectral (wavelength) dependence of each material is characteristic of its unique nature. When the photon characteristics are plotted as an *x-y* plot, the shapes of the varied distributions of photon levels produce patterns that further aid in identifying each material (with possibility of regrouping into a class or some other distinct physical feature).

9.1.3 Spectral signatures

For a given material, the amount of solar radiation that it reflects, absorbs, transmits, or emits varies with wavelength. When the amount (usually intensity, as a percent of its maximum value) coming from the material is plotted over a range of wavelengths, the values lie on a curve called the *spectral signature* or *spectral response curve* of the material. An example of the reflectance plot for some unspecified vegetation type, with the dominating factor influencing each interval of the curve, is shown in Fig. 9.4. Individual spectral signatures of different surface materials enable their identification (or of classes of materials), as shown in Fig. 9.5a. For example, for some wavelengths, sand reflects more energy than green vegetation cover but for other wavelengths it absorbs more (reflects less). Using differences in reflectance values, it is possible to distinguish four common surface materials from the above signatures (GL: grasslands; PW: pinewoods; RS: red sand; SW: silty water) simply by plotting the reflectance of each material at two wavelengths, as shown in Fig. 9.5b. In this case, the data points are sufficiently separated from each other to validate that just these two properly selected wavelengths permit markedly different

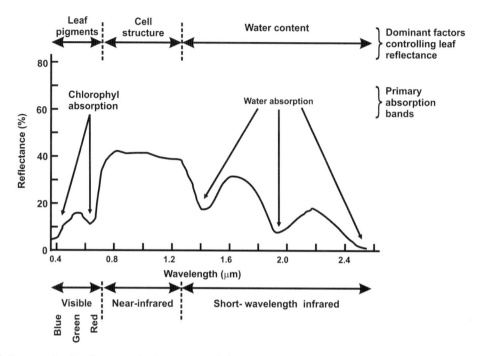

Fig. 9.4 An example of a reflectance plot for an unspecified vegetation type with the dominating factor influencing the entire reflectance curve. Redrawn from http://rst.gsfc.nasa.gov/Intro/Part2_5.html accessed on 18th Nov., 2008.

materials to be distinguished by their spectral properties. When we use more than two wavelengths, the plots in multi-dimensional space (up to three can be visualized; more than three are best handled mathematically) tend to produce more separation amongst different materials. This improved distinction amongst materials, due to use of extra wavelengths, forms the basis of *multispectral remote sensing*.

Spectral signatures for individual materials or classes can be determined best under laboratory conditions, where the sensor is placed very close

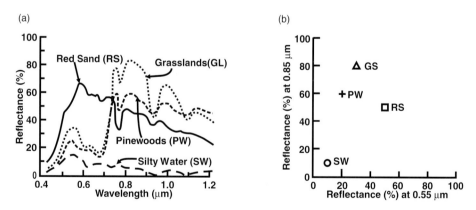

Fig. 9.5 (a) Spectral signatures, as variation of percent reflectance with wavelength, of some typical surface materials. Different materials reflect varying proportions of incident EMR at different wavelengths, which forms the basis of identifying materials through these signatures. (b) using differences in reflectance, it may be possible to distinguish the four common surface materials through the above signatures by plotting reflectances of each material at two wavelengths. Redrawn from http://rst.gsfc.nasa.gov/Intro/Part2_5.html accessed on 18th Nov., 2008.

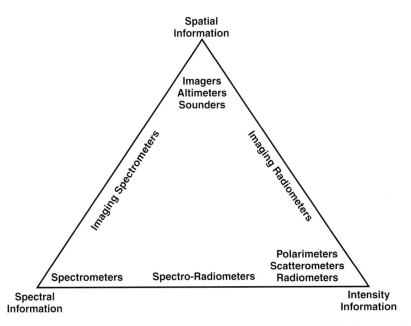

Fig. 9.6 A classification of remote sensing sensors based on the type of information provided by them, namely spatial, spectral, or intensity. Redrawn from http://rst.gsfc.nasa.gov/Intro/Part2_5a.html accessed on 18th Nov., 2008.

to the target. This results in a 'pure' *spectral signature*. But when the sensor is located well above the target, which is the case when a satellite remote sensing device looks down at the Earth, the telescope that examines the scene may cover a larger surface area than just the target at any given moment. Individual objects smaller than the field of view are not resolved and each object contributes its own spectral signature. In other words, for low resolution conditions, each one of the several different materials/classes sends (unresolved) radiation back to the sensor. The resulting spectral signature is a composite of all components in the sampled area. Analytical techniques (e.g. Fourier series analysis) can extract individual signatures under some special circumstances. But the sampled area is usually assigned a label equivalent to its dominant class.

9.1.4 Sensors

Most remote sensing measuring instruments (*sensors*) are designed to count the number of photons. As mentioned earlier, the basic principle underlying a sensor operation is based on the photoelectric effect. In this section, some of the main ideas on

the sensors that are used for this purpose are summarized. Three types of information are obtained using different kinds of sensors. Based on this criterion sensors may be classified, as shown in Fig. 9.6. An alternative and more elaborate classification of sensors based on the EM radiation measured is shown in Fig. 9.7.

Common types of a sensor system are:

- *Optical systems* – employing lenses, mirrors, apertures, modulators, and dispersion devices;
- *Detectors* – providing electrical signals proportional to irradiance on their active surface, generally being some type of semiconductor;
- *Signal processors* – performing specified functions on the electrical signal to provide output data in the desired format.

With reference to Fig. 9.7, two broad classes of sensors are: (i) *passive*, in which the energy giving rise to the radiation received comes from an external source, for example, the Sun; the MSS is an example, and (ii) *active*, in which energy generated from within the sensor system is beamed

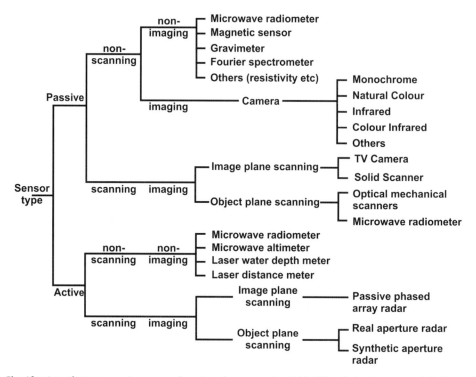

Fig. 9.7 Classification of remote sensing sensors based on the manner in which EM radiation is measured. Redrawn from http://rst.gsfc.nasa.gov/Intro/Part2_5a.html accessed on 18th Nov., 2008.

outwards and the fraction returned is measured; radar is an example. Sensors can be *non-imaging* or *imaging*. *Non-imaging* sensors measure the radiation received from all points in the sensed target, integrate it, and convert the result into an electrical signal or some other quantitative attribute, such as radiance. In the *imaging* sensors, electrons released are used to excite or ionize a substance such as silver (*Ag*) deposited on a film or drive an image-producing device such as a TV, computer monitor, cathode ray tube, oscilloscope, or an array of electronic detectors. As the radiation is related to specific points in the target, the end result is an image (picture) or a raster display (e.g. parallel horizontal lines on a TV screen).

Radiometer is a general term for any instrument that quantitatively measures the EM radiation in some interval of the EM spectrum. When the radiation is in the narrow spectral band including the visible light, the term *photometer* is used. If the sensor includes a component, such as a prism or

diffraction grating, that can break radiation extending over a part of the spectrum into discrete wavelengths and disperse (or separate) them at different angles to be detected by an array of detectors, it is called a *spectrometer*. One type of spectrometer (used in the laboratory for chemical analysis) employs multi-wavelength radiation passing through a slit onto a dispersing medium, which reproduces the image of the slit as lines at various fixed spacings onto a film plate. The term spectro-radiometer is reserved for sensors that collect the dispersed radiation in bands rather than in discrete wavelengths. Most air/space-borne sensors are spectro-radiometers.

Sensors that measure radiations coming from the entire scene instantaneously are called *framing systems*. The human eye, photo camera, and TV picture tube belong to this category. The size of the scene that is framed is determined by the aperture and optics employed in the system that defines the *field of view* (*FOV*). If the scene is sensed point by point (equivalent to small areas within the scene) along successive lines over a finite time, this mode of measurement makes up a scanning system. Most non-camera-based sensors operating from moving platforms image the scene by scanning.

Further, in the classification scheme the optical set-up for imaging sensors involves setting up either an image plane or an object plane, depending on whether the lens is before the point where the photon beam is made to converge at the focus or beyond it. For the *image plane arrangement*, the lens receives parallel light rays after these are deflected onto it by the scanner by the focusing device placed at its end. For the *object plane set-up*, the rays are focused at the front end (and have a virtual focal point behind the initial optical system) and are intercepted by the scanner before being finally focused at the detector.

Another attribute of this classification scheme is whether the sensor operates in *a scanning mode* or a *non-scanning mode*. A film camera held firmly in hand is a non-scanning device that captures light almost instantaneously when the shutter is opened for a short time and then closed. However, when the camera and/or the target move (as is the case with a movie camera), scanning is being performed in some sense. Conversely, the target can be static but the sensor sweeps across the sensed object/scene. This can be called scanning in that the sensor is designed for its detector(s) to move systematically in a progressive sweep, even as they move across the target. This is how a scanner operates in a computer; its flatbed platform (the casing and glass surface on which a picture is placed) also stays put. Scanning can also be carried out by putting a picture or paper document on a rotating drum (having two motions: circular and progressive shift in the direction of the drum's axis), in which the illumination is done by a fixed beam.

The sensors generally monitor the ground trace of the satellite over an area out to the sides of the trace. This is known as the swath width. The width is determined by part of the scene that encompasses full angular field of view of the telescope, which is actually sensed by a detector array. This is normally narrower than the total width of the object/scene from which light is made to pass through an external aperture (usually a telescope). With ref-

erence to the flight path, two principal modes of data gathering are shown in Fig. 9.8.

In the *cross-track mode* the data is gathered across the flight path by sweeping the object/scene along a line, or more commonly a series of adjacent lines, traversing the ground that is long (a few *km*) but very narrow (a few *m*). This is sometimes referred to as the 'whiskbroom mode' from the conceptual picture of sweeping a table from side to side with a small hand-held broom. Each line is subdivided into a sequence of individual spatial elements that represent a corresponding square, rectangular, or circular area (ground resolution cell) on the surface of the scene being imaged (or within, if the target to be sensed is the three-dimensional atmosphere). Thus, along any line there is an array of contiguous cells from each of which radiation emanates. The cells are sensed one after another along a line. In the sensor, each cell is associated with a pixel (picture element) that is coupled with/connected to a microelectronic detector. Each pixel is characterized for a brief time interval by some definite value of radiation (e.g. reflectance) that is converted into current (i.e. number of electrons flowing per unit time) using the photoelectric effect.

The areal coverage of the pixel (i.e., the ground cell area it corresponds to) is determined by the instantaneous field of view (IFOV) of the sensor system. The IFOV is defined as the solid angle extending from a detector to the area on the ground it measures at any instant (Fig. 9.8a). IFOV is a function of optical characteristics of the sensor, sampling rate of the signal, dimensions of any optical guides (such as optical fibres), size of the detector, and altitude above the target or scene at which the sensor is located.

The *along-track mode* of scanning has a linear array of detectors oriented normal to the flight path. The IFOV of each detector sweeps a path parallel to the flight direction. This type of scanning is also referred to as 'pushbroom' scanning (from the mental image of cleaning a floor with a wide broom through successive forward sweeps). The scanner consists of a line of small sensitive detectors stacked side by side, each being of very small dimension, on its plate surface. These detectors may number several thousands. Each detector is a charge-coupled device (CCD).

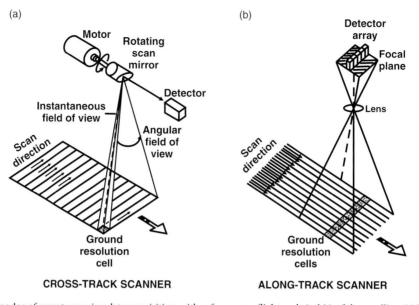

Fig. 9.8 Two modes of remote sensing data acquisition with reference to flight path (orbit) of the satellite. (a) The instantaneous field of view (FOV) of each detector in 'cross-track scanner' sweeps the scene across the flight path along a line or more commonly a series of adjacent lines traversing the ground that is very long ($\sim km$) but very narrow ($\sim m$). (b) Instantaneous FOV of each detector in 'along-track scanner' sweeps a path parallel to the flight direction. Redrawn from http://rst.gsfc.nasa.gov/Intro/Part2_5a.html accessed on 18th Nov., 2008.

9.1.5 Resolution

The obvious definition of spatial resolution can simply be stated as the smallest size of an object that can be picked out from its surrounding objects or features. This demarcation from neighbours or background may or may not be sufficient to identify the object, but high spectral resolution definitely aids material and class identification. Similar to spatial resolution, a price has to be paid to get better spectral resolution in terms of the number of detectors that must be appropriately increased. These must be physically placed suitably to capture the wavelength-dependent radiation spread over the desired part of the spectrum. This affects the process of signal handling onboard and data handling on the ground.

Field of view controls the swath width of a satellite image. The width, in turn, depends on: (i) optics of the observing telescope; (ii) inherent sampling limitations of the electronic sensor; and (iii) altitude of the sensor. Normally, the higher the orbit of the satellite, the wider is the swath width and consequently the lower will be the spatial resolu-

tion. Both altitude and swath width determine the 'footprint' of the sensed scene, such that it is across the track dimensions and frequency of repeat coverage. It may be noted that present-day popular digital cameras have much in common with some of the sensors described above. Light is recorded on CCDs after passing through a lens system and a device for splitting light into its primary colours. The light levels in each spectral region are digitized, processed by a computer chip, and stored (some cameras use removable disks) for immediate display on the camera screen or for downloading to a computer for processing to yield an image that can be reproduced on a printer connected to a computer. Alternatively, commercial processing facilities can produce high-quality pictures from the images recorded by the camera that are saved on its disk.

A summary description of some current major imaging Earth observation satellites available is given in Table 9.1. Major specifications of IRS series of Indian remote sensing satellites are given in Table 9.2.

Table 9.1 Salient features of currently available major imaging Earth observation satellites*. After Navalgund *et al.* (2007).

Country/organization	Satellite	Major sensors	Description (Spatial resolution, spectral bands)
USA/NASA	Terra	ASTER	15 m (VNIR), 30 m (SWIR), 90 m (TIR)' 14 XS (VNIR, SWIR, TIR)
	Terra/Aqua	MODIS	250–1000 m, 36 bands (VIS–TIR)
	NMP EO-1	ALI	10 m (Pan), 30 m (VNIR/SWIR)' Pan and 9 XS (VNIR/SWIR)
		Hyperion	30 m, 220 band, hyper spectral
USA/USGS	Landsat-7	ETM+	15 m (Pan), 30 m (Vis-SWIR), 60 m (TIR), Pan and 8 XS (VIS-TIR)
EU/ESA	Envisat	ASAR	C Band, multi-polarization, multi-imaging mode SAR
		MERIS	260–1200 m, 15 bands (VNIR)
Canada/CSA	Radarsat-1	SAR	C-Band SAR, HH polarization and multi-imaging modes
France/CNES	SPOT-5	HRG	2, 5 m (Pan), 10 m (MS)' Pan + 4 XS (VNIR/SWIR)
		HRS	High resolution (10 m, Pan) stereo 1.15 km,
		VEGETATION	3 XS(VNIR/SWIR)
China/CAST	CBERS-2	CCD	20 m, Pan + 4 XS (VNIR)
		IR-MSS	78 m (VNIR/SWIR), 156 m (TIR), 4 XS (Vis-TIR)
		WFI	258 m, 2 XS (VNIR)
Japan/JAXA	ALOS	AVNIR-2	10 m, 4 XS (VNIR)
		PALSAR	Multi-imaging mode L band radar
		PRISM	2.5 m panchromatic

* For detailed list, reference is made to CEOS (2005).
VNIR – visible and near infrared; SWIR – short-wave length infrared; TIR – thermal infrared; XS – multispectral bands.

9.2 Approaches to data/image interpretation

The ability to extract information from the data and interpret it depends not only on the capability of the sensor used but on how the data are processed to convert the raw data into images improving it for further analysis and applications.

As already mentioned, individual spectral signatures of different surface materials (or of classes of materials) enable their identification, as shown in Fig. 9.5a. Most sensors mounted on spacecraft divide the spectral response of ground scene into intervals or bands. Each band contains wavelength-dependent input that varies in intensity. The spectral measurements in each band characterise the interactions between the incident radiation and the atomic and molecular structures of the materials present on the ground. The measurements also reflect on the nature of the response of the detector system employed in the sensor. Two additional features, namely shape (geometric patterns) and use or context (sometimes including geographical locations), aid in distinguishing the various classes, even when the same material is being observed. Thus, one may assign a feature composed of concrete to the class 'street' or 'parking lot', depending on whether its shape is long and narrow or square/rectangular. Two features that have nearly identical spectral signatures for vegetation, could be assigned to the class 'forest' and 'crop', depending on whether the area in the images has irregular or straight (often rectangular, which is the case for most agricultural farms) boundaries. The task of any remote sensing system is to detect radiation signals, determine their spectral character, derive appropriate signatures, and compare the spatial positions of the classes they represent. This ultimately leads to some type of interpretable display

Table 9.2 Salient features of IRS series of Indian remote sensing satellites. After Navalgund *et al.* (2007).

Satellites (year)	Sensor	Spectral bands (μm)	Spatial res. (m)	Swath (km)	Radio-metric Res. (bits)	Repeat Cycle (days)
IRS-1A/1B (1988, 1991)	LISS I	0.45–0.52 (B) 0.52–0.59 (G) 0.62–0.68 (R) 0.77–0.86 (NIR)	72.5	148	7	22
	LISS-II	Same as LISS-I	36.25	74	7	22
IRS-P2 (1994)	LISS-II	Same as LISS-I	36.25	74	7	24
IRS-1C/1D (1995, 1997)	LISS-III	0.52–0.59 (G) 0.62–0.68 (R) 0.77–0.86 (NIR)	23.5	141	7	24
		1.55–1.70 (SWIR)	70.5 (SWIR)	148	7	24
	WiFS	0.62–0.68 (R) 0.77–0.86 (NIR)	188	810	7	24(5)
	PAN	0.50–0.75	5.8	70	6	24 (5)
IRS-P3 (1996)	MOS-A	0.755–0.768 (4 bands)	1570 × 1400	195	16	24
	MOS-B	0.408–1.010 (13 bands)	520 × 520	200	16	24
	MOS-C	1.6 (1 band)	520 × 640	192	16	24
	WiFS	0.62–0.68 (R) 0.77–0.86 (NIR) 1.55–1.70 (SWIR)	188	810	7	5
IRS-P4 (1999)	OCM	0.402–0.885 (8 bands)	360 × 236	1420	12	2
	MSMR	6.6, 10.65, 18, 21 *GHz* (V & H)	150, 75, 50 and 50 km	1360	–	2
IRS-P6 (2003)	LISS-IV	0.52–0.59(G) 0.62–0.68(R) 0.77–0.86 (NIR)	5.8	70	10(7)	24(5)
	LISS-III	0.52–0.59 (G) 0.62–0.68 (R) 0.77–0.86 (NIR) 0.77–0.86 (NIR)	23.5	141	7	24
	AWiFS	0.52–0.59 (G) 0.62–0.68 (R) 0.77–0.86 (NIR) 0.77–0.86 (NIR)	56	737	10	24(5)
IRS-P5 (Cartosat-1) (2007)	PAN Fore (+26°) & Aft (−5°)	0.50–0.85	2.5	30	10	5
Cartosat-2 (2007)	PAN	0.50–0.85	0.8	9.6	10	5

product, be it an image, map, or a numerical dataset, that reflects the reality of the surface in terms of nature and distribution of features present in the field of view.

Radiances (from the ground and intervening atmosphere) measured by various sensors, from hand-held digital cameras to distant orbiting satellites, vary in intensity. Thus reflected light at some wavelength or span of wavelengths (spectral region) can range in its intensity from very low value (dark portions in an image) to very bright (light toned). Each level of radiance can be assigned a quantitative value (commonly as a fraction of unity or as a percentage of the total radiance that can be handled by the range of the sensor employed). The values are converted to digital numbers (DNs) that consist of equal increments over the range. To generate an image, a DN is assigned some level of 'grey' (from all black to all white and shades of grey in between). When the pixel array acquired by the sensor is processed to show each pixel in its proper relative position and then the DN for the pixel is given a grey tone, a standard black and white image results.

Another vital parameter in most remote sensing images is colour. While variations in black and white imagery can be very informative and was the norm in the earlier aerial photographs, the number of different grey tones that the human eye can resolve is limited to about 20–30 steps (out of a maximum of ~250) on a contrasting scale. On the other hand, the human eye can distinguish 20 000 or more colour tints, so we can discern small but often important variations within the target materials or classes. Any three bands (each covering a spectral range or interval) from a multispectral set can be combined using optical display devices, photographic methods, or computer-generated imagery to produce a colour composite (simple colour version in natural colours, or quasi-natural or false colours).

New kinds of images can also be produced by making special datasets using computer processing programs. For example, one can divide the DNs of one band by those of another at each corresponding pixel site. This produces a band ratio image.

An important application of remote sensing data is in classifying the various features in a scene (usually presented as an image) into meaningful categories or classes. The image then becomes a thematic map (the theme is selectable, e.g. land use, geology, vegetation type, rainfall, etc.).

Another topic that is integral to effective interpretation and classification is commonly known as ground truth. This includes the study of maps and databases, test sites, field and laboratory measurements, and most importantly actual onsite visits to the areas being studied by remote sensing. The last of these has two main aspects: (i) to identify ground features in terms of their classes or materials so as to set up procedures for classification; and (ii) to revisit parts of a classified image area to establish the accuracy of identification in places that have not been visited for verification of ground situation.

9.2.1 Pattern recognition

Pattern Recognition (*PR*), also referred to as *Machine Learning* or *Data Mining*, uses spectral, spatial, contextual, or acoustic inputs to extract specific information from visual or sonic data sets. It involves techniques for classifying a set of objects into a number of distinct classes by considering similarities between objects belonging to the same class and the dissimilarities of objects belonging to different classes. A common example is the Optical Character Recognition (OCR) technique that reads a pattern of straight lines of different thicknesses called the bar code. An optical scanner reads the set of lines and searches a database for the exact matching pattern. A computer program compares patterns, locates it, and ties it into a database that contains information relevant to this specific pattern (e.g. in a grocery store, this would be the bar code that is printed on a package to show the current price of a product on display).

Pattern recognition has a definite role in remote sensing, particularly because of its effectiveness in geospatial analysis. It also plays an important role in Geographic Information Systems (GIS).

9.3 Radar and microwave remote sensing

9.3.1 Radar

Electromagnetic radiation at long wavelengths (0.1–30 *cm*) falls into the microwave region. Remote sensing employs passive microwaves, which are emitted by thermally activated bodies. But, in more common use is RADAR (acronym for Radio Detection and Ranging), an active (transmitter-produced) microwave system that sends out electromagnetic radiation, part of which is reflected back to the receiver. A radar system is a *ranging* device that measures distance as a function of round trip travel time (at the speed of light) of a directed beam of pulses (the signal whose strength is measured in decibels, *dB*) spreading out over specific distances. In this way, radar determines the direction and distance from the instrument (fixed or moving) to an energy-scattering object. We can also derive information about target shapes and certain diagnostic physical properties of materials at and just below the surface or from within the atmosphere by analysing modifications produced by the target in the back-scattered signal. The varying signal, which changes with the position and shape of a target body and is influenced by their properties, can be used to form images that

Table 9.3 Bands in the microwave region of the EM Spectrum.

Band	Frequency (MHz)	Wavelength (cm)
Ka	40 000–26 000	0.8–1.1
K	26 500–18 500	1.1–1.7
X	12 500–8000	2.4–3.8
C	8000–4000	3.8–7.5
L	2000–1000	15.0–30.0
P	1000– 300	30.0–100.0

Source: http://rst.gsfc.nasa.gov/Sect8/Sect8_1.html accessed on 18th Nov. 2008.

superficially resemble those recorded by imaging sensors. Commonly used frequencies and their corresponding wavelengths are specified by the nomenclature, as given in Table 9.3.

Radar has become increasingly important in various applications of remote sensing. Several operational satellites have radar as the principal sensor. By providing its own signal, a radar can function round the clock and, for some wavelengths, without significant interference from adverse atmospheric conditions (e.g. clouds). The most familiar civilian application of radar is in meteorology, mainly for tracking of storms, rainfall, and advancing weather fronts.

The ability of radar to mirror ground surfaces for displaying topography is its prime use for a variety of remote sensing applications. Some radars operate on moving platforms, others are fixed on the ground.

Each pulse emanating from a radar transmitter lasts only for a few microseconds (typically there are ~1500 pulses per second). The spatial resolution of a radar system is proportional to the length of its antenna, which for a *Side Looking Airborne Radar (SLAR)* is 5-6 *m* long and usually shaped as a section of a cylinder wall. To increase its effective length, and hence resolution, an electronic 'trick' of integrating the pulse echoes into a composite signal is performed in the *Synthetic Aperture Radar (SAR)* systems. These systems are associated with moving platforms. SAR utilizes both recording and processing techniques to generate a signal that acts as though it has an 'apparent' length greater than the antenna itself.

Another type of radar system is exclusive to conditions under which there is relative motion between the platform and the target. This system depends on the Doppler effect (apparent frequency shift due to relative motion between the target and the vehicle on which radar is mounted) that determines its azimuthal resolution. The Doppler shift occurs when a target is moving towards the observer and emits a signal such as sound; the frequency of the signal continues to increase as the target moves closer – the pitch of sound increases, whereas movement of the target away from the observer lowers its frequency.

Radar image tones may also vary in yet another systematic way that can be manipulated/controlled. When a pulse of photon energy leaves the transmitter, its electrical field vector can be made to vibrate in either a horizontal (H) or a vertical (V) direction, depending on its antenna design. There is no change of polarization in the reflected pulses, that is, they have the same direction of electric field vibration as in the transmitted pulse. Thus, we get either an HH or VV polarization pairing of the transmitted and returned signals. However, upon striking the target, the pulses can undergo depolarization to some extent so that reflections occur with different directions of vibration. A second antenna picks up the cross-polarization that is orthogonal to the transmitted direction, leading to either a VH or HV mode (the first letter refers to the transmitted signal and the second to the reflected signal). Some ground features appear nearly the same in either parallel or cross-polarized images. But vegetation, in particular, tends to show different degrees of image brightness in HV or VH modes, because of depolarization by multiple reflecting surfaces, such as branches and leaf cover of vegetation.

There are some other factors that contribute to the brightness or intensity of the returned signal. Two material properties provide clues about composition and surface condition from the manner in which these attributes interact with the incoming pulses. One of these properties is the dielectric constant. Radar waves penetrate deeper into materials with low dielectric constant and reflect more efficiently from those with high dielectric constant.

Values of dielectric constant range from 3 to 16 for most dry rocks and soils and up to 80 for water with impurities. Moist soils have values typically between 30 and 60. Thus, variation in reflected pulse intensities may indicate differences in soil moisture content, other factors being constant. Variations of the dielectric constant amongst rocks are generally too small to distinguish most types by this property alone.

The second property of materials is their surface roughness. Materials differ from one another in their natural or artificially altered state of surface roughness. Roughness, in this sense, refers to minute irregularities that relate either to textures of the surfaces or due to the presence of objects on them (such as closely-spaced vegetation that may have a variety of shapes). Examples include the surface character of pitted materials, granular soils, gravel, grass blades, and other covering objects whose surfaces have dimensional variability of the order of a couple of millimetres to a few centimetres. The vertical extent of an irregularity, together with radar wavelength and grazing angle at the point of contact, determines the behaviour of a surface as smooth (causing specular reflection), or intermediate or rough (diffuse reflection). To quantify the effect of different wave bands, a surface with a given small irregularity height (in cm) reflects Ka band ($\lambda = 0.85\ cm$), X band ($\lambda = 3\ cm$), and L band ($\lambda = 25\ cm$) radar waves as if it were a smooth, intermediate, and rough surface, respectively. This situation means that radar, broadcasting three bands simultaneously in a quasi-multispectral mode, can produce colour composites if a colour is assigned to each band. Patterns of relative intensities (grey levels) for images made from different bands may serve as diagnostic tonal signatures for diverse materials whose surfaces show contrast in their roughness.

Radar wavelengths also influence penetrability below target tops to ground surfaces. Depth of penetration increases with wavelength, λ. L and P band radar penetrate deeper than K or X bands. In forests, shorter wavelengths, such as the C band, reflect mainly from the leaves on the canopy tops that are encountered first. At longer wavelengths, most tree leaves are too small to have significant influence on backscatter, although branches do interact so that canopies can be penetrated to varying degrees.

Another active sensor system, similar in some respects to radar, is LIDAR (light detection and ranging). A Lidar transmits coherent laser light, at various visible or NIR (Near-IR) wavelengths, as a series of pulses (at the rate of a few hundreds per second) to the surface, from which some of the light is reflected. In this sense it is similar to radar. Round-trip travel times are measured. Lidars can be operated around the clock as profilers and as scanners. Lidar serves either as a ranging device to determine altitudes (topographic mapping), depths (in water bodies), or as a particle analyser (in the atmosphere). Light penetrates certain targets so that one of its prime uses is to assess condition of tree canopies.

9.3.2 Microwaves

Although active microwave systems, for example radar, are the more commonly used sensors, passive microwave sensors, both air- and space-borne, are also used. These directly measure the radiation emitted by thermal states of the Earth's surface, oceans, and atmosphere and hence are representative of natural phenomena innate to the materials (hence called passive).

The underlying principle is that emission of radiation from bodies depends on their temperature and at longer wavelengths extends into the microwave region. Though much weaker in intensity compared to shorter wavelengths, microwave radiation is still detectable by sensitive instruments and is not significantly attenuated by the Earth's atmosphere. The temperature measured by these instruments is brightness temperature, which is different for land areas, water, air, and ice, thus enabling their identification.

The wavelength region employed in passive microwave detectors is generally between 0.15 and 30 cm, equivalent to the frequency range between 1 and 200 GHz. Commonly used frequencies are centred on 1, 4, 6, 10, 18, 21, 37, 55, 90, 157, and 183 GHz (thus the multispectral mode is feasible) but the signal beam width is usually kept large in order to gather a sufficient amount of the weak radiation. The spatial resolution of the instrument also tends to be low (commonly in kilometres from space-borne and in metres from aircraft-mounted sensors), enabling the sensor to cover large sampling areas that provide enough radiation for ready detection. The sensors are typically radiometers that require large collection antennas (fixed or movable). On moving platforms, the fixed antenna operates along a single linear track so that it generates intensity profiles rather than images. Scanning radiometers differ by having the antenna move laterally to produce multiple tracking lines. The result can be a swath in which variations in intensity, when converted to photographic grey levels, yield images that resemble those formed in visible, near-IR, and thermal-IR regions.

On land, passive microwave surveys are particularly effective for detecting soil moisture, and temperature, because of sensitivity to the presence of water. Microwave radiation from thin soil cover (overburden) is contributed by near-surface bedrock geology. Assessing snowmelt conditions is another use of microwaves. Tracking distribution and conditions of sea ice is a prime oceanographic application. Yet another use is for assessing sea surface temperature. Passive microwave sensors are one of the important instruments on some meteorological satellites, being well suited to obtain water vapour and temperature profiles through the atmosphere, as well as ozone distribution, and precipitation conditions.

9.3.3 Gravity monitoring from space

Large regional scale changes in the Earth's gravitational field due to movement of water on or under the surface are being monitored from space using a coupled satellite system GRACE (Gravity Recovery and Climate Experiment). The two satellites of the GRACE mission were launched in 2002 and orbit the Earth 220 *km* apart and measure subtle variations in the gravitational pull by using microwaves to precisely gauge the changing distance between the two spacecrafts. As the lead spacecraft passes over a patch of anomalously strong/weak gravity field, it accelerates/decelerates ahead of the trailing spacecraft. Once past the anomaly, the lead satellite slows down/recovers speed. Then the trailing spacecraft accelerates/decelerates and again closes

on/trails the lead spacecraft. The changing distance between the two spacecrafts is a measure of the gravity anomaly. By making repeated passes over the same spot, GRACE measures changes in the Earth's gravity, which are mainly due to the changing load of water moving on or under the surface. GRACE has recorded shrinking of ice sheets, shifting ocean currents, desiccation of land surface due to droughts, and draining of large lakes.

9.4 Geographic Information Systems (GIS)

Modern-day computers, with their enhanced data handling capacity and powerful software, have greatly changed the manner in which multiple maps and other datasets can be merged, compared, and manipulated. This capability has led to emergence of a powerful new tool known as the Geographic Information System (GIS). Since its inception, GIS has blossomed into mainstream technology for using maps that are novel and practical in most applications focusing and relying on several kinds of geographic data. As remote sensing has routinely provided new images of the Earth's surface, it has become intertwined with GIS as a means to constantly update some of the GIS data (i.e. land use and land cover) in a cost-effective manner.

The objective in any GIS operation is to assemble a database that contains all the parameters to manipulate data through models and other decision-making procedures to yield a series of outputs to solve the given problem (Fig. 9.9). As an example, if one could relate information on the rainfall of a particular region to its aerial imagery, it might be possible to identify which wetlands dry up at a certain time of the year.

GIS uses information from multiple sources in different ways that can help in such analysis. The primary requirement for the source data consists of knowing the locations corresponding to the variables. Location may be denoted by x, y, and z coordinates of longitude, latitude, and elevation, or by other geo-coding systems such as ZIP or PIN codes or by highway milestone markers (*mile* or *km*). Any variable that can be located spatially can be fed into a GIS. Several computer databases that are compatible with the format

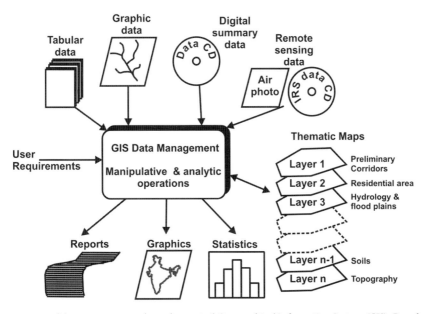

Fig. 9.9 Various aspects of data management through a typical Geographical Information System (GIS). Based on http://rst.gsfc.nasa.gov/Sect15/Sect15_4.html#anchor5681002 accessed on 18th Nov., 2008.

of GIS can be directly entered into it. Such databases are being routinely produced by various governmental – as well as non-governmental – organizations. Different kinds of data in the form of maps can be entered into a GIS as input. Some input data may already exist in digital form but one must digitize data, which are mostly given in the form of maps and tables or other sources. This can be done by scanning some of the maps or satellite/aerial imagery, thus obtaining the resulting products as digital inputs to individual pixels that are data cell equivalents. However, these values must be recoded (called *geocoding*) if they are to be associated with any specific attributes. Although recent technology has automated this conversion process, in many instances maps are still being digitized manually using a digitizing board or tablet. A GIS can also convert existing digital information, which may not be available in the form of maps or in a form it can recognize and use. For example, digital satellite images generated through remote sensing can be analysed to produce a map-like layer of digital information about vegetative cover. Likewise, census hydrologic tabular data can be converted to map-like form, serving as layers of thematic information in a GIS.

9.4.1 Data representation

GIS data represents real-world objects (roads, land use, elevation, etc.) in a digital form. Real-world objects/data can be divided into two abstractions: discrete objects (e.g. a house) and continuous fields (e.g. rainfall amount or elevation of a landform). There are two general methods, Raster and Vector, used for storing data in a GIS for abstractions.

9.4.1.1 Raster

A raster data type is, in essence, any type of digital image represented in grids (Fig. 9.10a), consisting of rows and columns of cells, with each cell storing a single value. Raster data can be images (raster images) with each pixel (or cell) containing a colour value. Additional values recorded for each cell may be a discrete value, such as land use; a continuous value, such as temperature; or a null value, if no data is available. While a raster cell stores a single value, it can be extended by using raster bands to represent RGB (red, green, blue) colours, colour maps (mapping between a thematic code and RGB value), or an extended attribute table with a given row having a unique value for each cell. The

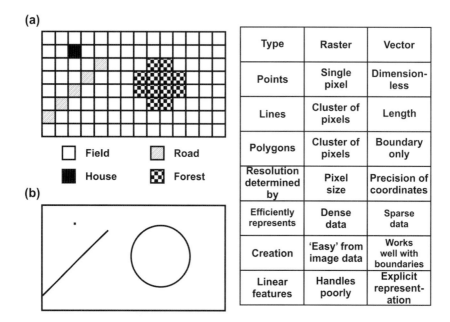

(a)

Type	Raster	Vector
Points	Single pixel	Dimension-less
Lines	Cluster of pixels	Length
Polygons	Cluster of pixels	Boundary only
Resolution determined by	Pixel size	Precision of coordinates
Efficiently represents	Dense data	Sparse data
Creation	'Easy' from image data	Works well with boundaries
Linear features	Handles poorly	Explicit represent-ation

Field　☐　Road ▨

House ■　Forest ▨

(b)

Fig. 9.10 Two methods of representing data in GIS: (a) Raster representation; (b) Vector representation. The table alongside compares the two methods of data representation and also lists some salient features of each method. Based on http://www.vcrlter.virginia.edu/~jhp7e/advgis/local/lectures/rastervsvector08.pdf?meid=51 accessed on 7th Nov. 2008.

resolution of the raster dataset is its cell width in units of the area represented by each cell on the ground.

Raster data is stored in various formats: from a standard file-based structure of TIF, JPEG, etc. to binary large object (BLOB) data stored directly in a relational database management system (RDBMS) similar to other vector-based feature classes. Database storage, when properly indexed, typically allows for quick retrieval of the raster data but may require storage of millions of large-sized records.

9.4.1.2 Vector

In a GIS, geographical features are often expressed as vectors, by considering those features as geometrical shapes (Fig. 9.10b). Different geographical features are expressed by different types of geometry:

- *Points*: Zero-dimensional points are used for geographical features that can best be expressed by a single point reference, for example, location of a place/feature. These could be locations of wells, peak elevations, features of interest or trailheads,

etc. Points can also be used to represent areas when displayed on a smaller scale. For example, cities on a world map would be represented by points rather than polygons. However, points convey least amount of information and no measurements are possible with point features.

- *Lines or polylines*: One-dimensional lines or polylines are used for linear features such as rivers, roads, railroads, trails, and topographic lines. As in the case of point features, linear features displayed on a smaller scale are represented as line features rather than as a polygon. Line features can measure distance.

- *Polygons*: Two-dimensional polygons are used to represent geographical features that cover a particular area on the Earth's surface. Such features include lakes, park boundaries, buildings, city boundaries, or land uses. Polygons convey the maximum amount of information on the file types. For example, by means of polygons, one can measure perimeter and area.

Each of these geometries is linked to a row in a database that describes their attributes. For example, a database that describes lakes may contain its depth, water quality, and pollution level. This information can be used to make a map to describe a particular attribute of the dataset. For example, lake waters could become coloured depending on the level of pollution. Different geometries can also be compared. For example, GIS could be used to identify the wells (point geometry) that are located within a given distance (say 1 *km*) of a lake (polygon geometry) that has a high level of pollution.

Vector features can be made to conform to spatial integrity through application of topological rules, such as 'polygons must not overlap with each other'. Vector data can also be used to represent continuously varying phenomena. Contour lines and triangulated irregular networks (TIN) are used to represent elevation or other continuously changing values. TINs record values at point locations, which are connected by lines to form an irregular mesh of triangles. The faces of triangles represent the terrain boundaries.

9.4.1.3 Non-spatial data

Additional non-spatial data can also be stored along with the spatial data represented by co-ordinates of vector geometry or position of a raster cell. In vector data, additional data contains attributes of the feature. For example, a forest inventory polygon may also have an identifier value and information about tree species. In raster data, the cell value can store attribute information and can also be used as an identifier that can relate to records presented in another table.

9.4.1.4 Raster-to-vector translation

Data restructuring can be performed by a GIS to convert data into different formats. For example, a GIS may be used to convert a satellite image map into a vector structure by generating lines around all cells with the same classification, while determining the spatial relationships between different cells, such as adjacency or inclusion.

9.4.1.5 Map overlay

A combination of several spatial data sets (points, lines, or polygons) creates a new output vector dataset, similar in appearance to stacking together several maps of the same region. A *union overlay*

combines the geographic features and attribute tables of both the inputs into a single new output. An *intersect overlay* defines the area where both inputs overlap and retains a set of attribute fields for each input. A *symmetric difference overlay* defines an output area that includes the total area of both inputs except for the overlapping area.

Data extraction is a GIS process similar to vector overlay, though it can be used in either vector or raster data analysis. Rather than combining the properties and features of both datasets, data extraction involves using a 'clip' or 'mask' to extract the features of one dataset that falls within the spatial domain of another dataset.

9.4.1.6 Graphical display techniques

Traditional maps are abstractions of the real world, a sampling of important elements displayed on a sheet of paper with symbols to represent physical objects. The symbols must be interpreted appropriately by the user of the map. Topographic maps show the shape of land surface with contour lines or with shaded relief.

Today, graphical display techniques, such as shading based on altitude in a GIS, can make relationships amongst map elements visible, enhancing one's ability to extract and analyse information.

9.4.1.7 GIS software

Geographic information can be accessed, transferred, transformed, overlaid, processed, and displayed using numerous software applications. Industry leaders in this field are companies such as Smallworld, Manifold System, ESRI, Intergraph, Mapinfo, and Autodesk, offering an entire suite of tools for GIS applications. Although free tools exist to view GIS datasets, public access to geographic information is limited by availability of online resources such as Google Earth and interactive web mapping.

9.4.1.8 Global Positioning System

The *Global Positioning System* (*GPS*) is a relatively recent technology, which provides unmatched accuracy and flexibility of positioning for navigation, surveying, and GIS data capture. The GPS provides continuous three-dimensional positioning round the clock throughout the world. The technology seems to benefit the GPS user community in terms of obtaining accurate data up to spatial resolution of about 100 metres for navigation, metre-level for mapping, and up to millimetre level for geodetic positioning. The GPS technology has tremendous applications in GIS data collection, surveying, and mapping.

The GPS uses satellites and computers to compute positions anywhere on the Earth's surface. It is based on satellite ranging, which means that the position on the Earth is determined by measuring the distance from a group of satellites in space. The basic principle underlying GPS is simple, even though the system employs some of the most hightech equipment ever developed.

To compute positions in three dimensions, one needs to have four satellite measurements. The GPS uses a trigonometric approach to calculate positions of objects/features on the Earth. The GPS satellites are positioned quite high up, as a result of which their orbits are quite predictable and each of the satellites is equipped with an accurate atomic clock.

9.5 Applications in hydrology

Both remote sensing and GIS are routinely used in areas such as applied hydrology, forestry, land-use dynamics, etc. Capabilities of remote sensing technology in hydrology include measurement of spatial, spectral, and temporal information and providing data on the state of the Earth's surface. It provides observations on changes in the hydrologic state of various reservoirs, varying over both time and space, that can be used to monitor prevailing hydrologic conditions and changes taking place in them over time. Sensors used for hydrologic applications cover a broad range of the electromagnetic spectrum. Both active and passive sensors are used. Active sensors (i.e. radar, lidar, etc.) send a pulse and measure time taken by the return pulse. Passive sensors measure emissions or reflectance from natural sources/surfaces, such as the Sun, thermal energy emitted by the human body, etc. Sensors can provide data on reflective, thermal, and dielectric properties of the Earth's surface.

Since remote sensing techniques indirectly measure hydrologic conditions using electromagnetic waves, these have to be related to hydrologic variables either empirically or by employing transfer functions.

Remote sensing applications in hydrology are mainly for:

- Estimation of precipitation
- Runoff computation
- Snow hydrology applications
- Estimation of evapotranspiration over land surface
- Evaluation of soil moisture content
- Water quality modelling
- Groundwater source identification and estimation
- Hydrologic modelling

GIS can play a crucial role in the application of spatially distributed data to hydrological models. In conventional applications, results either from remote sensing or from GIS analyses serve as input to hydrologic models. Land use and snow cover are the most commonly used input variables for modelling. The integration of GIS, database management systems and hydrologic models facilitates the use of remote sensing data in hydrologic applications.

9.5.1 Precipitation estimation

For a field hydrologist, satellite rainfall estimation methods are valuable when either few or no surface gauges are available for measuring rainfall (see also Section 2.5.1.3). However, direct measurement of rainfall from satellites for operational purposes has not been feasible as presence of clouds prevents direct observation of precipitation by visible, near infrared, and thermal infrared sensors. However, improved analysis of rainfall can be achieved by combining satellite and conventional rain gauge data. Useful data can be obtained from satellites employed primarily for meteorological purposes, including polar orbiters such as the National Oceanographic and Atmospheric Administration (NOAA) series, the Defense Meteorological Satellite Program (DMSP), and geostationary satellites such as Global Operational Environmental

Satellite (GOES), Geosynchronous Meteorological Satellite (GMS), and Meteosat, but their images in the visible and infrared bands can provide information only about the cloud top rather than cloud base or interiors. However, since these satellites provide frequent observations (even at night with thermal sensors), the characteristics of potentially precipitable clouds and rate of change in cloud area and shape can be observed. From these observations, estimates of rainfall can be made that relate cloud characteristics to instantaneous rainfall rates and cumulative rainfall over time.

Whereas the visible/infrared techniques provide only indirect estimates of rainfall, microwave techniques have great potential for measuring precipitation because the measured microwave radiation is directly related to the properties of raindrops themselves. Microwave techniques estimate rainfall in two ways: by their emission/absorption and scattering from the falling raindrops. Through the emission/absorption approach, rainfall is observed by the emission of thermal energy by raindrops against a cold, uniform background. A number of algorithms have been developed to estimate precipitation over the ocean by this method (Kummerow *et al.* 1989; Wilheit *et al.* 1977).

Using the scattering approach, the rainfall is observed through enhanced scattering primarily caused by frozen ice in clouds and not directly by the rain. Thus rainfall rates must be established empirically or with cloud models, but this method is not restricted to ocean background and may be the only feasible approach for estimating rainfall over land employing microwave radiometry. Spencer *et al.* (1988) have shown that the DMSP Special Sensor Microwave/Imager (SSM/I) data can identify rainfall areas. Adler *et al.* (1993) have used a cloud-based model with 85 and 37 *GHz* SSM/I data to estimate rainfall rates. It should be noted that these approaches provide instantaneous rainfall rates that are subsequently aggregated to yield monthly values.

9.5.2 *Snow hydrology*

Almost all regions of the electromagnetic spectrum provide useful information about snow packs. Depending on the need, one may like to know the areal extent of the snow, its water equivalent, its 'condition' or grain size, density, and presence of liquid water. Although no single region of the electromagnetic spectrum yields all these properties, techniques have been developed to estimate these to a certain degree (Rango 1993).

The water equivalent of snow can be measured from a low elevation aircraft carrying sensitive gamma radiation detectors (Carroll and Vadnais 1980). This approach is limited to low flying aircrafts (altitudes ~150 *m*), because the atmosphere attenuates a significant fraction of the gamma radiation.

Snow can be identified and mapped with the visible bands of satellite imagery, because of its high reflectance in comparison to non-snow covered areas. But at longer wavelengths, that is, in the near infrared region, the contrast between snow covered and non-snow covered areas is considerably reduced. However, the contrast between clouds and snow is higher in the infrared imagery, which serves as a useful discriminator between clouds and snow (Dozier 1984). Thermal data can also be useful in identifying snow/no-snow boundaries and discriminating between cloud and snow.

Use of satellite data for snow cover mapping has become operational in several regions of the world. Currently, NOAA has developed snow cover maps for about 3000 river basins in North America, of which approximately 300 are mapped according to elevation for use in stream flow forecasting (Carroll 1990). NOAA has also produced regional and global maps of mean monthly snow cover (Dewey and Heim 1981).

Microwave remote sensing data can provide information on snow pack properties that are of utmost interest to hydrologists, for example, area under snow cover, snow water equivalent (or depth), and presence of liquid water in a snow pack that signals the onset of its melting (Kunzi *et al.* 1982). An algorithm for estimating snow water equivalent for dry snow using satellite microwave data (Scanning Multichannel Microwave Radiometer (SMMR and SSM/I) has been developed by Chang *et al.* (1982). The depth and global extent of snow cover was mapped by Chang *et al.* (1987). Passive microwave systems are limited by their interaction with other media, such as forest areas, although a method to correct for the absorption of the snow signal by forest cover has also been developed (Chang *et al.*

1990). The spatial resolution achievable by the passive satellite systems is also a limitation, but Rango *et al*. (1989) have shown that reasonable snow water equivalent estimates can be made on basins with an area smaller than 10 000 km^2.

Aircraft SAR measurements have shown that SAR can discriminate between snow and glaciers from other targets and also discriminate between wet and dry snow (Shi and Dozier 1992; Shi *et al*. 1994). A number of models have been developed to use remote sensing data for predicting snowmelt runoff (Dillard and Orwig 1979; Hannaford and Hall 1980). NASA specifically developed (Martinec *et al*. 1983) the Snowmelt Runoff Model (SRM) for use in remote sensing of snow cover by taking the elevation zone as the primary input variable. SRM has been extensively tested on basins of different sizes in various regions of the world (Rango 1992; WMO 1992).

9.5.3 Soil moisture

Knowledge of soil moisture has many applications in hydrology, primarily in evaporation and runoff modelling. Studies have shown that soil moisture can be measured by a variety of remote sensing techniques. However, only microwave technology has demonstrated a quantitative ability to measure soil moisture under a variety of topographic and vegetation cover conditions (Engman 1990). Available passive systems do not have the optimum wavelengths for soil moisture research. But it seems that in areas with sparse vegetation cover, a reasonable estimate of soil moisture can be obtained (Owe *et al*. 1988). Experimental passive microwave systems using aperture synthesis such as the Electronically Steered Thinned Array Radiometer (ESTAR) hold promise for high resolution satellite systems. The airborne ESTAR has been demonstrated in the Walnut Gulch watershed in Arizona (Jackson 1993). The SAR systems offer perhaps the best opportunity to measure soil moisture routinely. Change detection techniques have been used to detect changes in soil moisture in a basin in Alaska (Villasenor *et al*. 1993). However, Merot *et al*. (1994) have shown that radar data becomes ambiguous when ponding of water occurs in variable source areas.

9.5.4 Evapotranspiration

In general, remote sensing techniques cannot measure evaporation or evapotranspiration (ET) directly. However, two potentially important roles of remote sensing in estimating evapotranspiration are: (i) extending point measurements or empirical relationships, such as the Thornthwaite (1948), Penman (1948), and Jensen and Haise (1963) methods, to much larger areas, including the areas where experimental meteorological data may be sparse; and (ii) measurement of variables in the energy and moisture balance models of ET. Some progress has also been made in direct remote sensing of the atmospheric parameters that affect evapotranspiration using Light Detection and Ranging (LIDAR) – which essentially is a ground-based technique.

Extrapolation of point ET measurements to a regional scale can be achieved by using thermal mapping data. Gash (1987) proposed an analytical framework for relating location specific changes in evaporation to corresponding changes in surface temperature. These concepts were used for an agricultural area under clear sky conditions by Kustas *et al*. (1990). Humes *et al*. (1994) proposed a simple model using remotely sensed surface temperatures and reflectances for extrapolating energy fluxes from a point measurement to a regional scale. However, other than for clear sky conditions, variations in incoming solar radiation, meteorological conditions, and surface roughness limit the applicability of this approach.

One formulation of potential ET that lends itself to suitability for remote sensing inputs was developed by Priestly and Taylor (1972) and was used for estimating ET with satellite data by Kanemasu *et al*. (1977). Barton (1978) used airborne microwave radiometers to remotely sense soil moisture in his study of evaporation from bare soils and grasslands. Soares *et al*. (1988) demonstrated how thermal infrared and C-band radar could be used to estimate bare soil evaporation. Choudhury *et al*. (1994) have shown a strong relationship between evaporation coefficients and vegetative indices. Another approach is to develop numerical models that simulate the heat content and water balance in the soil in relation to the energy balance at the surface (Camillo *et al*. 1983; Taconet *et al*.

1986). Taconet and Vidal-Madjar (1988) used this approach with Advanced Very High Resolution Radiometer (AVHRR) and Meteosat data.

9.5.5 Runoff and hydrologic modelling

Runoff cannot be directly measured by remote sensing techniques. However, remote sensing can be used in hydrologic and runoff modelling in two general ways: (i) for determining watershed geometry, drainage network, and other map-type information for distributed hydrologic models and for estimation of empirical flood peak, annual runoff, or for formulating low flow equations; and (ii) for providing input data such as soil moisture or delineating land-use classes that are used to define runoff coefficients.

In many regions of the world, remotely sensed data is the only source of good cartographic information. Drainage basin areas and the stream network are easily obtained from good quality satellite imagery. There have also been a number of studies to extract quantitative geomorphic information from Landsat imagery (Haralick et al. 1985). Remote sensing can also provide quantitative topographic information with appropriate spatial resolution, which is extremely valuable for model inputs. Digital Elevation Model (DEM) with high horizontal and vertical resolution approaching 5 m is possible. A new technology using SAR interferometer has been used to demonstrate similar horizontal resolutions with approximately 2 m vertical resolution (Zebker et al. 1992).

Quick estimate of peak flow when very little information is available is often made using empirical flood routing formulae. Applicability of these equations is generally limited to the basin and the climatic/hydrologic region of the world for which these were developed. Most such empirical flood routing formulae relate peak discharge to the drainage area of the basin (UN 1955). Landsat data have been used to improve empirical regression equations of various runoff characteristics. For example, Allord and Scarpace (1981) showed how the addition of Landsat-derived land cover data can improve regression equations based on topographic maps alone.

Most of the work on adapting remote sensing data to hydrologic modelling has involved the Soil Conservation Service (SCS) runoff curve number model (USDA 1972), for which remote sensing data are used as a substitute for land cover maps obtained by conventional means (Bondelid et al. 1982; Jackson et al. 1977).

The pixel format of digital remote sensing data makes it ideal for merging with GIS. Remote sensing can be incorporated into a system in a variety of ways: as a measure of land use and impervious surfaces, for providing initial conditions for flood forecasting, and for monitoring flooded areas (Neumann et al. 1990). Kite and Kouwen (1992) showed that Landsat-derived semi-distributed model at a sub-basin scale performs better than a lumped model. The GIS allows for combining of other spatial data forms such as topography, soils maps, etc. with hydrologic variables such as rainfall or soil moisture distribution. Fortin and Bernier (1991) proposed combining SPOT DEM data with satellite-derived land use and soil mapping data to define Homogeneous Hydrologic Units (HHU) in HYDROTEL. Ott et al. (1991) and Schultz (1993) defined Hydrologically Similar Units (HSU) by combining DEM data, soils maps, and satellite-derived land use in a study of the impact of land use changes on the Mosel River Basin. They also used satellite data to determine a Normalized Difference Vegetation Index (NDVI) and a leaf Water Content Index (WCI), which are combined to delineate areas where a subsurface source of water is available for growth of vegetation. Mauser (1991) showed how multi-temporal SPOT and TM data can be used to derive plant parameters for estimating ET in a GIS-based model.

9.5.6 Groundwater hydrology

As groundwater is subterranean, most remote sensing techniques, with the exception of airborne geophysics and radar, are incapable of penetrating beyond the uppermost layer of the Earth's surface. But the synergy between satellite imagery and airborne geophysical data can be advantageous in the early stages of groundwater exploration and also for groundwater modelling.

Long-wave radar can under suitable conditions (e.g. for coarse-grained deposits, dry vadose zone without vegetation, and with some *a priori* knowledge of the geology) detect groundwater at depths of a few metres and other subsurface features, such as buried channels (McCauley *et al*. 1982). Radar imagery has its use in hydrogeology for interpretation of geological structures (Drury 1993; Koopmans 1983). On the other hand, thermal imagery involves long-wave radiation emitted from the surface and its use largely pertains to detect temperature anomalies in surface water bodies. The surface features contained in the imagery of the reflected shorter wavelengths relate to the surface expression of geological and geomorphological features and land cover, yielding indirect hydrogeological information.

Groundwater exploration using photo-geology has been a major field of interest in the past and is still being done in areas inadequately covered by geological maps. Visual interpretation should make use of imagery that optimally displays the terrain information. To achieve this, the colour and intensity information of the multispectral data can be manipulated by image processing techniques. However, its interpretation needs experience and familiarity with the terrain. Image interpretation is the most important aspect of remote sensing studies of groundwater. In certain cases, the imagery contains features that have a direct link with the groundwater flow. The appearance of groundwater at or near the surface is caused by either the intersection of topographic depressions (blow-outs, silicate, karst depressions, back swamps, etc.) and the static phreatic groundwater level or the discharge zone of a groundwater flow system. In the latter case, the groundwater table is shallow and there is a pressure gradient that causes upwelling of groundwater. Such conditions were found in an area (Fig. 9.11) lying in an alluvial area in north India. Fig. 9.11a shows the image in band 7 and Fig. 9.11b is the interpretation for this example. In this case, the area has low relief and most of the area was fallow at the time of recording by Landsat Multi-Spectral Scanner (MSS). Often the exfiltration areas are less obvious on the imagery and have to be discerned from the vegetation pattern. Such direct links are, however, exceptional. Generally the in-

terpreted information from satellite imagery gives more of circumstantial evidence or proxy hydrogeological information.

The synergy between various types of imageries, such as multi-temporal short wave reflection, active microwave, and thermal imagery, can increase the discriminating capability for mapping of different vegetation types and wet soils in exfiltration areas. In addition, the classification may be superimposed on a digital elevation model (DEM) of the groundwater level, since the flow systems are driven by the groundwater potentiometric surface or by a DEM of the ground surface with detailed drainage included, because the water table generally follows the surface topography of the area. There is a close relationship between groundwater depth and stream network characteristics in permeable lowlands where the stream system can be considered an outcrop of groundwater flow systems. Hydrogeological image interpretation can be a cost-effective aid to prepare hydrogeological maps.

If upward flow of deep groundwater occurs, it will emerge in valleys or rivers and may be detected on a large scale (1:1000 or 1:2000) thermal imagery because of the thermal inertia of relatively warm groundwater emerging in cold rivers in mountainous terrain. Large head differences of the groundwater surface and the presence of fracture conduits contribute to the ease of its detection. Burger *et al*. (1984) presented examples in mountainous terrain, using enlargements of airborne thermal imagery on a scale of 1:2000.

Using imagery in the visual and thermal ranges together with some ground meteorological measurements, the evaporation in the central zone of the *sebkhas* (playas) that are important evaporative sinks of the partially leaking-out fossil groundwater in parts of northern Africa, was estimated as 2 *mm/day* (Menenti 1984).

Another field of application of thermal imagery is the detection of submarine groundwater discharge in the oceans. This is based on the temperature difference between the fresh and sea water, provided a sufficient amount of groundwater outflow exists to offset the effect of mixing in the depth interval between the sea bottom and the sea surface. Several winter time NOAA images of the Tunisian coast near Djerba show a strong anomaly

(a)   (b)

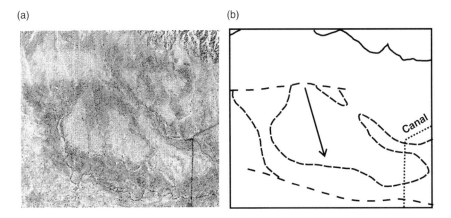

Fig. 9.11 An example of data extraction by visual interpretation aided by some field observations: (a) MSS image, band 7, after image sharpening, of a mainly alluvial area in north India with deposits of different ages and characteristics; and (b) interpreted exfiltration zones, partly influenced by faults and tilting. Reproduced from Meijerink (1996). Reproduced by permission of IAHS Press.

where there is significant temperature difference between the Mediterranean surface (with temperature $\sim 14°C$) and the groundwater (with temperature $>18°C$) and other suitable conditions exist (Meijerink 1996).

Vast areas of the world consist of hard rocks (basement complexes) or limestone terrain where groundwater flow occurs only through secondary permeability and is thus limited to fractures and weathered zones. Normally when the success ratio of drilling in hard rock terrain is low and the use of geophysical tools is too expensive, the study on imagery of lineaments and other structural and geometric controls favouring occurrence of groundwater is an attractive proposition. Photo-lineaments can be described as linear structural elements, which are thought to have developed over fracture zones. Interpreted lineaments can pertain to fractures of a different tectonic nature, with or without intrusive or secondary clay fillings. Reasonably straight low topographic corridors, formed by denudation initiated by open fracturing, have been identified as lineaments as well as sharp linear features in outcrop areas. Because of their varied nature, the hydrogeological significance of a lineament needs to be established by drilling and well testing. In some studies, high groundwater yields have been found to be associated with proximity to lineaments; in others this was, however, not found to be so.

In specific areas, valuable information on groundwater quality can be derived from imagery by making use of the relationship between water quality and vegetation. Kruck (1990) presented well-illustrated examples of the adjustment of vegetation to the salinity of the shallow, phreatic waters of the inland Okavango delta in Botswana and the pampas in Argentina. The local effects of hydrocarbon contamination on the vegetation have been described by Svoma (1990).

Imagery provides valuable information for groundwater resource evaluation as well as management. In certain cases, not only the likelihood of groundwater occurrence, risk of contamination, and saltwater intrusion, but also of the characteristics of the terrain for groundwater use, such as suitability for small-scale irrigation, location, and size of settlements and so on, are reflected in the imagery. Fig. 9.12 shows a typical scheme of preparing a groundwater distribution map using remote sensing data and GIS in conjunction with relevant ground truth information on geology, geomorphology, structural pattern, and recharge conditions that ultimately define a given groundwater regime. Such work has facilitated identification of sources of drinking water for water-scarce villages (NRSA 2003).

GRACE satellite mission has recorded the world's largest decline of gravity (other than that caused by post-glacial melting of ice sheets during the first

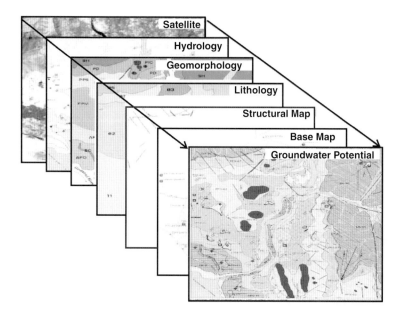

Fig. 9.12 Schematic showing a typical sequence of remote sensing and GIS application to groundwater. Starting with the satellite images and various derived geo-spatial inputs, a groundwater prospecting map is generated in the GIS environment. Redrawn from Navalgund *et al.* (2007) with permission from Current Science. See also plate 9.12.

6 years (2002–2008) of its operation across India in a 2.7×10^6 km^2 east-west swath extending from eastern Pakistan across northern India and Nepal upto Bangladesh, which has important implications for sustainability of groundwater resources. This decline in gravity is due to an estimated loss at the rate of 54 ± 9 km^3 a^{-1} of groundwater from this region. GRACE has also estimated that an adjacent area comprising states of Rajasthan, Haryana, Delhi, and Punjab in northwest India has lost groundwater to the extent of 17.7 ± 4.5 km^3 a^{-1} during the same period corresponding to a mean decline of the water table by 4 ± 1 cm a^{-1}. These losses of groundwater correspond to more than the total surface reservoir capacity created in India and point to unsustainable consumption of groundwater for irrigation and other human uses.

9.5.7 Urban issues

With the advent of remote sensing and GIS, city planners have begun to assess the resources through a multi-disciplinary approach for achieving quick results in a cost-effective manner and with minimal manpower support. Driven by technological advances and societal needs, remote sensing of urban areas is increasingly becoming a new application of geospatial technology, with additional applications in various socioeconomic sectors (Weng and Quattrochi 2006).

Urban landscapes are typically a complex combination of buildings, roads, parking lots, sidewalks, gardens, cemeteries, open lands with soil cover, surface and groundwater, and so on. Each of the urban component surfaces exhibits unique radiative, thermal, moisture, and aerodynamic regimes related to their respective surrounding environment that creates spatial complexity of ecological systems (Oke 1982). To understand the dynamics of patterns and processes and their interactions in heterogeneous landscapes that characterize urban areas, one must be able to precisely quantify the spatial pattern of the landscape and its temporal changes (Wu *et al.* 2000).

Remote sensing technology has been widely applied to urban land-use planning, land cover classification, and change detection. However, because several types of land use/land cover (LU/LC) are contained in one pixel, referred to as the mixed pixel problem, it is rare that a classification accuracy of more than 80% can be achieved by using pixel-wise classification (the so-called 'hard classification') algorithms (Mather 1999). The mixed pixel problem results from the fact that the scale of observation (i.e. pixel resolution) fails to represent the spatial characteristics of the area (Mather 1999). Alternative approaches have been suggested to overcome this problem, but analysis of urban landscapes is still based on 'pixels' or 'pixel groups' (Melesse *et al*. 2007).

Impervious surfaces such as roads, driveways, sidewalks, parking lots, rooftops, etc. are urban anthropogenic features through which water cannot infiltrate into the ground. In recent years, impervious surface cover has emerged, not only as an indicator of the degree of urbanization but also a major indicator of urban environmental quality (Arnold and Gibbons 1996). Various digital remote sensing approaches have been developed to estimate and map impervious surfaces (Weng 2007).

Since the launch of the first Earth Resources Technology Satellite, ERTS-1, in 1972, remotely-sensed data from different sensors have been used to characterize, map, analyse, and model the state of the land surface and sub-surface processes. This new information, together with better estimation of hydrological parameters, has increased our understanding of different hydrologic processes as an aid to quantification of rate and amount of water and energy fluxes in the environment. The ability of these sensors in providing data on various spatio-temporal scales has also increased our capability to grapple with one of the challenges of environmental modelling, such as mismatch between scale of environmental processes and the available data.

Remote sensing data is being widely used, particularly in developing countries for preparation of perspective development plans. Remote sensing and GIS are typically used to provide maps of existing land use, infrastructure network (roads, railways and human settlements), hydrologic features (river/stream, lakes) etc. that are useful for preparation of regional level landscape, updating of base maps, urban sprawl, land-use changes, population growth, and master plans for equitable development of environmental resources of the area. The data are being extensively used for infrastruc-

ture planning, ring road alignment, utility planning (e.g. site selection for locating garbage dump, water works, etc.), road network and connectivity planning, growth centre locations, etc. Several important applications of remote sensing and GIS can be found in the literature for almost every city in developed countries. Even in India, use of these technologies is growing rapidly for urban sprawl mapping, preparation of a master plans, pipeline surveys, planning of ring roads, port trusts, etc.

9.5.8 Monitoring of global change

In recent years, the term 'climate change' has found its way into the top agenda of scientific discussions. Muddled up with this is another term 'global change' that has evolved with its starting point of the concept for treating the Earth as an integrated system. The Earth system encompasses the climate system in addition to biophysical and anthropogenic processes that are important for its functioning. Some of the changes in the Earth system, natural as well as human-induced, could have significant consequences without involving any concurrent changes in the climate system. Global change should, therefore, not be confused with just climate change; it is considerably more comprehensive (Steffen *et al*. 2004). Much of global change study focuses on the vertical links between the atmosphere and the Earth's surface and its components, such as ecosystems on land and in the sea, which are also connected to each other laterally through dynamics of the Earth System – involving horizontal movement of water and other Earth materials, atmospheric transport of water vapour and pollutants, and deposition and migration of plants and animals. Moreover, they are connected through energy transfers and also through chemical and biological signatures that persist over time. It is essential to note that global changes in social, cultural, economic, and environmental systems converge in small localities. On the other hand, changes on the local scale contribute to global change with various linkages between systems and their spatial scales. Examples of global driving forces include the world population that has doubled since the decade of 1960s; the global economy has improved since the 1950s by more than a factor of 15 but the regional inequality is increasing and so on. It is a daunting task to identify, monitor, and predict the various effects and linkages between processes involved in global change.

On a global scale, the response of the Earth system to contemporary anthropogenic forcing can clearly be seen in biogeochemical cycles, in the hydrologic cycle, and recently in climate. The Earth's climate system responds to various direct and indirect human forcing in many ways, in addition to modification of the hydrologic cycle.

From space, the whole world unfolds every day. Orbiting the planet, the Earth Observing Satellite (EOS) sensors are uniquely able to make the kind of measurements that experts need to understand the systemic changes that are taking place on the Earth. Satellite remote sensing is an evolving technology with the potential for contributing to studies of the human dimensions of global environmental change by making comprehensive evaluations of various human activities from multispectral satellite images obtained from sensors deployed on satellites, such as Landsat and ASTER, both offering spatial resolution of 15 m.

Satellite images are expected to contribute to a wide array of global change-related application areas for studying vegetation and ecosystem dynamics, hazard monitoring, geology and soil analysis, land surface climatology, hydrology, land-cover changes, and generation of digital elevation models (DEMs).

For climatologists and meteorologists studying atmospheric and climatic variability on a global scale, monitoring of the snow cover component is a major challenge. Snow cover significantly influences the evolution of weather on a daily basis, as well as climate variability on longer time scales. For example, glaciers on the Himalaya and Tibetan Plateau, and ice caps and ice sheets in Antarctica, Greenland, and in the Arctic Circle, play an important role in the Earth's climate. This is because the snow cover modifies the radiative balance and energy exchanges between the land surface and the atmosphere significantly, thus indirectly affecting the atmospheric dynamics. Snow cover is also extremely sensitive to climatic fluctuations. Consequently, accurate observations of the spatial and temporal variability of snow cover could aid monitoring of global change. Presently, because of some limitations inherent in various data types employed in monitoring snow cover from space, the most favourable solution probably consists of jointly using multi-sensor satellite data (in the visible, near-infrared, thermal infrared, and microwave bands).

Remote sensing, particularly use of radar altimeters and imaging radiometers, offers a global perspective for some hydrologic parameters. Satellite radar altimetry, with its all-weather operation, can be used to derive water levels in lakes, rivers, and wetlands and infrared imagery can be used to demarcate lake and wetland areas. Remote sensing techniques can be applied to several problems, including determination of regional scale hydrologic variability and proxy monitoring of precipitation changes.

The Earth Sciences community has made extensive use of satellite imagery data for mapping land cover, estimating geophysical and biophysical characterization of terrain features, and monitoring changes in land cover with time. For example, Ehrlich et al. (1994) reviewed many of the reported findings associated with the capabilities of NOAA Advanced Very High Resolution Radiometer (AVHRR) 1-km resolution data to provide global land-cover information. Tucker et al. (1991) used coarse-resolution satellite image data for monitoring continental-scale climate-related phenomena in 'Expansion and Contraction of the Sahara Desert from 1980 to 1990'. Colwell and Sadowski (1993) used high-resolution satellite data for monitoring regional patterns and rates of forest resource utilization. Freeman and Fox (1994) discussed the semi-operational use of satellite imagery for several forest assessment programmes.

It is thus clear that evaluation of static attributes (type, amount, and arrangement) and the dynamic attributes (type and rate of change) on satellite images enable the types of change to be regionalized and the proximate sources of change to be identified or inferred. This information, combined with results of case studies or surveys, can provide helpful inputs for an informed evaluation of interactions amongst the various driving forces for global change.

10

Urban hydrology

Urbanization, which has of late become an important environmental issue, involves excessive physical growth of urban areas, in rural or natural setting,
as a result of population migration to and around
existing urban areas, mostly for generating livelihood. Effects of urbanization include changes in
the density of population and consequent pressure
on civic administration to provide various amenities. While the exact definition and population size
of urban areas vary in different countries, urbanization involves growth of cities. The United Nations
has defined urbanization as the migration of people
from rural to urban areas. People move to cities to
seek or further economic opportunities. In rural areas on small agricultural farms it is often difficult to
improve one's living standard beyond basic sustenance. Agricultural economy is largely dependent
on unpredictable environmental conditions and in
times of drought, flood, or epidemics, very survival
of affected population becomes difficult. Furthermore, as a result of industrialization, agricultural
operations have progressively become more mechanized, rendering many manual labourers jobless.
Many of the basic services as well as other specialized services are not available in rural areas. Also,
there are more job opportunities and a variety of
jobs available in urban areas. Health care is another
major concern in rural areas. People, especially the
elderly, are often forced to move to cities where
better medical facilities exist. Other factors include
more recreational facilities (restaurants, movie the-
atres, theme parks, etc.) and a better quality of

education, offered by universities as well as specialized institutions. These have come about due
to changes in the life style from a pre-industrial society to an industrial one, and more recently due
to the effect of globalization and the free market
economy. In addition, many new commercial en-
terprises have created new job opportunities in the
cities.

For the first time in the history of human civilization, almost half of the global population was
residing in cities by the end of 2008. In cities of the
developed world, urbanization traditionally manifests as a concentration of human activities and settlements around the downtown area, the so-called
in-migration. Recent developments, such as inner-
city redevelopment schemes, imply that new im-
migrants to cities do not necessarily settle in the
centre of the city. In some developed regions, however, the reverse process has also occurred, with
cities losing populations to rural areas. This is particularly true for affluent families and has become
possible because of improved communication network and also due to factors such as rising crime in
poor urban environments. When a residential area
expands outwards and new areas of population
concentration form outside the downtown areas,
a networked poly-centric concentration emerges.
Los Angeles in the United States and Delhi in India
are two examples of this type of urbanization. A
common problem often associated with urbanization is settlement of rural migrants in shanty towns
under conditions of extreme poverty, lacking even

Modern Hydrology and Sustainable Water Development, 1st edition. © S.K. Gupta
Published 2011 by Blackwell Publishing Ltd.

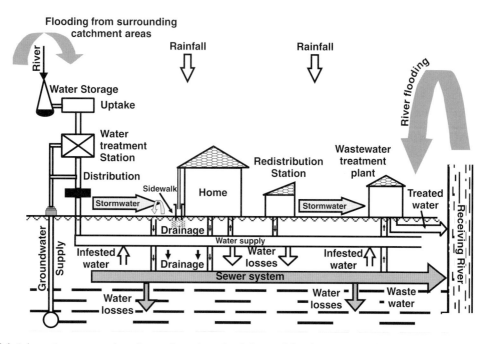

Fig. 10.1 Schematic representation of water formation, circulation, and distribution in an urban area. Redrawn from http://echo.epfl.ch/VICAIRE/mod_1b/ chapt_10/ main.htm.

basic amenities such as a water supply and sanitation. Urbanization in India and other developing countries is taking place at a faster rate than in the rest of the world. According to the UN report 'State of the World Population 2007', about 41% of India's population will be living in urban areas by 2030, compared to about 28% now. According to this report, over 90% of slum-dwellers live in developing countries, with China and India together accounting for 37% of them.

Urbanization reduces the cumulative volume of water storage in soils and vegetation, increases the fraction of rainfall that becomes surface runoff during storms, creates structures that accelerate the movement of runoff to streams, reduces evaporation, and increases the runoff yield of the watershed. Protecting water resources and water quality of streams in newly urbanized basins necessitates comprehensive implementation of structural as well as non-structural mitigation measures that are not limited to just retention of natural vegetation.

The obvious definition of *urban hydrology* is the study of hydrologic processes occurring within the urban environment, where a substantial part

of the area consists of nearly impervious surfaces and artificial land relief as a result of urban developments. However, further consideration of formation, circulation, and distribution of water in an urban area (Fig. 10.1) clearly reveals the inadequacy of this simplistic concept.

In urban areas, natural drainage systems are modified and supplemented by a sewerage network. The effects of flooding are mitigated by creating engineering structures, such as dams or storage ponds. In the initial stages of urban development, septic tanks are employed for disposal of domestic waste waters. As the urban area grows, sewerage systems are installed to divert sewage to treatment plants and the treated effluent is returned to local water courses or, in coastal locations, to the ocean. In the initial stages of development, water supplies are drawn from local surface and groundwater sources to minimize the cost. However, with increase in population and consequently the rise in demand for water, additional supplies can only be obtained from remote locations. Both waste water disposal and water supply, therefore, extend the influence of the urban area well beyond its immediate geographical boundaries. Most urban areas and their surroundings are not self-sufficient in terms of water availability and they can develop only within the framework of a well-organized water supply and flood prevention/mitigation systems. The bigger their size, the greater is their vulnerability and the need for integrated water management, therefore, becomes imperative.

The process of urbanization may thus be seen to create three major hydrologic problems: (i) provision of adequate water resources for urban areas, both in terms of quantity as well as quality; (ii) prevention of flooding within urban areas; and (iii) disposal of waterborne wastes from urban areas without impairing the quality of local water courses.

Of these three problems, water supply forms part of the wider subject of water resources development. Often sources of water for cities and heavily industrialized areas are from rivers and/or ground water that are generally not found in adequate quantities within urban areas and often water has to be imported from the adjoining areas to meet the demand. Although such water sources are not included within the scope of urban hydrology, their role cannot be overlooked when dealing with urban hydrology. The other two urban hydrologic problems, that is, prevention of flooding within urban areas and disposal of waterborne wastes from urban areas, are more specific to individual urban areas and some common links are considered in the following.

10.1 Water balance in urban areas

A major source of formation and circulation of natural water in an urban area is the local precipitation and storm water generated either locally or in the surrounding rural and suburban areas and flowing into the urban area. An additional source is flood water in rivers flowing through the area and storm water runoff that collects in lakes or ponds that exist inside the urban area or in its close proximity. Storm water generated from impervious surfaces (to which the snowmelt is added where applicable) is drained through pipes and canals that constitute the urban sewerage network. Part of the

precipitation falling on pervious areas (surrounding rural areas, urban yards, parks, stadiums, and sports fields) is absorbed by the soil and eventually recharges the groundwater and the remaining part either joins the sewerage network via drains or is evaporated.

Rivers and groundwater (especially in deep aquifers) are sources of water supply for meeting demands of domestic as well as industrial consumption and also for energy generation. Water is conveyed from the uptake points through canals and pipes to the water treatment plant for purification to meet the required quality standards for various end uses. The required quantity of surface water is met by reservoir regulation systems in the framework of integrated water engineering structures. In water supply from groundwater, the capacity may be enhanced by artificial recharging of water diverted from rivers, provided that the river water quality is good and does not cause groundwater pollution. From the treatment plant, water is conveyed to domestic and industrial users as well as for public uses (road washing, recreational parks irrigation, etc.).

The components that constitute the output term in the water balance equation for an urban area are the water: (i) carried through the sewerage network; and (ii) lost through evapotranspiration. Sewerage systems usually mix the waste water with storm water but in some cases there are two separate systems for conveying the waste water and the storm water. Schematic representation of the water balance of both natural water as well as waste water generated in a typical urban area is given in Fig. 10.2.

10.1.1 Influence of urbanization on the formation and circulation of water originating from rainfall and snowmelt

Modification of thermal and radiation balance in the atmosphere due to absorption of the long-wave radiation emitted by the ground is likely to result in an increase in the intensity of rainfall spells. This happens because the urban land cover is darker than the open spaces in rural areas. Furthermore, aerosols and trace gases generated by polluting

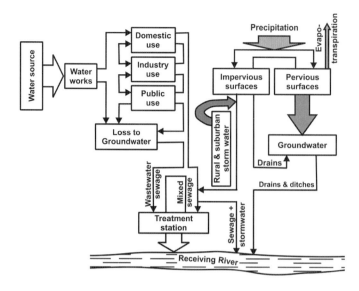

Fig. 10.2 Schematic showing water balance of natural and generated water supply and waste water. Redrawn from http://echo.epfl.ch/VICAIRE/mod_1b/chapt_10/ main.htm.

sources in the urban air also contribute to modification of the local micro-climate that may affect the rainfall over the area, particularly the number and severity of extreme events. Finally, the 'cupola effect' of the city leads to the increase of the convective motion of the air, which results in more intense cumulonimbus-type of cloud formation causing higher rainfall in urban zones as compared to nearby rural areas.

As already mentioned, due to the large fraction of land surface being increasingly covered with impervious materials (concrete, asphalt, stone-paved roads, roofs, etc.), the infiltration of rain water into the soil is reduced. This leads to an increase of the surface runoff as well as the peak discharge rate in urban areas. There is also a substantial increase in the velocity of the surface runoff as well as of the wastewater flow through the sewerage system, as compared to the velocity of overland flow in natural catchments or the flow in river channels. Thus the duration of flood wave propagation is shorter in urban areas compared to that in rural basins, leading to higher peak discharge in urban catchments.

The runoff generation process is impeded mainly due to water accumulation held in detention storages in surface depressions and also due to infiltration in pervious/semi-pervious zones of the urban area. The element that has a dominant effect on runoff is the infiltration that may be computed by

use of the well-known empirical equations of infiltration of the Horton, Philip, or Green-Ampt type (see Section 2.5.3.3). Another method for calculation of infiltration is by using runoff coefficients that express the ratio between the effective (or runoff equivalent) rainfall and the input rainfall, depending on the type of land cover and land use in the area. In practical situations, the runoff coefficient is taken to be constant during the entire duration of a rainstorm. The runoff coefficients for different types of land cover are given in Table 10.1 and as functions of land use in Table 10.2.

Table 10.1 Runoff coefficients vis a vis the nature of land cover. Adapted from http://echo.epfl.ch/VICAIRE/ mod_1b/chapt_10/main.htm.

Land cover type	Runoff coefficient
Dense pavement, asphalt or concrete	0.70–0.95
Ordinary pavement or brick	0.70–0.85
Roofs of buildings	0.75–0.95
Lawn on sandy terrain with slope <2%	0.05–0.10
Lawn on sandy terrain with slope >7%	0.15–0.17
Lawn on average and hard texture terrain with slope <2%	0.13–0.17
Lawn on average and hard texture terrain with slope >7%	0.25–0.35

Table 10.2 Runoff coefficient as function of land use in an urban area. Adapted from http://echo.epfl.ch/VICAIRE/mod_1b/chapt_10/main.htm.

Use in the urban area		Runoff coefficient
Commercial	Downtown	0.70–0.95
	Other urban areas	0.50–0.70
Residential	Buildings with yards	0.30–0.35
	Dwellings	0.50–0.70
	Suburban area	0.25–0.40
Industrial zones (average)		0.50–0.80
Heavily industrialized zones		0.60–0.90
Parks		0.10–0.25
Sport fields or arenas, recreational areas		0.20–0.35
Unused terrain		0.10–0.30

In the case when runoff is computed, starting with the rainfall averaged over the entire urban area encompassing all types of land-cover zones or different land uses, the runoff coefficient is considered as an average value $\bar{\alpha}$ over the entire area F:

$$\bar{\alpha} = \frac{1}{F} \sum_{i=1}^{n} \alpha_i \cdot f_i \qquad (10.1)$$

where a given value of f_i represents an area covered by a specific type of urban land use or land cover, with its respective runoff coefficient α_i.

Effective rainfall, h_{eff}, is computed as:

$$h_{eff} = \bar{\alpha} \cdot P_n \qquad (10.2)$$

where the net rainfall $P_n = P - INT$; P being the average rainfall over the entire urban area and INT (in mm) the amount of rainfall interception. The rainfall interception in an urban area plays a significant role only in the case of dense parks and trees along the streets. The equation given by Horton (1919) for accounting for the interception is:

$$INT = a + b \cdot P^n \qquad (10.3)$$

The parameters of the Horton equation for different land cover and trees are given in Table 10.3.

Furthermore, the effective rainfall is transformed into a runoff hydrograph by the use of an integration procedure. One of the commonly used integration procedures that is applied in urban areas is the 'unit hydrograph method' (see Section 2.7.1.3).

10.1.2 Hydraulic effects of changes in stream channels and floodplains

Development along stream channels and floodplains can modify the water-carrying capacity of a stream channel and can increase the height of the water surface (known as stage) for a given discharge passing through the channel. In particular, structures that encroach upon the floodplain, such as bridges, can increase upstream flooding by narrowing the width of the channel, which increases the resistance of the channel to flow. As a result, water flows at a higher stage as it flows past the obstruction, creating a backwater that inundates a larger area of the flood plain upstream.

Table 10.3 Parameters of the Horton equation (Eqn 10.3) for estimation of interception from a tree covered landscape.

Land cover	a [mm]	b	n
Orchards	1.016	0.18	1.0
Ash trees	0.508	0.18	1.0
Beech trees	1.016	0.18	1.0
Willow trees	1.27	0.18	1.0
Oak trees	1.016	0.18	1.0
Maple trees	0.508	0.40	1.0
Resinous trees (pine-tree, fir-tree, etc.)	1.27	0.2	0.5

Source: (VICAIRE, 2003).

Sediment and debris carried by flood waters deposited in the riverbed can further constrict a channel and increase flooding in the floodplain of the river. This hazard is greatest upstream of culverts, bridges, or other locations in the river channel where debris transported during high floods collects. For example, small stream channels can be filled with sediment or become choked with debris because of undersized culverts. This creates a closed basin with no outlet for the passage of flood waters. Although river channels can be trained to rapidly convey flood water and debris downstream, local benefits of this approach must be balanced against the possibility of increased flooding downstream.

Erosion caused by urban streams is yet another consequence of urbanization. Frequent flooding in urban streams increases both channel bed- and bank-erosion. Where channels are straightened and vegetation is removed from the catchment area, stream flow velocities increase, resulting in higher sediment loads carried by the stream. In many urban areas, stream-bank erosion, caused by higher stream flow velocities coupled with removal of vegetation from the banks, represents a serious threat to roads, bridges, and other structures that is difficult to control, even by strengthening the stream banks.

10.1.2.1 Reducing flood hazard in urban areas

There are many approaches for reducing flood hazard in urban basins under development. Areas that are identified as flood-prone can be used for construction of parks and playgrounds, as these can withstand occasional flooding. Buildings and bridges can be elevated, protected with floodwalls and levees, or designed to withstand temporary inundation. Drainage systems can be augmented to increase their capacity for detaining and conveying high stream flows, for example, by using rooftops and parking lots to store water. Techniques that promote infiltration and storage of water in the soil profile, such as infiltration trenches, permeable pavements, soil amendments, and reducing the proportion of impermeable surfaces can also be incorporated into new as well as existing residential and commercial complexes to reduce storm

runoff from these areas. As an example, wet-season runoff from a neighbourhood in Seattle, Washington, was reduced by 98% by reducing the width of the street and incorporating vegetated swales and native plants in the street right-of-way, that is, an area dedicated to public streets, sidewalks, and utilities (Konrad 2003).

10.2 Disposal of waterborne wastes

In addition to natural precipitation and the runoff generated, urban hydrology is mostly concerned with disposal of waterborne waste generated within urban and peri-urban areas, which in most cases gets mixed with the storm water. The main vehicle for wastewater collection and disposal is the sewerage system. Two types of sewerage systems in use are: (i) mix type, where the waste water is mixed with storm water; and (ii) separate type, where there are two separate pathways of conveying generated waste and storm waters towards the receiving river or any other water body.

Historically municipalities, consulting engineers, and individuals had the option of choosing between centralized or decentralized wastewater management and chose from a variety of collection and disposal technologies to implement their management strategy. The final choice was based on a combination of cost/benefit consideration, urban development patterns, its scientific basis, tradition, socio-cultural attitudes of the concerned populations, prevailing public opinion on sanitation, political environment of the region, and several other factors. Two main types of wastewater management strategies can be identified: (i) centralized, where all waste water generated is collected and conveyed to a central location for treatment and/or disposal; and (ii) decentralized, where the waste water is treated to the primary level or disposed off on-site or near the source itself.

10.2.1 Municipal waste water

Municipal waste water is a combination of liquid or water-borne wastes originating in the sanitary systems of households, commercial/industrial units, and institutions, in addition to any groundwater, surface water, and storm water that may be generated in the area and discharged into the sewerage system (Fig. 10.1 and Fig. 10.2). Untreated waste water generally contains high levels of organic material, variety of pathogenic microorganisms, as well as nutrients and toxic compounds. It thus entails environmental and health hazards and consequently must immediately be conveyed away from its source locations and treated appropriately before its final disposal. The ultimate goal of waste water management is protection of the environment in a manner commensurate with public health and socio-economic concerns.

Wastewater quality may be defined by its physical, chemical, and biological characteristics. Physical parameters include colour, odour, temperature, and turbidity. Included in this category are insoluble substances such as solids, oil, and grease.

Solids may be further subdivided into suspended and dissolved solids, as well as organic (volatile) and inorganic mineral fractions. Chemical parameters associated with the organic content of waste water include biochemical oxygen demand (*BOD*), chemical oxygen demand (*COD*), total organic carbon (*TOC*), and total oxygen demand (*TOD*). Inorganic chemical parameters include salinity, hardness, *pH*, acidity, and alkalinity, as well as ionized metals such as iron and manganese, and anionic species such as chlorides, sulphates, sulphides, nitrates, and phosphates. Bacteriological parameters include coliforms, faecal coliforms, specific pathogens, and viruses. Both types of constituents and their concentrations vary with time and local conditions. Typical ranges of concentration of various constituents in untreated domestic waste water are shown in Table 10.4. Sewage waters are

Table 10.4 Typical composition of untreated domestic waste water. Adapted from Metcalf and Eddy (1991).

Contaminants	Unit	Concentration		
		Weak	Medium	Strong
Total solids (TS)	$mg\,l^{-1}$	350	720	1200
Total dissolved solids (TDS)	$mg\,l^{-1}$	250	500	850
Fixed	$mg\,l^{-1}$	145	300	525
Volatile	$mg\,l^{-1}$	105	200	325
Suspended solids	$mg\,l^{-1}$	100	220	350
Fixed	$mg\,l^{-1}$	20	55	75
Volatile	$mg\,l^{-1}$	80	165	275
Settleable solids	$mg\,l^{-1}$	5	10	20
BOD_5, 20°C	$mg\,l^{-1}$	110	220	400
TOC	$mg\,l^{-1}$	80	160	290
COD	$mg\,l^{-1}$	250	500	1000
Nitrogen (as total N)	$mg\,l^{-1}$	20	40	85
Organic	$mg\,l^{-1}$	8	15	35
Free ammonia	$mg\,l^{-1}$	12	125	50
Nitrites	$mg\,l^{-1}$	0	0	0
Nitrates	$mg\,l^{-1}$	0	0	0
Phosphorus (as total P)	$mg\,l^{-1}$	4	8	15
Organic	$mg\,l^{-1}$	1	3	5
Inorganic	$mg\,l^{-1}$	3	5	10
Chlorides	$mg\,l^{-1}$	30	50	100
Sulphate	$mg\,l^{-1}$	20	30	50
Alkalinity (as $CaCO_3$)	$mg\,l^{-1}$	50	100	200
Grease	$mg\,l^{-1}$	50	100	150
Total coliforms	No. per/100 *ml*	10^{-6}–10^{-7}	10^{-7}–10^{-8}	10^{-7}–10^{-9}
Volatile organic compounds	$\mu g\,l^{-1}$	<100	100–400	>400

Table 10.5 Important contaminants in terms of their potential effects on receiving waters and their treatment concerns. Adapted from Metcalf and Eddy (1991).

Contaminant	Reason for importance
Suspended solids (SS)	Can lead to development of sludge deposits and anaerobic conditions when untreated waste water is discharged into an aquatic environment
Biodegradable organics	These are made up mainly of proteins, carbohydrates, and fats. They are commonly measured in terms of BOD and COD. If discharged into inland rivers, streams, or lakes, their biological stabilization can deplete natural oxygen levels and cause septic conditions that are detrimental to aquatic species
Pathogenic organisms	These can cause infectious diseases
Priority pollutants, including organic and inorganic compounds	These may be highly toxic, carcinogenic, mutagenic, or teratogenic
Refractory organics including surfactants, phenols and agricultural pesticides	These tend to interfere with conventional wastewater treatment
Heavy metals	These are usually contributed by commercial and industrial activities and must be removed for reuse of waste water as they tend to bio-concentrate in the food chain
Dissolved inorganic constituents such as calcium, sodium and sulphate	May be undesirable for many wastewater reuse applications at high levels of concentration

classified as strong, medium, or weak, depending on the level of their contaminant concentration.

The effects of discharging untreated waste water into the environment are manifold and depend on the type and concentration of various pollutants. Important contaminants in terms of their potential effects on receiving waters and treatment concerns are outlined in Table 10.5.

10.2.2 Wastewater treatment technologies

Physical, chemical, and biological methods are used to remove contaminants from waste water. In order to achieve different levels of contaminant removal, individual wastewater treatment procedures are combined into a variety of systems, classified as primary, secondary, and tertiary treatments (Fig. 10.3). More rigorous treatment of waste water includes removal of specific contaminants as well as removal and control of nutrients. Natural systems are also used for treatment of waste water in land-based applications. Sludge resulting from wastewater treatment operations is treated by various meth-

ods in order to reduce its water and organic content and make it suitable for final disposal and/or reuse. This section briefly describes various conventional and advanced technologies that are in current use and explains how these are applied for effective treatment of municipal waste water.

Preliminary treatment prepares influent waste water for further treatment by reducing or eliminating unfavourable wastewater characteristics such as the presence of large solids and rags, abrasive grit, odours, and sometimes unacceptably high hydraulic or organic loading. Preliminary treatment processes consist of physical unit operations, namely screening and comminution for removal of debris and rags, grit removal for elimination of coarse suspended matter, and flotation for removal of oil and grease. Other preliminary treatment operations include flow equalization, septage handling, and odour control methods.

Primary treatment involves partial removal of suspended solids and organic matter from waste water by means of physical methods such as screening and sedimentation. Pre-aeration or mechanical

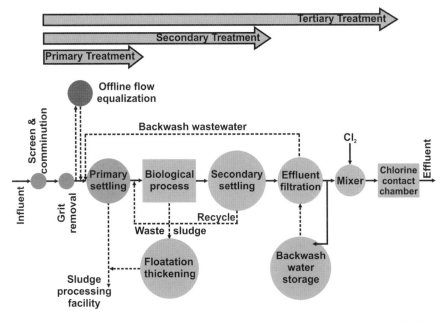

Fig. 10.3 Flow diagram showing various treatment levels in conventional wastewater treatment plants. Preliminary treatment operations include screening and comminution, grit removal, and flow equalization, and to prepare the influent wastewater for further treatment. Redrawn from (UNESCWA, 2003).

flocculation with chemical additives can be used to enhance the effect of primary treatment, which is a precursor to the secondary treatment. The effluent from primary treatment contains a good deal of organic matter and is characterized by relatively high *BOD*.

The purpose of *secondary treatment* is removal of soluble and colloidal organics and suspended solids that are not taken care of in the primary treatment. This is done typically through biological processes, namely treatment by activated sludge, fixed-film reactors, or lagoon systems and sedimentation.

Tertiary treatment is a step beyond the conventional secondary level treatment in removing significant amounts of nitrogen, phosphorus, heavy metals, biodegradable organics, bacteria, and viruses. In addition to biological nutrient removal processes, unit operations frequently used for this purpose include chemical coagulation, flocculation, and sedimentation, followed by filtration and passing the sewage through an activated carbon bed. Less commonly used processes include ion exchange and reverse osmosis for removal

of specific ion(s) or for reduction of dissolved solids.

As mentioned above and shown in Fig. 10.3, wastewater treatment involves a combination of several physical, chemical, and biological processes. Various unit operations included in each category are listed in Fig. 10.4.

10.2.2.1 Physical unit operations

10.2.2.1.1 Screening

One of the oldest treatment methods, such as *screening* of waste water, removes gross pollutants from the waste stream to protect downstream treatment plant equipment from damage, avoid interference with plant operations, and prevent undesirable floating material from entering the primary settling tanks. Screening devices may consist of parallel bars, rods or wires, grating, wire mesh, or perforated plates, to intercept large floating or suspended material. The openings may be of any shape, but are generally circular or rectangular. The material retained in the waste water, after manual

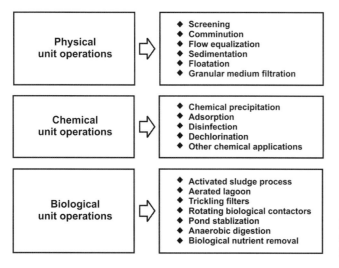

Fig. 10.4 Various unit operations and processes in a conventional wastewater treatment system. Redrawn from (UNESCWA, 2003).

or mechanical cleaning of bar racks and screens, is referred to as 'screenings' and is disposed off either by burial or by incineration or returned into the waste flow after grinding (Metcalf and Eddy 1991).

10.2.2.1.2 Comminution

Comminution is a process involving pulverization of large floating material in the wastewater flow where handling of the screenings is impractical, generally between the grit chamber and the primary settling tanks. For this, comminutors with either rotating or oscillating cutters are installed. Rotating-cutter comminutors either employ a separate stationary screen alongside the cutters or a combined screen and cutter rotating together. A different type of comminutor, known as a barminutor, involves a combination of a bar screen and rotating cutters (Liu and Lipták 2000). Comminution reduces odours, fly menace, and unsightliness.

10.2.2.1.3 Flow equalization

Flow equalization is a technique employed to improve the effectiveness of secondary and advanced wastewater treatment processes by levelling out operating parameters such as flow, pollutant levels, and temperature over a period of time. Fluctuations in flow are damped out until a near-constant flow rate is achieved, minimizing the downstream effects of these parameters.

10.2.2.1.4 Sedimentation

Sedimentation is a widely used unit operation in wastewater treatment and involves gravitational settling of heavy particles suspended in a mixture. This process is used for removal of grit and particulate matter in the primary settling basins, biological flocs in the activated sludge settling basins, and chemical agglomerates when a chemical coagulation process is used.

10.2.2.1.5 Flotation

Flotation is a unit operation employed to remove solid or liquid particles from the liquid phase by introducing a gas, usually air bubbles. The gas bubbles either adhere to the liquid or are trapped in the particulate structure of the suspended solids, raising the buoyant force of the combined particle–gas bubbles. Particles that have a higher density than the liquid can thus be made to rise and float on the surface of the liquid. In wastewater treatment, flotation is used mainly to remove suspended matter and to concentrate biological sludge. The main advantage of flotation over sedimentation is that very small or light particles can be removed more or less completely and in a shorter time. Once particles have been raised to the surface by flotation, they can be skimmed out. Flotation, as currently practised in municipal wastewater treatment, uses air exclusively as the floating agent. Furthermore, various chemical additives can be introduced to accelerate the removal process (Metcalf and Eddy 1991).

10.2.2.1.6 Granular medium filtration

Granular medium filtration of effluents as part of wastewater treatment processes is a relatively recent practice but has come to be widely used for supplemental removal of suspended solids from wastewater effluents of biological and chemical treatment processes, in addition to the removal of chemically precipitated phosphorus. The complete filtration process comprises two phases: filtration, and cleaning or backwashing. The waste water to be filtered is passed through a filter bed consisting of granular material (sand, anthracite, and/or garnet), with or without adding any chemicals. Within the filter bed, suspended solids contained in the waste water are removed by means of a complex process involving one or more removal mechanisms such as straining, interception, impaction, sedimentation, flocculation, and adsorption. The phenomena that occur during the filtration phase are basically the same for all types of filters used in wastewater filtration. The cleaning/backwashing phase, however, differs depending on whether the filtration process is continuous or semi-continuous. In semi-continuous filtration, filtering and cleaning operations occur sequentially, whereas in continuous filtration, filtering and cleaning operations occur simultaneously (Metcalf and Eddy 1991).

10.2.2.2 Chemical unit processes

Chemical processes employed in wastewater treatment are always used in conjunction with physical unit operations and biological processes. In general, chemical unit processes have an inherent disadvantage compared to physical operations in that they are additive processes. This means that there is usually a net increase in the dissolved constituents of waste water. This can become a significant factor if the waste water is to be reused.

10.2.2.2.1 Chemical coagulation

Chemical coagulation of raw waste water before sedimentation promotes flocculation of finely divided solids into more readily settleable flocs, thereby enhancing removal efficiency of suspended solids, BOD_5, and phosphorus, as compared to the basic process of sedimentation without coagulation. The degree of clarification of waste water achieved depends on the amount of chemicals used and the precision with which the process is controlled (Metcalf and Eddy 1991). Selection of the coagulant is based on considerations of its performance, reliability, and cost. Some commonly used chemical coagulants in wastewater treatment include alum ($Al_2(SO_4)_3$ 14.3 H_2O), ferric chloride ($FeCl_3$ $6H_2O$), ferric sulphate ($Fe_2(SO_4)_3$), ferrous sulphate ($FeSO_4$ $7H_2O$), and lime ($Ca(OH)_2$). Sometimes organic polyelectrolytes are also used as flocculation agents. Removal of suspended solids through chemical treatment involves a series of three unit operations: rapid mixing, flocculation, and settling. First, the chemical is added and completely dispersed in the waste water by rapid mixing for 20–30 seconds in a basin employing a turbine mixer. Coagulated particles are then brought together by flocculation by mechanically inducing velocity gradients within the liquid. Flocculation takes 15–30 minutes in a basin containing turbine or paddle-type mixers. The advantages of coagulation include greater removal efficiency, feasibility of using higher overflow rates, and consistent performance. On the other hand, coagulation results in a large mass of primary sludge that is often difficult to thicken and dewater. It also entails higher operational costs and demands care on the part of the plant operator.

10.2.2.2.2 Adsorption

Adsorption is the process of collecting dissolved substances in a solution on a suitable interface (see Section 6.1.10). In wastewater treatment, adsorption with activated carbon – a solid interface – usually follows normal biological treatment and is aimed at removing a part of the remaining dissolved organic matter. Particulate matter present in the waste water may also be removed in this process. Activated carbon is produced by heating charcoal to a high temperature and then activating it by exposure to an oxidizing gas at high temperatures ($600–1200°C$). The gas develops a porous structure in the charcoal and thus creates a large internal surface area. The activated charcoal can then be separated into various sizes with different adsorption capacities. The two most common types of activated carbon are granular activated carbon (GAC) with a grain size of more than $0.1\ mm$, and powdered activated carbon (PAC) with a grain size of less than 200 mesh (opening $0.0686\ mm$). A fixed-bed column is often used to bring the waste water in contact with GAC. The waste water is applied at the top of the column and withdrawn from the bottom with carbon held in its place. Backwashing and surface washing are applied to limit build-up of head loss. Expanded-bed and moving-bed carbon contactors have also been developed to overcome the problem of head loss build-up. In the expanded-bed system, the influent is introduced at the bottom of the column and is allowed to expand. In the moving-bed system, the spent activated carbon is continuously replaced with fresh carbon. Spent granular carbon can be regenerated by removal of the adsorbed organic matter from its surface through oxidation in a furnace. The capacity of the regenerated carbon is slightly inferior to that of the virgin carbon. Wastewater treatment using PAC involves addition of the powder directly to the biologically treated effluent or after the physico-chemical treatment process. PAC is usually added to waste water in a contacting basin for a certain length of time. It is then allowed to settle to the bottom of the tank and removed. Removal of the powdered carbon may be facilitated by addition of poly-electrolyte coagulants or filtration through granular-medium filters. A major problem with the use of powdered activated carbon is that the methodology of its regeneration is not well defined.

10.2.2.2.3 Disinfection

Disinfection refers to the selective destruction of disease-causing micro-organisms. This process is of importance in wastewater treatment owing to the nature of waste water, which harbours a number of human enteric organisms that are associated with various waterborne diseases. Commonly used means of disinfection include: (i) physical agents such as heat and light; (ii) mechanical means such as screening, sedimentation, filtration, etc.; (iii) radiation, mainly gamma rays; (iv) chemical agents including chlorine and its compounds, bromine, iodine, ozone, phenol and phenolic compounds, alcohols, heavy metals, dyes, soaps and synthetic detergents, quaternary ammonium compounds, hydrogen peroxide, and various alkalis and acids. The commonly used chemical disinfectants are oxidizing chemicals, and of these, chlorine is the most widely used (Qasim 1999). Disinfectants act through one or more of a number of mechanisms including damaging the cell wall, altering cell permeability as well as the colloidal nature of the protoplasm, and inhibiting enzyme activity of micro-organisms. In applying disinfecting agents, several factors need to be considered, that is, contact time, concentration and type of chemical agent, intensity and nature of physical agent, wastewater temperature, number of micro-organisms present, and nature of the suspending liquid (Qasim 1999).

10.2.2.2.4 Dechlorination

Dechlorination is removal of free and total combined chlorine residues from chlorinated wastewater effluent before its reuse or discharge to receiving waters. Chlorine compounds react with many organic compounds in the effluent to produce undesired toxic compounds that cause long-term adverse impacts on the water environment and potentially toxic effects on aquatic micro-organisms. Dechlorination may be brought about by use of activated carbon or by addition of a reducing agent such as sulphur dioxide (SO_2), sodium sulphite (Na_2SO_3), or sodium metabisulphite ($Na_2S_2O_5$). It is important to note that dechlorination does not remove toxic by-products that are already produced during the process of chlorination (Qasim 1999).

In addition to the chemical processes described above, various other methods are occasionally applied in wastewater treatment and disposal (ESCWA 2003).

10.2.2.3 Biological unit processes

Biological unit processes are employed to convert the finely dispersed as well as dissolved organic matter in waste water into flocculent settleable organic and inorganic solids. In these processes, micro-organisms, particularly bacteria, convert the colloidal and dissolved carbonaceous organic matter into various gases and into cell tissues that are removed in sedimentation tanks. Biological processes are used generally in conjunction with physical and chemical processes, the main objective being to reduce the organic content (measured as *BOD*, *TOC*, or *COD*) and nutrient content (notably nitrogen and phosphorus) of waste water. Five major biological processes used in wastewater treatment may be classified as: (a) aerobic processes; (b) anoxic processes; (c) anaerobic processes; (d) combined processes; and (e) pond processes. These processes are further subdivided, depending on whether the treatment takes place in a suspended-growth system, an attached-growth system, or a combination of both. Some of the commonly used biological systems are trickling filters, activated sludge process, aerated lagoons, rotating biological contactors, and stabilization ponds.

10.2.2.3.1 Activated sludge process

The *activated sludge process* is an aerobic, continuous-flow system containing a mass of activated micro-organisms that are capable of stabilizing organic matter. The process consists of delivering clarified waste water after primary settlement into an aeration basin where it is mixed with an active mass of micro-organisms, mainly bacteria and protozoa, which aerobically degrade organic matter into carbon dioxide, water, new cells, and other end products. The bacteria involved in activated sludge systems are primarily Gram-negative species, including carbon oxidizers, nitrogen oxidizers, floc formers as well as non-floc formers, aerobes, and facultative anaerobes. The protozoa, for their part, include flagellates, amoebas, and ciliates. An aerobic environment is maintained in the basin by means of diffused or mechanical aeration, which also serves to keep the contents of the reactor (or mixed liquor) well mixed. After a specific retention time, the mixed liquor passes into the secondary clarifier where the sludge is allowed to settle and a clarified effluent is produced for disposal. The process recycles a portion of the settled sludge back to the aeration basin to maintain the required activated sludge concentration (Fig. 10.5). In this process, a portion of the settled sludge is deliberately discarded to maintain the required solids retention time (SRT) for effective removal of organics. Control of the activated-sludge process is

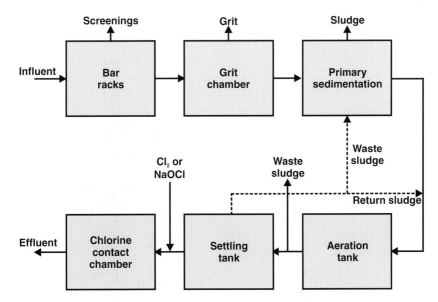

Fig. 10.5 Typical flow diagram of an activated-sludge process. Redrawn from (UNESCWA, 2003).

important to maintain a high treatment performance level under a wide range of operating conditions. Important factors in process control are: (a) maintenance of dissolved oxygen levels in the aeration tanks; (b) regulation of the amount of returned activated sludge; and (c) control of the waste activated sludge. The main operational problem encountered in such a system is sludge bulking, which can be caused by absence of phosphorus, nitrogen, and trace elements and large fluctuations in *pH*, temperature, and dissolved oxygen (*DO*). Bulky sludge has poor settleability and compatibility due to the excessive growth of filamentous micro-organisms. This problem can be controlled by chlorination of the returned sludge (Liu and Lipták 2000; Metcalf and Eddy 1991).

10.2.2.3.2 Aerated lagoons

An *aerated lagoon* is a basin of depth between 1 and 4 *m*, in which waste water is treated either on a flow-through basis or by recycling of solids. The microbiology involved in this process is similar to that of the activated-sludge process. However, differences arise because the large surface area of a lagoon may cause more temperature-dependent effects than are ordinarily encountered in conventional activated-sludge processes. Waste water is oxygenated by surface exposure, turbine aeration, or diffused aeration. The turbulence created by aeration is used to maintain contents of the basin in the state of suspension. Depending on the retention time, aerated lagoon effluents contain approximately one-third to one-half the incoming *BOD* value in the form of cellular mass. Most of the solids must be removed in a settling basin before the effluent is finally discharged (Fig. 10.6).

10.2.2.3.3 Trickling filters

A *trickling filter* is the most commonly encountered aerobic attached-growth biological treatment process used for removal of organic matter from waste water. It consists of a bed of highly permeable medium, which attaches micro-organisms, forming a biological slime layer through which waste water is made to percolate. The filter medium usually consists of rock or plastic packing material. The organic material present in the waste water is degraded by adsorption onto the biological slime layer. In the outer portion of this layer, waste water is degraded by aerobic micro-organisms. As micro-organisms grow, the thickness of the slime layer increases and the oxygen is depleted before it has penetrated the full depth of the slime layer. An anaerobic environment is thus established near the surface of the filter medium. As the slime layer increases in thickness, the organic matter is degraded before it reaches the micro-organisms present near the surface of the medium. Deprived of their external organic source of nourishment, the micro-organisms die and are washed off by the flowing liquid. A new slime layer grows in its place. This phenomenon is referred to as 'sloughing' (Liu and Lipták 2000; Metcalf and Eddy 1991).

After passing through the filter, the treated liquid is collected in an under-drain system, together with any biological solids that have detached from the medium (Fig. 10.7). The collected liquid then passes to a settling tank where solids are separated from the treated waste water. A portion of the liquid collected in the under-drain system or the settled effluent is recycled to dilute the strength of the incoming waste water and to maintain the biological slime layer in a moist condition (Fig. 10.8).

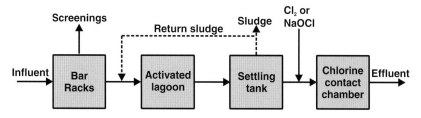

Fig. 10.6 Typical flow diagram of an aerated lagoon. Redrawn from (UNESCWA, 2003).

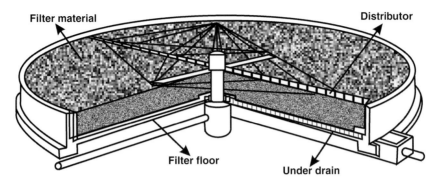

Fig. 10.7 Cutaway view of a trickling filter. Redrawn from (UNESCWA, 2003).

10.2.2.3.4 Rotating biological contractors

A *rotating biological contractor* (RBC) is an attached-growth biological process that consists of one or more basins in which large closely spaced circular disks mounted on horizontal shafts rotate slowly through waste water (Fig. 10.9). The disks, made of high density polystyrene or polyvinyl chloride (PVC), are partially submerged in the waste water so that a bacterial slime layer forms on their wetted surfaces. As the disks rotate, the bacteria present in the waste water are exposed alternately to waste water from which they adsorb organic matter and to air from which they absorb oxygen. Their rotary movement also allows excess bacte-

ria to be removed from surfaces of the disks, thus maintaining a suspension of sloughed biological solids. A final clarifier is needed to remove sloughed solids. Organic matter is degraded by means of mechanisms similar to those operating in trickling filters. Partially submerged RBCs are used for carbonaceous *BOD* removal, combined carbon oxidation, and nitrification of secondary effluents. Completely submerged RBCs are used for denitrification (Metcalf and Eddy 1991). A typical arrangement of RBCs is shown in Fig. 10.9. In general, RBC systems are divided into a series of independent stages or compartments by means of baffles in a single basin or separate basins arranged in stages. Compartmentalization creates a plug flow pattern,

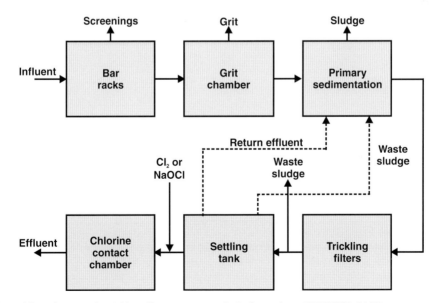

Fig. 10.8 Typical flow diagram of a trickling filter treatment unit. Redrawn from (UNESCWA, 2003).

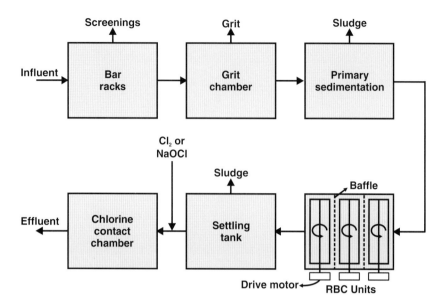

Fig. 10.9 Typical flow diagram of a RBC treatment unit. Redrawn from (UNESCWA, 2003).

increasing the overall removal efficiency of micro-organisms. It also promotes a variety of conditions where different organisms can flourish to varying degrees. As waste water flows through the various compartments, each subsequent stage receives influent with a lower organic content than the previous stage; the system thus enhances removal of the organics.

10.2.2.3.5 Stabilization ponds

A *stabilization pond* is a relatively shallow body of waste water contained in an earthen basin, using a completely mixed biological process without solids being returned. Mixing may be either natural (caused by wind, heat, or fermentation) or induced (due to mechanical or diffused aeration). Stabilization ponds are usually classified on the basis of the nature of the biological activity that takes place in them – as aerobic, anaerobic, or a combination of aerobic-anaerobic processes. Aerobic ponds are used primarily for treatment of soluble organic wastes and effluents from wastewater treatment plants. Aerobic-anaerobic (facultative) ponds are the most common type and are being used to treat domestic waste water and a wide variety of industrial wastes. Anaerobic ponds are particularly effective in bringing about rapid stabilization

of high concentrations of organic wastes. Aerobic and facultative ponds are biologically complex. The bacterial population oxidizes organic matter producing ammonia, carbon dioxide, sulphates, water, and other end products that are subsequently used by algae during sunlight hours to produce oxygen. Bacteria use this oxygen and also the oxygen provided by air flowing over the pond to break down the remaining organic matter. Wastewater retention time ranges between 30 and 120 days. This is a treatment process that is commonly employed in rural areas, because of its low construction and operating costs. Fig. 10.10 presents a typical flow diagram of stabilization ponds (Liu and Lipták 2000).

10.2.2.3.6 Completely mixed anaerobic digestion

Anaerobic digestion involves biological conversion of organic and inorganic matter in the absence of molecular oxygen to a variety of end-products, including methane and carbon dioxide. A host of anaerobic organisms work together to degrade organic sludge and waste in three steps comprising: (i) hydrolysis of high molecular mass compounds; (ii) acidogenesis; and (iii) methanogenesis. The process takes place in an airtight reactor. Sludge is introduced either continuously or intermittently and

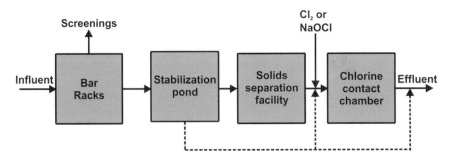

Fig. 10.10 Typical flow diagram of stabilization ponds. Redrawn from (UNESCWA, 2003).

retained in the reactor for varying periods of time. After withdrawal from the reactor, whether continuous or intermittent, the stabilized sludge has lower organic and pathogen contents and is non-putrescent. The two widely used types of anaerobic digesters are standard-rate and high-rate. In the high-rate digestion process, contents of the digester are heated, mixed completely, and retained typically for a period of 15 days or less. In the standard-rate digestion process, contents of the digester are usually not subjected to either heating or mixing and are retained for a period ranging from 30–60 days. A combination of these two basic processes is known as the two-stage process, which is used to separate the digested solids from the supernatant liquor. However, additional digestion and gas production may occur (Metcalf and Eddy 1991). Anaerobic digesters are commonly used for the treatment of sludge and waste water with high organic content. Disadvantages of a system of this kind, as compared to aerobic treatment, stem directly from the slow growth rate of methanogenic bacteria. A slow growth rate necessitates a relatively long retention time in the digester for adequate waste stabilization to occur. An advantage of this type of system is production of methane gas, which can be used as a fuel if produced in sufficient quantities. Furthermore, the system produces a well-stabilized sludge, which can be safely disposed off in a sanitary landfill after drying or dewatering. On the other hand, the fact that a high temperature is required for adequate treatment is a major drawback of this system.

10.2.2.3.7 Biological nutrient removal

Nitrogen and phosphorus are the two main nutrients of concern in waste water, as these may accelerate eutrophication of lakes and reservoirs and stimulate the growth of algae and rooted aquatic plants in shallow streams. A significant concentration of nitrogen may have other undesirable effects, such as depletion of dissolved oxygen in receiving waters, causing toxicity to aquatic life, adverse impact on chlorine disinfection efficiency, creation of public health hazard, and waste water that is less suitable for reuse. Nitrogen and phosphorus can be removed by physical, chemical, and biological methods. Biological removal of these nutrients is described below.

10.2.2.3.8 Nitrification-denitrification

Nitrification is the first step in the removal of nitrogen using a biological process. It is the working of two bacterial genera: (i) *Nitrosomonas*, which oxidizes ammonia to the intermediate product nitrite; and (ii) *Nitrobacter*, which converts nitrite to nitrate. Nitrifying bacteria are sensitive organisms and are extremely susceptible to a wide variety of inhibitors such as high concentrations of ammonia and nitrous acid, low *DO* levels (<1 mg l^{-1}), *pH* outside the optimal range (7.5–8.6), and so on. Nitrification can be achieved through both suspended growth and attached growth processes. In suspended growth processes, nitrification is brought about either in the same reactor that is used for carbonaceous *BOD* removal or in a separate suspended-growth reactor following the conventional activated sludge treatment process. Ammonia is oxidized to nitrate either by air or high-purity oxygen. Similarly, nitrification in an attached growth system may be brought about either in the same attached growth reactor that is used for carbonaceous *BOD* removal or in a separate reactor.

Trickling filters, rotating biological contactors, and packed towers can be used as nitrifying systems.

Denitrification involves removal of nitrogen in the form of nitrate by conversion to nitrogen gas under anoxic conditions. In denitrifying systems, *DO* is a critical parameter. Its presence suppresses the enzymatic action needed for denitrification. The optimal *pH* is in the range 7–8. Denitrification can be achieved through both suspended and attached growth processes. Suspended growth denitrification takes place in a plug flow type of activated sludge system. An external carbon source is usually necessary for micro-organism cell synthesis, since the nitrified effluent is low in carbonaceous matter. Some denitrification systems use the incoming waste water itself for this purpose. A nitrogen gas stripped reactor should precede the denitrification clarifier, because nitrogen gas hinders settling of the mixed liquor. Attached growth denitrification takes place in a column reactor containing stones or one of a number of available synthetic media upon which bacteria grow. Periodic backwashing and an external carbon source are necessary in a system of this kind.

10.2.2.3.9 Phosphorus removal

Phosphorus appears in water as orthophosphate (PO_4^{-3}), polyphosphate (P_2O_7), and organically bound phosphorus. Microbes utilize phosphorus for their cell synthesis and also for their metabolic processes. As a result, 10 to 30% of all influent phosphorus is removed during secondary biological treatment. More phosphorus can be removed if one of the several specially developed biological phosphorus removal processes is used. These processes are based on exposing microbes in an activated-sludge system to alternating anaerobic and aerobic conditions. This causes stress on the micro-organisms, as a result of which their uptake of phosphorus exceeds the normal levels.

10.2.3 Natural treatment systems

Natural systems for wastewater treatment are designed to take advantage of physical, chemical, and biological processes that occur in the natural environment when water, soil, plants, micro-organisms, and the atmosphere interact with each other (Metcalf and Eddy 1991). Natural treatment systems include land treatment, floating aquatic plants, and constructed wetlands. All natural treatment systems are preceded by some form of mechanical pre-treatment for removal of bulk solids. Where sufficient land suitable for the purpose is available, these systems can often be the most cost-effective option in terms of both construction and operation. They are generally well suited for small communities and rural areas (Reed *et al.* 1988).

10.2.3.1 Land treatment

Land treatment involves controlled application of waste water to the land at rates compatible with the natural physical, chemical, and biological processes that occur in the soil at a given location. Three main types of land treatment systems in use are: (i) slow rate (SR); (ii) overflow (OF); and (iii) rapid infiltration (RI) systems.

10.2.3.1.1 Slow rate systems

Slow rate (SR) land application systems are important amongst the various land treatment systems for municipal and industrial waste water. This technology incorporates wastewater treatment, water reuse, utilization of nutrients for growing crops, and wastewater disposal. It involves application of waste water to vegetated land by means of various techniques, including sprinklers or surface techniques such as graded border and furrow irrigation. Water is applied intermittently (every 4–10 days) to maintain aerobic conditions in the soil profile. The applied water is lost by evapotranspiration as well as by percolation vertically downward and also laterally through the soil profile. Any surface runoff is collected and reapplied to the system. Treatment occurs as the waste water percolates through the soil profile (Table 10.6). In most cases, the percolated water either joins the underlying groundwater or it may be intercepted by natural surface waters, or recovered by means of sub-surface drains or by recovery wells (Reed *et al.* 1988).

Table 10.6 Mechanisms of wastewater constituent removal by slow rate (SR) land treatment systems.

Parameter	Removal mechanism
BOD	Soil adsorption and bacterial oxidation
SS	Filtration through the soil
Nitrogen	Crop uptake, denitrification, ammonia volatilization, soil storage
Phosphorus	Chemical immobilization (precipitation and adsorption), plant uptake
Metals	Soil adsorption, precipitation, ion exchange, complexation
Pathogens	Soil filtration, adsorption, desiccation, radiation, predation, exposure to other adverse environmental factors
Trace organics	Photodecomposition, volatilization, sorption, degradation

Source: (UNESCWA, 2003).

SR systems can be classified into two types, Type 1 and Type 2, based on the design objectives. Type 1 systems are designed primarily for wastewater treatment itself rather than for crop production. Accordingly, in systems of this kind, the maximum possible amount of water is applied per unit land area. Type 2 SR systems, in contrast, are designed mainly with a view to reuse water for crop production, and consequently the amount of water applied in a system of this kind is just enough to satisfy the irrigation requirements of the crop being grown. Of all natural treatment systems, SR systems have the highest potential for land treatment of waste water.

10.2.3.1.2 Rapid infiltration systems

Rapid infiltration (RI) is the most intensive of all land treatment methods. Relatively high hydraulic and organic loadings are applied intermittently to shallow infiltration or spreading basins (Fig. 10.11). The RI process uses the soil matrix itself for physical, chemical, and biological treatment. Physical straining and filtration occur at the soil surface and also within the soil matrix. Chemical precipitation, ion exchange, and adsorption occur as water percolates through the soil. Biological oxidation, assimilation, and reduction occur within top one or two metres of the soil. Vegetation growth is not

allowed in systems of this kind. The RI system is designed to meet multiple performance objectives, including the following (Metcalf and Eddy 1991; Sanks and Asano 1976):

1. Recharging of streams by interception of groundwater;
2. Recovery of water by wells or sub-surface drains with subsequent reuse or discharge;
3. Groundwater recharge;
4. Temporary storage of renovated water in local aquifers.

As both unsaturated soil zone and the underlying phreatic aquifer are involved in this treatment process, rapid infiltration systems are also known as *soil aquifer treatment* (SAT) systems.

10.2.3.1.3 Overland flow systems

Overland flow (OF) is a treatment process in which waste water is treated as it flows down a network of vegetated sloping terraces. Waste water is applied intermittently to the top part of each terrace, which flows down the terrace to a runoff collection channel at the bottom of the slope (Fig. 10.12). Application techniques include high-pressure sprinklers, low-pressure sprays, or surface methods such as gated pipes. OF systems are normally used with relatively impermeable surface soils, since in contrast to SR and RI systems, natural infiltration through soil is limited. The effluent waste water undergoes a variety of physical, chemical, and biological treatment mechanisms while it flows as surface runoff. Overland flow systems can be designed for secondary treatment, advanced secondary treatment, or nutrient removal, depending on user requirements (Reed *et al.* 1988; Sanks and Asano 1976).

10.2.3.2 Constructed wetlands

Wetlands are inundated land areas with water depths typically shallower than 0.6 *m* that support growth of emergent plants such as cattail, bulrush, reeds, and sedges. The vegetation provides surfaces for attachment of bacteria, aids filtration and adsorption of wastewater constituents, transfers oxygen into the water column, and controls

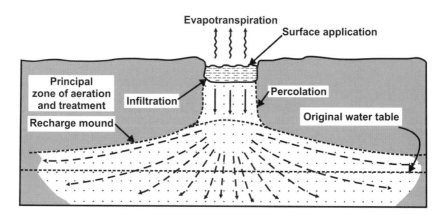

Fig. 10.11 Schematic diagram showing rapid infiltration treatment system. Redrawn from (UNESCWA, 2003).

growth of algae by restricting penetration of sunlight. Two types of constructed wetlands have been developed for wastewater treatment, namely: (i) free water surface (FWS) systems; and (ii) subsurface flow systems (SFS) (Reed *et al.* 1988; Sanks and Asano 1976).

10.2.3.2.1 Free water surface systems

Free water surface (FWS) systems consist of parallel shallow basins with depth ranging from 0.1–0.6 *m* or channels with relatively impermeable bottom soil or subsurface barrier and emergent vegetation (Fig. 10.13a). As a rule, pre-clarified waste water is applied continuously and undergoes quality improvement as it flows through the stems and roots of the emergent vegetation.

10.2.3.2.2 Subsurface flow systems

Subsurface flow systems (SFSs) consist of beds or channels filled with gravel, sand, or other permeable media planted with emergent vegetation (Fig. 10.13b). Waste water is treated as it flows horizontally through the media-plant filter. Systems of this kind are designed for secondary or higher levels of treatment.

10.2.3.3 Floating aquatic plants

This system is similar to the FWS system, except that the plants used are of the floating type, such as water hyacinths and duckweeds. Water depths, ranging from 0.5–1.8 *m*, are greater than in the

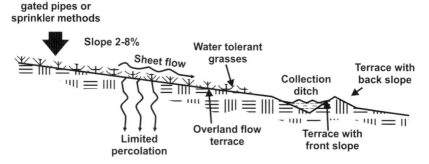

Fig. 10.12 Schematic showing cross-section across an overland flow system for wastewater treatment. Redrawn from (UNESCWA, 2003).

(a)   (b)

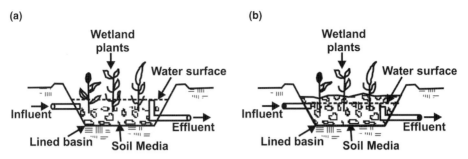

Fig. 10.13 Schematic of two types of constructed wetlands for wastewater treatment: (a) Free water surface (FWS) system; (b) subsurface flow system. The essential difference between the two is the level of water surface. Redrawn from Qasim (1999).

case of wetland systems. The floating plants shield the water from sunlight and reduce the growth of algae. Systems of this kind are effective in reducing *BOD*, nitrogen, metals, and trace organics and also in removing algae from lagoons and stabilizing pond effluents. Supplementary aeration has been used with floating plant systems to increase treatment capacity and to maintain the aerobic conditions necessary for biological control of mosquitoes (Metcalf and Eddy 1991).

10.2.4 Sludge treatment and disposal

Sewage sludge consists of organic and inorganic solids present in the raw waste water and removed in the primary clarifier, in addition to organic solids generated in secondary/biological treatment and removed in the secondary clarifier or in a separate thickening process. The generated sludge is usually in the form of liquid or semi-solid, containing 0.25 to 12% solids by weight, depending on the treatment method used. Sludge handling, treatment, and disposal are complex operations, owing to the presence of offensive constituents, depending upon the source of waste water and the treatment processes employed (Metcalf and Eddy 1991). Sludge is treated by means of a variety of processes that can be used in various combinations. Various unit sludge treatment operations and processes currently in use include thickening, conditioning, dewatering (primarily to remove moisture from sludge), digestion, composting, incineration, wet-air oxidation, and vertical tube reactors to treat or stabilize the organic matter present in the sludge.

10.3 New approaches and technologies for sustainable urbanization

The focus on urban hydrology discussed thus far has been on understanding the hydrologic impacts of urbanization and the conventional approaches to providing water supply, treatment, and disposal systems for storm and waste water. These approaches forming the basic framework of urban development and planning have made cities the world over more liveable with regard to the public health and recreational aspects and will no doubt continue to be relevant for a long time to come. But a price has to be paid – in terms of jeopardizing the environmental sustainability and consequently the well-being of future generations. Even as cities become more populous and their number increases, impacts of urbanization spread to surrounding regions as a result of which water supplies and sanitation services, including wastewater treatment and safe disposal, get stressed to meet demands of ever-increasing populations and correspondingly the financial resources. New approaches are, therefore, being formulated and practised in some areas, with the objective of ensuring environmentally sound and sustainable cities.

Broadly, the objectives of sustainable urban development are: (i) operational sustainability for public health, flood protection/mitigation, and structural integrity; (ii) environmental sustainability including pollution mitigation, efficient use of natural resources including reuse of wastes, and preservation of the natural hydrologic cycle; and (iii) economic sustainability, including cost-effective technology and cost recovery.

10.3.1 The need

As discussed above, concentrated human settlements with their propensity to create hard, impermeable surfaces for building houses and roads, and the need for water intake and outflow in a variety of forms, are not in harmony with the natural hydrologic cycle. The adverse effects of creating impervious surface cover in urbanized watersheds, reducing the groundwater recharge and consequent reduction in the base flow of the stream/river flowing through the area, are well documented (Cianfrani *et al.* 2007). Sewerage and water supply systems serving dense settlements can further interfere with groundwater and surface water hydrology. Urban settlements also create the 'heat island effect', reduce evapotranspiration, and modify local microclimates (Ward and Trimble 2003).

The historic shift from population settlement based on resource exploitation to one based on economy driven, transportation, and amenity-based settlement patterns poses even greater challenges for achieving sustainable water supplies and water management, as it puts increasingly larger population and land-use transformation in areas that previously served as 'water banks' for meeting the requirement of population residing in the area.

The major driving force for change is essentially population growth coupled with rising standards of living globally, a combination that has resulted in resource over-exploitation, including water. The current world population is about 6 billion, which is expected to grow to about 9 billion by the year 2050. When the population was much smaller (e.g. <2 billion) and when per capita use of resources was also much smaller, the traditional pattern of resource consumption, namely, 'take, make, waste' was sustainable. However, what is now needed is to recycle and reuse all types of resources (including water) and also increase the use of renewable resources. Water stress currently affects only a modest fraction of the human population, but it is expected to affect 45% of the population by 2025 (Daigger 2007; WRI 1966). This situation may be further exacerbated by global climate change, which may alter water supply and storage patterns in ways that would make the existing water-management infrastructure ineffective.

Recycling technologies can significantly reduce net water abstraction from the environment but many of these technologies require an increase in the use of other resources, especially energy. In our resource-constrained world, increasing the consumption of any resource, even for necessary functions such as water management, must be carefully considered.

Another aspect of water stress caused by urban water-management systems is the increase in the amount of nutrients, especially phosphorus, in the aquatic environment (Wilsenach *et al.* 2003). Mined as phosphate rock, phosphorus is used for manufacturing fertilizers that are used widely to increase crop production for human consumption. Phosphorus (and other nutrients) then passes through human metabolism and ends up in the wastewater discharge. When these effluents are discharged into the aquatic environment, the excess nutrients can cause eutrophication of surface water bodies. At the current rate of consumption, the supply of phosphate, an essential nutrient with no known replacement, is expected to be exhausted in about 100 years. Thus, there are at least two urgent reasons for us to recover phosphate from the wastewater stream.

Two other factors must be taken into consideration. First, although water supply is uniformly provided in the developed world, approximately 1 billion people in the developing world do not have access to safe drinking water, and more than 2.5 billion do not have access to adequate sanitation. Clearly to meet global needs, a more efficient urban water management system is needed.

10.3.2 Achieving sustainability goals

Urban water supply, storm water, and wastewater management is at a critical juncture all over the world. Methods must evolve in response to urban development, population growth, and diminishing natural resources. Based on information available in recent literature, three aspects of urban water management are becoming increasingly important and will continue to be important in the foreseeable future. These aspects are: (i) decentralized waste water management (DWM); (ii) wastewater reclamation and reuse; and (iii) increased attention to

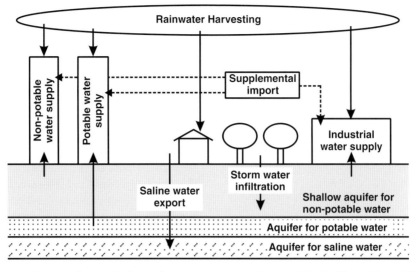

Fig. 10.14 Schematic diagram of a decentralized urban water management system. Modified from Daigger (2008).

wet-weather flow (WWF) management. Currently, consideration of these three aspects in urban water management planning is improving the functionality of wastewater systems and creating sustainable alternatives to the traditional centralized combined or segregated sewerage systems.

However, if sustainability goals as defined above are to be achieved, the current 'take, make, waste' approach to water and resource management needs to give way to a closed-loop approach, involving a combination of decentralized and centralized elements for recycling both liquid as well as solid waste material. Because of their superior environmental performance, closed-loop systems have the potential to meet the environmental, social, and economic goals of sustainability (Daigger 2007).

A closed-loop system for urban water management is schematically depicted in Fig. 10.14. The water supply, both domestic and commercial, is segregated into water for potable uses, such as direct consumption and bathing, and water for non-potable uses, such as toilet flushing, laundry, irrigation, and industrial uses. Overall, the requirement of potable water is quite small, though it should be of the highest quality in respect of its physical, chemical, biological, and radiological properties. The amount of water needed from the environment necessary for this purpose is much smaller

than the amount of domestic water that is being currently provided. In fact, the demand for potable water can be met either from local water supplies or by importing modest quantities of potable water. By separating potable and non-potable water, the net removal of water from the environment for potable uses can be dramatically reduced. The bulk of the domestic and commercial water requirement is of non-potable water quality, which can be supplied from a variety of local sources, including recycled water and captured rainwater, supplemented by modest import of water. As shown in Fig. 10.14, storage of non-potable water is a critical component of the system. Non-potable water can be stored either in an aquifer underlying the urban area (as shown) or in a surface storage facility if the requisite land area is available.

The repeated recycling of water may result in the build-up of dissolved solids, including salts, which must be managed to maintain the quality of the water for its intended use(s). Mixing with rainwater or employing reverse osmosis (RO) and other processes can dilute/remove the dissolved salts. The waste brine can be discharged into a saline-water aquifer (Fig. 10.14) or disposed off in some other way.

A question that is often asked relates to desalination of brackish/saline groundwater or surface source, or even sea water in water-scarce regions. Although this is technologically feasible, desalination does not always meet the environmental criteria for sustainability because of its significant energy requirements. Even though technological advances continue to reduce the energy requirement, it will always be higher than for treating waste water, because the content of dissolved solids in waste water (typically $\sim 1000\ mg\ l^{-1}$) is much smaller than that of sea water (35 000 $mg\ l^{-1}$). Therefore, depending on the need for desalination, water with the least salt content should be preferred.

10.3.2.1 Available tools/technologies

The main hydrologic consideration in achieving sustainability is conservation of local water resources to meet a variety of local needs without compromising living standards to a significant degree. Fortunately, tools and technologies are available that enable: (i) more efficient harvesting and local use of storm water; (ii) better water conservation by reducing the consumption; (iii) reclamation and reuse of waste water for a variety of applications; (iv) management and extraction of energy from the wastewater stream; (v) recovery of nutrients; and (vi) separation of specific wastewater sources that are difficult to treat by the commonly employed methods. Several of these technologies are presently in use, though not widely, and have demonstrated the ability to facilitate implementation of systems, such as the one shown in Fig. 10.14, and for improving decentralized and centralized water and resource management.

Technologies are available for managing storm water that can be collected and used either directly or treated by natural methods and infiltrated into the groundwater for subsequent use (see Section 10.2.3, and Chapter 6). Additional technologies include construction of: (i) permeable pavements to augment soil infiltration in urban landscapes; (ii) green roofs to reduce and in some cases completely eliminate the storm water contribution from the existing structures; and (iii) surface depressions that are planted as rain gardens with the objective of absorbing rainwater runoff generated from impervious urban areas such as roof tops, driveways,

walkways, and compacted lawn areas. In the past decade, as understanding of these systems has improved, storm water harvesting and treatment have become much more reliable and predictable.

Water- and wastewater-treatment technologies are crucial components of urban water management systems. Membrane technologies for removing particulate matter (micro- and ultra-filtration) and dissolved substances (nano-filtration and RO) are increasingly being used. When particle-removal membranes are coupled with biological systems, they can create membrane bioreactor (MBR) that is fast becoming an essential water-reclamation process (Judd 2006). Advanced oxidation processes include combinations of ozone, ultraviolet (UV) light, and hydrogen peroxide to create the highly reactive hydroxyl radical (OH). In addition, activated carbon is still widely used for water reclamation.

10.3.2.1.1 Tools that address environmental goals

Some other tools and technologies do not necessarily reduce the overall abstraction of water but do contribute significantly to meeting *environmental goals*, such as optimal use of energy and reduced nutrient dispersion. For example, as shown in Fig. 10.15, water from laundries and bathrooms (typically referred to as *grey water*) that contain few pollutants, constitute the largest component of urban waste water (Henze and Ledin 2001). Because of its low-pollutant content, grey water requires a minimal degree of treatment to become reusable for non-potable purposes. Recycling this large volume of waste water does not require much energy and thus needs fewer resources than recycling combined potable and non-potable waste water.

Organic matter in several components of the wastewater stream represents a significant source of energy. Most of the organic matter (quantified in terms of the five-day biochemical oxygen demand, or *BOD5*) is contained in toilet- and kitchen- waste (typically referred to as *black water*). The amount of waste water generated by these components is quite small, suggesting that the black-water fraction can be used efficiently for energy production. Technologies for energy generation from organic matter in black waters include thermal combustion

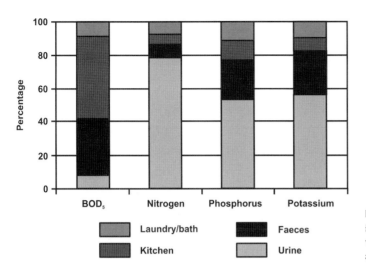

Fig. 10.15 Typical distribution of four important constituents in domestic waste waters of different sources. Source: Henze and Ledin (2001).

and anaerobic treatment for producing biogas (Grady *et al.* 1999), which can be used in combined heat and power systems. The microbial fuel cell is an emerging energy-production technology (Logan *et al.* 2006). A microbial fuel cell is a device that converts chemical energy to electrical energy by catalytic reaction of micro-organisms (Allen and Bennetto 1993). A typical microbial fuel cell consists of anode and cathode compartments separated by a cation-specific membrane. In the anode compartment, fuel is oxidized by micro-organisms, generating electrons and protons. Electrons are transferred to the cathode compartment through an external electric circuit and protons are transferred to the cathode compartment through the membrane. Electrons and protons are consumed in the cathode compartment, combining with oxygen to form water.

Several nutrients are found in human urine (typically referred to as *yellow water*). Development of urine-separating toilets and technologies for treating urine to produce fertilizer products hygienically is the key to managing nutrients with minimal requirements of external resources, particularly additional sources of energy. When energy management and nutrient recovery are combined with source separation, energy can be efficiently produced and extracted from the waste water stream along with nutrient recovery. A variety of technologies are available for nutrient recovery. For example, biosolids containing nitrogen and phosphorus produced from treatment and nutrient-recovery processes, can be applied directly to agricultural lands as fertilizers, as an alternative to phosphorus-based fertilizers. The storm water capture and water-reclamation technologies are most effective at the local (decentralized) level. Water-reclamation technologies result in reduced pumping requirements, because the reclaimed water is produced close to the place where it is used. In contrast, energy-management and nutrient-recovery technologies are most effective in large-scale centralized systems.

A recent survey by the *British Medical Journal* (BMJ 2007) found that modern water supply and sanitation has been the most significant contribution to public health in the past 150 years. The US National Academy of Engineering listed modern water supply and sanitation systems as one of the greatest engineering achievements of the 20th century (Constable and Sommerville 2003). Despite these developments, the present situation is grim and new approaches to water and sanitation systems are urgently needed. We are thus faced with many new, interesting, yet formidable challenges in water supply and sanitation management.

Fortunately, many technologies to meet these challenges already exist and work is continuing on refining them and integrating them into high-performance and more sustainable systems.

11 Rainwater harvesting and artificial groundwater recharge

Rainwater harvesting (RWH), in essence, involves collection, conveyance, and storage of rainwater. The scope, method, technologies, system complexity, purpose, and end uses vary widely – rainwater filled in barrels for garden irrigation in urban areas, large-scale collection of rainwater for various domestic uses, supplemental water for offices, industries and farm irrigation, aquifer recharge, and abatement of flooding caused by storm water. In view of depletion of groundwater resources and poor quality of some groundwaters, simplicity of rainwater harvesting systems and modern methods of treatment provide excellent avenues for harvesting rainwater for domestic use and other applications. Enlarging the scope of rainwater harvesting, from individual rooftops collection to water harvesting from a catchment for a variety of end uses, can lead to conjunctive use of surface- and ground-water for judicious management of water resources. In this chapter, a review is presented of the commonly used technologies of RWH with emphasis on domestic applications. In addition, technologies for various applications in irrigation and groundwater recharge are presented briefly.

The scope of RWH is wide and the technologies discussed can be applied, with or without minor modification and innovation, to any region facing water scarcity (for whatever reasons) but endowed with unutilized potential of rainwater.

11.1 Historical perspective

Archaeological evidence attests to the practice of rainwater harvesting in India as far back as 4500 years ago at the Harappan site of Dholavira in the Great Rann of Kachchh. In fact, rainwater harvesting in some form or the other has always been a part of Indian lifestyle and civilization. An excellent overview of the Indian traditions in water harvesting can be found in Agarwal and Narain (1997). During the 20th century, perhaps the best-known example of rainwater harvesting at the domestic scale comes from the ancestral home of Mahatma Gandhi at Porbandar in Gujarat State, India. The terrace of the house, which was thoroughly washed before the arrival of first monsoon showers, served as a rainwater catchment, and an underground tank with a capacity of ~90 m^3, constructed within the house itself, served as storage of the collected water for drinking and other domestic purposes throughout the year. A pipe with a heap of lime placed at its mouth served to convey filtered water from the terrace into the tank. However, with the convenience offered by piped water supply and the State assuming responsibility of supplying water through centralized systems, the practice of rainwater harvesting gradually declined almost everywhere. A renewed interest in this time-honoured approach of collecting rainwater has been rekindled in the last few decades because

Modern Hydrology and Sustainable Water Development, 1st edition. © S.K. Gupta
Published 2011 by Blackwell Publishing Ltd.

of escalating water scarcity (due to frequent occurrence of droughts, growing population, increasing pollution of both surface water and groundwater bodies) and increasing environmental and economic costs of providing water through centralized water systems or by drilling wells. These issues are forcing major departures from conventional water management strategies, leading to the realization that a single source of water can no longer ensure security of future water supplies. Successful water supply strategies must include rainwater harvesting as well as other sources of water that require a synergistic approach involving harnessing of multiple water resources.

The National Drinking Water Mission (NDWM) of the Government of India (NDWM 1990) took a more comprehensive view and defined '*water harvesting*' as 'collection and storage of rainwater and also other activities aimed at harvesting both surface water and groundwater, prevention of losses through evaporation and seepage, and all other hydrologic studies and engineering interventions aimed at conservation and efficient utilization of the available water in a physiographic unit such as a watershed or geomorphic basin.'

More recently, the Central Groundwater Board (CGWB) in India, spurred by success stories of farmers' movements for promoting artificial groundwater recharge has led to a vigorous movement involving various water users, including ordinary citizens, industries, offices, hotels, educational institutions, and hospitals, to take up recharging of groundwater by letting roof water to quickly recharge the aquifer(s) through abandoned dug wells, abandoned/working hand pumps, recharge pits, gravity-head recharge wells, recharge shafts, defunct bore wells, and digging trenches/pits around injection wells. This movement has gained such momentum that people started believing that rainwater harvesting is just for groundwater recharging and also a misconception that collected and stored rainwater is not for direct use such as domestic or potable consumption. Extensive research has been done by voluntary agencies, research organizations, and private companies on topics such as first-flush systems to improve the quality of rooftop water injected into wells to prevent contamination of groundwater and to improve its quality.

11.2 Rainwater harvesting – some general remarks

The interest in reviving the traditional technique of rainwater harvesting has grown recently due to the inherently good quality of rainwater in preference to treated waste water. Rainwater is valued for its chemical purity, particularly its low salt content. It has a nearly neutral pH, and is free from disinfection by-products, salts, minerals, and other natural and man-made contaminants. Plants/crops thrive on irrigation from stored rainwater. Appliances last longer when they use rainwater which is free from the corrosive or scaling effects of hard water. Rainwater is also preferred for potable purposes due to its superior taste and cleansing properties.

Perhaps one of the most important aspects of rainwater harvesting is the use of this natural resource at the very place it occurs. Rainwater harvesting also includes land-based systems with man-made landscape features to channel and concentrate rainwater in either storage basins or on planted areas. The main goals of water harvesting could be:

- to provide domestic water for people as well as for livestock and other animals;
- pasture improvement;
- to secure existing water supply where other water sources (surface or groundwater) are not available/polluted and/or uneconomical to develop;
- to increase the yield of rain fed agriculture;
- to minimize the risk of crop failure/low yield in drought-prone areas;
- reduction of soil salinity leading to increase in productivity of land;
- soil conservation (leading to reduced soil erosion);
- improved re-/afforestation (reducing desertification);
- groundwater recharge.

In any RWH method, rain and runoff from the *runoff area* (i.e. the catchment) is directed/concentrated and utilized and/or stored in the *run-on area*. The runoff contributing area comprises rooftops, courtyards, streets, and public spaces, part of which may be constructed (i.e. roads/pavements), or part of the ground surfaces and slopes in urban/rural areas that may be treated to enhance runoff. Runoff areas also include large natural catchment areas that feed seasonal water courses.

The equation of potential RWH supply is:

$$Pot.\ RWH\ (m^3) = Catchment\ area\ (m^2)$$
$$\times Rainfall\ (m) \times Runoff\ coefficient \qquad (11.1)$$

The runoff coefficient gives an estimate of the fraction of rainfall that can be harvested from a given surface area.

The equation for irrigation demand is:

$$Demand\ (m^3) = (ET_0\ (m) \times Plant\ factor)$$
$$\times Area\ (m^2) \qquad (11.2)$$

The ET_0 is the monthly potential evapotranspiration. The Plant factor represents the fraction of ET_0 needed by plants. This is determined by the type of plants existing in the area. Other types of demands are estimated using various empirical methods or, in established areas, by metering or otherwise estimating the actual demand.

The storage media in the run-on area could be over ground and/or subsurface. The over ground storage includes surface tanks, jars or ponds, and large reservoirs. The underground storage media could be soil, sediments, cisterns, and groundwater.

The regions where RWH has been/is being practised in a significant manner are arid/semi-arid areas where water demand exceeds supply because of: (i) poor and/or uneven seasonal distribution of rainfall; and/or (ii) high temperatures leading to high evapotranspiration, particularly in regions where rainfall occurs in the hot season.

11.2.1 RWH at the household scale

This is a special application of RWH because of the possibility of using the harvested water for drinking and other domestic uses and is, therefore, discussed here in some detail. In small-scale residential applications, rainwater harvesting can be as simple as channelling storm water running off an unguttered roof onto a planted area via a contoured landscape. More complex systems include gutters, pipes, storage tanks or cisterns, filters, pump(s), and water treatment for potable use. Regardless of the complexity of the system, the domestic rainwater harvesting system comprises six basic components (Fig. 11.1), as described below.

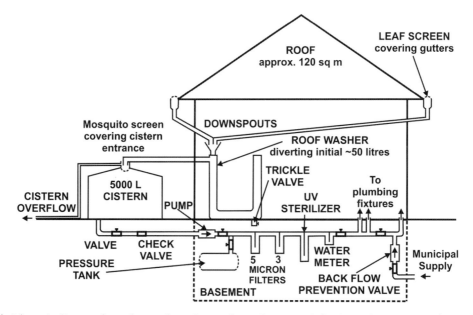

Fig. 11.1 Schematic diagram of a modern rooftop rainwater harvesting system indicating main components for potable end use. Redrawn from Errson (2006).

11.2.1.1 The catchment surface

This is the collecting surface from which rainfall runs off. The rooftop/terrace of a building or house is the obvious preferred choice for a catchment. Water quality from a given rooftop catchment is a function of the type of roofing material, climatic conditions, and the environment around the site (Vasudevan 2002). Commonly used roofing materials for rainwater harvesting include metal sheets, clay/slatestone/sandstone/concrete tiles, composite or asphalt shingles (rarely wood shingles are used), tar, and gravel. Caution should be exercised regarding roofing components to avoid toxins leaching from the tile sealant or from paint and other composite materials used.

11.2.1.2 Gutters and downspouts

Gutters and downspouts are used to channel water from the roof to the tank. The common materials employed for gutters and downspouts are half-round pipes made of PVC, vinyl, seamless aluminium, and galvanized steel. Regardless of the material used, other necessary components in addition to the horizontal gutters are the drop outlets that divert water from the gutters downwards and 45-degree elbows that allow the downspout pipe to run alongside a wall of the house. Additional components include the hardware, brackets, and straps to fasten the gutters and downspout to the fascia and the wall. Special attention is needed to determine where gutter overrunning areas may occur. At these locations, appropriate steps must be taken to minimize possible overrunning to improve catchment efficiency. Gutters should be installed sloping towards the downspout and also the outside face of the gutter should be lower than the inside face, to divert drainage away from the building wall.

11.2.1.3 Leaf screens

Leaf screens remove plant debris and dust from the captured rainwater before it enters the tank. Usually appropriate size mesh screens in wire frames that fit the length of the gutter are used to remove leaves and other debris. The size depends on the amount and type of tree litter, as well as dust accumulation. Often 5 *mm* ($^1/_5$-*inch*) is adequate. Leaf screens must be regularly cleaned

to function effectively. If not maintained properly, they can become clogged and prevent rainwater from flowing into the tank.

11.2.1.4 First-flush diverters

A roof collects dust, leaves, blooms, twigs, bodies of dead insects, animal faeces and bird droppings, pesticides, and other airborne particulate matter such as pollen grains. The first-flush diverters route the first flow of rain water from the catchment surface out of the storage tank. The simplest first-flush diverter is a PVC standpipe. During a rainfall event, standpipe fills with water first; the balance of water flows into the tank. The standpipe is drained continuously via a pinhole or by keeping the screw closure slightly loose. In any case, cleaning of the standpipe must be undertaken to remove collected debris after each rainfall event. There are several other types of first-flush diverters. The ball valve type consists of a floating ball that seals off the top of the diverter pipe when the pipe is filled up with water.

Opinions vary on the volume of rainwater to be diverted. The number of dry days, amount of debris, and type of roof surface are the variables to be considered. A rule of thumb for first-flush diversion is to divert a minimum of 50 *litres* of water for every 100 m^2 of collection surface. However, first-flush volumes vary with the amount of dust on the roof surface, which is a function of the number of dry days, the amount and type of debris collected on the roof, degree and type of tree overhang, and season.

11.2.1.5 Roof washers

A roof washer, placed just ahead of the storage tank, filters small particles of debris for potable systems and also for systems employing drip irrigation. Roof washers consist of a tank, usually between 100 and 200 *litre* capacity, with leaf strainers and a filter. All roof washers must be periodically cleaned. Without proper maintenance they not only become clogged and restrict the flow of rainwater but may themselves become breeding ground for pathogens.

11.2.1.6 Storage tanks

The storage tank is the most expensive component of a rainwater harvesting system. The size of storage

Table 11.1 Domestic scale RWH cistern materials, their features, and recommended precautionary measures to be observed. Adapted from Krishna (2005).

Material	Features	Recommended Precautions
Plastics		
Drum (80–250 *litres*)	Commercially available; inexpensive	Only new cans should be used
Fibreglass	Commercially available; alterable and moveable	Must be sited on smooth, solid, level footing
Polyethylene/polypropylene	Commercially available; alterable and moveable	UV-degradable, must be painted or tinted on the outside surface
Metals		
Steel drums (80–250 *litres*)	Commercially available; alterable and moveable	Must be checked for toxicity prior to use; prone to corrosion and rusting
Galvanized steel tanks (80–250 *litres*)	Commercially available; alterable and moveable	The inner surface must be lined for potable use in view of possible corrosion and rusting
Concrete and Masonry		
Ferro-cement	Durable and immoveable	Potential risk of breakage and structural failure
Stone, concrete block	Durable and immoveable	Difficult to maintain
Monolithic/Poured-in-place	Durable and immoveable	Potential for cracking/breakage
Wood		
Often wrapped with steel tension cables, and lined with plastic	Durable, can be dismantled and moved	Expensive, for potable use, a food-grade liner must be used

tank or cistern is dictated by several factors, such as local precipitation, demand for water, projected length of rainless spells, catchment surface area, aesthetic considerations, personal preference, and budgetary provision. Storage tanks must be opaque to light to inhibit growth of algae. For potable systems, storage tanks used must be made from virgin plastic and should not have been used to store toxic materials. Tanks must be covered on top and the air vents screened to ward off mosquitoes and prevent their breeding. Tanks used for potable systems must be provided with appropriate opening and should be accessible for cleaning. Commonly used storage tank materials for rainwater harvesting include: fibreglass, polypropylene, galvanized sheet metal, concrete, and ferro-cement (a low-cost steel and mortar composite material). RWH tanks can be located above ground level or partly/completely buried below ground level, depending on available space and convenience. For potable systems, it is essential that the interior of the tank is plastered with a high-quality material suitable for potable use. A summary of cistern materials, their features, and precautionary measures to ensure integrity of the stored water are provided in Table 11.1.

A frequently overlooked aspect of rainwater harvesting is the effect of storage itself on the water quality. As water is stored in a stagnant condition, several processes such as sedimentation, flotation, and bacterial die-off can take place, improving the water quality. Typical die-off behaviour for microorganisms in water follows the pattern in which for a short period their numbers either remain constant or increase (Droste 1997), followed by their eventual exponential decline. Adverse environmental factors outweigh favourable factors due to removal of organisms from their natural environment. Cleaning of the accumulated sediments at the bottom of the tank should be done periodically but taking care to leave the bio-films that develop on the inner side of the tank and in the sludge as its presence aids the die-off process of pathogens that may grow inside the tank. In addition, algae decompose due to lack of sunlight.

11.2.1.7 Delivery system

It could be either gravity-fed or pumped to the end user, according to the situation. Simple systems for limited potable use for a single household could comprise a tap from an over ground storage tank or a rope-and-bucket water drawing arrangement from a shallow underground cistern, although a hand pump fitted on top of the underground cistern would be preferable to avoid possibility of contamination. From practical considerations, it would, however, be more convenient to have a system comprising pumps and pressure tank/overhead tank with piped distribution network.

A typical automated pump-and-pressure tank/overhead tank arrangement consists of a $^3/_4$- or 1-*horsepower* pump, usually a shallow well jet pump or a multistage centrifugal pump, check valve, and a pressure switch. A one-way check valve between the storage tank and the pump prevents pressurized water from being returned to the underground tank. The pressure switch regulates operation of the pressure tank. The pressure tank/overhead tank, with a typical capacity of 100–200 *litres*, maintains pressure throughout the system. Power supply to the pump is cut off by a pressure switch when the pressure/level in the pressure tank/overhead tank reaches a preset threshold value. When there is a demand for household use, the pressure switch detects the drop in pressure in the tank and turns on the pump, drawing more water into the tank.

The new 'on-demand' type of pumps eliminate the need for a pressure tank. These pumps combine various components, such as pump, motor, controller, check valve, and pressure tank function, all contained in one unit. These systems are self-priming and are built with a check valve incorporated into the suction port.

11.2.1.8 Treatment and disinfection equipment

For potable water systems, treatment beyond the leaf screen and roof washer is necessary to remove suspended sediments and disease-causing pathogens from the stored water. Treatment generally consists of filtration and disinfection processes in series to ensure safe drinking water.

11.2.2 RWH for improving rain-fed agriculture and supplementary irrigation

Various RWH systems and techniques are summarized in Table 11.2. These are dependent on harvesting water flowing down-gradient in catchments and its subsequent use for agriculture/horticulture. The captured water is diverted to a cultivated area and part of it eventually permeates the ground.

Quite often, to improve the runoff efficiency of catchments, certain technical procedures are adopted. These procedures, known as catchment treatment, are generally specific to a given situation and include: (i) clearing the sloping surfaces of vegetation and loose earth material; (ii) smoothing and compacting of soil mechanically; (iii) applying sodium salt to make the surface impermeable; (iv) applying sprays containing bitumen or asphalt as chemical binders to seal the surfaces; (v) covering the catchment with paving material such as concrete, asphalt, corrugated iron sheet, flexible plastic films/sheets, etc.; and (vi) planting of shrubs/grasses, etc. to stabilize bunds or other earthen structures. These measures can increase runoff efficiency from 20–90%. An overview of the main types of water harvesting techniques for crop production is given in Table 11.3.

Depending on the size of catchment, RWH for irrigation and/or horticulture is often classified into three types: (a) micro-catchment water harvesting; (b) macro-catchment water harvesting (or medium-size catchment WH); and (c) floodwater harvesting (or large-size catchment WH).

11.2.2.1 Micro-catchment RWH

The catchment area is usually in the range 1–1000 m^2 and the runoff is essentially due to sheet and rill flow. Run-off (catchment) and run-on (storage) areas are generally adjacent to each other with a typical area ratio varying from 1:1 to 10:1. Storage of water is mostly in the root zone of soil where it is used for the planting of a few bushes/trees or

Table 11.2 Examples of different RWH systems dependent on capturing water flowing downhill in catchments used for agriculture/horticulture. These systems divert the harvested water to a cultivated area, which then permeates into the earth. After Food and Agriculture Organization of the United Nations (Prinz et al. 2000).

RWH Structure/s	RWH Type	Main Uses	Description	Where Appropriate	Limitations	Rough Sketch
Negarim micro-catchments	Micro-catchment (short slope catchment) technique	Trees and grasses	Closed grid pattern with diamond shapes or open-ended 'V's formed by small earth ridges, with infiltration pits	For tree planting in situations where land is uneven or only a few trees are planted	Cannot be easily mechanized, therefore, limited to small scale. Not practical to cultivate between rows of trees	
Contour bunds	Micro-catchment (small slope catchment) technique	Trees and grasses	Earthen bunds on contours spaced at 5–10 m apart, with furrow upslope and cross-ties	For large-scale tree plantation, especially when mechanized	Not suitable for uneven terrain	
Semi-circular bunds	Micro-catchment (short slope catchment) technique	Rangeland and fodder growing (and also trees)	Semi-circular shaped earthen bunds with tips on contour. In a series with bunds in staggered formation	Useful for grass reseeding, growing fodder, or trees in degraded rangelands	Cannot be mechanized, therefore, limited to areas where manual labour is available	
Contour ridges	Micro-catchment (short slope catchment) technique	Crops	Small earth ridges on contours at 1.5–5 m interval. Furrow upslope and cross-ties. Uncultivated catchment between ridges	For crop production in semi-arid areas, especially where soil is fertile and the area is easily accessible	Requires new technique of land preparation and planting, may, therefore, have problem with acceptance	

Technique	Catchment type	Use	Description	Suitability	Limitations	Diagram
Contour stone bunds	External catchment (long slope catchment technique)	Crops	Small stone-masonry bunds constructed on the contour at intervals of 15–35 m for slowing down and filtering runoff	Versatile system for crop production in a wide variety of situations. Easily constructed, even by resource-constrained farmers	Possible only where abundant loose stone available	
Permeable rock dams	Floodwater farming technique	Crops	Long low rock dams across valleys slowing and spreading floodwater as well as treating gullies	Suitable for situations where gently sloping valleys turn into gullies and better water spreading is required	Very site-specific and needs considerable stone material as well as availability of transportation	
Water spreading bunds	Floodwater farming technique	Growing of crops and rangeland	Earthen bunds set at a gradient, in a 'dog-leg' shape, to spread the diverted floodwater	For arid areas where water is diverted from a watercourse onto a growing crop or fodder	Does not impound much water and requires high degree of maintenance in early stages of construction	

Table 11.3 Overview of the main types of water harvesting methods for crop production.

Type of RWH	Type of Flow	Annual Rainfall	Treatment of Catchment	Catchment Size	Ratio: $\frac{\text{Catchment Area}}{\text{Storage Area}}$
Micro-catchment	Sheet and rill flow	>200 mm	Treated or untreated	~1000 m^2	1:1–10:1
Macro-catchment	Turbulent runoff +	>300 mm	Treated or untreated	1000 m^2–200 ha	10:1–100:1
Flood water harvesting	channel flow flood water	>150 mm	Untreated	200 ha–50 km^2	100:1–10 000: 1

Based on (Critchley et al., 1991).

annual crops. In urban areas underground tanks or infiltration wells may also store the generated runoff. Usually there is no provision for overflow. When used for agriculture or horticulture, micro-catchments mostly consist of a series of units, each bounded by earth dug out from the storage area, usually employing manual labour. A typical micro-catchment RWH system is schematically shown in Fig. 11.2. A variety of other shapes for runoff and storage areas are used – some of these are given in Table 11.2.

11.2.2.2 Macro-catchment RWH

The catchment area of a macro-catchment is usually 1000 m^2 to 200 ha, a large part of which lies outside the arable area. Typical ratios of catchment to storage areas vary from 10:1 to 100:1. The catchment may either be untreated or be treated to enhance runoff efficiency. The slope of a catchment area usually varies between 5% and 50%. Storage of water is mostly as soil moisture in a storage area, but due to larger flow volumes, a provision for overflow (spillway) during intense rain spells may be necessary. The cropping area could be terraced in a flat terrain (<10% slope). The preparation and construction could be done either manually or by employing mechanized farming equipment. A typical macro-catchment RWH system is schematically shown in Fig. 11.3.

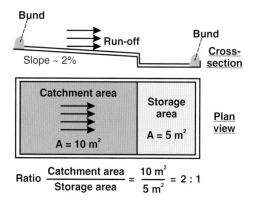

Fig. 11.2 Schematic drawing of a typical micro-catchment water harvesting system. The bunds are usually made with locally available excavated material such as soil, stone, etc. along a contour and can have shapes such as linear along a contour (contour bunds), semi-circular, triangular, V-shaped, and are arranged in staggered rows. If slope is high, terracing may also be resorted to.

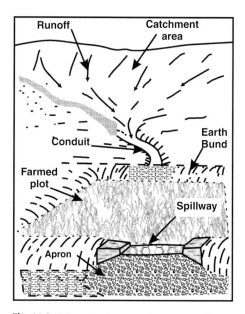

Fig. 11.3 Schematic diagram of a macro-catchment water harvesting system. Redrawn from http://www.fao.org/ag/AGL/AGLW/wharv/wh02/sld009.htm.

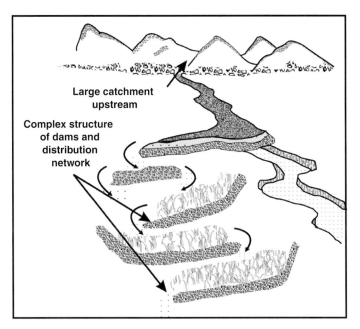

Fig. 11.4 Schematic of a typical floodwater harvesting system. Redrawn from http://www.fao.org/ag/AGL/AGLW/wharv/wh02/sld011.htm.

11.2.2.3 Floodwater harvesting

Floodwater harvesting involves collection and utilization of flood flow of a major seasonal river channel. External catchment upstream should be fairly large (200 ha–50 km^2) with catchment to storage area ratio, varying from 100:1 to 10 000:1. Special benefits of floodwater harvesting include: (i) supply of soil moisture for crop production; (ii) storage of water in ponds and reservoirs; (iii) recharge of groundwater; and (iv) minimizing damage due to flash floods. The system may comprise a complex structure of dams with provision for an overflow and distribution network. A typical flood harvesting system is schematically shown in Fig. 11.4.

Some of the limitations of RWH for crop production are: (i) climatic risk is not completely eliminated; (ii) there is no assurance of getting high yields; (iii) the methods are not yet scientifically well established; (iv) generally there is limited availability of extension services; (v) there is possibility of conflict between people living upstream and downstream in a catchment; (vi) larger schemes and structures are difficult to implement in view of arranging requisite financial support, and acceptance by stakeholders and political groups.

11.2.3 Rainwater harvesting for groundwater recharge

This is one of the more recent applications of water harvesting. Rainwater runoff from buildings, parking lots, etc., is diverted into seepage ponds or deep wells in order to directly recharge aquifers. This may sound overly simple, but it is not just putting the water back into the ground. Water that falls on the ground may, in the natural course, infiltrate and become part of the groundwater reservoir, but is likely to be redirected downstream to join a river or pond or enter the sewerage system (particularly in urban areas) and be transported far away. Surface water is likely to cause soil erosion and to pick up contaminants, e.g. fertilizers and chemicals. The recharge system puts water through a sand and gravel filter with possible biological components to clean it and then directs it down a deep well. This directly recharges aquifers without the long delay in reaching these depths in the natural course of percolation through the soil profile.

Rain/snow being the primary renewable sources of water; rainwater harvesting viewed in a wider perspective constitutes the primary source of water for *artificial groundwater recharge*. This includes water from perennial or ephemeral streams regulated by dams, storm runoff (including that generated from urban/peri-urban areas), ponds, and lakes; in fact any source of surplus water directly originating from rain/snow. Other sources of recharge water may originate from aqueducts or other water conveyance structures, irrigated areas, drinking water treatment plants, and sewage treatment plants.

A variety of methods have been developed to recharge the groundwater artificially. The widely practised methods that complement the RWH techniques involve some form of water spreading – releasing water over the ground surface in order to increase the quantity of water infiltrating into the ground, which then percolates to the water table. Spreading methods may be classified as basin, stream channel, ditch and furrow, flooding, and irrigation. These, together with techniques employing pits and recharge wells, are described below.

Artificial recharging of groundwater involves augmentation of groundwater resources using engineering systems where surface water is applied on the ground surface or diverted under ground for infiltration and subsequent movement to the aquifer for short- or long-term storage. Other objectives of artificial groundwater recharge often include preventing seawater intrusion into coastal aquifers, improving the quality of water through soil-aquifer treatment (geo-purification), using aquifers as subsurface water conveyance system, reducing land subsidence, and improving water quality in polluted aquifers through dilution.

Artificial groundwater recharge through enhanced infiltration requires permeable surface soils and/or sufficient land area for surface infiltration. Where these are not available, trenches, abandoned dug wells, tube wells, or shafts in the unsaturated zone and injection wells for aquifers that are deep and/or confined can be used.

Permeable soils, particularly coarse alluvial deposits have vertical hydraulic conductivity ranging from 1 m d^{-1} (for fine loamy sands) to 10 m d^{-1} (for sand and gravel mixes). However, because of clogging, the actual infiltration rate of recharge basins generally varies from 0.3 to 3 m d^{-1}. The quantity of infiltrated water obviously varies with the duration of infiltration, which may be limited to

a few days per year if no surface water storage facility exists. In any case, the total duration of the infiltration process seldom exceeds 100 $days$ per year since, if a surface reservoir exists to control floods, putting water in the ground would be worthwhile only if it can be achieved in 100 to 150 $days$. The amount that one can reasonably expect to be infiltrated in basins will therefore range approximately from 10^5 to 3×10^6 m^3 a^{-1} ha^{-1}.

The infiltration scheme may consist of basins, channels, or pits, depending on the local topography and on land use. The most common system consists of a number of basins each with an area ranging from 0.1 to 10 ha, according to availability of space. Each basin must have its own water supply and drainage so that it can be flooded, dried, and cleaned according to a predetermined schedule to achieve optimum renovation. Basins should not be constructed in series, because in such a system they cannot be dried and cleaned individually. Often the first basin is used as a pre-sedimentation facility for suspended solids present in wastewater.

The main problem in artificial groundwater recharge systems is inevitable clogging of the infiltrating surface (basin bottom, walls of trenches, vadose zone wells, and well–aquifer interface in recharge wells), which results in reduced infiltration rate. Clogging needs to be managed by applying desilting or other pre-treatment techniques for input water and remedial measures in the system, such as drying, scraping, disking, ripping, or other tillage methods. Recharge wells are required to be pumped periodically to backwash clogged surfaces.

As artificial groundwater recharge systems involve surface water and earth materials and their interaction, their design and management involve geological, geochemical, hydrologic, biological, and engineering aspects. Some of the artificial groundwater recharge structures in use are summarized in Table 11.4.

Two examples of successful groundwater recharge in Gujarat, India, are schematically shown in Fig. 11.5 and Fig. 11.6. In the first system, stream runoff water is recharged into a deeper confined aquifer using the head difference between the river water level and the piezometric level of the recharge tube well (Fig. 11.5). This

Table 11.4 Types of artificial groundwater recharge structures suitable for RWH in different geohydrological situations. Adapted from Gupta (2003).

Surface Infiltration

In-channel structures

Dams or weirs are placed across ephemeral or perennial streams to create backwater storage on the upstream side, thereby increasing the wetted area of the stream bed/floodplain so that more water infiltrates into the ground and percolates to the water table. The dams may be closely spaced low weirs or large dams spaced at greater distances. Where channels have small slope and shallow water depths, water is spread over the entire width of the channel or flood plain by placing T- or L-shaped earthen levees, generally <1 m high, in the channel itself. If the levees are washed out by high flows, they are restored again when the flood danger recedes. Check dams, gully-, or *Nallah*- plugs are some of the common types of 'in-channel' recharge structures used in India.

Off-channel structures

These are specially constructed infiltration basins, lagoons, old gravel pits, flood-irrigated fields, perforated pipes, or other such structures. To minimize land requirement, it may be useful to remove less permeable over-burden, if not very thick. The underlying aquifer should be unconfined, with sufficient transmissivity to accommodate lateral flow of infiltrated water away from the recharge area without forming high groundwater mound. Vadose zone should not have layers of clay or other fine-grained material that may unduly restrict downward flow and form perched water table and eventually cause water-logging in the recharge area. Soils in the vadose zone and aquifers as well as the recharged water should be free from undesirable contaminants. The figure alongside shows a cross-section across a typical infiltration basin.

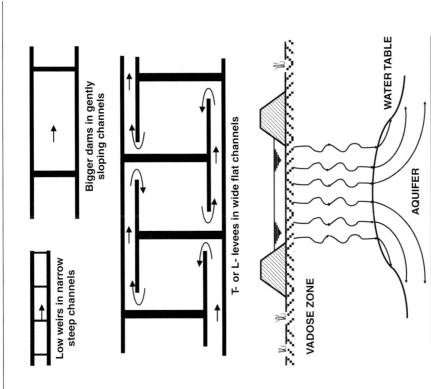

Low weirs in narrow steep channels

Bigger dams in gently sloping channels

T- or L- levees in wide flat channels

VADOSE ZONE

WATER TABLE

AQUIFER

A TYPICAL INFILTRATION BASIN GROUNDWATER RECHARGE SYSTEM

(*Continued*)

Table 11.4 (*Continued*)

Vadose Zone Infiltration

Where sufficiently permeable soils and/or sufficient land areas for surface infiltration systems are not available, vertical infiltration systems, such as trenches or wells dug in the vadose zone, can also be used for effecting groundwater recharge. Recharge trenches are typically dug up to ~5 m depth, whereas the wells may be 50–60 m deep. In both cases, water is normally applied through a perforated or screened pipe – 'horizontal' in trenches and 'vertical' in vadose zone wells placed in the centre of the well. Both are backfilled with coarse sand or fine gravel and covered properly to blend with the surroundings. The main advantage of recharge trenches or wells in the vadose zone is that they are relatively inexpensive. The disadvantage is that because of accumulation of suspended sediments and/or biomass, their permeable surface gets clogged up in the course of time.

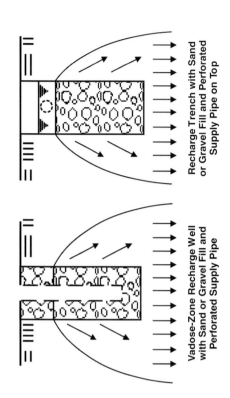

Vadose-Zone Recharge Well with Sand or Gravel Fill and Perforated Supply Pipe

Recharge Trench with Sand or Gravel Fill and Perforated Supply Pipe on Top

Wells

Direct recharge or injection wells are used where: (i) permeable soils and/or sufficient land area for surface infiltration are not available; (ii) the vadose zone is not suitable for digging trenches or wells; and (iii) aquifers are deep and/or confined. According to Bouwer (2002), even fully confined aquifers could also be recharged because such aquifers accept/yield water by expansion/compression of aquifer material and particularly from inter-bedded clay layers and aquitards. However, excessive compression of aquifer materials due to over-pumping, which may result in land subsidence, is mostly irreversible (Bouwer 1978). In the United States, the water used for well injection is usually treated to meet drinking water quality standards for two reasons. First, is to minimize the clogging of well–aquifer interface and second, to protect the quality of water in the aquifer, especially where the same water is pumped by other wells for potable uses. Also the water used for well injection in the United States is often chlorinated and has a chlorine residual of ~0.5 mg l^{-1} when it goes into the recharge well.

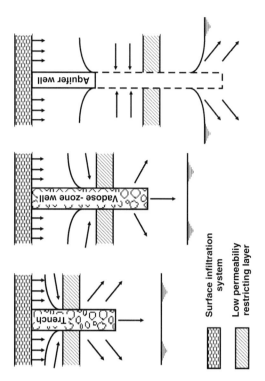

Trench

Vadose-zone well

Aquifer well

Surface infiltration system

Low permeability restricting layer

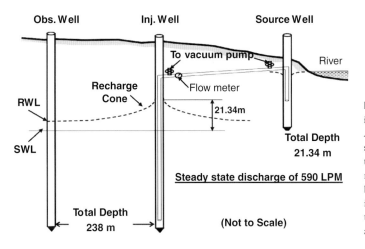

Fig. 11.5 Schematic illustration of an injection groundwater recharge system at Ahmedabad, India. The Sabarmati riverbed sand was used to filter the river runoff water through a shallow depth source well in the river bed. The hydrostatic head difference between the supply well and the deep injection well on the river bank was used to transfer water between the two wells using a siphon. See also Plate 11.5.

head difference not only conveys the water between the two wells through a siphon but also overcomes the formation and well losses between the two wells. The riverbed sand provides the necessary filtration to provide nearly silt-free water to the source well that gets transferred to the recharge well through the siphon. In the second system, runoff water collected in the village pond is recharged into the underlying aquifer to dilute the groundwater with high fluoride content. In this case, additional filtration systems are needed to minimize clogging of the recharge well (Fig. 11.6).

11.3 Watershed management and water harvesting

Watershed management offers an effective method to intercept dispersed runoff. Several techniques of water conservation have been developed along hill slopes with the purpose of preventing soil erosion and reducing surface runoff, as well as increasing infiltration into the ground, thus recharging the aquifers. Traditional terraced agriculture is certainly one of the most common water harvesting methods in arid areas. Where terraces are well maintained, they effectively control runoff and

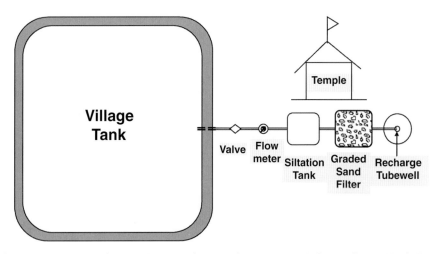

Fig. 11.6 Schematic arrangement of an injection groundwater recharge system at Balisana village in North Gujarat, India. An external siltation-cum-filtration system is required to minimize clogging up of the recharge well. The recharging of groundwater was required to dilute the high fluoride-containing groundwater in the area.

improve aquifer recharge but, once they fall into disuse, they progressively lead to gully erosion, collapse of the retaining walls, destruction of the entire system, and substantial modification of the hydrologic regime. Therefore, whatever the economic benefits of such terraces, it should be realized that their abandonment on a large scale can disturb the hydrologic conditions within a basin for a considerable period of time.

Because of siltation problems in surface reservoirs resulting from soil erosion in the upper catchment, large programmes of soil and water conservation as well as afforestation are being undertaken in several arid/semi-arid countries. Although the primary objective of watershed management is to limit soil erosion, thereby reducing sediment accumulation in surface reservoirs downstream, the effect of these practices may become significant for aquifer recharge when large areas are involved. However, there are a few examples of quantitative analysis of the modification of the water cycle in a catchment where soil and water conservation have been practised.

11.3.1 Conjunctive use of surface- and groundwater as part of integrated river basin management

The increasing severity of water scarcity, worldwide, requires the adoption of a dual approach of water supply and demand management.

Conjunctive use of surface and groundwater is one of the strategies of water supply management that has to be considered to optimize water resources development, management, and conservation within a basin, and artificial recharge of aquifers is certainly an important aspect in achieving these objectives.

Considering a river basin as a spatial unit for analysing the interactions and interrelations between the various components of the system, and for defining water management policy, is quite justified, and is increasingly a common practice in several countries:

- In China, river basin plans have legal status and development projects are required to be consistent with the provisions of the plans.

- In Indonesia, the Government recently adopted new water management policies in order to prepare spatial management plans and to link water and land use through river basin plans, to centralize water management at river basin level, and to assign water management responsibilities through more effective participation and collaboration of the various stakeholders.
- In Italy, a 1989 law introduced the river basin as a management unit to regulate programmes of the various sectoral and regional institutions.
- In the United Kingdom, Spain, France, and in most other European countries, water resources management is now essentially centred on river basins.

Adoption of an integrated river basin management approach to elaborate policies and strategies of water resources development, management, and conservation would regard water resources (both ground- and surface water) as one system and would avoid a water resources development approach focused only on surface water. This approach also facilitates management of the resource itself, allowing a better understanding, by water users, of the hydrologic issues involved.

Research is still needed, however, to understand better, and to quantify the role of watershed management and particularly of soil and water conservation practices in the water cycle. The development of GIS techniques could certainly help in assessing overall water resources within a river basin and the effect of various human interventions such as water conservation practices over large areas, large dams, or small dams.

11.4 Tutorial

Ex 11.1 Given average monthly rainfall and potential evapotranspiration, calculate monthly rainwater harvesting potential and demand for the whole year. Next calculate monthly storage requirement. The collecting area is a rooftop 15 m long, 10 m wide, sloping at an angle of 15° in the direction of width. The area to be irrigated is a nearby plot of 50 m^2. The runoff coefficient of the surface is 0.9.

Table 11.5

Month	Avg. Rainfall (mm)	ET_0 (mm)	Pot. harvested rainwater Availability (m^3)	Plant Water Demand (m^3)	Cum. harvested rainwater Avail. (m^3)	Cum. Demand (m^3)	Net Storage (m^3)
Jan	30.5	73.7	3.13	0.96	3.13	0.96	2.17
Feb	25.4	94.0	2.60	1.22	5.73	2.18	1.38
Mar	22.9	152.4	2.34	1.98	8.08	4.16	0.36
Apr	7.6	205.7	0.78	2.67	8.86	6.84	−1.89
May	7.6	248.9	0.78	3.24	9.64	10.07	−2.45
Jun	0.0	269.2	0.00	3.50	9.64	13.57	−3.50
Jul	33.0	243.8	3.39	3.17	13.02	16.74	0.22
Aug	45.7	205.7	4.69	2.67	17.71	19.42	2.01
Sep	25.4	185.4	2.60	2.41	20.32	21.83	0.19
Oct	17.8	149.9	1.82	1.95	22.14	23.77	−0.12
Nov	17.8	91.4	1.82	1.19	23.97	24.96	0.63
Dec	35.6	63.5	3.65	0.83	27.61	25.79	2.82
Total	269.2	1983.7	27.61	25.79			

Solution. The potential RWH supply is estimated using Eqn 11.1.

In this case, due to slope, the catchment intercepted by the rainfall is $30 \times 10 \times \cos(15°) = 113.85 \ m^2$.

The plant water demand is estimated using Eqn 11.2.

From Table 11.5 it is seen that even though the annual harvested rainwater supply is more than the demand, there are several months when the accumulated plant water requirement exceeds the accumulated harvested rainwater supply. This implies that it will be best to have the storage capacity $\geq$ the maximum deficit, which is 3.50 m^3 for the month of June. This is likely to ensure that the plantation can be supported from the harvested rainwater supply alone. During the first year there will be a deficit of harvested water, particularly in April, May, and June, because to start with there is no storage. However, beginning with the second year, the storage will build up and there will be progressively less deficit in subsequent years. To allow for variation in monthly precipitation, additional storage capacity may be required, depending on the coefficient of variation of rainfall for each month.

12

Water resource development: the human dimensions

Water is perhaps the most crucial element for evolution and sustenance of life on our planet. It is a natural resource that sustains the environment and supports livelihood. Availability of fresh water has a dramatic impact on the quality of human life, and the progress of mankind in diverse fields is critically dependent on it. Access to water fundamentally impacts on equality amongst nations and between rich and poor the same country. Access to water influences human behaviour, which includes adaptation to different environments. Human activity, in turn, impacts on the availability of good-quality water with its associated health implications. Water also has direct implications for poverty alleviation, economic growth, and overall development in general. Competition for water is intensifying day by day with the progressive collapse of water-based ecological systems, diminishing river flows, and groundwater depletion. All social systems have a hydrologic dimension and water has been part of the social fabric since time immemorial, for daily life activities and religious ceremonies. In fact, water has been a symbol of life, purity, and regeneration in many civilizations and religions around the world.

The human dimensions in water resource development and management encompass a diverse range of issues and policy domains. The concepts often transcend boundaries between pure and applied sciences, combining insights from fields such as physics, chemistry, microbiology, civil engineering, mathematics, economics, sociology, law, and psychology, and has led to new interdisciplinary approaches in exploration as well as management of water. Of the various social and natural resource crises facing humanity today, the water crisis is the one that lies at the heart of our survival and the future of our planet Earth itself.

12.1 The global water crisis

The amount of fresh water that exists on the Earth today is the same that existed 100 years ago or a million years ago. Only about 3% of all the water on Earth is fresh and about 2% is locked up in polar and mountain glaciers. Rivers and lakes contain only one-fiftieth of the remaining 1%, with only half of it being accessible (Table 1.3). This amount will be almost constant for all time. In the last century, the population of the world has tripled, but the use of annually renewable water has grown six-fold. In the next 50 years, the world population will increase by about 40–50%. This coupled with increased industrialization and urbanization, will result in an increasing demand for water and will have serious repercussions on the environment. Presently, more waste water is generated and discharged than at any other time in the history of the Earth. Nearly 1.1 billion persons, or 1 in every 6, lack access to safe drinking water; nearly 2.6 billion (i.e. more than 2 out of every 6 persons) lack adequate sanitation (WHO/UNICEF 2004). About 3900 children die every day from waterborne

Modern Hydrology and Sustainable Water Development, 1st edition. © S.K. Gupta
Published 2011 by Blackwell Publishing Ltd.

diseases. The issues are interlinked, since without an adequate amount of water needed for sewage disposal, cross-contamination of drinking water by untreated sewage is inevitable and is the main adverse outcome of inadequate water supply.

Despite food security having significantly improv in the past 30 years, water for irrigation accounts for 66% of the total withdrawals (up to 90% in arid regions). The remaining 34% is being used for domestic consumption (10%), in industry (20%), and evaporation from reservoirs (4%) (Shiklomanov 1999).

As the per capita use of water increases due to improved lifestyle and, further still, as the world's population increases, the proportion of water for human use is bound to increase. This, coupled with spatial and temporal variations in water availability, means that the water needed to produce food, industrial processes, and all other uses is becoming progressively scarce.

Vegetation and wildlife are main dependent upon adequate freshwater resources, with marshes, bogs, and riparian zones dependent upon a sustainable water supply. Forests and other upland ecosystems are equally at risk of significant productivity changes, as water availability is diminished. Large scale production of biofuels is set to further deplete the world's water supply.

There are approximately 260 international rivers, where conflicts exist on sharing of waters between the nations concerned. While international treaties, such as the Helsinki Rules, help to interpret intrinsic water rights amongst concerned countries, some issues related to basic human survival are so acute that conflict is inevitable. In many cases, water sharing disputes further exacerbate aleady-existing underlying tensions.

Water deficits are already spurring heavy grain imports in most of the small countries and may soon do the same in larger countries, such as China and India (Raja 2006). Water tables are constantly falling in several countries (including Northern China, the United States, and India) due to widespread over-pumping of groundwater, particularly in arid and semi-arid regions. Other affected countries include Pakistan, Iran, and Mexico. This may eventually lead to water scarcity and reduction in grain harvest. Even with over-pumping of aquifers, China is

developing into a grain-deficit country. When this happens, it will further escalate grain prices. Most of the three billion people projected to be added to the world's population by the mid 21st century will be born in countries already experiencing water shortages. After China and India, there is a second tier of countries with large water deficits – Algeria, Egypt, Iran, Mexico, and Pakistan. Some already import a large share of their food grain requirements, only Pakistan being self-sufficient. But with its population increasing by four million a year, Pakistan will probably soon turn to the world market to buy grain.

The Himalayan glaciers, sources of Asia's largest rivers – the Ganges, the Indus, the Brahmaputra, the Yangtze, the Mekong, the Salween, and the Yellow River, could disappear by 2035 as global temperature rises (Khadka 2004). Approximately 2.4 billion people live in the drainage basin of the Himalayan rivers. India, China, Pakistan, Bangladesh, Nepal, and Myanmar could experience floods alternating with droughts in the coming decades. In India alone, the Ganges provides water for drinking and farming to feed over 500 million people (Khadka 2004). The west coast of North America, which gets much of its water from glaciers in the Rocky Mountains and Sierra Nevada, could also be affected.

The precise nature of climate change induced by greenhouse gases (CO_2, CH_4, and others) and its impact on water resources is uncertain. Modelling studies suggest that precipitation will probably increase between the latitudes 30°N and 30°S, but many tropical and sub-tropical regions will probably receive lower and more erratic rainfall. With a discernible trend towards more frequent extreme weather conditions, it is likely that the frequency of occurrence of floods, droughts, mudslides, typhoons, and cyclones will increase. Stream flows in low-flow periods may well decrease and water quality is likely to deteriorate, because of increased pollution loads and concentrations and higher water temperatures (UNESCO–WWAP 2003).

These and many other problems, and a good deal about where they exist or are likely to spread, are now well known. There perhaps exists enough knowledge and expertise to tackle them. Concepts, such as equity and sustainability, have been developed, with the aim of ensuring that dis-advantaged people may participate in global de-velopmental issues, which does not happen at present.

12.2 Global initiatives

Perceptions of the global water crisis and under-standing of the requisite responses have progres-sively matured since the latter part of the 20th cen-tury. The International Conference on Water and the Environment held in Dublin in 1992 set out the four Dublin Principles that are still relevant today. These are: (i) fresh water is a finite and vulnera-ble resource, essential to sustain life, development, and the environment; (ii) water development and management should be based on a participatory approach, involving users, planners, and policy-makers at all levels; (iii) women play a central part in the provision, management, and safeguarding of water; (iv) water has an economic value in all its competing uses and should be recognized as an economic commodity.

The last of the Dublin Principles can also be seen as a starting point of another vital issue in the hu-man dimension of water, namely, the water ethic as an extension of the environmental ethics in relation to the interaction between ethics and economics (Sen 1987; Singer 1997), and similar aspects of the problem.

Those dealing with water problems are of the view that improvement in management and disas-ter amelioration through forewarning and advance preparedness is far more effective and economic than any *ad hoc* fire-fighting approach after a dis-aster has actually struck.

This raises two types of ethical issues. First, there is an obligation on the part of scientists, engineers, and other experts, trained and experienced in the relevant disciplines, to provide the best possible analysis of the problem, particularly concerning lo-cal vulnerabilities. Both reliable data and its coher-ent interdisciplinary interpretation are of utmost importance. Second, there is an obligation on local and national government authorities to disseminate information, making the public aware of the rele-vant conclusions of these expert studies.

12.3 Water and ethics

Sustainable development has often been described as 'development which meets the needs of the present without compromising the ability of fu-ture generations to meet their own needs.' With increasing stress on global freshwater resources, it becomes imperative to contemplate and evolve appropriate ideas, taking into account our moral obligations (defined simply as *ethics*), not just to-wards future generations but also addressing water resource development issues of the present day. The basic elements of any ethical approach are: (i) acceptance of a set of moral principles; (ii) an objec-tive perception of the ground situation; and (iii) a well-reasoned ethical judgment on the issue under consideration. There is considerable variety in sys-tems of ethical principles accepted by individuals, either on the basis of their authority or as a result of a wellthoughtout reflection or a combination of both. Certain moral injunctions are considered of vital importance and are widely adopted to become part of legislation governing a given society. Some of these have been liberally accepted amongst na-tions with different cultures that they became part of international or universal treaties binding on a wide comity of nations. Personal ethics is con-cerned with either problems that are not incorpo-rated into national or international law, or else with the application of such ethical codes (Priscoli *et al.* 2004).

12.3.1 Scope of water ethics

There are some basic differences between ethi-cal problems relating to water and those relating to the environment. First, the problems of wa-ter management are perceived largely in terms of such factors as human health, food security, eco-nomic development, quality of human life, dis-placement/migration of populations, and the asso-ciated economic losses. Accordingly, support for any action by the general public or by decision-makers does not depend on wide acceptance of extension of the concept of community to include land, plants, and animals as advocated by authors such as Leopold (1997) and other hard-core con-servationists. Individuals may differ on the moral theoretical basis for contributing to disaster aid (Arthur 1997; Singer 1997), but the public reaction to the request for humanitarian assistance is much stronger than for similar requests on conservation issues.

Many social groups are particularly disadvantaged in relation to water management. In the case of water-related disasters, Blaikie *et al.* (1994) list the five most vulnerable groups as: (1) the poorest third of all households; (2) women; (3) children and youth; (4) the elderly; and (5) some numbers of minority groups. In the whole gamut of water, women play a key role in many respects (Rodda 1991) and hence their involvement is of paramount importance in dealing with problems of water management. Women are more exposed to the immediate impacts of water problems, in relation to problems of household stability, and have less access to resources than men during recovery from water-related disasters.

Another group playing a key role in water resource development is the professional group, comprising scientists/engineers, social scientists, and legal experts, which has an important contribution to make in ethical management of water resources. The professional opinion is a vital component but does not constitute the totality of the required input.

The responsibilities of governments in relation to risk analysis in water resources management have been highlighted over the past decade. There is a growing realization that water scarcity is an important constraint in regard to economic development in many poor countries (Falkenmark *et al.* 1990). The Dublin Conference stressed the need to 'establish and maintain effective cooperation at the national level between the various agencies responsible for collection, storage and analysis of hydrological data.'

While ethical considerations may require reallocation of resources to reduce inequity within a given society, the position in regard to an existing inequality arising in a given location from scarce water resources or from higher disaster-proneness is not very clear. Sen (1970, 1973, 1981, 1982, 1987) has analysed the nature of inequality in society in some detail. It becomes clear that traditional methods of economic evaluation should be

widened to include indirect benefits and external values such as social justice. These kinds of ethical considerations require in-depth deliberations at all levels in society in order to reach a consensus solution to the water problems. This leads to the second of the four Dublin Principles, namely water development and management should be based on a participatory approach involving users, planners, and policy-makers at all levels. Clearly, participation should be substantial and not just marginal and should aim to bring about awareness of the importance of water amongst policy-makers as well as the general public. The participatory approach is equally applicable for all water related problems and is essential if the impact of such problems is to be minimized to the extent possible with minimum diversion of resources.

12.3.2 Ecological considerations

Nature is constantly changing and one must avoid taking its equilibrium and inviolability for granted when considering developmental issues. Indeed, very little of the Earth is left in its pristine state, completely unaffected by human intervention. Almost every eco-system has been encroached upon by humans, advertently or inadvertently. The need, therefore, is to strike a reasonable balance between ecological considerations and developmental needs, taking into account the symbiotic relationship between nature and the water resources on the one hand and the developmental issues on the other.

Ecological systems function as buffers between the Earth and the atmosphere, and can aid to achieve the most cost-effective solutions to water management. By the same token, water management can benefit ecological systems by regulating flows, sustaining fish and other aquatic organisms, etc. Ecological sustainability must be the major objective in all freshwater uses. Over-abstraction and water pollution must be minimized to maintain the sanctity of the ecological system of which water is a vital component. A renewed awareness of how people can conserve ecosystems along with development is needed. This requires better integration of ecological considerations with traditional economic values; the latter are often distorted by subsidies given to users. Claiming to preserve nature that would stall any development, or to separate human interventions from a perceived state of Nature, can be as unrealistic as ignoring human impact on the ecosystem.

12.3.3 Social context of water ethics

The social context of ethical questions concerning water tends to centre on notions of water as a common resource; its relationship with human well-being, and as a basic need for life; rights and responsibilities in regard to water access; social justice relating to water issues; and wealth-generating and developmental roles of water infrastructure.

Much of water management is about finding an ethical balance between water availability, use, and advertent and/or inadvertent modification of the hydrologic cycle and land resources. The general consensus on sustainable development can be seen as an ethical norm arising from this basic premise.

12.3.4 Gender related issues

Women, being economically more disadvantaged than men, carry a disproportionate burden of inequity. In most societies they often do not have access to property, land, or water rights. The failure to address this inequity in the developing world, especially in Africa and parts of Asia, is perhaps a major cause of social inequality. Gender-biased poverty is possibly the root cause of the population increase, which further exacerbates water scarcity resulting in the so-called water crisis. Promoting literacy, information, education, and jobs for women can go a long way in overcoming water scarcity in the long term.

Despite these considerations, women are *de facto* water managers in their own homes and in most small villages and communities. Further, they are central to maintenance and operation of water-related facilities. Women have a greater interest and are affected more by the impact of water-related issues. Nonetheless, it is ironic that women are rarely involved in crucial decision-making processes regarding water resources management. Studies have shown that increased participation of women is both an ethical requirement

and a pragmatic approach. Projects having active participation of women are more likely to be sustainable and to yield desired benefits. This was formally recognized in the Dublin Principles and also implied in many other UN declarations.

12.3.5 Floods and droughts

Floods and droughts are the most important water-related emergencies. These are generally triggered by natural variability of rainfall but their tragic consequences and coping mechanisms seem to be related to social and institutional systems prevailing in the affected areas. There is a growing realization to assess the vulnerability to such emergencies as well as assign the responsibility to take effective measures to prevent and mitigate the resulting disasters. Unfortunately, in many parts of the world, such natural calamities trigger wider human disasters. Often attitudes are formed and decisions made on the basis of perceived risks. Frequently, large gaps exist between perceived and real risk, especially in the case of extreme events, be it drought or flood. There is, thus, a need for formulating guidelines on the professional as well as ethical responsibilities of experts to define risk as clearly as possible, for decision makers to realize and communicate such risk, and for public and informed groups to actively participate in choosing solutions having acceptable levels of vulnerability. Nevertheless, the ground reality is that sometimes poverty gets into the way of flood management operations, because the poor cannot find alternative places to live.

Floods and droughts are extreme hydrologic events, bringing death and disease to thousands of humans and livestock, and substantial damage to property. Nevertheless, their prediction and mitigation procedures are different. Recently, flood forecasting has become somewhat more reliable than drought prediction. Furthermore, floods are usually of short duration (from hours to a few days), whereas droughts are a long-term phenomena, lasting from a few months to several years. Mitigation of floods is through structural solutions (afforestation in the catchment areas of rivers, construction of dams, dykes, etc.) as well as discouraging human encroachment onto river beds and floodplains. Drought protection, however, is often associated with non-structural approaches (water markets, insurance, restrictions and regulations on use of water, etc.) or conjunctive uses of surface- and groundwater. In some countries, failure to mitigate the severity of floods and droughts is largely due to corruption and lack of institutional support. Sometimes 'normal' floods and droughts (e.g. in the semi-arid countries) are declared as catastrophic in order to obtain more public funding to build large public works.

Investments in water policy offer great potential for enhancing the community infrastructure, which is crucial for keeping small natural hazards from becoming bigger emergencies. Indeed, there is considerable scope for redesigning natural water flow systems and water conservation structures at all levels. Many such projects are undertaken at public expense, from which private entrepreneurs reap the profits if there is no proper monitoring/appraisal system of such ventures and are, therefore, vulnerable to malpractices such as bribery and corruption.

The problems associated with droughts and floods arise in and around a river basin. However, institutional infrastructure meant to deal with them, even in the developed world, is fragmented. Thus, solutions tend to be *ad hoc*, mostly of the fire-fighting nature. A higher level of co-ordination between the monitoring agencies, planning agencies, and relief agencies will go a long way in alleviating human suffering and loss of property. The water prospecting and monitoring organizations ought to be seamlessly fused to the organizations responsible for planning, operations, and disaster relief.

12.3.6 Ethics and use of water

Agriculture is by far the largest consumer of water worldwide, yet the efficiency of water use in agriculture is very low. The world is dependent on irrigated agriculture that has become a major contributor to world food production. About 40% of global food production is from irrigated agriculture. Producing food to ensure global food security is a moral imperative as well as an ethical necessity. But ultimate food security comes from elimination of extreme poverty. Indeed, some of the poorest countries with low food security have hardly

been able to tap their water development potential properly.

Excessive attention on meeting short-term food requirements can, however, create severe, possibly irreversible, adverse impact on the environment and make it more difficult to meet food needs in the long run. On the other hand, excessive focus on preserving the natural environment and resource base can condemn millions to hunger and poverty. Therefore, the need is to develop a balanced approach to development of irrigation.

When water is used for irrigation, attention must also be paid to water quality, as pesticides, and fungicides can contaminate the surface runoff as well as infiltrating water from agricultural fields, impacting both aquatic life forms in surface water bodies and human users. The return seepage of applied irrigation water, which may contain high levels of nutrients (e.g. K, P, and N) as well as leaching of minerals from irrigated lands can be a significant source of groundwater pollution, particularly increasing its salinity. The problem may be further exacerbated by excessive use of irrigation water coupled with poor drainage of soils.

Attention should also be paid to the question whether, in times of scarcity, use of water for irrigation can be justified, particularly when the economic returns from industrial use of the same water may be many times higher, which would enable import of food grains.

According to experts, growing urban and industrial demands for water in the developing world will need to be met increasingly from irrigated agriculture, particularly from new water development projects. Therefore, promoting irrigation water efficiency together with appropriate conservation measures must be emphasized. Reducing poverty will also require supplementing agricultural yields from rain-fed agriculture with irrigation through various methods of water harvesting.

There is no doubt that, in general, industrial use of water gives more economic returns than agricultural use, which is also a necessity for poverty alleviation and development. But industrial waste waters can be toxic and highly polluting. Therefore, industries should use only the minimum required quantum of water and discharge their waste water only after treating it to environmentally safe

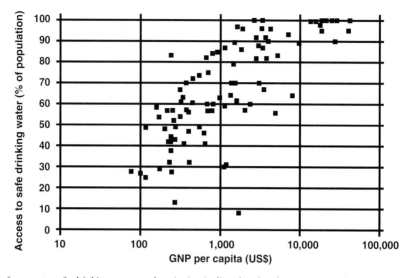

Fig. 12.1 Lack of access to safe drinking water and sanitation is directly related to poverty and indirectly to the inability of governments to invest in these systems. Redrawn from Priscoli *et al.* (2004) © UNESCO

levels of quality. Rather than waiting for irreversible damage to occur, pollution control should be exercised on a preventive basis employing scientific methods. During the last two to three decades, the water volumes used for industrial uses and the resulting pollution have dramatically decreased in many industrialized countries without causing significant adverse economic impact. In many developing countries, the same may be done through public awareness. Water is a precious resource and in the long run pollution levels should approach zero.

In 1995, approximately 55% of the global population was living in rural areas and about 45% in urban areas. By 2025, this scenario is expected to reverse, particularly in the developing world, with approximately 41% rural population and 59% urban. The rate of water supply and investment in nearly all the developing countries trails behind the urban growth. About 90% of the waste water in the developing world is discharged without any treatment. Thus, lack of access to safe drinking water and sanitation is directly related to poverty, as well as poor health status of the population. Fig. 12.1 shows that access to safe drinking water globally is directly related to the Gross National Product (GNP) and hence to investment in water. The situation calls for higher investment coupled with

establishing and enforcing of drinking water quality standards and protection of water sources from pollution and industrial effluents. Augmentation of water supply should occur with concomitant plans for sanitation. The quantum of finance for sanitation should be commensurate with the finance for water supply schemes. We must also look for innovative waste water treatment approaches beyond the conventional dilution systems.

In the prevailing schemes of water supply and sanitation services in urban areas, the poor actually pay higher price for water compared to the rich, though these costs are often hidden. It is also known that huge social dislocations can occur in cases where water prices are adjusted to reflect real costs, abolishing the subsidies. Political will coupled with ethical guidelines are necessary to deal with such issues.

12.3.7 Ethical aspects of groundwater use

The usual scale of surface water projects (dams, canals, etc.) is fairly large to warrant their design, finance, and construction by government agencies that normally manage and/or control the operation of large irrigation or urban public water supply systems. In contrast, groundwater development, which increased significantly during the past

50 years in most semi-arid or arid regions, has occurred through a number of small (largely private) entrepreneurs, with little administrative control of the government. This has resulted in largely uncontrolled development of groundwater that has led to problems such as decrease of well yields, deterioration of water quality, occurrence of land subsidence or collapse, interference with streams and surface water bodies, and ecological impact on wetlands or gallery forests.

Groundwater mining (or development of fossil water occurring in deep aquifers or of non-renewable groundwater resources) is controversial as it is not sustainable in the long run. However, under certain circumstances, it may be a reasonable option. The existence of fossil groundwater in a particular region by itself has no intrinsic value, except as a potential resource that can be tapped by future generations in times of scarcity. But this raises the question of how to determine whether fossil water will be needed more in the future than it is now.

The depletion of groundwater storage is not generally as serious a problem as groundwater quality degradation and may often be solved without much difficulty if water-use efficiency is improved. The crucial importance of preventing groundwater pollution in order to avoid a future water crisis to the extent of abandoning aquifers that are almost irreversibly polluted requires a concerted effort on the part of various concerned agencies as well as the users.

In general, groundwater development should not be rejected or seriously constrained if it is well planned and controlled. During the last few decades, groundwater withdrawal has undoubtedly made socio-economic benefits possible in addition to improving the quality of human life. Particularly in developing countries, it is a major source of potable drinking water, with 50% of municipal water supplies worldwide depending on it, as also in many rural areas and areas with sparse or scattered populations.

Real as well as perceived ecological impacts are becoming important constraints on groundwater development. These effects are mainly caused by water table depletion, which can culminate in decreasing flows or drying up of springs or reduced flow of streams, depletion of soil moisture to such an extent that it threatens the very survival of certain types of vegetation, and may lead to changes in local microclimates because of the decrease in evapotranspiration.

Irrigation with groundwater has made it possible to increase food production at a rate higherer than that of population growth; 70% of all groundwater withdrawals are used for this purpose, particularly in arid or semi-arid regions. It should also be noted that using groundwater for irrigating agriculture is often more expensive than using surface water, primarily because farmers generally bear all abstraction costs (development, maintenance, and operation) with little or no subsidy from the government. Groundwater usually produces significantly more income and jobs per cubic metre of water abstracted than surface water.

Despite the complexity of the problem and variety of responses that may be made based on location and time, there are several overbearing issues that have ethical implications in achieving sustainable, reasonable groundwater use. First, the subsidies (some being hidden) that have traditionally been a part of large hydraulic projects for surface water irrigation have led to the neglect of groundwater resources by water managers and decision makers alike. Careful consideration of cost-benefit could reveal that many proposed surface water projects are economically unviable, thus calling for serious consideration of groundwater planning, control, and management.

The question of groundwater ownership (public, private, or community owned) is also important. Some believe that the legal declaration of groundwater as a public domain resource is a necessary prerequisite for equiable groundwater development. This is far from being evident and there are examples where groundwater that had been in the public domain for many decades was subjected to chaotic management. Even so, promoting awareness in the use of groundwater for the common good is vital, particularly in view of the fact that thousands of stakeholders may subsist on a single aquifer of medium or large size. Groundwater management should be vested with the stakeholders, under supervision of an appropriate water authority.

12.3.8 Ethics and technology

Advances in technology may effectively eliminate much of the water supply problems which have resulted in public 'gloom and doom' scenarios. But technology does not completely remove the need for a multi-pronged management approach involving ethical issues related to flood management, agricultural uses, pollution of surface- and groundwaters, etc.

Therefore, it is undesirable to discard technological solutions altogether, just as it is to discard traditional approaches. In fact, both are needed. Water policy should not let the controversies over dams bias our perspective on the need for appropriate technology. As water systems begin to reach their limits, authorities face the daunting ethical imperative of managing risks to society that still persist. As societies begin to operate and manage their water, sanitation, and irrigation infrastructure, the limitations of their efficiency as well as technical competence may seriously affect their capacity to deal with extreme hydrologic events.

Hence new ethical norms may be needed for professional water resource engineers and managers. National, regional, and local governments, in partnership with consumers, users, and non-governmental organizations, together have a role to play.

12.4 Global water tele-connections and virtual water

In Section 12.3.6, the issue of the water use for irrigation in the scarcity situation was raised. It was noted that economic returns from the industrial use of the same water could alternatively enable import of food. We need to consider if it is appropriate from a global, regional, or national perspective in relation to water balance considerations, particularly choosing the option between depletion of local water resources resulting from irrigated agriculture against buying food from elsewhere where it is produced under water surplus or rain-fed conditions.

In the context of global trade, the water embedded in the production of goods or services is referred to as *virtual water*. Hoekstra and Chapagain (2007) have defined the virtual water content of a product (a commodity, good, or service) as 'the volume of fresh water used to produce the product, measured at the place where the product was actually produced.' It refers to the sum of the water used in various steps of the production chain. The water is said to be virtual because once a crop such as wheat (or any other produce) is grown, the real water used to grow it is no longer actually contained in the wheat or the produce.

Professor John Anthony Allan (1998) came up with the virtual water concept (Chenoweth 2008), for which he was awarded the 2008 Stockholm Water Prize. Virtual water has major impacts on global trade policy and research, especially in water-scarce regions, and explains how nations such as the United States, Argentina, and Brazil 'export' billions of litres of virtual water by exporting cereals, meat, fruits, etc. each year, while others such as Japan, Egypt, and Italy 'import' virtual water in billions of litres. The increasing inter-sectoral competition for water, the need to feed an ever-growing population, and increased water scarcity in many regions of the world, are some important considerations in rethinking the manner in which water is managed on our planet, considering its human dimensions.

The virtual water concept has opened the way to more productive uses of water. The importance of virtual water at the global level is likely to dramatically increase as projections show that food trade will increase rapidly: doubling for cereals and tripling for meat between 1993 and 2020 (Rosegrant and Ringler 1999).

One of the fundamental aspects of water management strategy is the ability to measure or evaluate fluxes and volumes of various goods, and virtual water is no exception. Its value is generally expressed as per unit volume (m^3), which results from multiplying the quantity of product (kg) by the unit value per unit mass of a product, expressed as volume of water per kg of product ($m^3 \, kg^{-1}$). Virtual water values for various food products, with reference to Californian production sites considering average productivity, are indicated in Fig. 12.2.

A breakdown of virtual water transferred around the world for main types of agricultural products

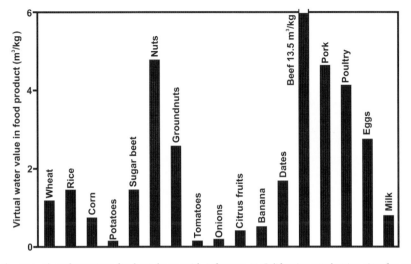

Fig. 12.2 Virtual water values for various food products, with reference to Californian production sites for average productivity. Redrawn from Renault (2003) © FAO.

is shown in Fig. 12.3. This figure is interesting because it shows that cereals, which have captured most of the attention in food security and virtual water studies, account for only for 24% of the total volume of virtual water exchanged. Of course, when it comes to nutritional values, and for arid regions, the importance of cereals is much higher than one-quarter. In 2000, cereals contributed to 40% of the food energy trade (Zimmer and Renault 2003).

12.4.1 The three colours of virtual-water content of a product

The virtual-water content of a product consists of three components: green, blue, and grey.

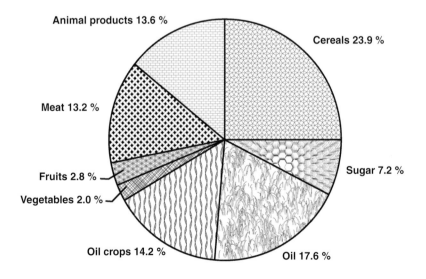

Fig. 12.3 Global virtual water trade partitioned into main types of agricultural food products. Redrawn from Renault (2003) © FAO.

The 'green' virtual-water content of a product is the volume of rainwater plus irrigation water that evapotranspires during the production process. This is mainly relevant for agricultural products, where it refers to the total water evaporated from a field during the growing period of a crop.

The 'blue' virtual-water content of a product is the volume of surface water or groundwater that evaporates as a result of the production. In crop production, the blue water content of a crop is defined as the sum of the evapotranspiration of irrigation water from the field and the evaporation of water from irrigation canals and artificial storage reservoirs created for the purpose of irrigation. In industrial production and domestic water supply, the blue water content of the product or service is equal to the part of the water withdrawn from ground or surface water that evaporates and thus does not return to the system where it originated.

The 'grey' virtual-water content of a product is the volume of water that becomes polluted during its production. This can be quantified by calculating the volume of water required to dilute the pollutants added to the natural water system during its production process, to such an extent that the quality of the ambient water remains inferior compared to the accepted water quality standards.

The distinction between green and blue water originates from the work of Falkenmark (2003). Evaporated water and polluted water have one thing in common, in that both are 'lost', that is, they become unavailable for other uses.

The concept of virtual water has its pros and cons. Preliminary studies show that improving our database on virtual water is likely to throw some light on the ongoing water management debate. Another important point related to production and trade, is that 'water' is not the only facet of the decision-making process. The issues of comparative advantages of considering land, jobs, rural development, and access to markets (Wichelns 2001) are also important. It is clear that looking at water in the food trade is not enough but at least it should be well understood, and this is one of the purposes of current research on virtual water.

Once all the virtual water consumed in the products that are bought is added up, along with the daily use of water coming out of the tap, a better idea of one's *water footprint* is obtained. Water footprints are used to give a better consumption-based indicator of water use on a national level.

13

Some case studies

In previous chapters we have discussed the science of hydrology, methods for studying the flow of water on the land surface or under ground, techniques for understanding various hydrologic systems, and processes on various spatial and temporal scales. The aim is to harness water resources of a given region taking into account several factors, including the societal needs along with environmental conservation.

In this chapter hydrologic developments in four regions from different parts of the world, currently facing severe water problems, are described as case studies. The objective is to demonstrate that while basic principles of hydrology are the same everywhere, for resource development they have to be region-specific. Solving a given problem requires considerable expertise and detailed understanding of local/regional hydrologic systems and, therefore, solutions obtained in one region, may or may not be directly applicable to other regions.

13.1 The Yellow River Basin, China

The Yellow River (Huang He) Basin, known as 'the cradle of Chinese civilization', has played a key role not only in the economic development of China but also in the historical and cultural identity of its people. The Yellow River has been rightly cast both as a source of great prosperity as well as of great disaster. Ironically, the Yellow River is also known as 'China's Sorrow' because the soils deposited by the river in its flood plain, which have fostered human civilization, are also associated with frequent, sometimes catastrophic floods.

In addition to the 'traditional' floods caused by heavy rain and consequent runoff, a second form of flooding, known as ice floods, occurs in winter and early spring, for quite different reasons. Ice floods occur when the river flow is south to north (Fig. 13.1; Plate 9), as happens in the Ningxia-Inner Mongolian area and in the lower reach starting near Kaifeng city. In winter, the higher latitudes at the downstream reaches are the first to freeze, blocking river flow from the lower latitude upstream regions, thereby causing floods. Similarly, downstream reaches at higher latitudes remain frozen longer into the spring than the upstream reaches, again blocking the flow and causing floods. It has been estimated that ice floods have been responsible for about one-third of all floods in the basin. Ice floods are notoriously difficult to control and the phenomenon in the lower reaches is a function of the particular course of the river. When the river's mouth was further south, as it was periodically before 1855, the occurrence of such floods was rare.

In addition to floods, droughts too have been a part of the Yellow River Basin. The catastrophic devastation brought about by floods and droughts forced successive Chinese administrations, from the legendary Xia Dynasty (*ca.* 2000 BC) through to the 20th century, to make flood control and water

Modern Hydrology and Sustainable Water Development, 1st edition. © S.K. Gupta
Published 2011 by Blackwell Publishing Ltd.

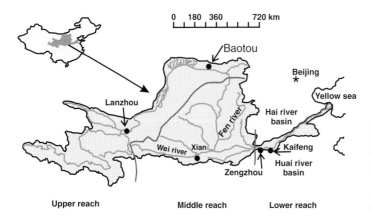

Fig. 13.1 The Yellow River Basin, China. By Chinese convention, the Yellow River is divided into three reaches, namely, upper, middle, and lower, as demarcated by red lines. Redrawn from Giordano *et al.* (2004) © Comprehensive Assessment Secretariat. See also Plate 13.1.

harnessing the utmost priority of Yellow River management. While the possibility of flooding is ever present and remains a key issue in basin management, major achievements in flood control have been made during the last 50 years. As a result of this success and the rapid economic and social changes that have taken place over the past few decades, new issues such as water scarcity, overuse of resources, and environmental degradation have become serious concerns of the river management agenda. In essence, a transition is now taking place in which the focus is shifting, from prevention of the river adversely affecting people to preventing the people from damaging the river ecology.

13.1.1 *Physical geography*

The Yellow River originates in the Qinghai-Tibetan plateau of Qinghai province, from where it flows across eight other provinces and autonomous regions before its outfall into the Yellow Sea north of the Shandong peninsula (Fig. 13.1). With a length of over 5400 *km*, it is the second longest river in China. The basin contains approximately 9% of China's population and 17% of its agricultural area. The Yellow River is often divided into three main reaches for purposes of analysis.

The upper reach of the Yellow River begins on the Qinghai-Tibetan Plateau where the terrain is mountainous with steep rock slopes, low evaporation, and high moisture retention, producing runoff coefficients estimated to range from 30%

(WB 1993) to 50% (Greer 1979). This, combined with relatively high precipitation levels, results in this westernmost region of the upper reach contributing 56% of the river's total runoff at Lanzhou gauging station (YRCC 2002). The evaporation increases, becoming several times that of the precipitation, as the river moves northwards into the Ningxia/Inner Mongolian plains and the Gobi Desert (WB 1993). As a result, in this section of the river, runoff is dissipated considerably and flow is greatly reduced. The spatial variation in flow contribution within the upper reaches is further accentuated by human usage. The westernmost regions of the upper reach have relatively low population densities, agricultural development, and industrialization that limit the local usage. Northwards from Lanzhou, the agricultural population, with its long history of irrigation, and a growing industrial base, leads to substantially increased water withdrawals from the river.

The middle reach of the Yellow River begins at the Hekouzhen gauging station downstream of the city of Baotou (YRCC 2002). This reach includes some of the Yellow River's major tributaries, such as the Fen and the Wei, which contribute substantially to the total flow. As the river course takes a southward turn, it cuts through the Loess Plateau and its potentially fertile but highly erodible loess soils. These soils contribute massive quantities of silt to the main stream and its tributaries, resulting in average sediment concentrations that are unprecedented amongst major waterways (Milliman and Meade 1983) and giving both the river and

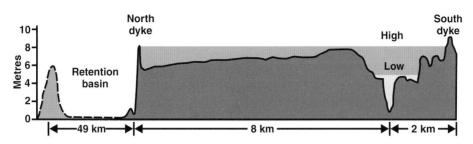

Fig. 13.2 A representative cross-section of the 'suspended' Yellow River. The diagram is not drawn to scale. Redrawn from Ronan (1995). © Cambridge University Press. See also Plate 13.2.

the sea into which it outfalls, their common name, 'Yellow'. Control of the potentially devastating Yellow River floods, which are greatly exacerbated by the high sediment loads generated in the middle reach, has formed a central theme in Chinese water management and politics for at least 3000 years. In addition, control of sedimentation to reduce the severity and frequency of flooding, accomplished through flushing, is now estimated to require about 25% of the total Yellow River flow and is, therefore, a major factor in current utilization of basin water.

The *lower reach* of the Yellow River commences at Huayuankou, near the city of Zhengzhou, and forms one of the most unique river segments in the world. Here the sediment transported from the middle reach begins to settle as the river spills onto the flat North China Plain, producing a consistently aggrading bed and a naturally meandering and unstable channel (Ren and Walker 1998). This instability has in fact been so severe that the Yellow River had undergone 6 major changes in its channel over the past 3500 years, in which its outfall point to the sea shifted by 400 *km* from one side of the Shandong Peninsula to the other (Greer 1979). These massive shifts in the river channel, as well as more frequent smaller movements, clearly create problems for millions of people who attempt to farm the fertile alluvial soils of the lower reach. As a mitigation measure, successive river managers down the millennia have constructed levees along the banks of the river in an attempt to stabilize the main channel. While such structures may contain the channel in the short term, their success depends on consistently raising levee walls as sediment elevates the level of the channel bed constrained by the levee walls. Over time, the process of levee raising

has contributed to a 'suspended' river, in which the channel bottom rises above ground level, sometimes by more than 10 *m* (Fig. 13.2; Plate 10). With the channel rising above ground level, the surrounding landscape cannot drain into the river nor can tributaries enter it. This essentially means that the river 'basin' becomes a narrow corridor no wider than a few *km* breadth of the embanked channel. With almost no inflow, the contribution of the lower reach is limited to only 3% of the total runoff. While much of the sediment is deposited in the lower reach, historically, approximately half has reached the river's outlet to the sea. These large deposits have, at least until recently, caused the river's delta to expand outward, creating substantial new farmland (Ren and Walker 1998).

When the Yellow River's channel shifts, typically after a flood event or from human intervention, it connects to either the Hai River system to the north or to the Huai River system to the south, resulting in an expansion of basin boundaries across various portions of the North China Plain. The last time such a change occurred was in 1938 when the Yellow River's south dyke was deliberately breached at Huayuankou to block advance of the Japanese army. The present course of the river was restored by engineering works in 1947. The imposition of the Grand Canal, which runs perpendicular to the generally east-west flowing rivers of eastern China and essentially links all the basins from Hangzhou north to Tianjin, further complicates the precise delineation of basin boundaries in the lower reach of the river.

Due to the highly engineered river systems of the region, there is considerable lack of congruence between the geographical extent of the basin, as commonly delineated, and the associated hydro-logic units. For example, in the lower reach of the basin, seepage from the suspended main stem of the river recharges groundwater aquifers in both the Hai and Huai basins, where it is extracted for crop production. Additional water is also trans-ferred out of the basin for industrial and domestic use, especially to the cities of Jinan, Qingdao, and Tianjin. Of potentially greater significance is the planned 'South waters North' engineering schemes, which may eventually transfer large amounts of water from the Changjiang Basin into the Yellow River, further blurring the relevance of the geographical definition of the Yellow River Basin (Biswas *et al*. 1983; Liu 1998).

13.1.2 Water supply and use

On average, the Yellow River Basin receives ap-proximately 450 *mm* of rainfall annually. Spatially, precipitation tends to be lowest in the north and west and highest in the south and east, with an average of 120 *mm* in the areas of Ningxia and Inner Mongolia and 800 *mm* in the hills of Shandong. Precipitation levels across much of the Loess Plateau, where soils are the most erodible, average less than 500 *mm*. Intra-annually, most rainfall occurs between June and September, with

almost no precipitation between November and March. In addition, summer rainfall, especially in the Loess areas of the middle reach, often occurs in intense spells of cloudbursts, which results in the soil infiltration capacity being exceeded quickly, causing substantial runoff and erosion and transportation of massive quantities of sediments into the Yellow River and its tributaries. It has been estimated that as much as 35% of precipitation occurs in the form of these sudden cloud bursts (Greer 1979). Inter-annual precipitation variability is such that, by some definitions, drought can be said to occur almost every two to three years.

The water resources of the Yellow River Basin for the year 2000 and the framework used for their accounting by the Yellow River Conservancy Com-mission (YRCC) are shown in Table 13.1.

Water used in the Yellow River Basin is currently considered to be contributed both by groundwater and surface water, and is used for agricultural, industrial, and domestic uses. Data for recent years on the use by various sectors and the contributing sources is shown in Table 13.2. As seen from the table, the average annual withdrawal from the Yellow River Basin is approximately 50×10^9 m^3, of which about 74% is from surface water and the remaining 26% from groundwater. Agriculture is by far the largest consumer of

Table 13.1 An estimate of the water resources of the Yellow River Basin (in 10^9 m^3) for the year 2000.

		Gauging Station				
		LZ	TDG	LM	SMX	HYK
(1) Surface runoff	(a) Measured flow	26.0	14.0	15.7	16.3	16.5
	(b) Depletion	2.7	13.0	13.6	17.0	18.4
	(c) Change in storage	−3.3	−3.3	−3.3	−3.2	0.1
	Surface runoff = (a) + (b) + (c)	25.4	23.7	26.0	30.1	35.0
(2) Groundwater	(e) Hilly area	12.6	13.1	15.3	19.7	22.6
	(f) Plain area	1.6	7.6	9.5	14.6	15.4
	(g) Double counting in (e) and (f)	0.7	1.3	1.8	3.8	4.1
	Groundwater = (e) + (f) − (g)	13.5	19.5	23.0	30.4	33.9
(3) Double counting in (1) and (2)		12.8	17.2	18.6	22.4	24.7
(4) Total water resources = (1) + (2) − (3)		26.0	26.0	30.4	38.1	44.1

Note: LZ = Lanshou; TDG = Toudaoguai; LM = Longmen; SMX = Sanmenxia; HYK = Huayuankou. *Source*: Giordano *et al*. (2004).
Source: (Giordano *et al*., 2004).

Table 13.2 An estimate of Yellow River Basin water sources and withdrawal for the period 1998-2000 (in 10^9 m^3).

	By source			By sector				
						Domestic		
	Surface water	Ground water	Total	Agri.	Ind.	Urban	Rural	Total
1998	37.0	12.7	49.7	40.5	6.1	1.6	1.5	49.7
1999	38.4	13.3	51.7	42.6	5.7	1.8	1.5	51.7
2000	34.6	13.5	48.1	38.1	6.3	2.1	1.6	48.1
Average	36.7	13.2	49.8	40.4	6.0	1.8	1.5	49.8
Share	74%	26%	100%	81%	12%	4%	3%	100%

Note: water withdrawal includes 2.7×10^9 m^3 pumping in the reach of the river downstream of Huayuankou.
Source: Giordano *et al*. (2004).

water, accounting for 80% of the total withdrawal, with industrial, urban, and rural domestic sectors sharing the remaining 20%.

13.1.3 Results of some important studies

In this section, some important studies and their results, based on various hydrologic investigation techniques discussed in the previous chapters, are described. These investigations have enhanced understanding of the development course and its impacts on the hydrologic system of the Yellow River Basin, enabling appreciation of the emerging issues

and thereby contributing to evolving appropriate mitigation strategies.

From the recorded data of rainfall, runoff, and rainfall/runoff ratios at various gauging stations located on the three reaches of the Yellow River (Table 13.3), it is clear that during the 1990s rainfall, and hence runoff, was substantially lower than in previous decades. A similar but apparently less severe dry spell had earlier been recorded in the decade 1922-1932. Therefore, it is pertinent to ask if the observed declines of the 1990s represent a short-term climatic cycle or have been caused by secular decline in long-term precipitation, probably brought about by global climate change.

Table 13.3 Rainfall-runoff in Yellow River Basin, 1956-2000.

Reach	Area ($\times 10^3$) km^2		Time period					1990s Change from average
			1956–1970	1971–1980	1981–1990	1991–2000	Avg.	
Upper	368	Rain (mm)	380	374	373	360	372	−3%
		Runoff ($\times 10^9$ m^3)	35	34	37	28	34	−16%
		Runoff yield (%)	25	25	27	21	24	−13%
Middle	362	Rain (mm)	570	515	529	456	523	−13%
		Runoff ($\times 10^9$ m^3)	29	21	23	15	23	−34%
		Runoff yield (%)	14	11	12	9	12	−25%
Lower	22	Rain (mm)	733	689	616	614	67	−8%
		Runoff ($\times 10^9$ m^3)	1.5	1.1	0.6	0.0	0.8	−100%
		Runoff yield (%)						
Basin	752	Rain (mm)	482	451	455	413	454	−9%
		Runoff ($\times 10^9$ m^3)	65	56	61	43	57	−24%

Source: Giordano *et al*. (2004).

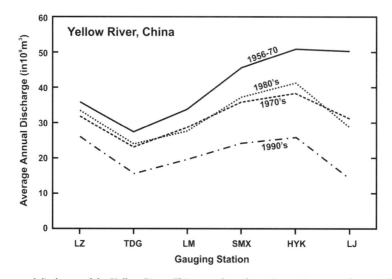

Fig. 13.3 Average annual discharge of the Yellow River, China, at selected gauging stations over the period 1956-2000. From left to right: the gauging locations are progressively in the downstream direction. Redrawn from Giordano *et al.* (2004). *Note*: LZ = Lanshou; TDG = Toudaoguai; LM = Longmen; SMX = Sanmenxia; HYK = Huayuankou; LJ = Lijin. Redrawn from Giordano *et al.* (2004) © Comprehensive Assessment Secretariat.

However, as shown graphically in Fig. 13.3, the runoff decline is not a phenomenon of the 1990s alone. Furthermore, decrease in rainfall does not fully explain the decline in the rainfall/runoff ratio. It is also noted that the relative decline in runoff seems to increase from upstream to downstream regions. This suggests that additional water detention and use in upstream regions also partly contribute to the runoff decline (Ongley 2000).

In response to declining supplies and increasing demand, groundwater pumping has also increased dramatically over the past two decades. From 1980 to 2002, groundwater abstraction increased by 5.1 km^3, that is, by 61% of pre-1980 values. In some regions of the basin, groundwater depletion is now a major problem. For example, in the Guanzhong plains located in the middle reach of the basin in the vicinity of Xian, the area with a groundwater depth of less than 4 m shrank from 85% of the total land in 1981 to 23% in 2000, while the area with a groundwater depth of more than 8 m increased by 24%, from 10 700 km^2 in 1981 to 13 200 km^2 in 2000 (Giordano *et al.* 2004). Groundwater overdraft is now causing significant land subsidence. For example, in the city of Xian, average

annual subsidence of the land surface is now as high as 35 mm in some areas and the largest accumulated subsidence has reached 1340 mm.

Due to declining supplies and increasing demand, the lower reach of the Yellow River has shown a seasonal desiccation of the river since the early 1970s. From 1995 to 1998 there was no flow in the lower reach for about 120 days in a year and in some cases the last 700 km stretch of the river dried up completely before it reached the sea, even before reaching Shandong province. Although it is perhaps the most visible manifestation of water scarcity in China, the drying up of the Yellow River is only one of many such signs. The Huai River, a smaller river situated between the Yellow and Yangtze rivers, also dried up during 1997 and failed to reach the sea for 90 days. Satellite images show hundreds of lakes disappearing and local streams drying up in recent years. As water tables fell and springs ceased to flow, millions of Chinese farmers found their wells going dry (Brown and Halweil 1998).

The dwindling of river flow has three important repercussions for the basin. First, it limits the availability of surface water for human use. Second, it limits the capacity of the river to carry its heavy sediment load and dump it into the sea, resulting in a more rapid aggradation of the already flood-prone channel. Third, it has deleterious consequences on the ecology of the downstream areas and, in particular, the Yellow River delta and coastal fisheries. After strengthening of the 1987 Water Allocation Scheme in 1998, the YRCC managed to marginally end the 'no-flow' condition since 1999, though flow levels were still far below those considered necessary from environmental considerations.

The primary environmental water use in the Yellow River, according to basin managers, is for sediment flushing to control potentially devastating floods. This requirement is estimated at $14 \times 10^9 \ m^3 \ a^{-1}$ and is about 25% more than the present flow. In the 'traditional' sense, from ecological considerations, lean-season flows are also required for biodiversity protection and sustenance of grasses, wetlands, and fisheries at the mouth of the river. To meet these needs, a minimum environmental flow requirement at the river mouth is estimated at $5 \times 10^9 \ m^3 \ a^{-1}$, along with a minimum continuous flow of $50 \ m^3 \ s^{-1}$ at the Lijin gauging station. The minimum flow requirement is also expected to partly meet requirements for sand flushing. Similarly, both the overall sediment flushing and minimum flow requirements are currently seen as adequate for the river to continue its function of diluting and mitigating the effect of human introduced pollutants. In view of these considerations, the ecological water requirements for the Yellow River Basin are currently estimated by the YRCC at $20 \times 10^9 \ m^3 \ a^{-1}$, a figure likely to remain relatively constant, as any reduction in sediment flushing requirement is offset by increase in erosion control requirement.

An estimate of regional scale annual ET in the Yellow River Basin has been made using NOAA-AVHRR NDVI satellite data on monthly mean air temperature and precipitation from 1982 to 2000. The model estimates were validated by the observed runoff data (Sun et al. 2004). Data of both pan evaporation and reference potential evaporation using climate data from 140 weather stations and hydrologic data from 71 stations in the Yellow River Basin analysed by Cong et al. (2008), show a significantly decreasing trend of evaporation. This is in conformity with several observations from all over the world that also show steadily decreasing values of pan evaporation. This observation is, however, contrary to the expectation that, as a result of global warming for the past few decades, the atmosphere has become drier and consequently evaporation should increase. The anomalous behaviour is called the 'evaporation paradox'. At the same time, actual evaporation estimated by water balance methods decreased with decrease in precipitation in most of the sub-basins. With increase of precipitation, potential evaporation and actual evaporation have an inverse relationship; while based on yearly results, potential evaporation and actual evaporation have a proportional relationship (Cong et al. 2008).

Liu and Zheng (2004) used linear regression and the Mann-Kendall method to detect trends in the hydrologic cycle components of the Yellow River and found that for the Lanzhou station, only surface runoff showed a decreasing trend. For Huayuankou station, the results show that both surface runoff and groundwater runoff have a significant decreasing trend. This study also suggests that increasing water resources exploitation and utilization are the two most important factors in causing frequent drying up of the main course of the Yellow River. But similarities of the trends in precipitation and natural runoff also suggest a link between climate and the hydrologic cycle.

Impact of climate change and human activity in the Yellow River Basin has been evaluated by quantitative analysis of the hydrologic trends obtained by incorporating historical meteorological data and available geographical information on the conditions of the landscape in a distributed hydrologic model to simulate the natural runoff, without considering the engineered water intake (Cong et al. 2009). By comparing the observed data and the results simulated by the model, it is found that the simulated natural runoff follows a similar trend as the precipitation in the entire area being studied during the last half century, implying thereby that changes in natural runoff are mainly controlled by climate change rather than by landuse change. Changes in actual evapotranspiration upstream of the Lanzhou gauge are controlled by changes – both in precipitation and potential evaporation, while changes of actual evapotranspiration downstream of the Lanzhou gauge are controlled mainly by changes in precipitation alone. The difference between the annual observed runoff and the simulated runoff indicates that there is little engineered water consumption upstream of the Lanzhou gauge, but it becomes larger downstream of the Lanzhou gauge. The engineered water consumption shows a significant increasing trend during the past 50 years and is the main cause of the drying up of the Yellow River. However, in contrast to the common perception that the near total drying up downstream of the river during the 1990s is caused by rapid increase in the engineered water consumption during the same period, it has been found that the main cause of this aggravation is the drier climate that has persisted since the 1990s. The main reason for the river flow situation improving in the 21st century is because of better water resources management since 2000 (Cong *et al*. 2009).

With a view to understanding the recent flow reduction of the Yellow River on a long-term basis, the tree-ring width indices of the six local *Juniperus przewalski* chronologies were constructed for the past 593 years (Gou *et al*. 2007) and were found to correlate significantly with the observed stream flow of the river recorded at the Tangnaihai hydrologic station. Several severe droughts and low-flow events are recognized in the decades 1920–1930, 1820–1830, 1700–1710, 1590–1600, and 1480–1490. The most severe droughts during 1480–1490 were also recorded in other studies on the Tibetan Plateau. The reconstructed increase in stream flow during much of the 20th century also coincides with generally wetter conditions in the Tienshan and Qilianshan mountains of China, as well as in northern Pakistan and Mongolia. After the 1980s, the tree-ring data agrees with the decreasing trend in stream flow. Presently, Yellow River stream flow is relatively low but lies within the range of stream flow fluctuations that occurred during the past six centuries (Gou *et al*. 2007).

Variation of $\delta^{18}O$, δD, and ^3H in the Yellow River were analysed on water samples from 18 sections of the Yellow River in rainy and dry seasons (Su *et al*. 2004). The results show an increasing trend of the ratios of the stable isotopes and a progressive decrease in the ^3H concentration from the river source to its estuary. The main factors affecting the isotopes in the river water are identified as:

(i) mixing of water from external water bodies; (ii) evaporation; and (iii) human activities. The primary river runoff arises mainly from the source in mountainous areas and the middle reaches of the river; (iv) in the river source above Lanzhou, the seasonal variation of $\delta^{18}O$ and δD in river water is different from that in precipitation, which indicates that the precipitation is not the only component of the river flow but groundwater is also an important component. The relative contribution of groundwater to the river flow in the dry season is larger than that in the rainy season. (v) In the northern segment of Jinshan gorge in the rainy season, the river water is recharged by karst groundwater and local precipitation with low $\delta^{18}O$ and tritium concentrations. In the dry season, the river is recharged by the karst water with low tritium concentrations. (vi) In the river segment between Wubao and Tongguan, the Yellow River is recharged by the tributaries with high isotope ratios in both seasons. (vii) In the segment between Lanzhou and Baotou and in the lower reaches, the evaporation from the river water surface has a large effect on the stable isotope ratios in the dry season, but in the rainy season, it is small. The evaporation of water from the agricultural fields irrigated by river water and subsequent return flow of this water joining the river are the two main factors that affect the isotopes in river water in these river segments (Su *et al*. 2004).

Groundwater levels of 39 observation wells, including 35 unconfined wells and 4 confined wells from North China Plain (NCP) monitored during 2004–2006 using automatic groundwater monitoring data loggers, were interpreted in conjunction with the major factors affecting the dynamics of groundwater flow, such as topography and landform, depth to water table, extent of groundwater exploitation, and proximity to rivers and lakes (Wang *et al*. 2009). This study identified six dynamic patterns of groundwater level fluctuations in NCP:

1. Discharge pattern in the piedmont plains, with steadily declining water level year after year with little seasonal fluctuation;
2. Lateral recharge–runoff–discharge pattern in the piedmont plains – quick rise of water table in July–August, due to lateral recharge from

mountain areas, lasting until March or April of the following year;

3. Recharge–discharge pattern in the central channel zone – the lowest water table from March to June/July caused by pumping for irrigation, which recovers after the end of the irrigation season due to the return seepage contributing to groundwater recharge. This persists until February or March of the following year;

4. Precipitation infiltration–evaporation pattern in the shallow groundwater region of the central plains, resulting in the lowest water table in the dry season with high evaporation and less precipitation. During the spring and early summer, the water table rises rapidly due to recharge by precipitation;

5. Lateral recharge–evaporation pattern in the recharge-affected area along the Yellow River – because of the high water level of the Yellow River, the river water recharges the groundwater in the vicinity, and the position of the water table changes in tune with the level of river stage;

6. Infiltration–discharge–evaporation pattern in the littoral plains – large pocket of saline water is distributed in the littoral plains, even though the degree of groundwater exploitation is low and in some regions there is no groundwater development at all. The precipitation can recharge the groundwater quickly and the evaporation is high because of the shallow water table. Also, the sea water may intrude and recharge the groundwater in the coastal aquifers.

Geological exploration and multiple well-pumping tests were conducted in the Yellow River terrace in the Zhengzhou area (Zisheng *et al.* 2004). Based on collected data, a groundwater flow net was constructed. Groundwater levels were measured and the water-bearing potential of the aquifers was evaluated. The results show that: (i) the alluvial sand deposit encountered within a depth of 80 m below the flood plain of the 'hanging river' in the lower reaches of the Yellow River constitutes a high water-bearing potential aquifer and its single well yield is about 5000–9000 $m^3 \, d^{-1}$; (ii) under natural conditions, the river discharges into the aquifer. The amount of this discharge increases

under large-scale groundwater exploitation. This indicates that large-scale waterworks can be developed in the region; (iii) stability of the Great Dyke of the Yellow River and runoff downstream may be influenced to some extent by groundwater exploitation along the river course (Zisheng *et al.* 2004).

Water samples from Qinghai show some unusual results. Normally the Qinghai samples are expected to show 'light' δ-values of both oxygen and hydrogen due to the altitude effect and some samples, in fact, do show 'light' δ-values. However, many other samples from Qinghai Lake show 'heavy' δ-values. This may be ascribed to a special condition such as low precipitation rates and high evaporation rates, where the evaporated 'light' vapour mass is quickly removed by the prevailing westerlies (west winds), which results in the 'heavy' water preferentially remaining in the lake environment (WIRG 2005).

Groundwater in deep confined aquifers is one of the major water sources for agricultural, industrial, and domestic uses in the North China Plain. Detailed information on groundwater age and recharge was obtained using ^{14}C of dissolved inorganic carbon and tritium in water (Zongyu *et al.* 2005). The isotopic data suggest that most groundwater in the piedmont region of the North China Plain is less than 40 years old and is recharged locally. In contrast, groundwater in the central and littoral portions of the North China Plain is 10 000–25 000 years old. $\delta^{18}O$ and δD values of this groundwater are 1.7‰ and 11‰ respectively, which are less than those in the piedmont plain groundwater and possibly reflect water recharged during a cooler climate during the Last Glaciation. The temperature of this recharge water, based on $\delta^{18}O$ values, lies in the range 3.7–8.4°C, compared to 12–13°C of modern recharge water. The combined isotopic dataset indicates that groundwater in the central and littoral part of the North China Plain is being mined under non-steady-state conditions (Zongyu *et al.* 2005).

In a study (Chen *et al.* 2002) from the lower reach of the Yellow River in Shandang province, comparison of groundwater potential and EC distribution on a regional scale indicates that the chemical solutes are transported by the groundwater flow system, and become concentrated along the flow path, with the lowest gradient and the maximum concentration occurring in topographically depressed areas. The unique linear relationships between the main anions are a result of mixing. The isotopic signature was used to calculate the contribution from three sources, that is, rainfall, old water, and diverted water from the Yellow River. The calculation indicated, on average, that the mixing rate from the above three sources is 18, 17, and 65%, respectively. The Yellow River is the dominant source for the local aquifer, and the amount contributed by rainfall is negligibly small for wells deeper than 30 m. The groundwater flow and chemistry are well integrated to form an interdependent system, which is driven primarily by the potential gradient. Because of this integration and surface–groundwater interaction, chemical components and the isotopic signature can be used as important indicators to trace the groundwater flow (Chen et al. 2002).

Geochemical analyses of the groundwater samples indicated that the groundwater in the North China Plain (NCP) has a two-layer structure, with a boundary at a depth of about 100–150 m. The two layers differ in pH, concentrations of SiO_2 and major ions, and isotopes (^{18}O, ^2H, and ^3H) (Fig. 13.4 and Fig. 13.5). Chemical components in the upper layer showed a wider range and higher variability

than those in the lower layer, indicating the impact of human activity. Groundwater in the upper layer indicated eastward flow, while the groundwater in the lower layer indicated northeastward flow. Three hydrogeological zones are identified: recharge (Zone I), intermediate (Zone II), and discharge (Zone III). The recharge zone was found to be low in chloride (Cl^-) but high in tritium. The discharge zone was found to be high in Cl^- and low in ^3H. This may be due to the difference in groundwater age. The discharge zone was subdivided into two sub-zones, Zone III1 and Zone III2, in terms of the impact of human activities. Zone III2 is strongly affected by water diversions from the Yellow River. As groundwater flows from the recharge zone to the intermediate and discharge zones, chemical patterns evolve in the order: Ca-HCO_3 > Mg-HCO_3 > Na-Cl > SO_4 (Chen et al. 2004).

A major economic activity throughout the Yellow River Basin is irrigated agriculture. The area under this has doubled in the last five decades. Data for major ions at 63 monitoring stations in the Yellow River system during the period 1960–2000 indicated that concentration of major ions in much of the basin, and especially downstream of the major irrigation areas of the upper basin, increased

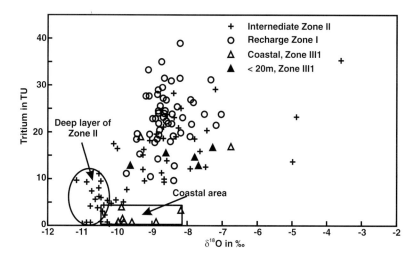

Fig. 13.4 Relationship between $\delta^{18}O$ and tritium values in the three hydrogeologic zones of the North China Plains. The recharge zone has high tritium levels of >10 TU, with $\delta^{18}O$ > −10‰, and the coastal zone (Zone III1) has a low tritium value (<5 TU), with $\delta^{18}O$ > −10.4‰, except for shallow wells with <20 m depth. The deeper layer of Zone II has tritium values of <10 TU and depleted $\delta^{18}O$ (< −10‰). Redrawn from Chen et al. (2004). © Wiley Interscience.

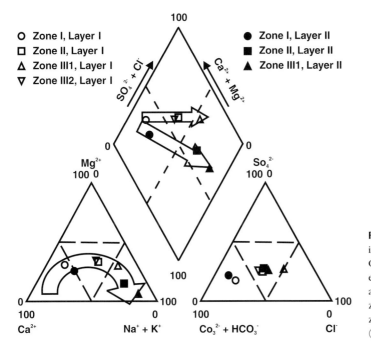

Fig. 13.5 Piper diagram of major ions found in the groundwater samples from the North China Plains. The arrows indicate the evolution of chemical facies for two layers as groundwater flows from the recharge zone to the intermediate and the discharge zones. Redrawn from Chen *et al*. (2004). © Wiley Interscience.

significantly between 1960 and 2000 (Chen *et al*. 2003). This indicates that the increasing trend is mainly a result of saline irrigation return waters, especially from irrigated lands in arid zones of northern China where sodium increases by 100% and total dissolved solids increases by 29% in the main stem of the river.

A narrow range of $\delta^{18}O$ from $-10\permil$ to $-8\permil$ and a wide range of δD, either highly enriched or depleted, were found in groundwater samples in the recharge zone (Fig. 13.6). The relatively enriched δD in springs or large diameter open wells of the recharge zone agrees well with the global meteoric line (GMWL) given by (Craig 1961a, b), implying that modern rainfall is the dominant component of such groundwater. The depleted δD locations are mainly found for wells near the Hutuo River, which is supplied mainly by the outflow from the Huangbizhuang reservoir. Though relatively enriched $\delta^{18}O$ and δD are found for Zone I and Zone III2, it was difficult to separate these zones based on the $\delta^{18}O$-δD relationship alone. The most depleted $\delta^{18}O$ and δD are found for the lower layer of the intermediate zone (Chen *et al*. 2004). Stable isotope data indicates that isotope values of groundwater tends to deplete systematically from discharge areas near the sea towards the recharge areas in mountains, as well as in deep groundwater in lowland areas. This suggests that shallow groundwater in the study area belongs to the local flow system and deep groundwater is part of the regional flow system (Aji *et al*. 2008).

In recent years, the earlier prevailing view of climate as static and unchanging on time scales important to river managers has given way to a new understanding that the gauged record represents only a small temporal window of the variability encompassing many centuries of river hydroclimate. River basin management decisions are inherently forward-looking and rely heavily on climate forecasts. These forecasts typically assume that past characteristics of the river basin system, as revealed by observations, will remain invariant in the future. However, the prospect of changing meteorological conditions and climate behaviour, associated with anthropogenic emissions of greenhouse gases, makes this assumption questionable. As a result, many water managers today are exploring innovative ways of water planning and management strategies.

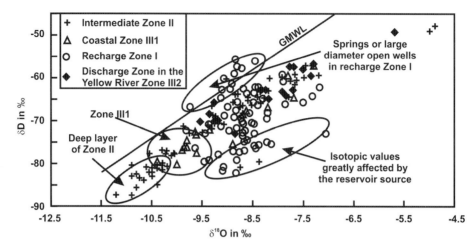

Fig. 13.6 Relationship between $\delta^{18}O$ and δD in groundwater samples from North China Plains. Redrawn from Chen *et al.* (2004). © Wiley Interscience.

Recent research on response of the Yellow River to climate change using numerical models of the global atmospheric general circulation, commonly known as General Circulation Models (GCMs), together with hydrologic models, predicts that by the year 2030 the runoff may decrease and the annual water deficit may be in the range of $1.9 \times 10^9 \ m^3$ to $12 \times 10^9 \ m^3$ (Liu 1997). Ye Aizhong and Wang (2006) and Xia *et al.* (2005) suggest that climatic factors greatly influence the course of hydrology in the Yellow River, and increase in precipitation by 10% exerts greater influence on runoff than the impact of a decrease by 10%. Wang and Shi (2000) analysed the response of hydrology in the upstream reach of the Yellow River and concluded that any change in precipitation has a major impact on the hydrology of the upper reach, while change in temperature exerts a relatively minor impact. Changes of runoff and soil moisture in response to climate change in flood seasons are greater than those in non-flood seasons. In terms of regional distribution, the middle reach is more sensitive to climate change than the upper reach (Wang *et al.* 2002). Hao *et al.* (2006), by using results from climate models and employing a large-scale distributed hydrologic model, conclude that in the next 100 years, the trend in the Yellow River source region will be in tune with global warming; temperatures will continue to increase, evaporation will significantly increase even with an in-

crease in precipitation, and the degree of future climate change to a certain extent will cause reduction in the volume of water resources as a whole. In addition, the inter-annual distribution will be increasingly uneven, so the threat of severe droughts and floods will increase.

Various investigations involving physical geography of the basin, discharge measurement over several decades, rainfall-runoff analyses, trend analyses and modelling of surface flow system, climate variability studies, evolution of water use pattern, and chemical quality indicate the following features:

- Since the 1990s, runoff of the Yellow River source region has reduced substantially. There is no consensus on causes for reduction in the runoff, but it is generally believed that it is mainly because of reduction in rainfall.
- Water and sediment in the middle reach of the Yellow River show a decreasing trend and annual analysis indicates that rainfall, runoff, and sediment load are limited to few events. The reason is that meteorological conditions change significantly in different reaches of the river and probability of occurrence of drought is high in the middle reach of the river.
- Both number and area of lakes in the lower reach of the Yellow River have reduced significantly and changes in runoff in the lower reach of the

river have a close relation with water delivery from the upper and middle reaches and also from the key tributaries. Thus, reduction of runoff in the lower reaches is obvious.

- For the whole basin, surface run-off has reduced significantly in the Lanzhou control section. Natural runoff (estimated without anthropogenic influence), surface runoff, and groundwater runoff have reduced remarkably in Huayuankou control section, and the trend of runoff reduction in the lower reaches is more obvious than in the upper reaches. Both climatic factors and increased water consumption due to human activities may be responsible for this.

- The alluvial sand deposit encountered within a depth of 80 m underneath the area of the 'hanging river' in the lower reaches constitutes a high water-bearing potential aquifer and its single well yield is about 5000–9000 $m^3\ d^{-1}$. Under natural conditions, the river discharges into the surrounding aquifers.

- Isotope data suggests that groundwater in the piedmont part of the North China Plain is less than 40 years old and is recharged locally. In contrast, groundwater in the central and littoral portions of the North China Plain is 10 000 to 25 000 years old and was possibly recharged during the colder climate of the Last Glaciation that occurred about 20 000 years ago.

- The geochemical analyses of groundwater samples indicated that the groundwater in the North China Plain (NCP) has a two-layer structure, with the boundary at a depth of about 100–150 m. Shallow groundwater belongs to the local flow system and deep groundwater is part of the regional flow system.

- Concentration of major ions in much of the basin, and especially downstream of the major irrigation areas of the upper basin, increased significantly between 1960 and 2000. This increasing trend is mainly a result of saline irrigation return waters.

13.1.4 Current critical issues

While flood control has historically been the primary issue in Yellow River management, water stress has now emerged as the most important issue for basin authorities as well as users. The increase in water stress as a critical issue has been caused by three factors: (i) a recent decline in water supplies due possibly related to climate change; (ii) an increase in demand; and (iii) growing awareness of environmental water needs.

Since 1949, flood protection works have greatly reduced the incidence of flooding but the threat of flooding has not disappeared altogether. Large-scale dam and reservoir construction in the middle and upper reaches has reduced the probability of the large flows that have historically caused floods, but the channel in the lower reach is now more constricted by deposition of sediment in the channel, reducing its capacity to carry a flood. Furthermore, the channel below Huayuankou continues to rise above the surrounding countryside as sediments are deposited and is now about 20 m higher than the floodplain at Xinxiang City, 13 m higher at Kaifeng City, and 5 m higher at Jinan City, aggravating the impact of a flood if it occurs. The basic problem of flooding in the Yellow River is not simply one of large flood peaks but rather a combination of flow coupled with sediment transport and deposition arising from changes in the river channel geometry. The middle reach of the river runs through the highly erodible Loess Plateau, which at some places produces 20 000 $tons\ km^{-2}\ a^{-1}$ of sediment (Giordano *et al.* 2004). About one half of all sediment in the river is estimated to be deposited in the lower main channel and the delta region.

The ultimate causes of soil erosion in the Loess Plateau are debated but are generally thought to be associated with both human action and climate change. Much of the plateau region contains deep soils and temperature regimes favourable for crop production, given sufficient water. Agricultural expansion and associated activities, especially from the time of the Qin Dynasty (221 BC to 206 BC) onwards, appears to have contributed to the loss of vegetation cover, which exposed the already fragile soils to erosion. The problem seems to have exacerbated with population growth, which eventually led to extensive land clearance and cultivation of highly sloping, and hence highly erodible lands. While soil conservation strategies may help arrest erosion, these also deplete water, for example, tree plantation adds green cover on land but also consumes water. A key policy issue for the future is whether or not the reduction in sediment transport and ever increasing agricultural activity upstream are sufficient to justify reduction in the runoff downstream.

With decreasing water supplies and increasing demand, the waters of the Yellow River are now almost fully allocated, suggesting that reduction in allocations to some sectors must be enforced if additional new demand is to be met. Further complicating the matters, it is now clearly established that environmental water demands have not been adequately accounted for in existing allocation schemes.

While the growing imbalance between supply and demand and the threat of flooding, exacerbated by erosion, are still considered to be the primary issues facing Yellow River water managers, rapid degradation of water quality is increasingly becoming a key factor in basin water management. Due to a sharp increase in pollution levels over the past few decades, degraded water quality is aggravating the existing water supply problems.

The water quality problem in the Yellow River is caused by a combination of various factors. Most importantly, large quantities of waste are discharged into the main stem of the river and its tributaries. Industrial activity, municipal wastes, and non-point pollution arising from agricultural lands are important factors in water quality deterioration and play a significant role in substantially increasing heavy metals in the basin as well as BOD in the river discharge. In the upper reach, large quantities of agricultural return flow drains directly to the main channel, while in the middle stream, most of the return flow enters major tributaries such as the Weihe, Fenhe, and Qinhe. In the flooding season, pollutants are taken up from the large flood plain and enter the river, providing a second non-point pollution source.

13.1.5 Proposed solutions

The Yellow River Basin has played, and continues to play, a role in Chinese society, disproportionate to either its land area or available water resources. For thousands of years, the challenge in the Yellow River has been to control flooding and bring

more area under irrigation. Flooding is now largely under control and irrigation growth coupled with rapid increases in industrial and household water use now renders the river bed completely dry in some years. Because of these fundamental changes, the challenge in modern Yellow River management is no longer one of engineering to control the river, but one of management, particularly how to allocate water between various competing uses while still meeting ecological requirements.

Since 2000, in tune with the most recent approach adopted by the Ministry of Water Resources, water management and related development activities in the Yellow River Basin aim to integrate the interests of various regions and sectors. Consequently, to balance available water supply and the demand of various sectors, the YRCC developed a water use plan based on medium- to long-term supply and demand patterns. Annual water use plans are issued to users to assure adequate supply for priority areas, especially during occurrence of droughts. Furthermore, the YRCC formulated regulations encouraging household users to install water-saving devices, farmers to adopt water-efficient irrigation practices, and industry to promote techniques minimizing water use as well as waste discharge. It also established a market pricing system for water.

On a long-term basis, the South-to-North Water Transfer Project has been conceived for transferring surplus water from the upper, middle, and lower reaches of the Yangtze River in the southern part of China to the Huang-Huai-Hai River basins in North China. This project involves three routes, East Route, Middle Route, and West Route, which can be connected to a huge network with 'four transverse and three longitudinal waterways' with the four rivers, namely, Yangtze, Yellow, Huai, and Hai.

13.2 The Colorado River Basin, United States

The Colorado River (Aha Kwahwat in Mojave or the Red River) has long been considered important in the exploration, development, and culture of the western United States. It descends

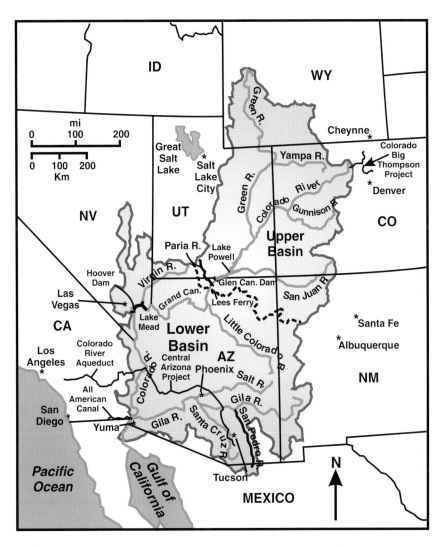

Fig. 13.7 Colorado River Basin, Southwestern United States. Redrawn from NRC (2007)
© International Mapping Associates. See also Plate 13.7.

from frozen heights of 4300 m in the Rocky Mountains, carves the mile-deep Grand Canyon, and ends its journey of 2334 km in the Gulf of California in Mexico (Fig. 13.7; Plate 11). The Colorado is a desert river. For millions of years it has shaped landforms and in the Grand Canyon has exposed geological formations that are half as old as the Earth itself. The river flows through the states of Colorado, Utah, and Arizona and serves as the boundary between Arizona and Nevada and California. It also separates the Mexican states of Sonoran and Baja California. The lower Colorado River separates two great deserts, the Mojave on the California (western) side and the Sonoran on the Arizona (eastern) side. The Gila River drains the Sonoran desert. South of the Mojave Desert lies the Salton Basin, a large structural depression about 70 m below sea level, extending 240 km northwest from the head of the Gulf of California.

Tributaries of the Colorado River include the Green, Gunnison, Dolores, San Juan, and Little

Colorado rivers. The entire drainage basin, including parts of Wyoming and New Mexico, occupies $632\,000\ km^2$ – about 7% of the area of United States. In its course to the sea, the river passes through sparsely populated arid and semi-arid land, mainly plateaus and deserts. In much of the middle part of its course, it flows through deep canyons with steep walls, the most prominent of which is the Grand Canyon, 349 km long and 6.4–9 km wide, with an average depth of 1600 m.

Numerous dams in the Colorado Basin provide hydroelectric power, flood control, and water for irrigation, recreation, and municipal use. The Hoover Dam is the largest producer of hydroelectric power and also the main flood-control structure. Amongst the canals diverting water from the river are: (i) the Colorado River Aqueduct, which extends to the southern Californian coast; (ii) the All American Canal, which carries water to California's Imperial Valley; and (iii) the Gila Canal in Arizona. The Central Arizona Project, a system of dams, canals, and aqueducts, brings large amounts of water to the Phoenix and Tucson areas. Tunnels dug through the Continental Divide divert water from the Upper Basin to cities and farmlands on the high plains.

The allocation of water of the Colorado River has long been a major problem. The 1922 Colorado River Compact divided the river between the upper and lower basins and reserved unused water for future development in the four upper basin states. Thus by treaty, Congressional Acts, and agreements amongst the states concerned, the Upper Basin and Lower Basin are each entitled to $9.25 \times 10^9\ m^3$ of water annually and Mexico to $1.85 \times 10^9\ m^3$. These commitments, however, often exceed the river's annual flow, which has caused disputes amongst the concerned states and also with Mexico. The decreased flow has caused the lower course of the river to become salty, endangering fish and damaging irrigated farmland.

Given the Colorado River's importance, the record of variations in its flow has long been of keen interest to the parties involved. Direct stream flow measurements date back to the late 1890s when gauging stations were established at a few sites along the river course. As the river flow was measured over the 20th century, and as the network of stream gauging stations grew, a better understanding of Colorado River flows and variability emerged. For example, it is known today that the Colorado River Compact of 1922 – the water allocation compact that divides Colorado River flows between the upper and lower Colorado River basin states – was signed during a period of relatively high annual flows. It is also accepted that the long-term mean annual flow (15 *million acre-feet* = $18.5 \times 10^9\ m^3$) of the river is less than the 16.4 *million acre-feet* ($20.3 \times 10^9\ m^3$) assumed when the Compact was signed – a hydrologic fact of no small importance with regard to water rights agreements and subsequent allocations.

13.2.1 Water supply and use – recent developments

The Boulder Dam (later known as the Hoover Dam) in the early 1920s marked the start of a time of rapid development of the Colorado River, and completion of the Glen Canyon Dam in 1964 signalled its end. Most of the basin's large dams were either planned or constructed during this period. A key factor affecting Colorado River basin water demands from the mid-1980s to the present has been the rapid increase in population in many areas of the western United States served by Colorado River water. The population growth is driven by a combination of factors such as migration from other US states, immigration, and the normal growth rate of the local population.

Increasing rates of population growth and water demand in the 1990s and early 2000s has prompted many water users and managers to consider nontraditional means to augment water supplies. For example, groundwater sources have been increasingly tapped over the past few decades; water tables have dropped precipitously in many areas, and the limits of groundwater resources are being reached in some areas (primarily Tucson and Phoenix metropolitan areas, the Las Vegas region, and in peripheral areas along the Colorado River basin falling in both Southern California and New Mexico). A prominent development on this front has been the sale, lease, and transfer of agricultural water to meet the needs of growing urban populations (NRC 2007).

Today, these agricultural-urban transfers are taking place in many sites across the region, including Colorado's South Platte River Basin (Denver), Las Vegas, and the Phoenix and Tucson metropolitan areas. In strict monetary terms, sale, lease, and transfer of water from agricultural to urban users often represent 'win-win' situation, both for the buyer and seller, as water typically shifts from lower-value agricultural uses to higher-value urban uses. Municipalities and industries generally have a greater inclination to pay for a given unit of water than irrigators or ranchers, and such transfers may offer a cost-effective method for cities to meet their increasing water demands. They also often prove to be profitable to individual farmers or ranchers (NRC 2007).

Another important consideration is that such changes in points of diversion and water uses nearly always entail 'third-party' effects beyond those that accrue to the buyers and sellers of water rights. Examples of these effects include reduced agricultural return flows that support riparian ecosystems, and loss of business suffered by local merchants as a result of reduction in the areas of irrigated cropland. The various costs that may be borne by such third parties are well recognized (NRC 1992). Another factor that may go against long-distance transfer is limited infrastructure, especially water storage and conveyance facilities, needed to enable such transfers. Several innovative and useful practices — especially water banking and aquifer storage — have been developed to help obviate the need for new storage and conveyance facilities. Nevertheless, there will invariably be some infrastructural limitations on potential transfers across the basin.

A severe drought across much of the western and southwestern United States for a number of years in succession had substantial impact on Colorado River Basin water supplies from 2000–2006. By one measure, as reckoned from inflow values into Lake Powell, drought conditions existed from 2000 to 2004 (Piechota *et al.* 2004). By other measures, however, drought conditions persisted beyond 2004 and affected portions of the basin until late 2006.

Steadily rising population and urban water demands in the Colorado River region inevitably result in increasingly costly, controversial, and un-

avoidable trade-off choices made by water managers, politicians, and their constituents. The increasing demand for water is also impeding the region's ability to cope with droughts and water shortages. The drought of the early 2000s brought climate-related concerns to the forefront across the Colorado River region. Not only did the drought result in numerous, direct hydrologic impacts, it raised questions about what climate trends and future conditions across the region and the planet might portend for Colorado River flows.

13.2.2 Climate and hydrology of Colorado River Basin

The Colorado River Basin comprises climate zones ranging from alpine to desert and exhibits significant climate variability on various time scales. These variations have important implications for snowmelt and river hydrology in the Colorado River region and are thus of interest to both researchers and water managers.

13.2.2.1 Precipitation patterns and sources

The Colorado River is primarily a snowmelt-driven system, with most precipitation in the basin falling as winter snowfall in higher elevations of Colorado, Utah, and Wyoming. Cold temperatures prevailing at high elevations cause precipitation to occur mainly as snow that remains frozen during the winter months. This snow drapes the mountainous terrain during winter months and survives throughout the summer at the highest altitudes. Some of the water in this snow pack is lost to the atmosphere through sublimation (change from solid phase directly to vapour) during the cool season. Most of the snow, however, survives and as the snow pack warms in the spring, meltwater steadily enters the soil. This process lasts for several weeks to months at higher elevations, and melting occurs slowly enough to recharge the soil and allow water to flow to the myriad channels that feed the Green and Colorado rivers. For these reasons, winter precipitation over the high-elevation portion of the upper basin plays a dominant role in generating runoff and stream flow.

Warm season precipitation occurs during the North American rainfall season, transporting moisture into the region from sources in the subtropical Pacific and the Gulf of Mexico. The rain falls more intensely, often in localized convective thunderstorms, and is mostly lost by evapotranspiration. A relatively small fraction of summer precipitation finds its way into aquifers and streams. In the high-elevation headwaters of the basin, summer precipitation amounts are generally less than winter values. The dominance of winter precipitation at high elevations in the annual precipitation is more pronounced in the Green River drainage system than in the Colorado River headwaters in central Colorado. In the lower and drier reaches of the basin, summer precipitation can account for a larger share of total annual precipitation, but because of high evaporation and transpiration rates, it is less effective in contributing to stream flow. In the hottest and lowest portions of the basin, summer precipitation is of considerable importance to local vegetation and to small runoff channels, but contributes insignificantly to the main Colorado River and its major tributaries. In the summer months, on the other hand, there is higher water demand. If, because of drought conditions, there is a deficit of snowfall in winter over high mountainous reaches, there is little opportunity to replenish water storage until the following winter.

13.2.2.1.1 The Tropical Pacific and the ENSO

Surface ocean temperature patterns in the equatorial Pacific between Peru and the International Date Line significantly influence the Colorado River Basin climate. At irregular intervals, typically ranging between 2 and 7 years, sea surface temperatures (SSTs) in this region are warmer compared to long-term averages. This phenomenon, called El Niño, is part of a complex ocean–atmosphere oscillation. A climatic counterpart of El Niño, called La Niña, is characterized by below-average SSTs (usually with smaller departures from average SST than during El Niño events). The two terms, El Niño and La Niña, refer to ocean temperature changes only in this geographical domain and not in any other region.

Another atmospheric feature relates to barometric pressure gradients in the South Pacific.

An inverse relationship in atmospheric surface pressure has been found to exist between Tahiti and Easter Island in the tropical Pacific, and over Darwin in northern Australia (Walker 1925). Thus, when atmospheric pressure is high in one of these locations, it tends to be low in the other region, and vice versa. Walker termed this phenomenon the Southern Oscillation. It is solely an atmospheric phenomenon. The Darwin-Tahiti pressure difference (normalized for variability over the past century) is the basis of the Southern Oscillation Index (SOI). Furthermore, when Tahiti has lower than average pressure and Darwin has higher than average pressure (negative SOI), a strong tendency exists for El Niño to occur. Conversely, there is a tendency for La Niña conditions to occur with higher pressure in Tahiti and lower pressure in Darwin. The oceanic (SST) and atmospheric (SOI) measures are usually highly correlated and these terms are sometimes used interchangeably (McCabe and Dettinger 1999). For historical reasons these phenomena are often lumped together and referred to (although somewhat oddly) as the El Niño-Southern Oscillation (ENSO). The ENSO phenomenon owes its existence to a coupled ocean–atmosphere interaction over the equatorial Pacific and is an important contributor to inter-annual global climate variability. The ENSO cycle has impacts on climate over large areas of both the tropical and extra-tropical regions (see Section 8.6, and Fig. 8.14, for its impact over Indian summer monsoon). The winter storm track over the eastern Pacific Ocean shifts southwards during El Niño episodes, often causing wet winters in the southwestern United States and dry winters in the Pacific Northwest and northern Rockies. La Niña winters tend to give rise to the opposite pattern, and moderately positive values of the SOI in the pre-summer/autumn invariably result in a nearly dry winter in the southwestern United States (Redmond and Koch 1991). These patterns are accentuated in stream flow, particularly under extreme high and low stream flows (Cayan et al. 1999).

Another pattern of regional scale climate variability related to SST variations in the Pacific area is the Pacific Decadal Oscillation (PDO). The PDO describes concurrent variations in SST, atmospheric pressure, and wind in the central and eastern

Pacific poleward of 20°N (Mantua *et al.* 1997). Historically, the warm as well as cool phases of the PDO have lasted for two to three decades, over a period of about a half-century. An abrupt change in Pacific-wide environmental conditions, known as the '1976 shift' (Trenberth and Hurrell 1994), was identified subsequently, leading to identification of the PDO. This pattern appears to alternately accentuate and counteract the effects of ENSO in the Pacific Northwest and the southwestern United States and manifests most strongly in winter. The origin of this oscillation has not been identified unambiguously. It is linked to periods of higher and lower frequency of El Niño and La Niña at equatorial latitudes, even though the PDO index has only a modest correlation with the SOI (Mantua *et al.* 1997). Although there are intriguing statistical relationships associated with the PDO, the causative mechanisms that underlie the PDO behaviour, leading to its manifestation within the Colorado River Basin (and primarily in the lower basin, as is the case with ENSO), have not been fully explained.

In recent years, yet another pattern has been identified that appears to influence the Colorado River Basin. Atlantic Ocean SSTs exhibit a mode of variability that has similar departures from their average values for one to two decades over an area spanning low to high latitudes; this feature is known as the Atlantic Multi-decadal Oscillation

(AMO). The fact that the AMO has effects on climate and stream flow in the eastern United States is understandable. Surprising, however, is the fact that when the North Atlantic is warmer for a decade or more, stream flow in the upper Colorado River Basin, governed by winter precipitation, tends to be lower than average, and vice versa (McCabe and Palecki 2006). The evidence so far is statistical and largely dependent on a small number of AMO cycles.

13.2.2.1.2 Precipitation

Inter-annual precipitation variability across the upper Colorado River Basin (spatially averaged over the basin upstream of Lees Ferry) is shown in Fig. 13.8. It is seen that after a period of small variability for several decades in the mid-20th century, there has been a tendency towards large variability towards the later part of the 20th century. The past 30 years of data include the highest and lowest annual precipitation in the 100-year record, and there has been a tendency towards multi-year episodes of both wet and dry conditions. Some of the years in the early- and mid-1980s were at least as wet as the period that preceded the signing of the Colorado River Compact of 1922. Prior to the early 21st century drought, the comparable drought for five consecutive years was in the 1950s. The

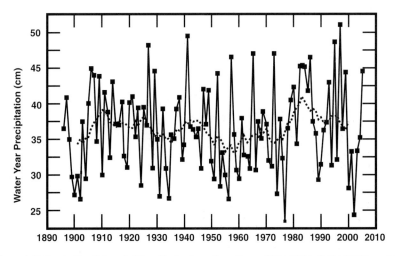

Fig. 13.8 Annual precipitation in the Colorado River Basin above Lees Ferry, 1895–2005. Full solid line: annual values. Dotted line: 11-year running mean. Redrawn from NRC (2007) © Western Regional Climate Center.

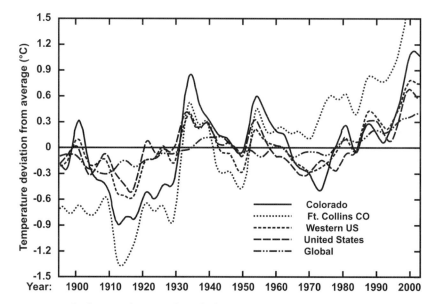

Fig. 13.9 Temperature trends, shown as deviations from the long-term average. For different regions of the globe since 1900. The Western United States has warmed by ~0.5°C, Colorado by ~0.8°C, and Fort Collins by ~2°C during the last 100 years. Redrawn from Udall and Hoerling (2005) with permission from California Department of Water Resources.

only other comparable five-year dry period was at the end of the 19th century and beginning of the 20th century. Despite these variations, there is no significant trend in inter-annual variability of precipitation over the past 110 years.

13.2.2.2 Temperature

Temperature trends for different regions of the globe since 1900 are shown in Fig. 13.9. Upper and lower basin temperature trends are similar and bear a strong resemblance to the record of temperature across the entire western United States, as well as to the mean global surface temperature trends. Fig. 13.9 also shows that since the late 1970s the Colorado River region has exhibited a steady upward trend in surface temperatures. The most recent 11-year average exceeds any previous values observed in the instrumental record of over 100 years. One striking aspect of Fig. 13.9 is how much warmer the region has been during the drought of the early 2000s as compared to previous droughts. For example, temperatures across the basin today are at least 0.8°C warmer than those that prevailed during the drought of the 1950s. Increasing tem-

peratures in the region have many important hydrologic implications.

13.2.3 Projecting future climate scenario

Many of the studies concerning prediction of future climatic and hydrologic conditions across the western United States are based on results of computer-based numerical models (GCMs) of the global atmosphere. On the whole, the future projections and past trends point to a strong likelihood of warmer future climate across the Colorado River Basin. Key indicators of warmer temperatures in western North America include a shift (towards earlier than the usual melting season) in the peak seasonal runoff generated by snowmelt, increased evaporation, and correspondingly less runoff.

13.2.4 Instrumental record of Colorado River stream flow

The best-known Colorado River stream gauge record is from Lees Ferry, Arizona, where the US Geological Survey (USGS) has been operating a gauging station since May 8, 1921. Discharge readings at Lees Ferry measure combined runoff

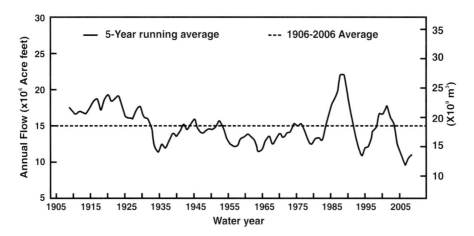

Fig. 13.10 Natural Colorado River annual flows at Lees Ferry, AZ, 1906–2006. Values for 2004–2006 are estimates. Redrawn from NRC (2007) © National Academies Press.

from the upper part of the Colorado River Basin, which includes the upper Colorado, Green, and San Juan rivers. The 1922 Colorado River Compact designated Lees Ferry as the hydrologic dividing point between the upper and lower basins. The record from Lees Ferry is the most important instrumental record of Colorado River flows. Its five-year moving average and long-term mean (1906–2006) are depicted in Fig. 13.10. The mean annual flow value is roughly 15 million *acre-feet* (1.85 million *ha m*; broken line). The drought of the late 1990s and early 2000s – which began in the fall of 1999 (water year 2000) – clearly stands out within the past century, as it represents the lowest five-year running average discharge in this record.

Several noteworthy hydrologic periods are reflected in the Lees Ferry gauge record. The time period used in Colorado River Compact negotiations, 1905–1922, included some particularly wet years. As of now, in the documented gauged record at Lees Ferry, the 1905–1922 period had the highest long-term annual flow volume in the 20th century, averaging 16.1 million *acre-feet* (2.04 million *ha m*) per year.

To examine the issue of how well the historical, gauged record represents long-term flow patterns, tree-ring-based reconstruction of past hydrologic conditions employing the method of dendrochronology have been done for the Lees Ferry gauging station, in the Colorado River Basin (Fig. 13.11). This reconstruction used conifer-

ous tree species growing at lower elevations on well-drained south-facing slopes and have been shown to be well suited for reconstruction of annual stream flow (Woodhouse and Lukas 2006; Woodhouse and Meko 2007). Dendrochronology-based reconstructions have inherent uncertainties as these are derived from trees that are not perfect recorders of hydrologic variability, as is evidenced by the fact that the tree-ring-based models do not account for the entire (100%) variance in the gauge record. There are other sources of uncertainties in these reconstructions, including those in the quality of the gauge record used for calibration. In spite of the various sources of error, the reconstructions shown in Fig. 13.11 reveal the following:

• Long-term Colorado River mean flows calculated over periods of hundreds of years are significantly lower than both the mean of the Lees Ferry gauge record upon which the Colorado River Compact was based and the full 20th-century gauge record.

• High flow conditions in the early decades of the 20th century were one of the wettest in the entire period of reconstruction.

• The longer reconstructed record provides a more reliable basis from which to assess the drought characteristics that have been experienced in the past, revealing that considerably longer droughts had occurred prior to the 20th century.

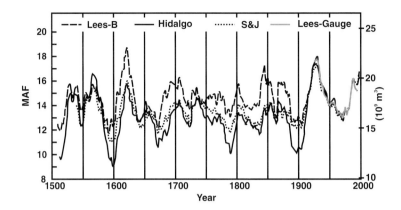

Fig. 13.11 Colorado River annual streamflow reconstructions, Lees Ferry, AZ (smoothed with a 20-year running mean). *Note*: year plotted is the last year in the 20-year mean. *Sources*: Lees-B from Woodhouse *et al*. (2006) and Hidalgo *et al*. (2000); S&J from Stockton and Jacoby (1976); Lees Gauge is gauge record, 1906–1995. Redrawn from NRC (2007) © National Academies Press.

- These reconstructions, along with temperature trends and GCM-based projections for the region, suggest that droughts will recur in future as well and that their severity may exceed that of historical droughts, such as the drought of the late 1990s and early 2000s.

13.2.5 Conserving and augmenting water supplies

In response to population growth and economic development, the traditional strategy has been to construct storage reservoirs and appropriate conveyance facilities to provide adequate water supplies to cope with occasional drought. This was complemented by a variety of means for increasing supplies and better management of demands: groundwater supplies have been tapped, irrigation practices have been modified, and improved in respect of efficiency of water utilization, some states and cities have modified landscaping practices, and there have been some efforts on weather modification.

In addition to environmental and other concerns related to large dams, traditional water projects today face a more stringent planning and feasibility studies and other obligations than in the past, which can literally entail decades of project planning and related activities.

An interesting and ambitious project to augment the Colorado River Basin water supply storage in-

volved various plans to transfer large quantities of water from the water-rich regions of Alaska and the Canadian Yukon to the water-deficient arid western United States through a complex system of reservoirs, tunnels, pumping stations, and canals. Dams were also meant to generate hydropower, the revenue from which was to be used to finance project construction. The cost estimate of this project in 1964 was $80 billion, revised upwards to $130 billion in 1979. Similarly, prospects of towing icebergs from south of Alaska or other arctic regions to augment Colorado River water supplies are equally unrealistic. Declining prospects for traditional water supply projects are to be viewed perhaps more realistically, not as an end to 'water projects' but as part of a shift towards innovative non-structural measures such as water conservation, water use technologies, xerophytic landscaping, groundwater storage, and demand management through changes in water pricing policies.

13.2.5.1 Cloud seeding

Weather modification, including cloud seeding to increase rainfall and suppress hail formation, has been studied and practised in the United States for at least five decades. There is still no convincing scientific proof of the efficacy of intentional weather modification efforts. However, in addition to the Colorado River Basin states, user agencies such as municipalities and the skiing industry are interested in the prospects of augmenting water supplies and snow packs by cloud seeding.

In assessing the efficacy of cloud seeding operations, one should keep in mind the experience of six decades of experiments and applications which did not produce clear evidence whether cloud seeding can definitely enhance water supplies on a large scale. Of course, unambiguous evidence is difficult to produce in cloud seeding experiments, as they are not amenable to studies under controlled conditions. Furthermore, such experiments are seen by many as being relatively inexpensive even if they do not positively result in enhanced precipitation. Given the ever increasing demand for water across the Colorado River Basin, cloud seeding is likely to continue to be pursued as a means for augmenting water supply.

13.2.5.2 Desalination

There have been steady scientific and engineering advances in the technologies for converting salt water into fresh water, and several desalination facilities have been constructed. Advances in technology have led to cost reduction, improved efficiency, and a worldwide increase in the number of desalination plants. More than half the world's desalinated water production today is in the Middle East; about 15% of the world's desalinated water is produced in North America.

Recent improvements in desalination technology have led to energy cost reduction per unit of fresh water produced. There is, for example, a variety of membrane technologies, such as reverse osmosis, nano-filtration, and ultra filtration, which remove salts, dissolved organics, bacteria, and other dissolved chemical constituents from salt water. There is also a range of thermal technologies that boil or freeze water, then capture the purified water, leaving behind the contaminants. Maximum attention has been directed to converting sea water to potable fresh water, while comparatively less attention has focused on subterranean and surface brackish water desalination. There is some interest in coupling future desalination plants with new power plants to reduce the energy cost of desalination. Rising energy costs, however, raise doubts as to whether this trend will continue.

Besides energy costs, desalination entails several environmental implications. A key barrier to economically viable desalination is disposal of the brine water that is an inherent by-product of the process. This is especially a problem in areas that do not have access to the sea, but it can also be problematic for coastal locations. For example, native species in bays and estuaries are impacted by large seawater influx and by discharge of concentrated brine. Drawdown of brackish water in subterranean reservoirs can lead to ground subsidence and/or a lowering of the water table.

Technical, economic, and environmental issues notwithstanding, desalination offers the Colorado River Basin states an option for actually increasing water supplies. This option is limited primarily to areas with access to water derived from the Pacific Ocean, although there may be other, select Colorado River Basin sites at which desalination facilities may be feasible (e.g. Yuma, Arizona). With increasing regional water demands, and increase in technical efficiency, desalination is likely to be perceived as an increasingly attractive option for augmenting supply, which will be especially true for economically sound communities (NRC 2007).

13.2.5.3 Removing water-consuming invasive species

Since settlers began moving into the southwestern United States in the mid-19th century, many invasive species have been advertently or inadvertently introduced in the region. These include cheat grass, camelthorn, ravenna grass, Russian olive, tamarisk, and salt cedar. These species are capable of surviving in a variety of habitats and have been identified by the National Park Service as the greatest threat to native species in the Grand Canyon National Park (NPS 2005). Today they are proliferating in the park, representing about 10% of the vegetation (USGS 2005). Tamarisk (*Tamarix ramosissima*) is an invasive plant of special concern across the basin. Tamarisk consumes large quantities of water, overwhelms native riparian species, and can lead to changes at the very ecosystem level. It forms dense stands and is difficult to eradicate. Tamarisk is a dominant riparian plant species today across the basin, consuming considerable amounts of water that would otherwise be available to downstream states or to support ecosystem goods and services. For example, in Colorado it is estimated that tamarisk occupies roughly 55 000 *acres* (~2.23 × 10^8 m^2) and consumes 170 000 *acre-feet* (~2.8 × 10^8 m^3) of water per year more than that used by the native replaced vegetation (NRC 2007). Several efforts have been made to remove tamarisk, including herbicide injection, stump removal, deliberate flooding, and the use of a leaf beetle (*Diorhabda elongata*) and its larvae that feed on tamarisk leaves, but their success is at best marginal.

13.2.5.4 Agricultural water conservation

Globally, irrigation is the largest consumer of water. As a result, increased efficiency of its application can save considerable amounts of water. Farm-level irrigation efficiency is of great importance to individual farmers, but unused irrigation water on a given farm usually generates return flow that is reused downstream by other farmers and also helps sustain ecological habitats. Farmers often periodically flush out water from the soil profile in order to prevent excessive salt accumulation in the crop root zone – especially if irrigation water is of marginal quality. The salt leaching practice is essential to successful irrigated agriculture (English *et al.* 2002). Leaching of salts from the root zone can, however, increase salt load to streams and aquifers. Increased on-farm efficiency of irrigation water can reduce production costs, improve water distribution amongst farmers and other users, and reduce negative, off-farm effects of irrigation. There is also potential to employ different cropping patterns, such as rotations employing both sensitive and salt-tolerant crops. It may be possible to grow a salt-sensitive crop, followed by a salt-tolerant crop, before salinity reaches unacceptable levels and soils must be leached (English *et al.* 2002).

There also exists the possibility of innovation in managing irrigation systems, taking advantage of remotely sensed soil moisture content and the state of crop growth – this latter approach includes deficit irrigation in which plants can be stressed during their growing phase but are properly watered at critical stages – flowering and fruiting (English and Raja 1996). In addition, some improvements in ir-

rigation water productivity are likely to result from better agronomic and cultivation practices, perhaps employing genetic modification.

13.2.5.5 Urban water conservation

Increased urbanization in the Colorado River states and rise in water demands have also resulted in new water conservation technologies (e.g. lower-flow plumbing fixtures and more efficient irrigation systems), market incentives, regulatory policies, new landscaping techniques and the use of drought-resistant (xerophytic) plants, and public education encouraging urban water conservation. Use of reclaimed (or 'grey') water for landscaping, golf course irrigation, and augmenting return flows to the Colorado River has also increased.

13.2.5.6 Off stream water banking and reserves

Water banking and groundwater recharge has been used for several decades in the western United States, and there has been a strong interest in these programmes during the past decade. The term 'water bank' generally applies to two different types of arrangements: (i) groundwater storage projects; and (ii) arrangements to facilitate voluntary water transfers through rental markets. Water banking and groundwater storage programmes have multiple objectives, which include: (i) creation of reliable supplies during extended drought; (ii) promotion of water conservation by encouraging 'deposits' into groundwater storage; and (iii) recharging of groundwater tables and reduction of evaporation from surface water bodies and reservoirs. From a geological perspective, large amounts of water can often be infiltrated, under gravity, into underground aquifers in favourable locations (Chapter 11). However, during drought conditions, large amounts of water may need to be withdrawn in a short period of time, which often entails significant pumping costs. Groundwater storage projects aim to facilitate water transfers in response to short-term changes in supply-and-demand conditions, with the aim to bring together people seeking to purchase water from people interested in selling water entitlements (Frederick 2001).

The State of Arizona created its first framework for water banking in 1986, with the passing of legislation to authorize underground storage and recovery projects. In 1996, the Arizona Water Banking Authority (AWBA) was established. The AWBA focused on storing surplus Colorado River water through groundwater recharge and on protecting Arizona's part of the Colorado River. This groundwater supply could subsequently be drawn upon when needed. By 2000, the AWBA was recharging about 294 000 *acre–feet* (36 265 *ha m*) per year of water.

A related concept being explored in the basin involves creation of water reserves, which is not to be confused with the concept of reserved (exclusive) rights that can exist under the principle of prior appropriation. The concept of water reserves generally entails storage of water, either from excess flows in wet periods or through water rights sales, leases, or transfers, to be used at a later date for specific purpose(s). It is a variant of the water banking concept in that it is not necessarily fully market-based and may be designed to benefit public and non-market values (e.g. in stream flows).

Water conservation programmes for the city of Tucson, Arizona, offer an interesting case study for several reasons. First, some of the programmes date back to the 1970s and are amongst the oldest urban water conservation efforts in the region. Second, these programmes have incorporated several different aspects, including public education, water-saving technologies, water pricing, and regulation. Third, the Tucson Water Department has maintained an excellent database on its conservation programmes and after a period of four decades has a compendium of valuable information. Finally, Tucson has constantly revised its policies and strategies to incorporate newer, more efficient technologies, and updated water pricing. The revised regulations include, for example, codes for low-water-use landscaping (xeriscaping) and drip irrigation systems. The city of Tucson has been delivering reclaimed water since the mid-1980s, from which a large percentage of parks, golf courses, and other public spaces today are irrigated with reclaimed water (NRC 2007).

Given the projections of both growing regional population and increasing regional temperatures,

along with tree-ring-based reconstructions that reveal the recurrence of severe drought conditions across the Colorado River region in the past, there has been a wide range of engineering and political efforts designed to overcome the water supply limitations. Technological options are available to extend water supplies in the Colorado River Basin and elsewhere, but these have limitations. Although limited opportunities do exist to construct additional reservoirs or to implement inter-basin water transfers into the Colorado River Basin, these are not the favoured solutions at the present time. Consequently, new water project prototypes that emphasize conservation, landscaping, new technologies, and other measures are being promoted. Although useful and necessary, technological and conservation options may not constitute a panacea for coping with the reality that water supplies in the Colorado River Basin are limited and that demand is inexorably rising.

13.3 The Murray-Darling River Basin, Australia

Deriving its name from the two major rivers draining a 1 061 469 km^2 area, the Murray-Darling Basin (MDB) is geographically located in the interior of southeastern Australia and spans parts of the states of Queensland, New South Wales, Victoria, and South Australia (Fig. 13.12; Plate 12). The basin is 3375 km long (the Murray River is 2589 km long), and drains one-seventh of the Australian landmass, and is currently by far the most productive agricultural area in Australia, accounting for around 41% of the national gross value of agricultural production. The basin supports almost one-third of the nation's cattle population, half of the sheep, half of the cropland, and almost three-quarters of the nation's irrigated land. While the MDB system is large in terms of catchment area and length, it is small in terms of surface runoff. Most of the basin is flat, low-lying, and far inland, and receives scanty rainfall. The multitude of rivers that drain it are long and slow-flowing. Just three rivers, the Upper Murray, Murrumbidgee, and Goulburn, account for over 45% of the basin's total runoff. Overall, about 86% of the basin contributes virtually no runoff to

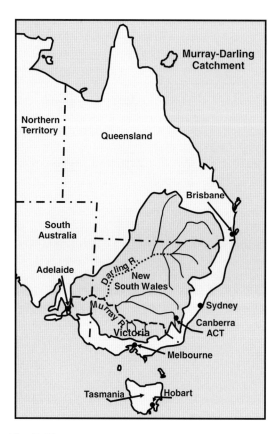

Fig. 13.12 Location map of the Murray-Darling Basin. Redrawn from http://en.wikipedia.org/wiki/Murray-Darling_Basin. See also Plate 13.12.

the river systems, except during floods. In general, the climate is hot and dry in summer and mild in winter. Much of the terrain is semi-arid and nearly all of it is only a few tens of metres above sea level.

Most of the agricultural land in the MDB is earmarked for grazing and about 12%, or 9.8×10^6 *ha*, for raising crops. While the landscape is dominated by dryland agriculture, irrigated agriculture also plays an important role in the economic development of the basin. Most irrigated land is used for pasture, but there are also significant areas of irrigated crops such as cereals, cotton, rice, fruit, vegetables, grapes, oilseeds, and legumes.

13.3.1 Surface water

Rainfall is unpredictable and varies from place to place, as well as from year to year. In the south-

east, average temperatures are low, elevations comparatively high, and rainfall occurs frequently. The spatially averaged annual average rainfall over the basin is about 480 *mm*, giving a total of just under 500 *billion cubic metres* of water, most of it falls in winter and spring. Along the southern and eastern borders of the basin are the inland slopes of the Great Dividing Range. It is here that most of the water in the rivers of the basin originates either as rainfall or, in the case of the Australian Alps which straddle the New South Wales–Victoria border, as winter snowfall. Over 90% of precipitation is consumed as evapotranspiration (ET) by native vegetation, forests, crops, and pastures. Total water flow in the Murray-Darling Basin in the period since 1885 has averaged around 24 billion m^3 per year (24 $km^3\ a^{-1}$). Water in the lakes and rivers evaporates and drains into groundwater systems as well as to the sea, although in most years only half of it reaches the sea and in dry years much less. Estimates of total annual flows for the basin ranged from 5 km^3 in 1902 to 57 km^3 in 1956.

The hydrology of the streams within the basin is varied, even considering its size. There are three main types (Brown 1983):

1. *The Darling and Lachlan basins*: These have extremely variable flows from year to year, with the smallest annual flow being typically as little as 1% of the long-term mean and the largest often more than ten times the mean. Periods of zero flow in most rivers can extend to months and in the drier parts (Warrego, Paroo, and Lower Darling basins) to several years in succession. Flows in these rivers are not strongly seasonal. However, in the north most floods occur in the summer from monsoon rains; in most of the Darling and Lachlan catchments, typically high or low flows begin in winter and extend to the following autumn.

2. *The southwestern basins (Campaspe, Loddon, Avoca, Wimmera)*: These have a marked winter rainfall maximum and relatively lower precipitation variability than the Lachlan or Darling. The variability of runoff is very high and most of the terminal lakes found in these basins frequently dry up. Almost all runoff occurs in the winter and spring and, in the absence of large dams

Table 13.4 Diverted water in the Murray-Darling Basin and the Murray Basin (the southern, more intensively developed, one-third).

Diversions	Murray-Darling Basin		Murray Basin	
	Volume (*km³*)			
Period	Total	Irrigation	Total	Irrigation
1994–2003 (annual average)	11.3	10.7	9.6	9.0
1996–1997 (wet year)	12.3	11.8	10.3	9.9
2002–2003 (dry year)	8.1	7.4	6.7	6.1

Source: (Giordano *et al.*, 2004) citing (MDBC, 2004).

for regulation, these rivers are often or usually seasonally dry during summer and autumn.

3. *The Murrumbidgee, Murray, and Goulburn basins* (except the Broken River Basin, which resembles the southwestern basins): Because these catchments have headwaters in alpine region with relatively young peaty soils, the runoff ratios are much higher than in other parts of the basin. Consequently, although gross precipitation variability is no less than in the Lachlan or Darling basins, runoff variability is significantly less than in other parts of the basin. Typically these rivers are perennial and the smallest annual flow is typically around 30% of the long-term mean and the highest value around three times the mean. In most cases the flow peaks vary strongly with the arrival of spring snowmelt and is the lowest in mid-autumn.

About half of average runoff is diverted for various uses, as shown in Table 13.4. Irrigation accounts for about 95% of the water extracted for raising various crops. Growing of rice and cotton is becoming highly controversial amongst the scientific community in Australia, owing to their high water consumption in a region with extreme scarcity of water (as much due to exceptionally low run off coefficients as due to low rainfall).

The region has been facing prolonged drought, almost since the beginning of the 21st century. In April 2007, the Australian prime minister announced that the region was facing an unprecedented water shortage and that water might have to be reserved for critical urban water supplies.

Basin-scale observations of this multi-year drought, integrated by choosing a grid with elements of 1 degree latitude and 1 degree longitude, to assess the response of water resources and the severity of the drought were undertaken using a combination of Gravity Recovery and Climate Experiment (GRACE) data with *in-situ* and modelled hydrologic data (Leblanc *et al.* 2009). The study shows propagation of the water deficit through the hydrologic cycle and occurrence of different types of drought. These observations revealed rapid drying of soil moisture and surface water storages, which reached near-stationary low levels only about two years after the onset of the drought in 2001. The multi-year drought has led to almost complete drying up of surface water resources, which account for most of the water used for irrigation and domestic purposes. High correlation between observed groundwater variations and GRACE data substantiates the observed persistent reduction in groundwater storage, with groundwater levels still declining even six years after the onset of the drought (groundwater loss of ~104 km^3 between 2001 and 2007). The hydrologic drought continues, even though the region had average annual rainfall during 2007.

Since 1997 a cap/limit on the diversion of water from the basin's river system has been in place. The introduction of the cap was seen as an essential first step in establishing management systems to rejuvenate rivers and ensure sustainable consumptive uses. The cap is a key policy decision to support the goal of the Murray-Darling Basin Initiative 'to promote and co-ordinate effective planning and

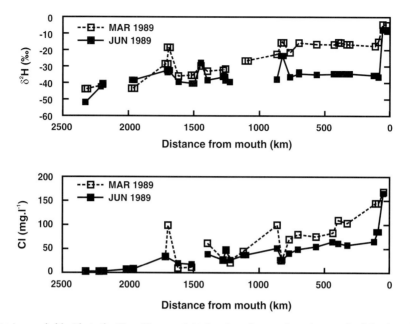

Fig. 13.13 Deuterium and chloride in the River Murray, plotted against distance from the mouth of the river. Differences in response between the two types of measurements are evident over certain reaches of the river, leading to conclusions concerning the relative balance of the various sources of salinity. For example, the large change in δ^2H observed between 1600 and 700 km during March 1989 is consistent with a substantial flux of irrigation drainage waters. Downstream of 700 km distance from the mouth of the river, the influx of Cl^- is not accompanied by large changes in δ^2H, and thus appears to result from a higher proportion of saline groundwater relative to irrigation drainage. Redrawn from (Simpson and Herczeg (1991). © Elsevier.

management for equitable, efficient, and sustainable use of the water, land, and other environmental resources of the Murray-Darling Basin.' Though implemented in 1997, it capped diversions at 1993 levels, and therefore irrigators lost a certain amount of water. The cap made water in the basin a more valuable resource and gave entitlements to its diversion more economic value and saw increased trade in these entitlements.

Soil salinity is a major environmental issue in Australia. It is a problem in most states, particularly in the southwest of Western Australia. The problem is basically of land-water management. Salinity in the Murray Basin is a natural consequence of the evaporative concentration of salts present in rainfall due to the semi-arid climate. However, land clearance, irrigation, and large-scale water development schemes that began in the 1880s have significantly altered the natural equilibrium, giving rise to a net movement of saline groundwaters to the surface and a discharge of Cl^- ions to the river several times

the value contributed by rainfall. Chemical and stable isotope data (Fig. 13.13) indicate that, while groundwater delivers 75% of chloride to the river, reactions occurring in the soil zone deliver 90% of the bicarbonate. Over time this process caused the thin top-soil layers to become irreversibly saline, unsuitable for agriculture. The area of salt-affected soils in 2002 was around 20 000 km^2.

13.3.2 Groundwater

The surface resources constitute the main source of water within the basin, but they are only part of the much larger hydrologic system that includes groundwater. Within this system, water moves back and forth between surface- and groundwater sub-systems and their dependent ecosystems, affecting the quantum available for use as well as its quality. Furthermore, it is increasingly evident that many of the problems related to water resource and environmental degradation of the MDB are linked to its groundwater and the movement of water and the dissolved salts between the surface and sub-surface systems.

There are large resources of groundwater in the MDB, categorized on the basis of electrical conductivity (unit: $\mu S\ cm^{-1}$) as fresh (EC < 325), marginal (325 < EC < 975), brackish (975 < EC < 3250), and saline (EC > 3250). Groundwater occurs in all three types of aquifers, namely: surficial (essentially unconsolidated clay, silt, sand, gravel, and limestone formations, mainly of Quaternary age); sedimentary (consolidated sediments, such as porous sandstones and conglomerates); and fractured rock (igneous and metamorphosed hard rocks that have been subjected to disturbance, deformation, or weathering). Covering the largest area are the sedimentary basins, in particular the Great Artesian Basin and the Murray Basin. Estimates of exploitable groundwater in MDB, together with abstractions during 1983–1984, are given in Table 13.5. In spite of their size, the groundwater resources are not unlimited. Many of the potentially high-yielding aquifers receive a relatively low rate of natural recharge compared with the volume of groundwater they store. For example, the volume of water stored in the alluvial sediments in the lower Namoi Valley is about 20 km^3, but the average yearly natural recharge is only about 0.03 km^3. Recharge areas for the various high-yielding alluvial aquifers are the river beds and flood plains. Several lenses of fresh groundwater also form in alluvial aquifers that receive recharge during high river levels. For the Great Artesian Basin, the most important recharge areas are the wetter areas along the Great Dividing Range. For the Murray groundwater basin, recharge of the deeper confined aquifers occurs around the margins of the basin; the shallower unconfined aquifers also receive recharge over most of their surface areas.

Most of the attention given to groundwater over recent years has been in response to water- and land-salinity problems. However, research is also increasingly highlighting the fact that considerable quantities of good quality groundwater are present in many parts of the basin, which represents a substantial economic resource.

In terms of the hydrogeology of the Murray-Darling basin, there are a number of distinct groundwater systems (Fig. 13.14; Plate 13):

- the Murray geological or groundwater basin, known as the Murray Basin;
- the Great Artesian Basin;
- the shallow aquifers of the Darling River Basin; and
- the local groundwater systems found in areas of fractured rocks of the Great Dividing Range and other areas.

As indicated by Fig. 13.14, the boundaries of the groundwater areas do not coincide with those of the Murray-Darling Basin, which is defined on the basis of its surface water resources. The different groundwater regions behave somewhat independently of each other, with only relatively small amounts of groundwater directly exchanged between them. However, water from different aquifer systems is transferred across boundaries as surface water that constitutes the base flow of streams. Through this process, a substantial volume of groundwater enters the surface streams in the upper and middle catchments as base flow and then re-enters the groundwater systems further down through seepage from stream beds.

The Great Artesian Basin (GAB) is one of the largest basins in the world, with a total area of 1.7 $\times 10^6\ km^2$, covering 22% of the area of Australia. The basin extends under the northern part of the MDB in Queensland and New South Wales. It has a multi-layered aquifer system, consisting mainly of sandstones alternating with impermeable siltstones and mudstones, and is up to 3000 m thick. The GAB contains an estimated 8700 km^3 of water.

The GAB comprises predominantly arid and semi-arid areas, where surface water resources

Table 13.5 Estimate of groundwater resource in the MDB.

Fresh	Marginal	Brackish	Saline	Total	Abstraction (1983/1984)
Volume (km^3)					
0.97	0.94	0.91	0.85	3.68	0.62

Source: http://www.mdbc.gov.au/nrm

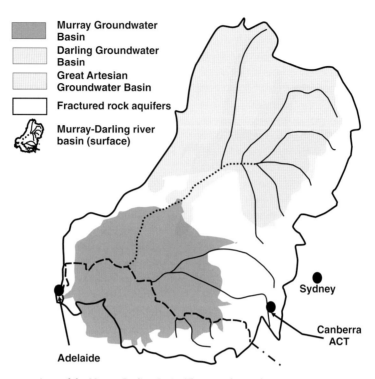

Fig. 13.14 Groundwater regions of the Murray-Darling Basin. The map shows the Murray Groundwater Basin, the portion of the Great Artesian Groundwater Basin within the MDB, the shallow aquifers of the Darling River Groundwater Basin, and the areas of fractured rock aquifers. *Source*: MDBC (1999). See also Plate 13.14.

are limited and extremely unreliable. As a result, groundwater is the only significant source of water for towns, farms, and livestock, as well as for mining and tourism. Without groundwater, these activities would not be possible. However, the quality of groundwater is generally unsuitable for irrigation because of its high sodium content, which makes it chemically unsuitable for cultivation. Overlying parts of the GAB are large alluvial fan aquifers of Tertiary age associated with the major rivers, the Macquarie, Gwydir, Namoi, Border, and Condamine. Increased attention is being paid by the government to the integrated management of land and water resources of the basin.

In the northern parts, high variability of surface flows has caused a greater demand for groundwater than further south. As a result, in many areas groundwater levels and water pressure are falling. In contrast, under current agricultural practices, the southern part of the Murray-Darling Basin continues to experience a steady rise in groundwater levels and in the mid-1990s it was estimated that about 1.3×10^6 *ha* of the irrigated land of the region would become saline or waterlogged by the year 2040. In addition, the rising groundwater levels will contribute ever increasing amounts of saline water to rivers throughout the Murray-Darling Basin.

13.3.3 Climate change and water resources of MDB

Severe drought conditions in 2006 and 2007 reduced inflows of water to the Murray-Darling River system to the lowest level recorded so far. Climate models suggest that rainfall in the basin will decline because of climate change, and as a result inflows to the system will also be reduced. In particular, climate change is likely to decrease the average rainfall and increase the frequency of occurrence of droughts and, therefore, the reliability of water supply.

A MDBC-sponsored study by Earth Tech, reviewed the potential future impact of eight important factors on flows in the River Murray System for the worst-case scenario on stream flow for 20 and 50 *years* from today. These factors are: (i) climate change; (ii) reforestation; (iii) groundwater extraction; (iv) changes in drainage and return flows from irrigation areas; (v) farm dams; (vi) vegetation regrowth in the upper catchment as a sequel to the 2003 bushfires; (vii) changes in industrial activities; and (viii) water trade.

It is concluded (MDBC 2003) that while the relative significance of these factors varies, collectively they are likely to have a substantial impact on stream flows in the River Murray System. Climate change and the construction of farm dams appear to be factors with a high degree of uncertainty and have been identified to have worst-case scenario impacts on stream flows in future. The impacts of reforestation and forest fires are rather well understood. However, ongoing research and careful monitoring are required to identify areas where reforestation is taking place within the basin. Groundwater extraction and irrigation return flow have the potential to significantly impact the river flow. However, both have a considerable level of uncertainty concerning their impacts on the River Murray System. With the volumetric cap on diversions in place, water trade and change in industrial activities are unlikely to impact on the volume of water in the River Murray system. However, changes in trade and industry are likely to change the location where water is used and have the potential to change the timing at which application of irrigation water is required. Both of these have the potential to impact on the capacity of the MDBC to meet the peak demand. In the context of the Living Murray initiative, the eight factors investigated have the potential to substantially reduce the volume of water within the River Murray system by 5–20% over the next 20–50 *year* period.

A recent investigation by the CSIRO Murray-Darling Sustainable Yield Project (Crosbie *et al*. 2008) on the impact of a future climate change on the diffuse groundwater recharge of the MDB using the WAVES model, groundwater recharge at selected points was modelled for a variety of soil and vegetation types. Recharge scaling factors were calculated for each climate scenario at the selected points. The point scale estimates of the recharge scaling factors were then up-scaled to the entire MDB using soil type, vegetation type, and change in rainfall as covariates to create rasters of recharge scaling factors for each scenario. The climate scenarios investigated involved a climate sequence based upon the data of the previous 10 years and the 3 future climate scenarios (dry, medium, and wet) as predicted by 15 different global climate models. The outputs of this report are a series of rasters for the change in recharge throughout the MDB at a resolution of a $0.05° \times 0.05°$ grid for each of these 46 climate scenarios. Under the scenario based on the last 10 years of climate observations, recharge decreases by up to 50% in the southern parts of the MDB and in the Condamine, while in the remainder of the northern parts of the MDB, recharge increases by up to 20%. Under the dry climate change scenario, recharge reduces in all parts of the MDB, though not as extremely as under the climate scenario based on the last ten years. Under the mid-climate change scenario, little change in recharge is observed throughout the MDB, with small decreases in the south and small increases in the north. Under the wet climate change scenario, recharge increases in all regions with 50% increase in the north and 5% increase in the south.

The Murray-Darling Basin is a vital region of Australia – producing a significant proportion of the nation's food, supporting rural populations, and providing many recreational and cultural avenues. Water is central to wealth generation and regional growth, and indeed all economic activities in the basin. However, rising environmental awareness in society along with scientific evidence indicate that irrigation and dry land salinity are major threats to sustainable use of natural resources in the basin. Also, there is concern that climate change may reduce the quantity of water in the rivers and the groundwater aquifers. State and Federal Governments through the Murray Darling Basin Commission (MDBC) are responding to these emerging issues. Since 1989 they have established a limit on water diversions (the cap), promoted salinity management and greater technical efficiency on farms, and established environmental flows through the recent Living Murray initiative of the Murray-Darling Basin Ministerial Council, which is aimed at restoring the health of the River Murray and the Murray-Darling Basin.

13.4 The North Gujarat–Cambay region, Western India

The North Gujarat–Cambay (NGC) region of Gujarat State in Western India is located between the 21.5°–24.5°N latitudes and 71.5°–74°E longitudes. The water resource situation of Gujarat State and its four major sub-divisions is shown in Fig. 13.15; Plate 14. The annual per capita natural endowment of water in North Gujarat is meagre – much below the internationally accepted comfort norm of more than 1000 m^3 – and most of this is already being exploited. The important rivers and other landmarks

in the region are shown in Fig. 13.16a. These rivers carry runoff only during the southwest monsoon season (June–September) and for some time thereafter. The prominent geographical features of the study area include the Nalsarovar (NS), the Little Rann of Kachchh (LRK), and the Gulf of Cambay (GC), also known as the Gulf of Khambhat.

The region is endowed with one of the richest alluvial aquifers of India but its uncontrolled exploitation for irrigation has resulted in many undesirable consequences. The region's groundwater resources are now under great stress, as evidenced by rapidly falling water levels and deteriorating water quality. The reason for the genesis of the groundwater crisis of North Gujarat has been its intensive use for irrigated agriculture, which is out of balance with its natural replenishment from rainfall. The tube wells pump out about $3 \times 10^9 \ m^3$ of groundwater annually against an annual replenishment of only about $2.4 \times 10^9 \ m^3$, resulting

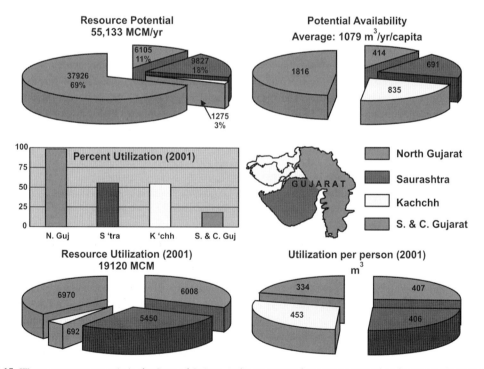

Fig. 13.15 Water resource scenario in the State of Gujarat, India. In terms of per capita natural endowment of water, North Gujarat is the most stressed region, with nearly 100% of its resource already being utilized. The South and Central regions, with the Narmada, Mahi, and Tapi rivers passing through them, have surplus water that is being or will be transferred to water-deficient regions through a network of interlinked canals. See also Plate 13.15.

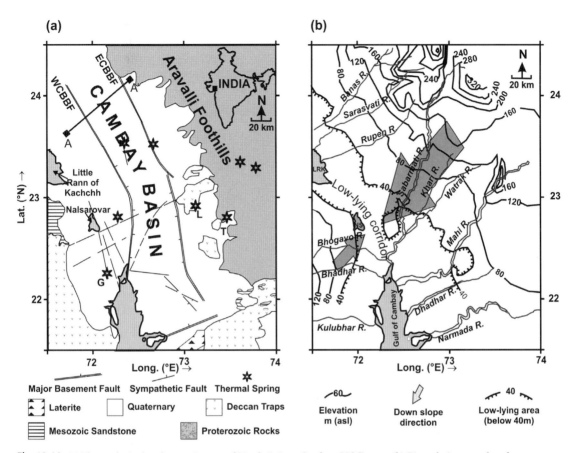

Fig. 13.16 (a) The geological and tectonic map of North Gujarat Cambay (NGC) area. (b) River drainage and surface elevation map of the study area. A large part of the area is covered by Quaternary alluvium. Proterozoic rocks (mainly granitic) lie along the Aravalli foothills in the east and the upper Cretaceous Deccan Traps on the southwest and southeast corners. A small patch of Mesozoic sandstone is also seen towards the west. A sub-surface lithological section along AA' is shown in Fig. 13.17. Traces of two major subsurface faults (ECBBF and WCBBF) and several other sympathetic faults are also seen. ECBBF = Eastern Cambay Basin Bounding Fault; WCBBF = Western Cambay Basin Bounding Fault. The low-lying belt linking LRK-NS-GC is less than 40 *m* in elevation and forms the zone of convergence for surface drainage from both sides. Redrawn from Deshpande (2006).

in an annual deficit of about 0.6×10^9 m^3. The direct implication of groundwater depletion to the economy in terms of increasing cost of electricity for lifting groundwater from progressively deeper levels is an area of serious concern to the state government, which subsidizes electricity used in well irrigation. For farmers too, agriculture is becoming increasingly uneconomic because of the high cost of groundwater irrigation, in spite of the highly subsidized electricity used for lifting water.

This dismal groundwater situation is accentuated by the inherently meagre surface water resources of the region. Rivers in North Gujarat flow from the hilly region in the northeast towards the Gulf of Cambay (now called Gulf of Khambhat) or the Rann of Kachchh. They are seasonal in nature, carrying stream flows only during three to four rainy months of the year. The rivers experience high variability in their annual flows and are nearly dry during low rainfall years. The major- and medium-irrigation schemes in North Gujarat were designed for runoff with very low dependability and hence their performance is highly susceptible to runoff variability. Due to over-appropriation of runoff, in most years there is no spillover from these reservoirs into the downstream. In April 2000, as against the total capacity of $2.2 \times 10^9 \ m^3$ of 13 major and minor dams of North Gujarat, the storage was only about $6 \times 10^6 \ m^3$. To sum up, any further surface water development of significant scale in the region using native water does not seem possible, and judicious management of groundwater seems to be the only alternative to ensure sustained availability of water in the region.

The region is characterized by a unique combination of geological, hydrologic, tectonic, and climatic features, namely: (i) two major bounding faults, defining the Cambay Graben, and several other sympathetic faults parallel as well as orthogonal to these (Fig. 13.16a); (ii) more than 3 km thick sedimentary succession, formed by syndepositional subsidence in the Cambay Graben, acting as a reservoir for the oil and gas at deeper levels and a regional aquifer system at shallower depths; (iii) higher than average geothermal heat flow; (iv) intermittent seismicity; (v) emergence of thermal springs (Fig. 13.16a); and (vi) arid climate with high rate of evapotranspiration.

The surface soils in the region are loamy to sandy loam type and agricultural productivity is high when water is available. As the streams flowing through the region are seasonal, groundwater mining has been resorted to in some parts. This has led to a decline in piezometric levels – at a rate more than 3 $m \ a^{-1}$ during the last couple of decades.

During the same period, it has also been noticed that fluoride concentrations in groundwater in some parts have progressively increased, leading to endemic fluorosis. Assuming that deep groundwaters have been in contact with the aquifer material for relatively long periods, it has been argued that high groundwater fluoride of the region is due to its slow leaching from the mineral grains comprising the aquifer matrix (Patel 1986). Some studies have also indicated that thermal springs in the region contain high concentration of fluoride and dissolved helium (Datta et al. 1980b). Therefore, it has been suggested that high concentrations of

fluoride may be related to subsurface injection of thermal waters (Chandrasekharam and Antu 1995).

It has long been recognized that helium and radon produced by radioactive decay of U and Th in rocks and minerals are being steadily released from grains by etching, dissolution, fracturing, and alpha recoil during weathering and subsequently released into the atmosphere due to diffusion and temperature variations. As already mentioned, anomalously high amounts of dissolved helium and radon in groundwater from some parts of the region have been reported earlier (Datta et al. 1980a).

There are several known thermal springs and flowing artesian wells in and around the Cambay Basin. The geothermal gradient in parts of the Cambay Basin is known to be in excess of $60 °C$ km^{-1} (Gupta 1981; Negi et al. 1992; Panda 1985; Ravi Shankar 1988). Based on reports of groundwaters with high temperature, anomalous helium, and high fluoride, it is believed that these might be related to each other (Chandrasekharam and Antu 1995; Gupta and Deshpande 2003b; Minissale et al. 2000).

The average annual relative humidity at present is only about 50% in the northern part of Cambay Basin; therefore water loss by evapotranspiration is expected to be high. The Late Quaternary sedimentary record of the region, however, suggests episodes of both wetter and drier periods (Juyal et al. 2003; Pandarinath et al. 1999; Prasad and Gupta, 1999; Wasson et al. 1983).

13.4.1 Groundwater hydrology

The groundwater occurrence in the NGC area can be grouped under three physiographic settings, namely: (i) hilly area of the Aravalli foothills in the northeast and east; (ii) alluvial plains including the Cambay Basin and both its flanks; and (iii) part of the Saurashtra uplands in the NGC area.

In the hilly region of the Aravalli foothills, where hard rocks are either exposed at the surface or are at a shallow depth of a few metres, groundwater is found in the secondary porosity zones resulting from weathering, joint planes, cracks, and fissures. The water table in this terrain is generally at relatively shallow depths of less than 10 m.

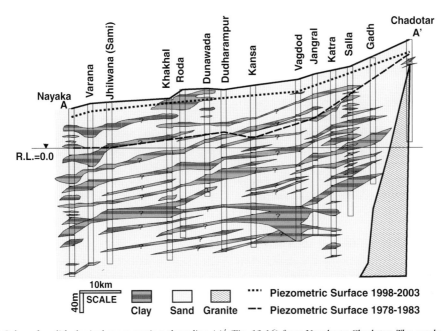

Fig. 13.17 Sub-surface lithological cross-section along line AA′ (Fig. 13.16) from Nayaka to Chadotar. The sandy layers forming aquifer horizons are seen to be laterally continuous and vertically interspersed with thin semi-permeable clay/silt layers that may not have lateral continuity over a large area. The uncertainty in the continuity of horizons in view of their large separation is indicated by the symbol "?". Tube wells tap all the water-bearing horizons up to their maximum depth. Redrawn from Gupta *et al.* (2005a). Reproduced with kind permission of Springer Science+Business Media. See also Plate 13.17.

The groundwater level rises considerably in the post-monsoon period, due to direct infiltration of rainwater in the secondary porosity aquifers.

The sediments in the foothill region are relatively coarse and this zone between the Aravalli hills in the east and the alluvial plains in the west forms the principal recharge area for the confined groundwater in the Cambay Basin area. Using tritium tagging of soil moisture, annual recharge to groundwater from this part of the NGC area has been estimated (Gupta and Sharma 1984) as approximately 15% of average annual precipitation.

A thick succession of Quaternary alluvial deposits comprising multilayered sequences of sediments of fluvio-marine and fluvio-aeolian origin forms the regional aquifer system, which extends from the Aravalli foothills in the northeast and east to the Little Rann of Kachchh and Saurashtra highlands in the west (Fig. 13.16a). A subsurface lithological cross-section across the Cambay Basin from Nayaka to Chadotar (along the line AA′), based on drilling logs obtained from the Gujarat

Water Resources Development Corporation Ltd (GWRDC), is shown in Fig. 13.17; Plate 15.

As can be seen, thickness of the alluvium rapidly increases westwards; away from the Aravalli hills, with a sand-clay/silt succession replacing the coarse sediments of the foothills. The sandy layers forming aquifer horizons are seen to be laterally continuous and vertically interspersed with thin semi-permeable clay/silt layers that may not have lateral continuity over a large area. It is also seen that various sub-aquifers are inclined nearly parallel to the ground surface so that, at different locations, a given depth below ground level (bgl) corresponds approximately to the same sub-aquifer within the basin. According to Patel (1986), the deeper aquifers are under artesian condition. Towards their western extension, the deeper aquifers abut against a thrust plane a little to the north of LRK. Along the LRK-NS-GC belt, tube wells tapping the deeper aquifers exhibit free-flowing condition and have high water temperature and saline water.

The shallow aquifers are under semi-confined condition. They receive recharge: (i) directly by seepage from the overlying unconfined aquifer; and (ii) by lateral flow from the recharge zone of the Aravalli foothills in the east. The shallow unconfined aquifer receives direct recharge from: (i) rainfall infiltration; (ii) nearby stream flow; and (iii) return flow from irrigation.

In recent years, the deep groundwater from this regional aquifer system, particularly in the Cambay Basin region, is being exploited to meet agricultural, domestic, and industrial water demands. Since extensive withdrawal of deep groundwater is not made up by natural or artificial groundwater recharge, water levels in most of the study area have declined at a rate of 3-4 $m\ a^{-1}$ during past decades (see piezometric surfaces in Fig. 13.17). In some parts of the NGC region, the groundwater levels have declined to more than 400 m depth

and the water is being pumped at progressively increasing energy cost.

13.4.2 Fluoride content and EC of groundwater

The geographical distribution of fluoride in groundwater from the NGC region is shown in Fig. 13.18a. It is seen that pockets of high groundwater fluoride concentration in the NGC region are approximately aligned around four linear belts marked PP', QQ', RR', and SS'. Pockets of very high fluoride concentration values (4-8 ppm) are aligned around PP', linking LRK-NS-GC, roughly in the north-south direction. To the east of PP', there is another linear belt around QQ' within the Cambay Basin with patches of high fluoride concentration values (1.5-4 ppm). On the easternmost part lies the linear belt around RR', almost in the recharge area

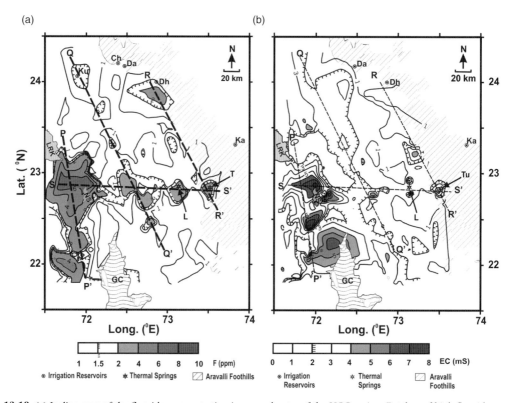

Fig. 13.18 (a) Isoline map of the fluoride concentration in groundwater of the NGC region. Patches of high fluoride concentration (>1.5 ppm) appear to be aligned around four lines (PP', QQ', RR', and SS') separated by areas with low fluoride. (b) Isoline map of EC of groundwater from the NGC region. Redrawn from Gupta et al. (2005a). Reproduced with kind permission of Springer Science+Business Media.

in the Aravalli foothills and roughly parallel to it, with several small and isolated pockets of fluoride concentration of more than 1.5 *ppm*. The fourth linear belt of high fluoride concentration (4–8 *ppm*) is around SS', roughly in the east-west direction, linking the region of thermal springs of Tuwa and Lasundra in the east to the Nalsarovar in the west.

An isoline map of the electrical conductivity (EC) of groundwater samples from the NGC region is shown in Fig. 13.18b. The EC values range from 0.3–8 *mS cm*$^{-1}$, with values of more than 5 *mS cm*$^{-1}$ in the LRK-NS-GC belt and the lowest values in isolated pockets along the Aravalli foothills. In general, areas with high EC overlap with those having high fluoride concentration (Fig. 13.18a). However, a pocket of high EC (>3 *mS cm*$^{-1}$) in the northwest part of the study area has low fluoride concentration (<1.5 *ppm*). Another pocket of EC of ~1 *mS cm*$^{-1}$, around 24°N latitude in the Aravalli foothills, on the other hand, has high fluoride concentration (>1.5 *ppm*). In spite of these differences, it can be seen from Fig. 13.18a that high EC

regions are aligned approximately along the same lines PP', QQ', RR', SS' that show high fluoride concentrations around them. The central belt (QQ') of both high fluoride concentration and EC is flanked on either side by belts of relatively lower values.

13.4.3 Groundwater dating

The estimated groundwater ^{14}C ages are seen (Fig. 13.19) to increase progressively from less than 2 *kaBP* (along the Aravalli foothills) to more than 35 *kaBP* in the low-lying tract linking LRK-NS-GC. Further west, lower ^{14}C ages are found. From hydrogeological considerations (Fig. 13.17), it seems that confinement of the regional aquifer in the NGC region becomes effective near the ECBBF around the ^{14}C age contour of about 2 *ka BP*. Within the Cambay Basin, the age isolines are nearly parallel to each other and the horizontal distance between the successive 5 *ka BP* isolines is nearly constant, giving a regional flow velocity in the range 2.5–3.5 *m a*$^{-1}$ for the prevailing hydraulic gradient of

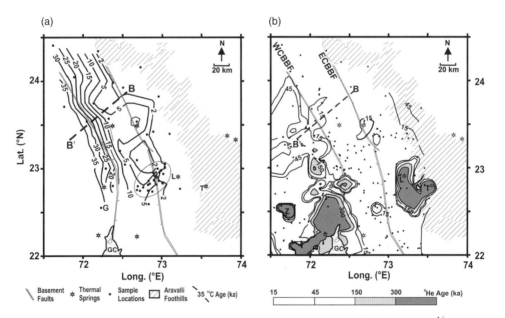

Fig. 13.19 (a) Isoline map of groundwater radiocarbon ages. Within the Cambay Basin, the groundwater ^{14}C ages increase progressively towards the WCBBF, beyond which a limiting ^{14}C age of >35 *kaBP* is observed. Dots indicate sampling locations. The ellipse encloses sampling locations. From Borole *et al.* (1979). Letters L, T, and G, respectively indicate the locations of thermal springs at Lasundra, Tuwa and a free-flowing thermal artesian well at Gundi. (b) Isoline map of estimated ^4He ages of groundwater (for helium release factor; $\Lambda_{He} = 1$). Redrawn from Agarwal *et al.* (2006). © Elsevier.

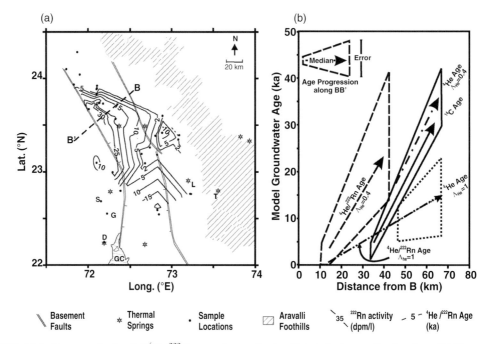

Fig. 13.20 (a) Isoline map of estimated $^4He/^{222}Rn$ ages of groundwater. Dots indicate sampling locations. (b) The groundwater age progression and gradient along BB′ for ^4He ages and $^4He/^{222}Rn$ ages for various assumed values of Λ_{He}. Also shown for comparison are the age progression and gradient for ^{14}C groundwater ages. Redrawn from Agarwal *et al.* (2006). © Elsevier.

1 in 2000 (GWRDC, unpublished data), which is comparable to an earlier estimate of approximately $6 \; m \; a^{-1}$ (Borole *et al.* 1979) for a small part of the Vatrak-Shedhi sub-basin (marked by an ellipse in Fig. 13.19). The Vatrak-Shedhi sub-basin is closer to the recharge area; therefore, both the permeability and the hydraulic gradient are expected to be relatively higher than that for the regional estimates of flow velocity (2.5–$3.5 \; m \; a^{-1}$) obtained from Agarwal *et al.* (2006).

The isoline map of estimated ^4He ages of the groundwater in the NGC region is shown in Fig. 13.19b. The isoline map of estimated $^4He/^{222}Rn$ ages of the groundwater in the NGC region is shown in Fig. 13.20. For details of ^4He and $^4He/^{222}Rn$ dating of groundwaters, reference is made to Agarwal *et al.* (2006). The $^4He/^{222}Rn$ ages are independent of porosity, density, and U concentration but depend on Th/U ratio in the aquifer material. In addition, the $^4He/^{222}Rn$ ages depend on the release factor ratio ($\Lambda_{Rn}/\Lambda_{He}$).

It is thus seen that the groundwater ages (employing ^{14}C, ^4He, and $^4He/^{222}Rn$ methods) of the

confined aquifers of the NGC region progressively increase with distance from the recharge area but the dissolved fluoride concentration does not increase correspondingly. Instead, as already mentioned, alternating bands of high and low fluoride concentration (and also EC) in the groundwater from confined aquifers are observed. This observation negates the possibility of a continuous and constant source of dissolved salts/fluoride in the recharged groundwater of the confined aquifers of the NGC region and suggests additional time varying control on EC/fluoride concentration in groundwaters of NGC region.

The confined groundwater in the central high fluoride groundwater belt (QQ′) within the Cambay Basin corresponds to groundwater ^{14}C age in the range 15–25 *kaBP* (Fig. 13.19a). The period around 20 *kaBP*, corresponding to the Last Glacial Maximum (LGM), is known to be a period of enhanced aridity in the NGC region (Juyal *et al.* 2003; Pandarinath *et al.* 1999; Prasad and Gupta 1999; Wasson *et al.* 1983). The enhanced aridity is generally associated with: (i) increased evaporation;

(ii) decreased rainfall; and (iii) increased dry deposition. Some imprints of evaporation and dry deposition, even in the present climate, were seen in the ionic concentration and stable isotopic composition of modern rainfall. Significant control of dry deposition is seen in the variation of fluoride and EC of fortnightly accumulated rainwater samples. It is, therefore, inferred that groundwater recharged around LGM in the Aravalli foothills during the period of enhanced aridity has since moved to its present position in the Cambay Basin and corresponds to the central high fluoride groundwater belt (QQ'). Groundwater, with relatively low fluoride concentration, on either side of QQ', suggests recharge during less arid climatic regime.

13.4.4 Distribution of stable isotopes in groundwater

From the geographical distribution of $\delta^{18}O$ and d-excess, it is also seen (Fig. 13.21) that groundwater around a linear belt (QQ') is characterized by relatively low values of d-excess and high values of $\delta^{18}O$. Since evaporation of water results in low values of d-excess and high $\delta^{18}O$ (see Section

7.4), the groundwater around QQ' represents relatively high evaporation (indicative of higher aridity) either during rainfall or during groundwater recharge, compared to that on either side of this belt. As already mentioned, ^{14}C age of the confined groundwater around QQ' has been estimated to be in the range 15–25 $kaBP$, corresponding to the known arid phase in the past. Taking a holistic view of the data, that is, 15–25 $kaBP$ groundwater age, its lower d-excess, and higher $\delta^{18}O$, it is concluded that groundwater recharged in the Aravalli foothills around the LGM with signatures of higher aridity has since travelled to its present position. This corroborates enhanced aridity in the past as one of the important causes for the occurrence of high values of fluoride as well as EC of groundwater around QQ'.

13.4.5 Outlook

Severe depletion of aquifers in the North Gujarat–Cambay region has led to the failure of a large number of wells in some areas. Farmers who invested heavily in newer technology and deeper wells went bankrupt when their wells ran dry. In

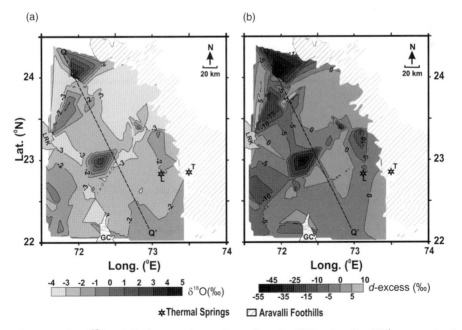

Fig. 13.21 Isoline map of (a) $\delta^{18}O$ and (b) d-excess of groundwater from the NGC region. Line QQ' representing the central linear belt of high fluoride (Fig. 13.18a) and high EC (Fig. 13.18b) is superposed for comparison. Redrawn from Deshpande (2006).

the quest for more water, farmers drilled deeper. The mounting costs involved in this operation resulted in the formation of wells owned by a group of farmers, who pooled their financial resources and sold the surplus pumped groundwater to other farmers. This laid the foundation for water markets in this region.

It is seen from Fig. 13.15 that, while the natural endowment of water in North Gujarat is the lowest, that in the neighbouring south and Central Gujarat is the highest. The major rivers of Gujarat, namely Narmada, Tapi, and Mahi, flow through this region. Therefore, to ensure equitable distribution of water within different regions and to utilize their full development potential, inter-basin transfer of water through a network of interlinking canals is already in progress. The Sardar Sarovar Dam on the

Narmada River, with its network of canals, is the most ambitious of these projects, providing assured water to parched lands in North Gujarat, Kachchh, and Saurashtra. The infrastructure created for this purpose is also being used for rejuvenation of the aquifers of North Gujarat, by diverting the uncommitted surplus flows during years of excess rainfall through a scheme known as 'Sujalam Sufalam', launched by the State Government. This project has three components: (i) pumping excess water from the Narmada canal into nine North Gujarat dams by laying a dozen-odd pipelines, each about 100-km-long; (ii) constructing an unlined spreading canal, 337 *km* long, from the Kadana dam in the Panchmahals to the Banas River in Banaskantha, connecting 21 rivers in North Gujarat; and (iii) building 200 000 farm ponds.

14

Epilogue

As the plan for writing this book was underway, it became obvious that hydrology forms part of almost every aspect of life on our planet and it would be unrealistic to address this diverse subject in a single book. It was, therefore, important to limit the scope of the book to the common issues, namely, understanding of the fundamental principles, methods, and problems encountered in the field, with emphasis on the underlying science and its applications to real life situations in the field of water. The study of modern hydrology, therefore, involves, in addition to conventional surface- and groundwater hydrology, the application of new environmental and artificial tracers, application of remote sensing, analytical and numerical models, and human dimensions.

The material of the book is drawn from basic concepts in geology, soil science, hydraulics, physics, chemistry, mathematics, engineering, and related disciplines as relevant to hydrology. The aim has been to provide adequate understanding of the various facets of modern hydrology in relation to our environment, ecology, and life on the Earth from local, regional as well as global perspectives, so as to present a book that can serve as a comprehensive textbook on modern hydrology at the graduate level and also serve as a reference book for researchers and professionals in diverse disciplines concerned with the various facets of water on the Earth.

The aim is to enable students as well as professionals to make informed analyses of any hydrologic dataset and plan additional investigations, if needed, to adequately address the hydrologic problem at hand, while keeping in mind the issues of water ethics and the larger issue of global change and the central role of water therein.

The first 12 chapters of the book are organized under three broad themes with the first 6 chapters addressing 'water, its properties, and its movement and modelling'. The next 3 chapters deal with the broad theme of 'studying the distribution of water in space and time'. The next 3 chapters are intended to address the broad theme of 'water resource sustainability'. The last chapter (Chapter 13) presents a few case studies linking many of the aspects discussed in the book with some real field situations. This chapter is the wrap-up summary to conclude in a comprehensive manner how hydrologic investigations and analyses enable one to study the local, regional, and global water cycle and manipulate and manage it, while keeping in mind considerations of sustainability and human welfare.

14.1 Water and its properties, quality considerations, movement, and modelling of surface- and groundwater

Many of the unique properties of water are largely due to its V-shaped molecule with one atom of oxygen bound to two atoms of hydrogen. Perhaps the most important physical property of water is that at the temperatures normally found on the Earth, it exists in all three states – liquid, solid (ice), and gas (vapour) – and moves around the globe continually changing its state. As it moves around in the 'global water cycle', transfer of energy is caused by the high values of its specific heat and the latent heat involved in its change of state.

Even though called the 'water' planet, the terrestrial component of the Earth contains only about 4% of all water on the Earth, with fresh water in lakes and rivers being about 0.007% and in groundwater being only about 0.76% of the total water on the Earth. Yet rivers, lakes, and groundwater are the water sources we are most familiar with and use in our daily lives. Due to a variety of geographic, geomorphic, and meteorological factors, the distribution of terrestrial water on the Earth is uneven. This gives every place on the Earth a unique character in almost every aspect, including landscape, ecology, and environment and has governed the evolution of human societies and cultures throughout the world.

The concept of the *hydrologic cycle* is fundamental to efficient management of water resources. When the flow of water is manipulated to meet human needs, it is necessary to understand how it affects the local and regional hydrologic cycle and, ultimately, the availability and quality of water to downstream users. To ensure availability of adequate water for human use, water managers need to be able to estimate the amounts of water that enter, flow through, and leave a given watershed.

Of all the components of the hydrologic cycle, the most commonly measured element is precipitation, both as liquid rain and snow. Several methods exist for measuring the magnitude and intensity of precipitation that include: (i) ground-based measurements at point sites; (ii) ground-based remote sensing measurements on a regional scale (weather radar); and (iii) aircraft and satellite-based sensors over still larger regions. Each measurement differs in respect of temporal and spatial scales and the measurement technique employed.

Evapotranspiration (*ET*) involves the vapour phase transfer of water from the land surface to the atmosphere, through a combination of evaporation from open water surfaces (e.g. lakes, rivers, puddles) and transpiration by plants. Three major factors that limit *ET* are: (i) input of solar energy (particularly at high latitudes, and in winter season); (ii) water availability (in dry soils, and in the absence of open water bodies); and (iii) turbulent transport of vapour (under low winds, sheltered areas). Perhaps the simplest way of measuring evaporation is with an evaporation pan. But hydrologists and engineers are not really interested in what evaporates from a pan; instead they wish to know the regional evaporation from a given land surface or from a nearby lake. Unfortunately, pan evaporation is often a poor indicator of these variables. *ET* can be estimated indirectly by measuring percolation using lysimeters and subtracting it from precipitation or by measuring soil water depletion during a given time interval. Some other methods are purely theoretical in nature. One such widely-used empirical method is the Thornthwaite method that predicts the monthly potential *ET* (*PE*; in *mm*) based on mean monthly air temperature (*T*; in $°C$).

As precipitation reaches the ground surface it begins to infiltrate into soil. Infiltration can be defined as the downward flux of water from the soil surface into the soil profile via pore spaces. The process becomes more complex, both with increasing non-uniformity of soil and rainfall rate. Infiltration through a soil profile is governed by two major forces opposing each other, namely capillary force and gravity. It is affected by several other factors that include: (i) water input from rainfall, snowmelt, irrigation, and ponding; (ii) soil profile properties, such as porosity, bulk density, conductivity; (iii) antecedent soil moisture content and its profile; (iv) soil surface topography and roughness; and (v) soil freeze and thaw conditions. The infiltration process is characterized by a general decrease in infiltration rate as a function of time and a progressive downward movement and diffusion of an infiltration front into the soil profile. The infiltration rate is determined by measuring the time it takes for a layer of water on a soil surface to drop by a fixed distance. In watershed studies for hydrologic modelling, empirical and physical equations enable the estimation of infiltration as a function of time for a given set of soil properties. Empirical methods rely on regression equations based on large datasets. Use of these equations enables determination of infiltration with limited measurements. Some of the commonly used equations are the Green-Ampt, Richards, and Horton equations.

Runoff occurs when the portion of rainfall that is not absorbed by the ground flows down-gradient. Part of the land from where runoff water collects and joins a particular river or stream is known as the watershed (or catchment) of the stream. The stream hydrograph is a primary observable in a river network draining a basin. Hydrologists use long term measured stream discharge and attempt to relate it to rainfall characteristics in the basin. A time-series of precipitation in a watershed is called a rainfall hyetograph and a similar time-series of stream discharge from the watershed is called a stream flow hydrograph. There are a number of methods to measure the amount of water flowing in a stream or a canal. The objective is to estimate the volume of water moving per unit time through a given cross-section of a river channel. This is usually calculated using the 'area-velocity method'. Flumes and weirs are constructed as 'in stream' structures to make water flow through a well-defined cross-section so that flow velocity and hence discharge estimation can be made by measuring overflow height above a standard fixed reference level.

Another commonly employed method to monitor surface water flow is to use a water depth-flow rate rating curve, which correlates the depth of flow to the flow rate based on measurements of both water depth and flow rate at the location of interest over a range of flows. The other approach, called an index velocity determination, uses a velocity meter, either magnetic or acoustic, to measure the index velocity of the flow at a gauging station. This index velocity is used to calculate the average velocity of the flow in a stream. A rating curve, similar to that used for a stage-discharge relationship, is constructed using discharge determination to relate the indicated index velocity with the stream discharge. As with other water cycle components, empirical equations obtained by regression analysis using a large dataset are used for discharge estimation, particularly in ungauged or inaccessible reaches of streams.

Rainfall-runoff analysis and modelling includes the physical processes operating in a watershed leading to the transformation of rainfall into stream runoff. The runoff hydrograph of a stream provides a reasonable indication of the processes that are op-

erative in a basin — perhaps more realistically than any other measurement. There are a number of ways to estimate the amount of water that runs off a given surface. In addition to field-based methods, one can also use computer models and simulations to estimate runoff volumes.

Transport of runoff through a channel system also gives rise to erosion and transport of sediments (as suspended and bed loads) and nutrients (as dissolved load). In fact, transport of runoff water, together with dissolved, suspended, and bed load transported materials, is a major agent for landscape evolution. Hill slopes, streams, and drainage basins together form the fluvial network that transports water, nutrients, and sediment through a given landscape that itself changes continuously with time due to concomitant processes of erosion and deposition occurring in its different parts. In its down-gradient journey, runoff produced at individual points is routed through a stream network. Several robust network laws relating to stream order and their number, length, and drainage area of a basin have been identified through characterization of river channel morphology. The near universality of these empirical stream networking 'laws' seem to indicate that some basic properties of fluvial geomorphology and the processes that shape the landscape may be invariant over a wide range of spatial and temporal scales and that these empirical equations, in some way, exhibit the underlying scaling laws.

In volume terms, groundwater is the most important component of the active terrestrial hydrologic cycle. But being subsurface, the main features of groundwater systematics are poorly known and these can only be inferred indirectly. Below the ground surface, water is contained: (i) in the top soil as soil moisture; (ii) in the intermediate unsaturated zone below the soil; and (iii) in the capillary fringe as pellicular water and below it in the aquifer as groundwater. Some amount of water occurs under the Earth's surface almost everywhere. This gives a unique advantage to groundwater in terms of its almost ubiquitous availability as compared to surface water, which can only be harvested at certain favourable sites and needs to be conveyed to other places by constructing appropriate engineering structures.

Groundwater and surface water are basically interconnected. In fact, it is often difficult to separate the two because they 'feed' each other. The aquifers are often partially replenished by seepage from streams and lakes. In other locations, the same aquifers may discharge through seeps and springs to feed the streams, rivers, and lakes. In fact, for perennial inland rivers of non-glacial origin in arid/semi-arid areas, the lean season base flow is contributed by groundwater discharge from the aquifers adjacent to the river banks.

Geological formations through which water can pass easily are said to be *permeable* and those that scarcely allow water to pass through or only with difficulty are described as *semi-permeable* or *impermeable*, depending on the degree of permeability. Saturated and permeable formations are called *aquifers*. An *unconfined aquifer* contains a phreatic surface (water table) as an upper boundary, being at the atmospheric pressure, that fluctuates in response to recharge and discharge of water. If, on the other hand, the effective thickness of an aquifer lies between two low permeability or impermeable layers, it is called a *confined aquifer*. An imaginary surface joining the water level in boreholes tapping a confined aquifer is called the *potentiometric* or *piezometric surface*.

External forces, which act on water in the subsurface, include gravity, atmospheric pressure, and hydraulic pressure due to the overlying water column, and molecular forces acting between aquifer solids and water. The driving force for groundwater flow is known as *hydraulic head* – being the sum of gravitational and pressure potentials. Groundwater flows in the down-gradient direction – from higher to lower head. The flow of groundwater through an aquifer is governed by *Darcy's Law*, which states that the rate of flow is directly proportional to the hydraulic gradient. The coefficient of proportionality defines an intrinsic property of porous media and is known as the *hydraulic conductivity* (*K*). This parameter is a measure of the ease with which water flows through the various materials that form aquifers and is a function of properties of the medium, collectively defining the intrinsic permeability (*k*), as well as the properties of the fluid. The overall permeability of a rock mass or sediment depends on a combination of the size of the pores and the degree to which the pores are interconnected. For well-sorted, granular materials, hydraulic conductivity increases with grain size.

The *specific discharge* per unit area or *Darcy flux* ($q = Q/A$) gives the apparent volumetric flow velocity through a given cross-section of an aquifer that includes both solids and voids. The average velocity through the pores (the *pore water velocity*) is estimated by dividing specific discharge by the *effective* or *dynamic porosity* (n_e). The effective porosity may be significantly less than the *total porosity, n*, of the media; the latter defined simply as the ratio of void space (V_V) to the total volume (V_T) of the material. *Transmissivity, T*, is a measure of the amount of water that can be transmitted horizontally through a unit width by the fully saturated thickness of an aquifer under a unit hydraulic gradient. Transmissivity is equal to the hydraulic conductivity multiplied by the saturated thickness of the aquifer through which the flow takes place.

The total load above an aquifer is supported by a combination of the solid skeleton of the aquifer matrix as well as by the hydraulic pressure exerted by the water in the aquifer. When an aquifer is pumped, water is released from the storage due to release of hydrostatic pressure within the pore spaces and compression of the solid skeleton by the overburden resulting from loss of buoyancy. The *storage coefficient* or *storativity S* (dimensionless) is the volume of water that a permeable unit will take into storage, or release from storage, per unit surface area per unit change in head. A related term, *specific storage S_s* [L^{-1}], is the amount of water per unit volume of a saturated formation that is taken into or released from the aquifer storage owing to compression of the mineral skeleton and expansion of the pore water per unit change in the hydraulic head.

The various flow parameters controlling the flow through the aquifer are defined for homogenous isotropic aquifers – an ideal situation rarely encountered in nature. However, this problem is often circumvented by considering mathematically equivalent properties for homogenous isotropic media. The two physical principles, namely Darcy's law and the mass conservation principle, define the groundwater flow equations. The most general form of the saturated flow equation is (Eqn. 3.36), which describes flow in three dimensions, transient flow ($\partial h/\partial t \neq 0$), heterogeneous conductivities (e.g. K_x being a function of x), and anisotropic hydraulic conductivities ($K_x \neq K_y \neq K_z$). This equation simplifies to the familiar Laplace equation (Eqn. 3.39) for steady flow with homogenous, isotropic K.

In saturated flow through porous media, velocities vary widely across pore spaces of different sizes, shapes, and orientations. As a consequence, in addition to molecular diffusion, an irreversible hydrodynamic mixing process occurs during laminar flow of groundwater between its various elements. This phenomenon is known as dispersion.

Many of the measurements related to groundwater and its flow characteristics are indirect because most of the subsurface is inaccessible to direct observation. But groundwater levels, temperature, and quality parameters can be determined *in situ* through wells to determine groundwater flow direction and velocities for resource management and pollution risk assessment.

Wells are used to extract groundwater and sometimes also for recharging aquifers. Well hydraulics deals with the process of groundwater flow to wells tapping an aquifer system. Hydraulic properties of aquifers that control groundwater flow are: (i) hydraulic conductivity or permeability; (ii) transmissivity; (iii) storativity; and (iv) hydraulic gradient. Response of a well to any discharge of water from or recharge into the well is a function of the aquifer properties.

Under isotropic, homogeneous aquifer conditions, groundwater flows radially towards a pumping well from all directions. When water is pumped, the level of the water table in the vicinity of the well lowers in the shape of an inverted cone, known as the *cone of depression*. The drawdown curve defines the shape of the depressed potentiometric surface/water table in three dimensions. During steady state conditions, the head and cone of the depression are in equilibrium between the pumping rate and aquifer properties, in contrast to unsteady flow when the head of the drawdown curve changes continuously with time.

A number of equations have been derived to describe the flow of water to wells, using calculus and application of Darcy's Law to groundwater flow from the surrounding aquifer to a pumping well. Implicit in these derivations are assumptions that the pumping well is 100% efficient in extracting water, it fully penetrates the aquifer, the water table or potentiometric surface has zero slope initially, and laminar flow conditions (characterized by a low Reynolds number) prevail.

Pump tests are employed to determine performance characteristics of a well and to determine the hydraulic parameters T, K, and S of an aquifer. Another use of the pump tests is to assess the performance of a well by monitoring the drawdown and yield to derive the specific capacity of the well defined as a ratio of the yield to drawdown. This is then used to choose the appropriate size of the pump relative to the production capacity of the well. Two types of commonly used aquifer pump tests are: (i) *the constant-rate test* involving pumping the well for a significant length of time at a uniform discharge rate to allow the aquifer to come to equilibrium with the well assembly; and (ii) the *step-drawdown test* wherein the well is pumped at successively higher discharge rates for relatively short duration intervals. When only a single well is available for recording of pumping rates and the corresponding drawdown, the pump test method can estimate only the transmissivity and specific capacity of the well and not the storativity or the geometry of the cone of depression.

Multiple well pump tests involving a pumping well and one or more observation wells are required to estimate storativity, and the three-dimensional geometry of cone of depression. Initial data analyses involve making semi-log plots of drawdown versus time since the beginning of pumping and matching the field curve with standard model curves.

Hydrologic systems are complex, with processes occurring over different spatial scales corresponding to the size and topography of geographical areas. In this situation, mathematical models are valuable tools that enable one to make assessments, investigate alternative scenarios, and assist in developing effective management strategies. Models are essentially aids to describe and evaluate the performance of relevant systems under various real or hypothetical constraints and field situations. A hydrologic model may be defined as a simplified conceptual representation of a part of the hydrologic cycle of a real-world system (here, a surface- or a groundwater, or a combined system) that approximately simulates the relevant input-output response of the system. Two major types of hydrologic models can be distinguished: (i) *stochastic models*, that use mathematical and statistical concepts to link a certain input (e.g. rainfall) to the model output (e.g. runoff); and (ii) *process-based models*, that mathematically attempt to simulate the physical processes of surface runoff, subsurface flow, evapotranspiration, and channel flow that can be quite complex. These models are known as deterministic models and can be subdivided into single-event models, and continuous simulation models. *Hydrological modelling* can be undertaken either as distributed or lumped-parameter simulation, differentiated by whether or not spatial variation of hydrologic parameters is accounted for.

Surface water hydrologic modelling is perceived to meet two basic requirements: (i) to determine the magnitude and frequency of flood flows; and (ii) to determine the long-term availability of water for consumption. The two requirements, however, involve different modelling approaches.

A groundwater model may be defined as a simplified, mathematical description of a real groundwater system, coded in a programming language. The accuracy of a model is dependent upon the level of understanding of the system and its realistic conceptualization consistent with modelling objectives and availability of data describing the physical system that the model is meant to represent. Any groundwater model essentially solves the governing equation of groundwater flow and storage.

The three common methods of solution used in groundwater modelling are: analytical, finite difference, and finite element. Each method differs in its approach, assumptions, and applicability to real-world problems.

Analytical methods use classical mathematical approaches to solve differential equations and obtain exact solutions. These provide quick and reliable results to simple problems but require assumptions of homogeneity and are essentially limited to one- and two-dimensional problems. However,

these can provide rough approximations for most problems with little effort.

Finite difference methods solve the partial-differential equations describing the system by using algebraic equations to approximate the solution at discrete points in a rectangular grid. The grid can be one-, two-, or three-dimensional. The points in the grid, called nodes, represent the average of the surrounding rectangular block (cell). Although adjacent nodes have an effect on the solution process, the value for a particular node is distinct from its neighbouring nodes. Grids used in finite difference codes generally require far less setting-up time than those of finite element codes, but have less flexibility in individual node placement. Many common codes, such as MODFLOW, use the finite difference solution method.

Finite element methods differ from finite difference methods in that the area (or volume) between adjacent nodes forms an element over which exact solution values of the input parameters are defined everywhere by means of basis functions. The essential difference is that finite element codes allow for flexible placement of nodes, which can be important in defining irregular boundaries. However, defining a unique location for each finite element node requires more effort in setting up the grid than that of a finite difference code. FEMWATER is a common code using the finite element solution method.

Some pre-processors allow superposition of the grid and the site map, and then interactive assignment of boundary conditions, aquifer properties, etc. Post-processors allow the numerical output to be presented as contour maps, raster plots, flow path plots, or line graphs. Choosing a code that does not have, or cannot easily be linked to, pre- and post-processors, should be avoided.

After selection of the modelling software, features of the conceptual model are transferred to an input file that defines the mathematical model. Features such as boundary conditions, grid dimensions and spacing, initial aquifer properties, and time-steps are specified according to the requirements of the selected code.

Groundwater models are useful in predicting the effects arising from specific recharge and withdrawal stresses, usually employing injection and extraction wells that cause a relatively large volume of water exchange in a relatively small area. These analyses can predict general aquifer response to such stresses.

Combining surface water and groundwater models to represent a real-life situation is a difficult task. First, surface- and groundwater tend to operate on very different temporal scales. In addition, in various numerical models, not only is the space discretized (e.g. employing layers and grid in MODFLOW the), the computations are performed at discrete time intervals, called the operating time-steps. The operating time-steps in a typical groundwater model and a typical surface water model are quite different. Groundwater models tend to run with time-steps of the order of a few months to a year. For a surface water system, for example, flood or reservoir routing, the time-step of a few minutes to a few hours would be more appropriate. While it may take several days for the surface water flow to make its way through the section, it is required to be tracked at much shorter time-steps. Because of these and other difficulties in solving groundwater–surface water–unsaturated zone flow in an integrated manner, individual existing sectoral models are often coupled.

Model coupling conceptually requires a realistic description of the hydrologic process and combined calibration of, for example, inflows, outflows, recharges, discharges, and aquifer parameters. Coupling, therefore, provides a means to realistically identify groundwater inflow into and outflow from the catchment and their coupling processes, for example, groundwater recharge/discharge. In practice, however, these potential benefits of coupled models are often offset by high demand on computation time and required computer storage capacity. Often results obtained from the coupled models might be more unrealistic than those obtained from the stand-alone sectoral models that require inputs from hydrochemistry, natural chemical, and isotopic tracers to achieve meaningful results.

One of the commonly-used codes is GSFLOW, which is a coupled groundwater–surface water FLOW model based on integrating the US Geological Survey Precipitation-Runoff Modelling System (PRMS) and the US Geological Survey Modular Ground-Water Flow Model (MODFLOW-2005).

Chemical composition of natural waters is derived from many different sources of solutes, including gases and atmospheric aerosols, weathering and erosion of rocks and soils, dissolution or precipitation reactions occurring below the land surface, and more recently anthropogenic processes. Application of principles of chemical thermodynamics can help in discerning broad interrelationships amongst these processes and their effects on the surface water–groundwater system. Some of the processes, for example, dissolution or precipitation of minerals, can be closely evaluated by means of principles of chemical equilibrium, including the law of mass action. Other processes are irreversible and require knowledge of relevant reaction mechanisms and their rates.

Basic data used in the determination of water quality are obtained by chemical analysis of water samples in the laboratory or onsite measurements in the field. Most of the measured constituents are reported in gravimetric units, usually milligram per litre ($mg\,l^{-1}$) or milli-equivalent per litre ($meq\,l^{-1}$). Chemical analyses may be grouped and statistically evaluated by determining the mean, median, and frequency distribution or ion correlation that help to consolidate and derive useful information from large volumes of data. Graphical methods of analyses or groups of analyses aid in identifying and showing chemical relationships amongst different waters, probable sources of solutes, regional water quality relationships, temporal and spatial variations, and water resource evaluation. Graphical methods may enable identification of water types based on chemical composition, relationships amongst ions or groups of ions in individual waters, or waters from multiple sources considered simultaneously. Relationship of water quality to hydrogeologic characteristics, such as stream discharge rate or groundwater flow pattern, can be represented by mathematical equations, graphs, and maps.

Adverse human impacts on water quality arise from contamination and resulting pollution of the water sources from diverse anthropogenic activities generating waste products. Transport and attenuation of point and non-point sources of pollutants and basic aspects of numerical modelling of solute transport also need to be considered.

Water quality standards for domestic, agricultural, and industrial uses have been published by various agencies. Irrigation and industrial requirements for water quality are particularly complex. Basic knowledge of processes that govern composition of natural waters is required for judicious management of water quality.

14.2 Distribution of water in space and time

To obtain direct insight into the dynamics of surface and subsurface water flow, hydrologists use tracers – substances that tag water because of their unique properties and follow its movement. Some of the commonly used tracers are dyes, solutes (e.g. chloride), radioactive and stable isotopes, dissolved gases (e.g. helium, *CFC*s), and some physical parameters (e.g. temperature).

When tracers are introduced artificially into a system for carrying out a tracer based study, these are known as artificial or injected tracers. However, if a tracer substance is already present in a system before the start of a tracer study, it is known as an environmental tracer. Depending on the method of analysis, artificial tracers can be classified into four broad groups – chemical, radioactive, activable, and particulate tracers. In recent years, *gaseous tracers* have also been used for various applications. These include dissolved inert gases used as geochemically conservative tracers in groundwater systems. When using *environmental tracers*, hydrologists exploit variations in the composition of a large number of substances, elements, or their isotopes within and/or across hydrologic reservoirs. A special class of environmental radioisotopes comprises tracers of cosmogenic origin. These isotopes are produced in nature by cosmic radiation entering the Earth's atmosphere. Cosmic rays produce nine radio-nuclides with half-lives ranging between 10 *years* and 1.5 *Ma* and five with half-lives between 2 weeks and 1 year. These have been used as tracers for measuring groundwater movement on timescales ranging from a few weeks to millions of years.

Groundwater age is generally considered as the average travel time for a water parcel from either the ground surface or from the water table of the unconfined aquifer in the recharge zone to a given point along the aquifer length. Various methods

for dating of young (<50 years) and old groundwaters are described. It is shown that, depending on the conceptual mathematical model employed to describe the aquifer system, the groundwater age estimation can yield additional information such as velocity, residence time, dispersion coefficient, and influx of young shallow unconfined aquifer water into the underlying semi-confined aquifers.

Tracers, in particular Cl^- and ^3H, have also been extensively used to estimate direct recharge of groundwater.

Amongst various isotopes used in hydrology as tracers, stable isotopes of oxygen (^{18}O) and hydrogen (^2H or D) are the most commonly used. Since these form an integral part of the water molecule, they are ideally suited to trace the movement of water in the hydrologic cycle. In hydrologic parlance, the two isotopes are also commonly referred to as *water isotopes*.

In addition, several chemical and gaseous tracers are available to investigate the hydrologic processes operating on different spatial and temporal scales.

Hydrologic data essentially represent random phenomena - temperature, rainfall, wind, etc. – measured using various instruments or derived from other measurements. All measurements inherently contain errors, both random and those arising from the measurement process itself. Statistics is the science of understanding and quantifying the uncertainty. The questions often relate to the outcome of a random event. The expected value, equivalent to the mean or average, and other descriptive parameters, such as variance, skewness, and kurtosis, summarize and describe the observed distributions quantitatively. Data is skewed when there is an imbalance in the number of high and low values of observations. The fourth-moment about the mean, or kurtosis, is used to describe the frequency of low probability events – both with extremely high and extremely low magnitudes. The coefficient of variation of a dataset is the ratio of the standard deviation to the mean. A hydrologic example of the coefficient of variation is streamflow data – low flow variations are probably much smaller than under high flow conditions, but their coefficients of variation may be of a similar magnitude.

A frequency distribution represents the distribution of observed values, while a probability distribution is a mathematical function that predicts the expected likelihood of an unknown variable. A discrete distribution is used when only a countable number of outcomes are possible, such as in the tossing of a coin, or the number of students in a class. An example of discrete distribution is the Binomial, which is used to predict the probability of the number of heads when a coin is repeatedly tossed 'n' times. Continuous distributions are used for describing outcomes that can have any fractional value, such as the monthly or annual rainfall values. Examples of continuous distributions include the Uniform, Normal, Exponential, Gamma, and Gumbel extreme value distributions. One can create additional distributions by taking the logarithm of the random variable, resulting in distributions such as the log-normal, log-gamma, etc.

One reason for applying statistical methods is to be able to make a definitive statement about occurrence of a situation – whether a chance occurrence is sufficiently unlikely that we can reasonably say how improbable it is. There are two types of errors when this method is used. Even an unbiased tossing of a coin has a small chance of yielding a rare outcome. So by rejecting the coin itself, one may be making a mistake; the rejection of a fair coin may be called a type-1 error. On the other hand, a coin that would appear to give a biased outcome may still give normal results and yet not be detected. The failure to reject such a coin is called a type-2 error.

In statistical tests, one generally assumes that observations follow a Normal distribution – giving the familiar bell-shaped curve. If a particular observation is too far from the mean, then one might think that it is fundamentally different from the other observations. To check this, one should first find the standard normal variable (Eqn. 8.47) and then use this variable to make a decision. Using the normal distribution, one can calculate the likelihood of this occurrence (variable). If the probability is quite small, one might infer that observations do not follow the Normal distribution.

One generally uses the least squares method to fit a distribution to a series of data points (x_i, $y_i \pm \sigma_i$) under the assumption of negligible un-

certainty in the variable x. In this method, values of arbitrary constants corresponding to chosen curve/distribution are obtained by minimizing the weighted sum of squares (S) of deviations between observed values of y_i and their computed values for each x_i, using equations similar to (Eqn. 8.119 and Eqn. 8.120). The least squares method is quite general and is applicable to any type of distribution, for instance, quadratic or second-degree polynomial, harmonic, exponential, etc. The Chi-Square (χ^2) test is used to estimate the probability that S_{min} is exceeded for the given degrees of freedom by less than 5% or 1% and the hypothesis that the chosen distribution can be accepted at 95% or 99% confidence levels, respectively.

Time series (TS) data are used to describe many aspects of the hydrologic cycle. TS data contain several pieces of information that can be utilized by a user for various analytical purposes. The data is usually collected at regular intervals, referred to as the time-step. Hydrologic models can also generate time series data. Methods for TS analyses can be divided into two broad classes: (i) *time-domain methods*, and (ii) *frequency-domain methods*. A time domain analysis aims to describe the pattern of the series over time. A frequency domain analysis, on the other hand, aims to determine the strength/power of periodicity(ies) inherent in the series within each given frequency band over a range of frequencies.

Forecasting with classical time domain, TS methods may be viewed as an attempt to decompose the series into component parts and then predict the future pattern of each part. The component parts are the trend, cycle, seasonal, and irregular components. These procedures require some knowledge of the mathematical model of the hydrologic process. However, in real-life situations, patterns of the data are not clear, as individual observations involve considerable errors. The Auto-Regressive Integrated Moving Average (ARIMA) methodology developed by Box and Jenkins enables one to uncover the hidden patterns in the data and generate forecasts. *Lag*s of the differenced series appearing in the forecasting equation are called 'auto-regressive' terms, *lag*s of the forecast errors are called 'moving average' terms, and a time series which needs to be differenced to make it stationary is said to be an 'integrated' version of a stationary series.

Spectrum analysis is a frequency-domain method of TS analysis and is concerned with the exploration of periodicities inherent in the data. The purpose of the analysis is to decompose a complex time series with periodic components into the inherent sinusoidal (sine and cosine) functions of particular wavelengths. Employing spectrum analysis, one might uncover recurring cycles of different periodicities in the time series of interest, which at first would appear more or less like random noise.

It is sometimes of interest to investigate the joint structure of two series, that is, the dependence or degree of coherence between the two series. This is achieved by examining coherency and phase relationships between the two series.

The hydrologic cycle integrates atmospheric, hydrospheric, cryospheric, and biospheric processes over a wide range of spatial and temporal scales and thus lies at the heart of the Earth's climate system. Studies of the integrated, global nature of the hydrologic cycle are crucial for a proper understanding of natural climate variability and prediction of climatic response to anthropogenic forcing. Only in recent years, particularly with the advent of satellite remote sensing, the required global data seems to be within reach.

Although most hydrologists believe that remotely sensed data is valuable for global hydrologic studies and even for regional hydrologic modelling and field operations, these data are rarely used in practice, possibly due to: (i) lack of necessary technical expertise in processing/interpreting the data; and (ii) the form of emitted and reflected radiances not being the type of data traditionally used to run and calibrate models. Remote sensing data also represent averages over finite areas, or pixels, and thus mask much of the detail at individual points to which most hydrologists are accustomed. In addition, current remote sensing observations are not optimized to provide the temporal resolution needed to measure certain changes in hydrologic processes. Furthermore, algorithms for converting these reflectances into physical quantities are often empirical in nature and are subject to noise present in the calibration data.

Nevertheless, remote sensing is beginning to prove its usefulness in providing hydrologic information. Hydrologic remote sensing can reveal complex spatial variations that cannot be readily obtained through traditional *in-situ* approaches. Development of such datasets and models in which these data can be used requires field experiments that combine appropriate remote sensing measurements with traditional *in-situ* measurements in regions that are hydrologically well understood. Once hydrologic models are developed for use with remote sensing data in well-monitored basins, they can possibly be extended to regions where little or no *in-situ* measurements exist.

Different remote sensing satellites carry sensors of varied characteristics. Often data are complementary in nature, for example panchromatic data have high spatial resolution and multispectral data have low spatial resolution. Fine spatial resolution is necessary for an accurate description of shapes, features, and structures, whereas fine spectral resolution enables better discrimination between attributes (e.g. for classification of land cover). Hence, merging of these two types of data to form multi-spectral images with high spatial resolution is useful for various applications such as vegetation mapping, land cover classification, precision farming, and urban management.

Geographical Information System (GIS) is a computer-assisted system for capturing, storage, retrieval, analysis, and display of spatial data and data with non-spatial attributes. Some of these involve reclassification, aggregation, overlays, suitability analysis, network, and route analysis, optimization, allocation, etc. The data can be derived from alternative sources such as survey, geographical/topographical/aerial maps, or archived data. Data can also be in the form of location data (i.e. latitude/longitude) or tabular (attribute) data. Applications of GIS range from simple database query systems to complex analysis and decision support systems. Areas of application range from natural resources management to near real-time application such as flood forecasting. GIS techniques are playing an increasing role in facilitating integration of multi-layer spatial information with statistical attribute data to arrive at alternative developmental scenarios.

Combining the ability of RS to measure spatial, spectral, and temporal information on the state of the Earth's surface and that of GIS to handle and manipulate geospatial data for a multitude of applications in resource management, quantification, and process understanding, have become possible. These include understanding and simulating landscape changes and hydrology, urban flood modelling, urban environment and impact assessment, generic ecosystem patterns, regional climate models, and a host of other fields.

14.3 Water resource sustainability

The historic shift from population settlement based on resource exploitation to economy-driven, transportation, and amenity-based settlement patterns poses great challenges for achieving sustainable water supplies and water management. The amenity-based settlement patterns put increasingly larger population and land-use pressure on areas that previously served as 'water banks' for meeting the requirement of urban population inhabiting the area.

Concentrated human settlements with their propensity to create hard, impermeable surfaces for building houses and roads and the need for water intake and outflow in a variety of forms, are not in harmony with the natural hydrologic cycle. The adverse effects of creating impervious surface cover in urbanized watersheds, reducing the groundwater recharge and the consequent reduction in the base flow of the stream/river flowing through the area, are well documented. Sewerage and water supply systems serving dense settlements can further interfere with groundwater and surface water hydrology. An urban settlement also creates a 'heat island' effect, reduces evapotranspiration due to reduction in the vegitative cover, and modifies the local microclimate.

The major driving force for change is essentially population growth coupled with a rising living standard globally, a combination that has resulted in over-exploitation of resources, including water. The current world population is about 6 billion, which is expected to grow to about 9 billion by the year 2050. When the population was much smaller (e.g. <2 billion) and the per capita use of resources was also much smaller, the traditional pattern of resource consumption, namely, 'take, make, waste' was sustainable. However, what is needed is to recycle and reuse all resources (including water) and also increase the use of renewable resources. Water stress currently affects only a modest fraction of the human population, but it is expected to affect 45% of the population by the year 2025. This situation may be further exacerbated by global climate change, which may alter water supply and storage patterns in ways that could render existing water-management infrastructure ineffective.

Recycling technologies can significantly reduce the net water abstraction from the environment, but many of these technologies require an increase in the use of other resources, especially energy. In our resource-constrained world, increasing the consumption of any resource, such as water, must be carefully considered.

Another aspect of water stress caused by urban water-management systems is increased load of nutrients, especially phosphorus, entering the aquatic environment. Mined as phosphate rock, phosphorus is used for manufacturing fertilizers that are widely used to increase the yield of crops for human consumption. Phosphorus and other nutrients then pass through the human body metabolism and end up in the wastewater discharge. When these effluents are discharged into the aquatic environment, the excess nutrients can cause eutrophication of surface water bodies. At the current rate of consumption, the supply of phosphate, an essential nutrient for crops with no known replacement, is expected to be exhausted in about 100 years. Thus, there is an urgent need to recover phosphate from wastewater.

Two other factors must be taken into consideration. First, although water supply is uniformly provided in the developed world, approximately 1 billion people in the developing world do not have access to safe drinking water, and more than 2.5 billion do not have access to adequate sanitation. Clearly, to meet global needs, more efficient urban water and waste management systems are needed.

Some of the aspects of urbanization that exert the most obvious influence on hydrologic processes are the increase in population density and the proportion of built-up areas within urbanized areas. With an increase in population, water demand begins to rise. The increase in water demand is accelerated with rising living standards, which further compounds the problem of developing adequate water resources – the first of the major urban hydrologic problems.

Due to increased urbanization and with the installation of sewerage systems for both domestic and storm water, the amount of water-borne waste load also increases in proportion to population growth. The resultant water quality changes are intimately linked to the increase in population density. As the latter rises, the extent of impervious built-up area also increases, the natural drainage system gets modified, and the local microclimate changes. Owing to the larger proportion of area becoming impervious, a higher fraction of the incident rainfall appears as runoff compared to the situation when the catchment was in its pristine state. Furthermore, laying of storm sewers, realignment of natural stream channels, and construction of culverts result in more rapid transmission of runoff to the drainage network. The increase in inflow velocities directly affects the nature of the runoff hydrograph. Since a large volume of runoff is discharged within a short time interval, peak rates of flow inevitably increase, giving rise to flash floods, the second of the major urban hydrologic problems, such as water-logging during heavy rain spells.

Municipal wastewater is a combination of liquid- or water-transported wastes originating in the sanitary systems of dwellings, commercial or industrial units, and institutions, in addition to any groundwater, surface water, and storm water that may be present. Untreated wastewater generally contains high levels of organic matter, nutrients, and toxic compounds as well as numerous pathogenic micro-organisms. It thus entails environmental and health hazards and, consequently, must immediately be conveyed away from its source locations and treated appropriately before its final disposal. The ultimate goal of wastewater management is protection of the environment in a manner commensurate with public health and socio-economic concerns of the area.

Physical, chemical, and biological methods are used to remove contaminants from waste water. In order to achieve different levels of contaminant removal, individual wastewater treatment procedures are combined into a variety of systems, classified as primary, secondary, and tertiary treatments. Physical unit operations, in which physical methods are applied to remove contaminants, include screening, comminution, flow equalization, sedimentation, flotation, and granular medium filtration. Chemical processes used in wastewater treatment are designed to bring about some form of change in the redox conditions by means of chemical additives and/or reactions. These are invariably used in conjunction with physical unit operations and biological processes and include chemical precipitation, adsorption on activated carbon, disinfection, dechlorination, etc. Biological unit processes are used to convert the finely divided and dissolved organic matter in waste water into flocculent settleable organic and inorganic solids. In these processes, micro-organisms, particularly bacteria, convert the colloidal and dissolved carbonaceous organic matter into various gases and build their cell tissues that are subsequently removed in sedimentation tanks. Biological processes are generally used in conjunction with physical and chemical processes, with the main objective of reducing the organic and nutrient loads of waste water. Biological processes used for wastewater treatment may be classified under five major categories, namely: (i) aerobic processes; (ii) anoxic processes; (iii) anaerobic processes; (iv) combined processes; and (v) pond processes. The commonly used biological processes include trickling filters, activated sludge process, aerated lagoons, rotating biological contactors, and stabilization ponds.

Natural systems for wastewater treatment are designed to take advantage of physical, chemical, and biological processes that occur in the natural environment where water, soil, plants, microorganisms, and the atmosphere are in constant interaction with each other. Natural treatment systems include land treatment, floating aquatic plants, and artificially created wetlands. All natural treatment systems are preceded by some form of mechanical pre-treatment for removal of gross solids. Where a sufficient land area suitable for this purpose is available, these systems can often be the most cost-effective option in terms of their construction as well as operation. They are generally well suited to small communities and rural areas.

Even though modern water supply and sanitation is considered the most significant contribution to public health in the past 150 years, rapid urbanization is challenging the sustainability of the development process and new approaches to water and sanitation management are urgently needed. Three aspects of urban water management are emerging as increasingly significant and will continue to be important in the foreseeable future. These aspects are decentralized wastewater management (DWM), wastewater reclamation and reuse, and increased attention to wet-weather flow (WWF) management.

Harvesting of rainwater, either directly from house rooftops, the runoff from private/public land, or natural water collection areas, is an option that holds significant promise in any sustainable water resource management strategy. In addition, groundwater, another natural resource representing natural subsurface accumulation of rainwater over timescales ranging from a few minutes to centuries to millennia, provides an important reserve to be exploited during periods of failure/ deficit of rain. Due to its ubiquitous nature and rather simple technology required for its exploitation, the groundwater resource has been widely over-exploited in the last few decades, while at the same time rainwater harvesting at household and community levels has been neglected. Several technological tools and practice of rainwater harvesting for three important applications, namely potable use at household and community levels, agriculture including horticulture at a farm scale, and for artificial groundwater recharge, are available. The scope of RWH is wide and the technologies discussed can be applied, with or without minor modification and innovation, in any region facing water scarcity (for whatever reason), but still having unutilized potential of rainwater.

Fortunately, most water resources are renewable (except some groundwaters), albeit with huge differences in availability in different parts of the world and wide variations in seasonal and annual precipitation in many places. Human influence on useable water is now a global phenomenon and plays a significant role in the hydrologic cycle.

Per capita use of water is increasing (with better lifestyles) and the world population is also constantly growing. Human activity, in turn, impacts the availability of clean water with its concomitant health implications. Access to water has also direct implications on poverty alleviation, economic growth, and development in general. Competition for water is intensifying day by day with the progressive collapse of traditional water-based ecological systems, diminishing river flows, and groundwater depletion arising from over-exploitation. Thus water as a symbol of life, purity, and regeneration is under threat in large parts of the world.

The human dimensions in water resource development and management encompass a diverse range of issues and policy domains. Ensuring sustainable and equitable resource development requires consideration of ethical principles involving questions of right or wrong from socio-economic as well as moral perspectives. Some of the important water-related challenges, with significant ethical considerations, have been identified as:

- *meeting basic needs* – for safe and adequate water supply and sanitation;
- *securing the food supply* – especially for the poor and vulnerable section of the population through effective use of water;
- *protecting ecosystems* – ensuring their integrity through sustainable water resource management;
- *sharing water resources* – promoting cooperation between different users of water and between concerned states/countries, through approaches such as sustainable river basin management;
- *managing risks* – to provide security from a range of water-related hazards, such as droughts, floods, pollution, etc.;
- *valuing water* – to manage water in the light of its different values (economic, social, environmental, and cultural) and to move towards pricing of water to recover the costs of providing the services, taking into account the equity and the needs of the poor and vulnerable sections of the population;

- *managing water prudently* – involving the interests of the public as well as various stake holders;
- *water and industry* – promoting cleaner industrial environment, particularly with regard to water quality and the needs of various users;
- *water and energy* – assessing the key role of water in energy production to meet rising energy demands;
- *ensuring the accessibility of knowledge base* – so that water knowledge becomes more easily available globally to all concerned;
- *water and cities* – recognizing the distinctive challenges posed by increase in the number of urbanized regions throughout the world.

In addition to 'water ethics', the amount of water embedded in the production of goods or services, referred to as 'virtual water', is another recent concept that may significantly influence the regional and global commodity trade and water allocation for the various competing demands. It is likely to lead to more productive uses of water, even though water is not the only component of any decision-making process. As with 'carbon footprints', adding up all virtual water in the products that are used in daily life and the water coming out of a tap leads to the idea of one's 'water footprints' – a concept that helps to understand the impact of an activity, individuals, communities, and nations on limited freshwater resources of the Earth.

Lastly, four case studies from three continents covering regions with high water stress have been described in Chapter 13, to highlight the various issues related to understanding of the hydrologic system, technology, and societal concerns and how these are being addressed to ensure sustainable development of water resource of each of these regions. Evolving region-specific adaptation measures to mitigate the situation of water stress are also described. Sound knowledge of fundamental principles and advancements in various disciplines provide important clues to ensure the survival of humankind on the Earth.

Bibliography

Abramowitz, M. and Stegun, I.A. (eds) (1972) *Handbook of Mathematical Functions with Formulas, Graphs and Mathematical Tables*, Dover Publications, New York Dover Publications, ISBN 978-0-486-61272-0, p. 1006.

Acharya, G.D., Hathi, M.V., Patel, A.D. and Parmar, KC. (2008) Chemical properties of groundwater in Bhiloda Taluka region, North Gujarat, India. *E-Journal of Chemistry*, **5**, 792–796.

Adamowski, K. (1989) A Monte Carlo comparison of parametric and nonparametric estimation of flood frequencies. *Journal of Hydrology*, **108**, 295–308.

Adamowksi, K. (2000) Regional analysis of annual maximum and partial duration flood data by nonparametric and L-moment methods. *Journal of Hydrology*, **229**, 219–231.

Adler, R.F., Huffman, G.J. *et al.* (2003) The version-2 global precipitation climatology project (GPCP) monthly precipitation analysis (1979–Present). *Journal of Hydrometeorology*, **4**, 1147–1167.

Adler, R.F., Negri, A.J., Keehn, P.R. and Hakkarinen, I.M. (1993) Estimation of monthly rainfall over Japan and surrounding waters from a combination of low-orbit microwave and geosynchronous IR data. *Journal of Applied Meteorology*, **32**, 335–356.

Aeschbach-Hertig, W., Peeters, F., Beyerle, U. and Kipfer, R. (1999) Interpretation of dissolved atmospheric noble gases in natural waters. *Water Resources Research*, **35**, 2779–2792.

Aeschbach-Hertig, W., Peeters, F., Beyerle, U. and Kipfer, R. (2000) Palaeotemperature reconstruction from noble gases in ground water taking into account equilibration with entrapped air. *Nature*, **405**, 1040–1044.

Aeschbach-Hertig, W., Stute, M., Clark, J., Reuter, R. and Schlosser, P. (2002) A paleotemperature record derived from dissolved noble gases in groundwater of the Aquia Aquifer (Maryland, USA). *Geochimica et Cosmochimica Acta*, **66**, 797–817.

Agarwal, A. and Narain, S. (1997) Dying Wisdom: Rise, Fall and Potential of India's Traditional Water Harvesting Systems. *State of India's Environment, 4th Citizen's Report*, Centre for Science and Environment, New Delhi.

Agrawal, V., Vaish, A.K. and Vaish, P. (1997) Groundwater quality: Focus on fluoride and fluorosis in Rajasthan. *Current Science*, **73**, 743–746.

Agarwal, M., Gupta, S.K., Deshpande, R.D. and Yadava, M.G. (2006) Helium, radon and radiocarbon studies on a regional aquifer system of the North Gujarat–Cambay region, India. *Chemical Geology*, **228**, 209–232.

Aji, K., Tang, C. *et al.* (2008) Characteristics of chemistry and stable isotopes in groundwater of Chaobai and Yongding River basin, North China Plain. *Hydrological Processes*, **22**, 63.

Allan, J.A. (1998) Virtual water: a strategic resource. Global solutions to regional deficits. *Ground Water*, **36**, 545–546.

Allen, R.M. and Bennetto, H.P. (1993) Microbial fuel cells – electricity production from carbohydrates. *Applied Biochemistry and Biotechnology*, **39/40**, 27–40.

Alley, W.M. and Smith, P.E. (1982) Distributed Routing and Rainfall-Runoff Model – Version II: *US Geological Survey Open-File Report 82-344*, US Geological Survey, p. 201.

Allison, G.B. and Hughes, M.W. (1978) The use of environmental chloride and tritium to estimate total recharge to an unconfined aquifer. *Australian Journal of Soil Research*, **16**, 181–195.

Allison, G.B., Barnes, C.J. Hughes, M.W. and Leaney, F.W.J. (1984) Effects of climate and vegetation on oxygen-18 and deuterium profiles in soils. *Isotope Hydrology 1983*, IAEA Symposium 270: IAEA, Vienna, pp. 105–123.

Allison, J.D., Brown, D.S. and Novac-Gradac, K.J. (1991) MINTEQA2/ PRODEFA2, a geochemical assessment model for environmental systems. *Version 3.0 User's Manual*: US EPA, Athens, GA, p. 104.

Allord, G.J. and Scarpace, F.L. (1981) Improving streamflow estimates through use of Landsat, in *Satellite Hydrology* (eds M. Deutsch, D.R. Wiesnet and A. Rango), 5th Annual William T. Pecora Memorial Symposium on Remote Sensing, Sioux Falls, South Dakota, June 10–15, 1979, American Water Resources Association, Minneapolis, pp. 284–291.

Anderson, M.A. (1984) Movement of contaminants in groundwater: Groundwater transport, advection and dispersion, in *Groundwater Contamination, Studies in Geophysics*, National Academy Press, Washington DC, pp. 37–45.

Andrews, J.N. (1977) Radiogenic and inert gases in groundwater. *Second International Symposium on Water-Rock Interaction*, pp. 334–342.

Andrews, J.N. and Lee, D.J. (1979) Inert gases in groundwater from the Bunter Sandstone of England as indicators of age and palaeoclimatic trends. *Journal of Hydrology*, **41**, 233–252.

Andrews, J.N., Giles, I.S. *et al.* (1982) Radioelements, radiogenic helium and age relationships for groundwaters from the granites at Stripa, Sweden. *Geochimica et Cosmochimica Acta*, **46**, 1533–1543.

Andrews, J.N., Goldbrunner, J.E. *et al.* (1985) A radiochemical, hydrochemical and dissolved gas study of groundwater in the Molase basin of Upper Austria. *Earth & Planetary Science Letters*, **73**, 317–332.

Andrews, J.N., Davis, S.N. *et al.* (1989) The *insitu* production of radioisotopes in rock matrices with particular reference to the Stripa granite. *Geochimica et Cosmochimica Acta*, **53**, 1803–1815.

Araguas-Araguas, L., Froehlich, K. and Rozanski, K. (1998) Stable isotope composition of precipitation over Southeast Asia. *Journal of Geophysical Research*, **103**, 28721–28742.

Arnold, C.L. and Gibbons, C.J. (1996) Impervious surface coverage: the emergence of a key environmental indicator. *Journal of the American Planning Association*, **62**, 243–258.

Aron, G. and Egborge, C.E. (1973) A practical feasibility study of flood peak abatement in urban areas, Sacramento, CA. US Army Corps of Engineers, Sacramento District.

Aron, G., Smith, T.A. and Lakatos, D.F. (1996) *Penn State runoff model, PSRM C96, User Manual*. Environmental Resources Research Institute, Penn State University, Penn State, PA, p. 54.

Arthur, J. (1997) Rights and the duty to bring aid, in *Ethics in Practice* (ed H. La Follette), Blackwell, Oxford, pp. 596–604.

Bagdasaryan, G. (1964) Survival of viruses of the Enterovirus group (Poliomyelitis, Echo, Coxsackie) in soil and on vegetables. *Journal of Hygiene, Epidemiology, Microbiology and Immunology*, **8**, 497–505.

Ballentine, C.J. and Burnard, P.G. (2002) Production, release and transport of noble gases in the continental crust, in *Noble Gases in Cosmochemistry and Geochemistry: Reviews in Mineralogy and Geochemistry*, vol. **47** (eds D. Porcelli, C. Ballentine and R. Wieler), Mineralogical Society of America, Geochemical Society, Washington DC, pp. 481–538.

Barsukov, V.L., Varshal, G.M. and Zamokina, N.S. (1984) Recent results of hydrogeochemical studies for earthquake prediction in the USSR. *PAGEOPH*, **122**, 143–156.

Barton, I.J. (1978) A case study comparison of microwave radiometer measurements over bare and vegetated surfaces. *Journal of Geophysical Research*, **83**, 3513–3517.

Bauer, S., Fulda, C. and Schafer, W. (2001) A multi-tracer study in a shallow aquifer using age dating tracers H-3, Kr-85, CFC-113 and SF6 – Indication for retarded transport of CFC-113. *Journal of Hydrology*, **248**, 14–34.

Bayer, R., Schlosser, P., Bönisch, G., Rupp, H., Zaucker, F. and Zimme, G. (1989) *Performance and Blank Components of a Mass Spectrometric System for Routine Measurement of Helium Isotopes and Tritium by the ^{3}He Ingrowth Method*, Springer-Verlag, New York, p. 42.

Bear, J. (1979) *Hydraulics of Groundwater*, McGraw-Hill, New York, p. 569.

Bear, J., Beljin, M.S. and Ross, R.R. (1992) Fundamentals of Ground-Water Modeling: Ground Water Issue, United States Environmental Protection Agency, EPA/540/S-92/005, pp. 1–11.

Bear, J. and Dagen, G. (1965) The relationship between solutions of flow problems in isotropic and anisotropic soils. *Journal of Hydrology*, **3**, 88–96.

BEIR (1999) The health effects of exposure to indoor radon, biological effects of ionizing radiation (BEIR), *VI Report*, National Academy of Sciences, USA, pp. 1–500.

Benda, L., Leroy Poff, N. *et al.* (2004), The network dynamics hypothesis: how channel networks structure riverine habitats. *BioScience*, **4**, 413–427.

Bennett, G.D. and Patten, E.P. (1962) Constant-head pumping test of a multiaquifer well to determine characteristics of individual aquifers. *US Geological Survey Water-Supply Paper 1545-C*, pp. 181–203.

Benson, M.A. (1968) Uniform flood-frequency estimating methods for federal agencies. *Water Resources Research*, **4**, 891–908.

Bethke, C.M. (1996) *Geochemical Reaction Modeling - Concepts and Applications*, Oxford University Press, New York, p. 397.

Bethke, C.M. and Johnson, T.M. (2002) Paradox of groundwater age. *Geology*, **30**, 107–110.

Bewers, J.M. (1971) North Atlantic fluoride profiles. *Deep Sea Research*, **18**, 237–241.

Beyerle, U., Purtschert, R. *et al.* (1998) Climate and groundwater recharge during the last glaciation in an ice-covered region. *Science*, **282**, 731–734.

Beyerle, U., Aeschbach-Hertig, W., Imboden, D.M., Baur, H., Gra, T. and Kipfer, R. (2000) A mass spectrometric system for the analysis of noble gases and tritium from water samples. *Environmental Science and Technology*, **34**, 2042–2050.

Bhandari, N., Gupta, S.K., Sharma, P., Sagar, P., Ayachit, V. and Desai, B.I. (1986) Hydrogeological investigations in Sabarmati and Mahi Basins and Coastal Saurashtra using Radio-isotopic and Chemical Tracers. Roorkee, High Level Technical Committee on Hydrology, Ministry of Water Resources, Government of India, National Institute of Hydrology, p. 115.

Bhattacharya, S.K., Froehlich, K., Aggrawal, P.K. and Kulkarni, K.M. (2003) Isotope variation in Indian monsoon precipitation: records from Bombay and New Delhi. *Geophysical Research Letters*, **30**, 2285.

Bhattacharya, S.K., Gupta, S.K. and Krishnamurthy, R.V. (1985) Oxygen and hydrogen isotopic ratios in groundwaters and river waters from India: *Proceedings of the Indian Academy of Sciences. Earth and Planetary Science*, **94**, 283–295.

Bicknell, B.R., Imhoff, J.C., Kittle, J.L., Donigian, A.S. and Johanson, R.C. (1993) Hydrological Simulation Program – Fortran Users Manual for Release 10: EPA-600/R-93/144, Environmental Research Laboratory, US Environmental Protection Agency, Athens, GA, p. 660.

BIS (1990) *Drinking Water Specifications, IS:10500*. Bureau of Indian Standards, New Delhi.

Biswas, A.K., Dakang, Z., Nickum, J.E. and Liu, C. (eds) (1983) Long-distance water transfer: A Chinese case study and international experiences, United Nations University Press, Water Resources Series, **6**. Tokyo, Japan, p. 432.

Blaikie, P., Cannon, T., Davis, I. and Wisner, B. (1994) *At Risk: Natural Hazards, People's Vulnerability and Disasters*, Routledge, London, p. 284.

Blank, L. (1980) *Statistical Procedures for Engineering, Management, and Science*, McGraw-Hill, New York, p. 649.

Blöschl, G. and Sivapalan, M. (1995) Scale issues in hydrological modelling: a review. *Hydrological Processes*, **9**, 251–290.

BMJ (2007) Medical milestones. *British Medical Journal*, **334**, s1–s20.

Bondelid, T.R., Jackson, T.J. and McCuen, T.H. (1982) *Estimating Runoff Curve Numbers Using Remote Sensing Data*. Proceedings of the International Symposium on Rainfall-Runoff Modeling, Applied Modeling in Catchment Hydrology, Water Resources Publications, Littleton, CO, pp. 519–528.

Boning, C.W. (1973) Index of Time-of-travel Studies of the US Geological Survey, US Geological Survey Water Resources Investigations 73–34, p. 71.

Borole, D.V., Gupta, S.K. Krishnaswami, S., Datta, P.S. and Desai, B.I. (1979) *Uranium Isotopic Investigations and Radiocarbon Measurement of River-Groundwater Systems, Sabarmati Basin, Gujarat, India, Isotopic Hydrology*. IAEA, Vienna, pp. 181–201.

Boulton, N.S. (1954) The drawdown of the water-table under non-steady conditions near a pumped well in an unconfined formation. *Proceedings Institute of Civil Engineers*, **3** (III), 564–579.

Boulton, N.S. (1963) Analysis of data from non-equilibrium pumping tests allowing for delayed yield from storage. *Proceedings Institute of Civil Engineers*, **26**, 469–482.

Boulton, N.S. and Streltsova, T.D. (1975) New equations for determining the formation constants of an aquifer from pumping test data. *Water Resources Research*, **11**, 148–153.

Bouwer, H. (1963) Theoretical effects of unequal water levels on the infiltration rate determined with buffered cylinder infiltrometers. *Journal of Hydrology*, **1**, 29–34.

Bouwer, H. (1978) *Groundwater Hydrology*, McGraw-Hill, New York, p. 480.

Bouwer, H. (2002) Artificial recharge of groundwater: hydrology and engineering. *Hydrogeology Journal*, **10**, 121–142.

Bowen, H.J.M. (1966) *Trace Elements in Biochemistry*, Academic Press, London and New York, p. 241.

Box, G.E.P. and Jenkins, G.M. (1976) *Time Series Analysis: Forecasting and Control*, Holden Day, San Francisco, CA, p. 187.

BR (1977) Groundwater Manual, Bureau of Reclamation, US Department of the Interior, p. 480.

Bradford, G.R. (1963) Lithium survey of California's water resources. *Soil Science*, **96**, 77–81.

Brown, J.A.H. (1983) *Australia's Surface Water Resources*, Government Publishing Service, Canberra, Australia, p.177.

Brown, L.R. and Halweil, B. (1998) China's water shortage could shake world food security. *World Watch*, **11**, 10–18.

Bruns, D.A., Minshall, G.W., Cushing, C.E., Cummins, K.W., Brock, J.T. and Vannote, R.C. (1984) Tributaries as modifiers of the river continuum concept: analysis by polar ordination and regression models. *Archiv für Hydrobiologie*, **99**, 208–220.

Brzezinski, M.A., Jones, J.L., Bidle, K.D. and Azam, F. (2003) The balance between silica production and silica dissolution in the sea: insights from Monterey Bay, California, Applied to the Global Data Set. *Limnology and Oceanography*, **48**, 1846–1854.

Buechler, S. and Devi, G.M. (2003) *The Impact of Water Conservation and Reuse on the Household Economy*. Proceedings of the Eighth International Conference on Water Conservation and Reuse of Wastewater, September 13–14, 2003, Mumbai.

Bugmann, H. (1997) Scaling issues in forest succession modelling, in *Elements of Change 1997* (eds S.J. Hassol and J. Katzenberger), Aspen Global Change Institute, Aspen, CO, pp. 47–57.

Bullister, J.L., Wisegraver, D.P. and Menzia, F.A. (2002) The solubility of sulfur hexafluoride in water and seawater: *Deep-Sea Research*, Part 1, **49**, 175–187.

Burger, A., Recordon, E., Bover, D., Cotton, L. and Saugy, F. (1984) *Thermique des nappes souterraines*, Presses Polytechniques Romandes, Lausanne, Switzerland.

Burgy, R.H. and Luthin, J.N. (1956) A test of the single and double ring type infiltrometers. *Transactions of American Geophysical Union*, **37**, 189–191.

Busenberg, E. and Plummer, N.L. (2000) Dating young ground water with sulfur hexafluoride: natural and anthropogenic sources of sulfur hexafluoride. *Water Resources Research*, **36**, 3011–3030.

Butler, J.N. (1998) *Ionic Equilibrium: Solubility and pH Calculations*, John Wiley & Sons, New York, p. 576.

Camillo, P.J., Gurney, R.J. and Schmugge, T.J. (1983) A soil and atmospheric boundary layer model for evapotranspiration and soil moisture studies. *Water Resources Research*, **19**, 371-380.

Carpenter, R. (1969) Factors controlling the marine geochemistry of fluorine. *Geochemica et Cosmochemica Acta*, **33**, 1153-1167.

Carroll, T.R. and Holroyd, E.W. (1990) Operational Remote Sensing of Snow Cover in the US and Canada, *Proceedings of the International Symposium of Haydrulics and Irrigation Division, American Society of Civil Engineers*, pp. 296-291.

Carroll, T.R. and Vadnais, K.G. (1980) Operational Airborne Measurement of Snow Water Equivalent using Natural Terrestrial Gamma Radiation, *48th Annual Western Snow Conference*, pp. 97-106.

Castro, M.C., Stute, M. and Schlosser, P. (2000) Comparison of ^{4}He ages and ^{14}C ages in simple aquifer systems: implications for groundwater flow and chronologies. *Applied Geochemistry*, **15**, 1137-1167.

Cayan, D.R., Redmond, K.T. and Riddle, L.G. (1999) ENSO and hydrologic extremes in the western United States. *Journal of Climate*, **12**, 2881-2893.

Chahine, M.T. (1992) The hydrological cycle and its influence on climate. *Nature*, **359**, 373-380.

Chandrasekharam, D. and Antu, M.C. (1995) Geochemistry of Tattapani thermal springs, Madhya Pradesh, India - Field and experimental investigations. *Geothermics*, **24**, 553-559.

Chandrasekharan, H., Sunandra Sarma, K.S., Das, D.K., Datta, D., Mookerjee, P. and Navada, S.V. (1992) Stable isotope contents of perched water under different land use and salinization conditions in the canal command of arid western Rajasthan. *Journal of Arid Environments*, **23**, 365-378.

Chang, A.T.C., Foster, J.L. Hall, D.K., Rang, A. and Hartline, B. (1982) Snow water equivalent estimation by microwave radiometry. *Cold Regions Science and Technology*, **5**, 259-267.

Chang, A.T.C., Foster, J.L. and Hall, D.K. (1987) Nimbus-7 derived global snow cover parameters. *Annals of Glaciology*, **9**, 39-44.

Chang, A.T.C., Foster, J.L. and Hall, D.K. (1990) *Effect of Vegetation Cover on Microwave Snow Water Equivalent Estimates*. Proceedings of International Symposium on Remote Sensing and Water Resources, pp. 137-145.

Chen, J., He, D. and Cu, S. (2003) The response of river water quality and quantity to the development of irrigated agriculture in the last four decades in the Yellow River Basin, China. *Water Resources Research*, **39**, 1047.

Chen, J.Y., Tang, C.Y., Sakura, Y., Kondoh, A. and Shen, Y.J. (2002) Groundwater flow and geochemistry in the lower reaches of the Yellow River: a case study in Shandang Province, China. *Hydrogeology Journal*, **10**, 587-599.

Chen, J., Tang, C. *et al.* (2004) Spatial geochemical and isotopic characteristics associated with groundwater flow in the North China Plain. *Hydrological Process*, **18**, 3133-3146.

Chenoweth, J. (2008) Looming water crisis simply a management problem. *New Scientist*, **199**, 28-32.

Chilton, J. and Seiler, K.-P. (2006) Groundwater occurrence and hydrogeological environments, in *Protecting Groundwater for Health* (eds O. Schmoll, G. Howard, J. Chilton and I. Chorus), IWA Publishing, London, pp. 21-47.

Chin, D.A. (2000) *Water Resources Engineering*, Prentice Hall, Upper Saddle River, NJ, p. 750.

Choudhury, B.J., Ahmed, N.U., Idso, S.B., Reginato, R.J. and Daughtry, C.S.T. (1994) Relations between evaporation coefficients and vegetation indices studied by model simulation. *Remote Sensing of Environment*, **50**, 1-17.

Cianfrani, C.M., Hession, W.C. and Rizzo, D.M. (2007) Watershed imperviousness impacts on stream channel condition in southeastern Pennsylvania. *Journal of the American Water Resources Association*, **42**, 941-956.

Clark, I.D. and Fritz, P. (1997) *Environmental Isotopes in Hydrogeology*, Lewis, Boca Raton, FL, p. 328.

Clark, J.F., Davisson, M.L., Hudson, G.B. and Macfarlane, P.A. (1998) Noble gases, stable isotopes, and radiocarbon as traces of flow in the Dakota aquifer, Colorado and Kansas. *Journal of Hydrology*, **211**, 151-167.

Clarke, W.B., Jenkins, W.J. and Top, Z. (1976) Determination of tritium by mass spectrometric measurement of ^{3}He. *International Journal of Applied Radiation Isotopes*, **27**, 515-522.

Clever, H.L. (ed) (1979) Helium and neon – gas solubilities: Solubility data series, v. 1: Oxford, Pergamon Press, 393 p.

Collins, W.D. (1923) Graphic representation of analyses. *Industrial and Engineering Chemistry*, **15**, 394-394.

Colwell, J. and Sadowski, F. (1993) Past patterns as a guide for future forest management: using landsat photographic imagery. *Remote Sensing of Environment*, **13**, 291-300.

Cong, Z., Yang, D., Sun, F. and Ni, G. (2008) *Evaporation Paradox in the Yellow River Basin, China: Hydrological Research in China: Process Studies, Modelling Approaches and Applications*. Proceedings of Chinese PUB International Symposium, Beijing, September 2006. IAHS Publ. 322, pp. 3-8.

Cong, Z., Yang, D., Gao, B., Yang, H. and Hu, H. (2009) Hydrological trend analysis in the Yellow River Basin using a distributed hydrological model. *Water Resources Research*, **45**, W00A13, doi:10.1029/2008WR006852.

Constable, G. and Sommerville, B. (2003) *A Century of Innovation: Twenty Engineering Achievements that Transformed our Lives*, Joseph Henry Press, Washington, DC, p. 256.

Cook, P.G. and Böhlke, J.K. (2000) Determining timescales for groundwater flow and solute transport, in *Environmental Tracers in Subsurface Hydrology* (eds P.G. Cook and A.L. Herczeg), Kluwer Academic Publishers, Boston, pp. 1-30.

Cooper, H.H. (1959) A hypothesis concerning the dynamic balance of fresh water and salt water in a coastal aquifer. *Journal of Geophysical Research*, **64**, 461-467.

Cooper, H.H.J. and Jacob, C.E. (1946) A generalized graphical method for evaluating formation constants and summarizing well-field history. *Transactions of American Geophysical Union*, **27**, 526-534.

Corbett, D.M. *et al.* (1945) Stream-gaging Procedure, *US Geological Survey Water-Supply Paper 888*, p. 245.

Craig, H. (1961a) Isotopic variations in meteoric waters. *Science*, **133**, 1702-1703.

Craig, H. (1961b) Standard for reporting concentrations of deuterium and oxygen-18 in natural water. *Science*, **133**, 1833-1834.

Craig, H. and Lupton, J.E. (1976) Primordial neon, helium and hydrogen in oceanic basalts. *Earth and Planetary Science Letters*, **31**, 369-385.

Craig, H. and Weiss, R.F. (1971) Dissolved gas saturation anomalies and excess helium in the ocean. *Earth and Planetary Science Letters*, **10**, 289-296.

Cresswell, R. Wischusen, J., Jacobson, G. and Fifield, K.L. (1999a) Assessment of recharge to groundwater systems in the arid southwestern part of Northern Territory, Australia, using Chlorine-36. *Hydrogeology Journal*, **7**, 393-404.

Cresswell, R.G., Wischuse, J., Jacobson, G. and Fifield, K.L. (1999b) Ancient groundwaters in the Amadeus Basin, Central Australia: evidence from the radioisotope ^{36}Cl. *Journal of Hydrology*, **233**, 212-220.

Critchley, W., Siegert, K. and Chapman, C. (1991) A Manual for the Design and Construction of Water Harvesting Schemes for Plant Production, Rome, Food and Agriculture Organization of the United Nations.

Crosbie, R.S., McCallum, J.L., Walker, G.R. and Chiew, F.H.S. (2008) Diffuse Groundwater Recharge Modelling across the Murray-Darling Basin – A report to the Australian government from the CSIRO Murray-Darling Basin Sustainable Yields Project, CSIRO, Australia, p. 108.

Cserepes, L. and Lenkey, L. (1999) Modelling of helium transport in groundwater along a section in the Pannonian basin. *Journal of Hydrology*, **225**, 185-195.

Dagan, G. (1967) A method of determining the permeability and effective porosity of unconfined anisotropic aquifers. *Water Resources Research*, **3**, 1059-1071.

Daigger, G.T. (2007) Wastewater management in the 21st century. *Journal of Environmental Engineering*, **133**, 671-680.

Daigger, G.T. (2008) *Creation of Sustainable Water Resources by Water Reclamation and Reuse*. Third International Conference on Sustainable Water Environment: Integrated Water Resources

Management - New Steps, Sapporo, Japan, October 24-25, 2007, pp. 79–88.

Dansgaard, W. (1964) Stable isotopes in precipitation. *Tellus*, **16**, 436-438.

Das, B.K., Kakar, Y.P. and Stichler, W. (1988) Deuterium and oxygen-18 studies in ground water of the Delhi area, India. *Journal of Hydrology*, **98**, 133-146.

Datta, P.S., Bhattacharya, S.K., Mookerjee, P. and Tyagi, S.K. (1994) Study of groundwater occurrence and mixing in Pushkar (Ajmer) valley, Rajsthan, with d^{18}O and hydrochemical data. *Journal Geological Society India*, **43**, 449-456.

Datta, P.S., Deb, D.L. and Tyagi, S.K. (1996) Stable isotope (^{18}O) investigations on the processes controlling fluoride contamination of groundwater. *Journal of Contaminant Hydrology*, **24**, 85-96.

Datta, P.S., Desai, B.I. and Gupta, S.K. (1979) Comparative study of groundwater recharge rates in parts of Indo-Gangetic and Sabarmati alluvial plains. *Mausam*, **30**, 129-133.

Datta, P.S., Gupta, S.K., Jaisurya, A., Nizampurkar, V.N., Sharma, P. and Plusnin, M.I. (1980a) A survey of helium in ground water in parts of Sabarmati basin in Gujarat state and in Jaisalmer district, Rajasthan. *Hydrological Sciences Bulletin*, **26**, 183-193.

Datta, P.S., Gupta, S.K. and Sharma, S.C. (1980b) A conceptual model of water transport through unsaturated soil zone. *Mausam*, **31**, 9-18.

Datta, P.S., Tyagi, S.K. and Chandrasekharan, H. (1991) Factors controlling stable isotopic composition of rainfall in New Delhi, India. *Journal of Hydrology*, **128**, 223-236.

Davis, S.N. and DeWiest, R.J.M. (1966) *Hydrogeology*, John Wiley and Sons, New York, p. 463.

Dawdy, D.R. and Matalas, N.C. (1964) Statistical and probability analysis of hydrologic data, Part III: Analysis of variance, covariance and time series, in *Handbook of Applied Hydrology, a Compendium of Water-Resources Technology* (ed V.T. Chow), McGraw-Hill Book Company, New York, pp. 8.68–8.90.

DeCoursey, D.G. (1996) Hydrological, climatological and ecological systems scaling: a review of selected literature and comments. USDA, Great Plains Systems Research Unit, Fort Collins, CO, p. 120.

DeGroot, M.H. (1975) *Probability and Statistics*, Addison-Wesley, Reading, MA, p. 816.

Desai, B.I., Gupta, S.K. and Sharma, S.C. (1979) Hydrochemical evidence of seawater intrusion in Mangrol-Chorwad coast of Saurashtra. *Hydrological Sciences Bulletin*, **24**, 71-82.

Deshmukh, A.N., Gajbhiye, R.D. and Ganveer, J.B. (1993) Hydrochemical impact of natural recharge on groundwater quality of fluorosis endemic Dongargaon village, Tehsil Warora, District Chandrapur, Maharashtra. *Gondwana Geological Magazine*, **6**, 16-26.

Deshmukh, A.N., Shah, K.C. and Appulingam, S. (1995a) Coal ash: a source of fluoride pollution, a case study of Koradi thermal power station, district Nagpur, Maharashtra. *Gondwana Geological Magazine*, **9**, 21-29.

Deshmukh, A.N., Wadaskar, P.M. and Malpe, D.B. (1995b) Fluorine in environment: a review. *Gondwana Geological Magazine*, **9**, 1-20.

Deshpande, R.D. (2006) Groundwater in and around Cambay Basin, Gujarat: Some geochemical and isotopic investigations, MS University of Baroda, Vadodara, India, p. 158.

Deshpande, R.D., Bhattacharya, S.K., Jani, R.A. and Gupta, S.K. (2003) Distribution of oxygen and hydrogen isotopes in shallow ground waters from southern India: influence of a dual monsoon system. *Journal of Hydrology*, **271**, 226-239.

Dev Burman, G.K., Singh, B. and Khatri, P. (1995) Hydrogeochemical studies of ground water having high fluoride content in Chandrapur district of Vidarbha region, Maharashra. *Gondwana Geological Magazine*, **9**, 71-80.

Devore, J.L. (1999) *Probability and Statistics for Engineering and Sciences*. Brooks/Cole Publishing, Monterey, CA, p. 775.

Dewey, K.F. and Heim, R. (1981) *Satellite Observations of Variations in Northern Hemisphere Seasonal Snow Cover*, NOAA, Washington, DC, p. 83.

Dillard, J.P. and Orwig, C.E. (1979) *Use of Satellite Data in Runoff Forecasting in the Heavily Forested, Cloud Covered Pacific Northwest*. Final Workshop on Operational Applications of

Satellite Snow Cover Observations, NASA CP-2116, pp. 127-150.

Dincer, T., Al-Mugrin, A. and Zimmermann, U. (1974) Study of the infiltration and recharge through the sand dunes in arid zones with special reference to the stable isotopes and thermonuclear tritium. *Journal of Hydrology*, **23**, 79-109.

Diskin, M.H. (1970) On the computer evaluation of Thiessen weights. *Journal of Hydrology*, **11**, 69-78.

Divine, C.E. and McDonnell, J.J. (2005) The future of applied tracers in hydrogeology. *Hydrogeology Journal*, **13**, 255-258.

Doherty, J. (2000) *Manual for PEST2000*, Watermark Numerical Computing, Brisbane, Australia, p. 122.

Domenico, P.A. and Schwartz, F.W. (1998) *Physical and Chemical Hydrogeology*, John Wiley & Sons, New York, p. 506.

Douglas, M. (1997) Mixing and Temporal Variation of Ground Water Inflow at the Con Mine, Yellowknife, Canada: an Analogue for a Radioactive Waste Repository. Unpublished thesis, University of Ottawa, Canada.

Dowson, D.C. and Wragg, A. (1973) Maximum entropy distribution having prescribed first and second moments. *IEEE Transaction on Information Theory*, **19**, 689-693.

Dozier, J. (1984) Snow reflectance from Landsat-4 thematic mapper. *IEEE Transactions on Geoscience and Remote Sensing*, **GE-22**, 323-328.

Drane, J.W., Cao, S., Wang, L. and Postelnicu, T. (1993) Limiting forms of probability mass function via recurrence formulas. *The American Statistician*, **47**, 269-274.

Drever, J.I. (1997) *The Geochemistry of Natural Waters (Surface and Groundwater Environments)*, Prentice Hall, Upper Saddle River, NJ, p. 436.

Droste, R.L. (1997) *Theory and Practice of Water and Wastewater Treatment*, John Wiley & Sons, New York, p. 800.

Drury, S.A. (1993) *Image Interpretation in Geology*, Chapman & Hall, London, p. 283.

Dudewicz, E. (1976) *Introduction to Statistics and Probability*, Holt, Rinehart & Winston, New York, p. 528.

Dupuit, J. (1863) *Études théoretiques et pratiques sur le mouvement des eaux dans les canaux découverts et á travers les terrains permeables*, Dunod, Paris.

Eaton, F.M. (1950) Significance of carbonates in irrigation waters. *Soil Science*, **69**, 127-128.

Eaton, F.M. (1954) Formulas for estimating leaching and gypsum requirements of irrigation waters. *Texas Agriculture Experiment Station Miscellaneous Publication 111*, p. 18.

Ehrlich, D., Estes, J.E. and Singh, A. (1994) Applications of NOAA-AVHRR 1 km data for environment monitoring. *International Journal of Remote Sensing*, **15**, 145-161.

Eilon, M., Leibundgut, A. and Leibundgut, C. (eds) (1995) *Application of Tracers in Arid Zone Hydrology*, IAHS Publishing, Wallingford, UK, p. 450.

Engesgaard, P. and Molson, J. (1998) Direct simulation of groundwater age in the Rabis Creek Aquifer, Denmark. *Ground Water*, **36**, 577-582.

Endreny, T.A. (2007) Simulation of soil water infiltration with integration, differentiation, numerical methods and programming exercises. *International Journal of Engineering Education*, **23**, 608-617.

English, M. and Raja, S.N. (1996) Perpective on deficit irrigation. *Agricultural Water Management*, **32**, 1-14.

English, M., Solomon, K.H. and Hoffman, G.J. (2002) A paradigm shift in irrigation management. *Journal of Irrigation and Drainage*, **128**, 267-277.

Engman, E.T. (1990) Progress in microwave remote sensing of soil moisture. *Canadian Journal of Remote Sensing*, **16**, 6-14.

Eriksson, E. and Kunakasen, V. (1969) Chloride concentration in groundwater, recharge rate and rate of deposition of chloride in the Israel coastal plain. *Journal of Hydrology*, **7**, 178-197.

Errson, O. (2006) *Rainwater Harvesting and Purification System, Oregon*, http://www.rwh.in/ (accessed 17 July 2007).

ESCWA (2003) *Waste-Water Treatment Technologies: a General Review*, United Nations, Economic and Social Commission for Western Asia, New York, p. 121.

Espeby, B. (1990) Tracing the origin of natural waters in a glacial till slope during snowmelt. *Journal of Hydrology*, **118**, 107–127.

FAA (1970) Circular on Airport Drainage. Report A/C 150-5320-5B, Federal Aviation Administration, US Department of Transportation, Washington, DC.

Fair, G.M. and Hatch, L.P. (1933) Fundamental factors governing the streamline flow of water through sand. *Journal of American Water Resources Association*, **25**, 1551–1565.

Falkenmark, M. (2003) Fresh water as shared between society and ecosystems: from divided approaches to integrated challenges. *Philosophical Transactions of the Royal Society of London B*, **358** (1440), 2037–2049.

Falkenmark, M., Lundquist, J. and Widstrand, C. (1990) *Water Scarcity – an Ultimate Constraint in Third World Development*. Tema V, Report 14. Department of Water and Environmental Studies, University of Linkoping, Linkoping, Sweden.

Faucher, D., Rasmusse, P.F. and Bobee, B. (2001) A distribution function based bandwidth selection method for kernel quantile estimation. *Journal of Hydrology*, **250**, 1–11.

Faure, G. (1986) *Principles of Isotope Geology*, New York, John Wiley & Sons, Inc., 589 p.

Filippo, M.D., Lombardi, S., Nappi, G. and Reimer, G.M. (1999) Volcano-tectonic structures, gravity and helium in geothermal areas of Tuscany and Latium (Vulsini volcanic district), Italy. *Geothermics*, **28**, 377–393.

Fitts, C.R. (2002) *Groundwater Science*, Academic Press, New York, p. 450.

Fleischer, M. and Robinson, W.O. (1963) Some problems of the geochemistry of fluorine, in *Studies in Analytical Geochemistry Toronto* (ed D.M. Shaw), University of Toronto Press, Toronto, pp. 58–75.

Fontes, J.C. and Garnier, J.M. (1979) Determination of the initial ^{14}C activity of the total dissolved carbon: A review of the existing models and new approach. *Water Resources Research*, **15**, 399–413.

Forchheimer, P. (1886) Uber die Ergiebigkeit von Brunnen-Anlagen und Sickerschlitzen. *Z. Archietekt. Ing. Ver. Hannover*, **32**, 539–563.

Forstel, H. (1982) ^{18}O/^{16}O ratio of water in plants and their environment in stable isotopes, in *Stable Isotopes* (eds H.L. Schmidt, H. Forstel and K. Heinzinger), Elsevier, Amsterdam, pp. 503–516.

Fortin, J.-P. and Bernier, M. (1991) *Processing of Remotely Sensed Data to Derive Useful Input Data for the Hydrotel Hydrological Model, Remote Sensing: Global Monitoring for Earth Management, IGARSS 91*. International Geoscience and Remote Sensing Symposium, Espoo, Finland. IEEE, New York, pp. 63–65.

Frederick, K.D. (2001) Water Marketing: Obstacles and Opportunities. *Forum for Applied Research and Public Policy,* **Spring**, pp. 54–62.

Freeman, P.H. and Fox, R. (1994) *Satellite Mapping of Tropical Forest Cover and Deforestation: A Review with Recommendations for USAID*. Environment and Natural Resources Information Center, DATEX, Arlington, VA. http://dlc.dlib.indiana.edu/archive/00002945/01/Reproduced.pdf

Freeze, R.A. and Cherry, J.A. (1979) *Groundwater*, Prentice Hall, Englewood Cliffs, NJ, p. 604.

Fröhlich, K. and Gellermann, R. (1987) On the potential use of uranium isotopes for groundwater dating. *Chemical Geology*, **65**, 67–77.

Fröhlich, K., Ivanovich, M. *et al.* (1991) Application of isotopic methods to dating of very old ground waters – Milk River Aquifer, Alberta, Canada. *Applied Geochemistry*, **6**, 465–472.

Gambolati, G. (1973) Equation for one-dimensional vertical flow of groundwater: 1. The rigorous theory. *Water Resources Research*, **9**, 1022–1028.

Gambolati, G. (1974) Second-order theory of flow in three-dimensional deforming media. *Water Resources Research*, **10**, 1217–1228.

Garrels, R.M. and Mackenzie, F.T. (1967) Origin of the chemical compositions of some springs and lakes, in *Equilibrium Concepts in Natural Waters* (ed W. Stumm), American Cancer Society, Washington, DC, pp. 222–242.

Gash, J.H.C. (1987) An analytical framework for extrapolating evaporation measurements by remote sensing surface temperature. *International Journal of Remote Sensing*, **8**, 1245–1249.

Gat, J.R. and Matsui, E. (1991) Atmospheric water balance in the Amazon Basin: an isotopic evapotranspiration model. *Journal of Geophysical Research*, **96**, 13179–13188.

Gelhar, L.W., Welt, C. and Rehfeldt, K.R. (1992) A critical review of data on field-scale dispersion in aquifers. *Water Resources Research*, **28**, 1955–1974.

Giordano, M., Zhu, Z., Cai, X., Hong, S., Zhang, X. and Xue, Y. (2004) *Water Management in the Yellow River Basin: Background, Current Critical Issues and Future Research Needs, Comprehensive Assessment Secretariat,* Colombo, Sri Lanka, p. 48.

Gleeson, C. and Gray, N. (1997) *The Coliform Index and Waterborne Disease*, E&FN SPON (Chapman & Hall), London, p. 194.

Gleick, P.H. (1996) Water resources, in *Encyclopedia of Climate and Weather*, vol. **2** (ed S.H. Schneider), Oxford University Press, New York, pp. 817–823.

Gmelin, L. (1959) *Handbuch der anorganischen Chemie*, VerlagChemie, GmbH, Weinheim, p. 136.

Gonfiantini, R. (1978) Standards for stable isotope measurements in natural compounds. *Nature*, **271**, 534–536.

Gonfiantini, R. (1986) Environmental isotopes in lake studies, in *Handbook of Environmental Isotope Geochemistry, vol. 2: The Terrestrial Environment* (eds P. Fritz and J.-C. Fontes), B. Elsevier, Amsterdam, pp. 113–168.

Goode, D.J. (1996) Direct simulation of groundwater age. *Water Resources Research*, **32**, 289–296.

Goode, D.J. (1999) Age, Double Porosity, and Simple Reaction Modifications for the MOC3D Ground-Water Transport Model. *US Geol. Survey Water-Res. Inv. Rept. 99-4041*, p. 34.

Goode, D.J. and Konikow, L.F. (1990) Apparent dispersion in transient groundwater flow. *Water Resources Research*, **23**, 2339–2351.

Gou, X., Chen, F., Cook, E., Jacoby, G. and Yan, M. (2007) Streamflow variations of the Yellow River over the past 593 years in western China reconstructed from tree rings. *Water Resources Research*, **43**, W06434, doi:10.1029/2006WR005705.

Grady, C.P.L., Daigger, G.T. and Lim, H.C. (1999) *Biological Wastewater Treatment*, Marcel Dekker, New York, p. 1076.

Green, W.H. and Ampt, G.A. (1911) Studies in soil physics: 1. The flow of air and water through soils. *Journal of Agricultural Science*, **4**, 1–24.

Greer, C. (1979) *Water Management in the Yellow River Basin of China*, University of Texas Press, Austin and London, p. 174.

Gröning, M. (1994) Edelgase und Isotopentracer im Grundwasser: Paläo-Klimaänderungen und Dynamik regionaler Grundwasserfliesssysteme. Ph.D. Dissertation thesis, Universität Heidelberg, Heidelberg, p. 136.

Grove, D.B. (1976) Ion exchange reactions important in groundwater quality models, in *Advances in Groundwater Hydrology* (ed Z.A. Saleem), American Water Resources Association, Minneapolis, MN, pp. 144–152.

Gulec, N., Hilton, D.R. and Mutlu, H. (2002) Helium isotope variations in Turkey: relationship to tectonics, volcanism and recent seismic activities. *Chemical Geology*, **187**, 129.

Guo, S.L., Kachroo, R.K. and Mngodo, R.J. (1996) Nonparametric kernel estimation of low quantiles. *Journal of Hydrology*, **185**, 335–348.

Gupta, M.L. (1981) Surface heat flow and igneous intrusion in the Cambay basin, India. *Journal of Volcanology and Geothermal Research*, **10**, 279–292.

Gupta, S.K. (1983) An isotopic investigation of a near-surface groundwater system. *Hydrological Sciences Journal*, **28**, 291–272.

Gupta, S.K. (2001) Modelling advection dispersion process for dual radiotracer dating of groundwater with an example of application to a ^{14}C and ^{36}Cl data set from Central Australia, in *Modelling in Hydrogeology* (eds L. Elango and R. Jayakumar), Allied Publishers Ltd, Chennai, India, pp. 169–190.

Gupta, S.K. and Deshpande, R.D. (1998) *Depleting Groundwater Levels and Increasing Fluoride Concentration in Villages of Mehsana District, Gujarat, India: Cost to Economy and Health*, Ahmedabad, Water Resources Research Foundation (WRRF), PRL, Ahmedabad, p. 74.

Gupta, S.K. and Deshpande, R.D. (2003a) Origin of groundwater helium and temperature anomalies in the Cambay region of Gujarat, India. *Chemical Geology*, **198**, 33–46.

Gupta, S.K. and Deshpande, R.D. (2003b) Synoptic hydrology of India from the data of isotopes in precipitation. *Current Science*, **85**, 1591–1595.

Gupta, S.K. and Deshpande, R.D. (2005a) Groundwater isotopic investigations in India: what has been learned? *Current Science*, **89**, 825–835.

Gupta, S.K. and Deshpande, R.D. (2005b) Isotopes for water resource management in India. *Himalayan Geology*, **26**, 211–222.

Gupta, S.K. and Deshpande, R.D (2005c) The need and potential applications of a network for monitoring of isotopes in waters of India. *Current Science*, **88**, 107–118.

Gupta, S.K. and Polach, H.A. (1985) *Radiocarbon Dating Practices at ANU, Handbook*, Radiocarbon Laboratory, Research School of Pacific Studies, ANU, Canberra, Australia, p. 173.

Gupta, S.K. and Sharma, P. (1984) Soil moisture transport through unsaturated zone: tritium tagging studies in Sabarmati basin, western India. *Hydrological Sciences Journal*, **29**, 177–189.

Gupta, S.K., Lal, D. and Sharma, P. (1981) An approach to determining pathways and residence time of ground waters: dual radiotracer dating. *Journal of Geophysical Research*, **86**, 5292–5300.

Gupta, S.K., Bhandari, N., Thakkar, P.S. and Rengarajan, R. (2002) On the origin of the artesian groundwater and escaping gas at Narveri after the Bhuj earthquake in 2001. *Current Science*, **82**, 463–468.

Gupta, S.K., Deshpande, R.D., Agarwal, M. and Raval, B.R. (2005a) Origin of high fluoride in ground water in the North Gujarat-Cambay region, India. *Hydrogeology Journal*, **13**, 596–605.

Gupta, S.K., Deshpande, R.D., Bhattacharya, S.K. and Jani, R.A. (2005b) Groundwater $\delta^{18}O$ and δD from central Indian peninsula: influence of Arabian Sea and the Bay of Bengal branches of summer monsoon. *Journal of Hydrology*, **303**, 38–55.

Guymon, G.L. (1972) Note on finite element solution of diffusion-convection equation. *Water Resources Research*, **8**, 1357–1360.

Haan, C.T. (1974) *Statistical Methods in Hydrology*, Iowa State University Press, Ames, IA, p. 496.

Haan, C.T. (1977) *Statistical Methods in Hydrology*, Iowa State University Press, Ames, IA, p. 378.

Hack, J.T. (1957) Studies of longitudinal stream profiles in Virginia and Maryland. *US Geological Survey Prof. Paper*, **294-B**, 45–97.

Haderlein, S. (2000) Underground chemical spies. *EAWAG News*, **49e**, 21–22.

Haitjema, H.M. (1995) *Analytic Element Modelling of Groundwater Flow*, Academic Press, San Diego, CA, p. 421.

Hamid, S., Dray, M., Fehri, A., Dorioz, J.M., Normand, M. and Fontes, J.C. (1989) Etude des transfers d'eau a l'interieur d'une formation morainique dans le Bassin du Leman-transfer d'eau dans la zone non-saturee. *Journal of Hydrology*, **109**, 369–385.

Handa, B.K. (1977) *Presentation and Interpretation of Fluorine Ion Concentrations in Natural Waters*. Symposium on Fluorosis, pp. 317–347.

Hannaford, J.F. and Hall, R.L. (1980) *Application of Satellite Imagery to Hydrologic Modeling Snowmelt Runoff in the Southern Sierra Nevada*. Final Workshop on Operational Applications of Satellite Snow Cover Observations, NASA CO-2116, pp. 201–222.

Hantush, M.S. (1956) Analysis of data from pumping tests in leaky aquifers. *Transactions of American Geophysical Union*, **37**, 702–714.

Hantush, M.S. (1964a) Drawdown around wells of variable discharge. *Journal of Geophysical Research*, **69**, 4221–4235.

Hantush, M.S. (1964b) Hydraulics of wells, in *Advances in Hydroscience*, vol. 1 (ed V.T. Chow), Academic Press, New York, pp. 281–432.

Hantush, M.S. (1966) Wells in homogeneous anisotropic aquifers. *Water Resources Research*, **2**, 273–279.

Hantush, M.S. (1967) Flow to wells in aquifers separated by a semipervious layer. *Journal of Geophysical Research*, **72**, 1709–1720.

Hantush, M.S. and Jacob, C.E. (1955) Non-steady radial flow in an infinite leaky aquifer. *Transactions of American Geophysical Union*, **36**, 95–112.

Hantush, M.S. and Papadopulos, I.S. (1962) Flow of ground water to collector wells. *Journal*

Hydraulics Division. American Society of Civil Engineers, **88**, 221-244.

Hantush, M.S. and Thomas, R.G. (1966) A method for analysing a drawdown test in anisotropic aquifers. *Water Resources Research*, **2**, 281-285.

Hao, Z., Wang, J. and Li, L. (2006) Impact of climate change on runoff in source region of Yellow River. *Journal of Glaciology and Geocryology*, **28**, 1-6.

Haralick, R.M., Wang, S., Shapiro, L.G. and Campbell, J.B. (1985) Extraction of drainage networks by using a consistent labeling technique. *Remote Sensing of Environment*, **18**, 163-175.

Harbaugh, A.W. (2005) MODFLOW-2005, The US Geological Survey Modular Ground-Water Model – the Ground-Water Flow Process, US Geological Survey.

Harriman, R., Ferrier, R.C., Jenkins, A. and Miller, J.D. (1990) Long- and short-term hydrochemical budgets in Scottish catchments, in *The Surface Waters Acidification Programme* (ed B.J. Mason), Cambridge University Press, Cambridge, pp. 31-45.

Harvey, L.D.D. (1997) Upscaling in global change research, in *Elements of Change 1997* (eds S.J. Hassol and J. Katzenberger), Aspen Global Change Institute, Aspen, CO, pp. 14-33.

Heath, R.C. (1983) Basic Groundwater Hydrology, US Geological Survey Water-Supply Paper 2220, p. 86.

Heaton, T.H.E. (1981) Dissolved gases: Some applications to groundwater research. *Transactions Geological Survey Society South Africa*, **84**, 91-97.

Heaton, T.H.E. and Vogel, J.C. (1981) 'Excess air' in ground water. *Journal of Hydrology*, **50**, 201-216.

HEC (1990) *HEC-1 Flood Hydrograph Package User's Manual: US Army Corps of Engineers*. Hydrologic Engineering Center, Davis, CA, p. 410.

Hem, J.D. (1985) Study and interpretation of the chemical characteristics of Natural Water, US Geological Survey, *Water Supply Paper 2254*, p. 225.

Henze, M. and Ledin, A (2001) Types, characteristics and quantities of classic, combined domestic wastewater, in *Decentralised Sanitation and Reuse: Concepts, Systems and Implementation*

(eds P. Lens, G. Zeeman and G. Lettinga), IWA Publishing, London, pp. 57-72.

Hidalgo, H.G., Piechota, T.C. and Dracup, J.A. (2000) Alternative principal components regression procedures for dendrohydrologic reconstructions. *Water Resources Research*, **36**, 3241-3249.

Hill, M.C. (1990) Preconditioned Conjugate-gradient 2 (PCG2), a Computer Program for Solving Ground-Water Flow Equations, *Water Resources Investigation Report 90-4048*, US Geological Survey, p. 4.

Hill, M.C. (1992) A Computer program (MODFLOWP) for Estimating Parameters of a Transient, Three-Dimensional Ground-Water Flow Model Using Nonlinear Regression. *Water Resource Investigation report 91-484*, US Geological Survey.

Hinsby, K., Edmunds, W.M., Loosli, H.H., Manzano, M., Melo, M.T.C. and Barbecot, F. (2001) The modern water interface: recognition, protection and development – advance of modern waters in European coastal aquifer systems, in *Palaeowaters in Coastal Europe – Evolution of Groundwater since the Late Pleistocene*, vol. Special Publication 189 (eds W.M. Edmunds and C.J. Milne), Geological Socety, London, pp. 271-288.

Hoekstra, A.Y. and Chapagain, A.K. (2007) Water footprints of nations: water use by people as a function of their consumption pattern. *Water Resources Management*, **21**, 35-48.

Holocher, J., Peeters, F. *et al.* (2002) Experimental investigations on the formation of excess air in quasi-saturated porous media. *Geochimica et Cosmochimica Acta*, **66**, 4103-4117.

Holzbecher, E. and Sorek, S. (2005) Numerical models of groundwater flow and transport, in *Encyclopedia of Hydrological Sciences* (ed M.G. Anderson), John Wiley & Sons, Ltd, New York, pp. 2401-2414.

Hooper, R.P. and Shoemaker, C.A. (1986) A comparison of chemical and isotopic hydrograph separation. *Water Resources Research*, **22**, 1444-1454.

Horton, R.E. (1919) Rainfall interception. *Monthly Weather Review*, **47**, 603-623.

Horton, R.E. (1933) The role of infiltration in the hydrologic cycle. *Transactions of American Geophysical Union*, **14**, 446-460.

Horton, R.E. (1940) An approach towards a physical interpretation of infiltration capacity. *Soil Science Society of America Proceedings*, **5**, 399–417.

Horton, R.E. (1945) Erosional development of streams and their drainage basins: hydrophysical approach to quantitative morphology. *Geological Society of America Bulletin*, **56**, 275–370.

Hsieh, P.A. and Freckleton, J.R. (1993) Documentation of a Computer Program to Simulate Horizontal Flow Barriers using US Geological Survey's Modular Three-Dimensional Finite-Difference Ground-Water Flow Model. *Open File Report 92-477*, US Geological Survey, p. 32.

Huber, W.C. and Dickinson, R.E. (1988) Storm Water Management Model Version 4, Part A: User's Manual. *Report EPA/600/3-88/001a*, US Environmental Protection Agency, p. 569.

Hubner, M. (1969) Geochemische Interpretation von Fluorid/Hydroxid-Austauschversuchen an Tonmineralen Deutsch. *Ges. Geol. Wiss. B. Miner. Lagerstattenforsch*, **14**, 5–15.

Humes, K.S., Kustas, W.P. and Moran, M.S. (1994) Use of remote sensing and reference site measurements to estimate instantaneous surface energy balance components over a semi-arid rangeland watershed. *Water Resources Research*, **30**, 1363–1373.

Hurst, C.J., Gerba, C.P. and Cech, I. (1980a) Effects of environmental variables and soil characteristics on virus survival in soil. *Applied and Environmental Microbiology*, **40**, 1067–1079.

Hurst, C.J., Gerba, C.P., Lance, J.C. and Rice, R.C. (1980b) Survival of enteroviruses in rapid-infiltration basins during the land application of wastewater. *Applied and Environmental Microbiology*, **40**, 192–200.

Hurst, C.J., Wild, D.K. and Clark, R.M. (1992) Comparing the accuracy of equation formats for modelling microbial population decay rates, in *Modelling the Metabolic and Physiologic Activities of Micro-organisms* (ed C.J. Hurst), John Wiley & Sons, New York, pp. 149–175.

Huston, M.A. (1994) *Biological Diversity: Coexistence of Species on Changing Landscapes*, Cambridge University Press, Cambridge UK, p. 550.

IPCC (2001) *Third Assessment Report Working Group I: The Scientific Basis. Chapter 2. Observed Climate Variability and Change*: Cambridge University Press, Cambridge, UK, pp. 99–181.

Ivanovich, M. (1992) *Uranium Series Disequilibrium: Applications to Earth, Marine and Environmental Sciences*: Clarendon Press, Oxford, p. 910.

Izzard, C.F. (1946) Hydraulics of runoff from developed surfaces. *Proceedings of the Highway Research Board*, **26**, 129–150.

Jacks, G., Rajagopalan, K., Iveteg, T. and Jonsson, M. (1993) Genesis of high F ground waters, southern India. *Applied Geochemistry*, Suppl Issue 2, 241–244.

Jackson, T.J. (1993) Measuring surface soil moisture using passive microwave remote sensing. *Hydrological Processes*, **7**, 139–152.

Jackson, T.J., Ragan, R.M. and Fitch, W.N. (1977) Test of Landsat-based urban hydrologic modeling. *Journal of Water Resources Planning and Management Division (ASCE)*, **103**, 141–158.

Jacob, C.E. (1947) Drawdown test to determine effective radius of artesian well. *Transactions of American Society of Civil Engineers*, **112**, 1047–1070.

Jardine, P.M., Wilson, G.V., Luxmore, R.J. and McCarthy, J.F. (1989) Transport of inorganic and natural organic tracers through an isolated pedon in a forested watershed. *Soil Science Society of America Journal*, **53**, 317–323.

Jensen, M.E. and Haise, H.R. (1963) Estimating evapotranspiration from solar radiation: Proceedings of the American Society of Civil Engineers (ASCE). *Journal of the Irrigation and Drainage Division*, **89**, 15–41.

JMP (2008) Joint Monitoring Programme for Water Supply and Sanitation, WHO/UNICEF, http://www.wssinfo.org/en/142_currentSit.html (accessed 4 March 2010).

Jones, V.T. and Drozd, R.J. (1983) Predictions of oil and gas potential by near-surface geochemistry. *American Association of Petroleum Geologists Bulletin*, **67**, 932–952.

Judd, S. (2006) *Principles and Applications of Membrane Bioreactors in Water and Waste Water Treatment*, Elsevier, Oxford.

Julian, P.Y. and Saghafian, B. (1991) CASC2D: A Two-Dimensional Watershed Rainfall-Runoff

Model, CASC2D User's Manual. Report CER90-91PYJ-BS-12, Colorado State University, Fort Collins, CO, p. 66.

Julian, P.Y., Saghafian, B. and Ogden, F.L. (1995) Raster-based hydrologic modeling of spatially-varied surface runoff. *Water Resources Bulletin*, **31**, 523–536.

Juyal, N., Kar, A., Rajaguru, S.N. and Singhvi, A.K. (2003) Luminescence chronology of aeolian deposition during the Late Quaternary on the southern margin of Thar Desert, India. *Quaternary International*, **104**, 87–98.

Kanemasu, E.T., Stone, L.R. and Powers, W.L. (1977) Evapotranspiration model tested for soybean and sorghum. *Agronomy Journal*, **68**, 569–572.

Kaufman, W.J. and Orlob, G.T. (1956) Measuring ground water movement with radioactive and chemical tracers. *American Water Works Association Journal*, **48**, 559–572.

Kendall, C. and McDonnell, J.J. (eds) (1998) *Isotope Tracers in Catchment Hydrology*, Elsevier, Amsterdam, p. 839.

Kennedy, V.C., Jackman, A.P., Zand, S.M., Zellweger, G.W. and Avanzino, R.J. (1984) Transport and concentration controls for chloride, strontium, potassium, and lead in Uvas Creek, a small cobblebed stream in Santa Clara County, California – Part 1: Conceptual model. *Journal of Hydrology*, **75**, 67–110.

Kerby, W.S. (1959) Time of concentration studies. *Civil Engineering*, **29** (3), 60.

Kerr, R.S. (1992) *Quality Assurance and Quality Control in the Application of Ground-Water Models*, US Environmental Protection Agency, Environmental Research Laboratory, Ada, OK, EPA/600/R-93/011.

Khadka, N.S. (2004) Environmentalists are warning that the melting of glaciers in the Himalayas could spell disaster for millions of people living in the region, *BBC News*.

Kielland, J. (1937) Individual activity coefficients of ions in aqueous solutions. *Journal of the American Chemical Society*, **59**, 1676–1678.

Kipfer, R., Aeschbach-Hertig, W., Peeters, F. and Stute, M. (2002) Noble gases in lakes and ground waters, in *Noble Gases in Geochemistry and Cosmochemistry: Reviews in Mineralogy and Geochemistry*, Vol. **47** (eds D. Porcelli, C. Ballentine and R. Wieler), Mineralogical Society of America, Geochemical Society, Washington DC. pp. 615–700.

Kipp, K.L. (1973) Unsteady flow to a partially penetrating, finite radius well in an unconfined aquifer. *Water Resources Research*, **9**, 448–452.

Kirby, M., Qureshi, M.E., Mainuddin, M. and Dyack, B. (2006) Catchment behavior and countercyclical water trade: an integrated model. *Natural Resource Modeling*, **19**, 483–510.

Kirpich, Z.P. (1940) Time of concentration in small agricultural watersheds. *Civil Engineering*, **10**, 362.

Kitching, R., Edmunds, W.M., Shearer, T.R., Walton, N.R.G. and Jacovides, J. (1980) Assessment of recharge to aquifers. *Hydrological Sciences Bulletin*, **25**, 217–235.

Kite, G.W. (1988) *Frequency and Risk Analyses in Hydrology*, Water Resources Publication, Littleton, CO, p. 257.

Kite, G.W. and Kouwen, N. (1992) Watershed modeling using land classifications. *Water Resources Research*, **28**, 3193–3200.

Klemes, V. (1986) Dilettantism in hydrology: Transition or destiny? *Water Resources Research*, **22**, 177s–188s.

Kobayashi, D., Suzuki, K. and Nomura, M. (1990) Diurnal fluctuations in stream flow and in specific electric conductance during drought periods. *Journal of Hydrology*, **115**, 105–114.

Konrad, C.P. (2003) Effects of Urban Development on Floods, *Fact Sheet 076-03*, US Geological Survey.

Koopmans, B.N. (1983) Side looking radar, a tool for geological surveys. *Remote Sensing Reviews*, **1**, 19–69.

Koritnig, S. (1951) Ein Beitrag zur Geochemie des Fluor. (A contribution to the geochemisty of fluorine). *Geochemica et Cosmochemica Acta*, **1**, 89–116.

Krishna, H.J. (2005) *The Texas Manual on Rainwater Harvesting*, Texas Water Development Board, Austin, TX, p. 56.

Krishnamurthy, R.V. and Bhattacharya, S.K. (1991) Stable oxygen and hydrogen isotope ratios in shallow ground waters from India and a study of the role of evapotranspiration in the Indian monsoon. *The Geochemical Society*, Special Publ. 3, 187–193.

Krishnaswami, S., Bhushan, R. and Baskaran, M. (1991) Radium isotopes and ^{222}Rn in shallow brines, Kharaghoda (India). *Chemical Geology*, **87**, 125–136.

Krishnaswami, S. and Seidemann, D.E. (1988) Comparative study of ^{222}Rn, ^{40}Ar, ^{39}Ar and ^{37}Ar leakage from rocks and minerals: implications to the role of nanopores in gas transport through natural silicates. *Geochimica et Cosmochimica Acta*, **52**, 655–658.

Kruck, W. (1990) *Application of Remote Sensing for Groundwater Prospection in the Third World*. International Symposium on Remote Sensing and Water Resources, Enschede, The Netherlands, IAH/Netherlands Society of Remote Sensing, pp. 455–463.

Krumbein, W.C. and Pettijohn, F.J. (1938) *Manual of Sedimentary Petrography*, Appleton-Century, New York, p. 549.

Kruseman, G.P. and Ridder, N.A. (1970) Analysis and Evaluation of Pumping Test Data. Wageningen, International Institute for Land Reclamation and Improvement, *Bulletin 11*, p. 200.

Kueper, B.H., Redman, D., Starr, R.C., Reitsma, S. and Mah, M. (1993) A field experiment to study the behavior of tetrachloroethylene below the water table: spatial distribution of residual and pooled DNAPL. *Ground Water*, **31**, 756–766.

Kulongoski, J.T., Hilton, D.R. and Izbicki, J.A (2003) Helium isotope studies in the Mojave Desert, California: implications for groundwater chronology and regional seismicity. *Chemical Geology*, **202**, 95–113.

Kulongoski, J.T., Hilton, D.R. and Izbicki, J.A (2005) Source and movement of helium in the eastern Morongo groundwater basin: the influence of regional tectonics on crustal and mantle helium fluxes. *Geochemica et Cosmochemica Acta*, **69**, 3857–3872.

Kumar, K.K. (2007) PRL Lecture – Some New Perspectives on the Variability and Prediction of Indian Summer Monsoon Rainfall, Pune, Indian Institute of Tropical Meteorology.

Kumar, B., Athavale, R.N. and Sahay, K.S.N. (1982) Stable isotope geohydrology of the Lower Maner

Basin, Andhra Pradesh, India. *Journal of Hydrology*, **59**, 315–330.

Kummerow, C., Mack, R.A. and Hakkarinen, I.M. (1989) A self-consistency approach to improve microwave rainfall estimates from space. *Journal of Applied Meteorology*, **28**, 869–884.

Kunzi, K.F., Patil, S. and Rott, H. (1982) Snow-covered parameters retrieved from NIMBUS-7 SMMR data. *IEEE Transactions on Geoscience and Remote Sensing*, **GE-20**, 452–467.

Kustas, W.P., Moran, M.S. *et al.* (1990) Instantaneous and daily values of the surface energy balance over agricultural fields using remote sensing and a reference field in an arid environment. *Remote Sensing of Environment*, **32**, 125–141.

Kwan, J.Y., Riley, J.P. and Amisal, R.A. (1968) A digital computer program to plot isohyetal maps and calculate volumes of precipitation. *The Use of Analog and Digital Computers in Hydrology*, IAHS Publ. No. 80, 240–248.

Lakshminarayana, V. and Rajagopalan, S.P. (1978) Type-curve analysis of time-drawdown data for partially penetrating wells in unconfined anisotropic aquifers. *Ground Water*, **16**, 328–333.

Lal, D. (1999) An overview of five decades of studies of cosmic ray produced nuclides in oceans. *The Science of Total Environment*, **237/238**, 3–13.

Lal, D. and Peters, B. (1967) Cosmic ray produced radioactivity on the Earth, in *Handbuch der Physik*, vol. **46**, Springer Verlag, Berlin, pp. 551–612.

Lal, D., Nijampurkar, V.N., Somayajulu, B.L.K., Koide, M. and Goldberg, E.D. (1976) Silicon-32 specific activities in coastal waters of the world oceans. *Limnology and Oceanography*, **21**, 285–293.

Langbein, W.B. and Durum, W.H. (1967) The Aeration Capacity of Streams, US Geological Survey, Circular 542, p. 6.

Langmuir, D. and Mahoney, J. (1984) Chemical equilibrium and kinetics of geochemical processes in groundwater studies, in *First Canada-American Conference on Hydrology* (eds B. Hitchon and E.I. Walleck), National Water Well Association, Dublin, Ohio, pp. 69–95.

Leadbetter, M.R., Lindgren, G. and Rootzen, H. (1983) *Extremes and Related Properties of Random Sequences and Processes*, Springer-Verlag, New York, p. 335.

Leavesley, G.H., Lichty, R.W., Troutman, B.M. and Saindon, L.G. (1983) Precipitation-Runoff Modeling System – User's Manual, *US Geological Survey Water Resources Investigation Report 83-4238*, p. 207.

Leblanc, M.J.P., Tregoning, G., Ramillien, S., Tweed, O. and Fakes, A. (2009) Basin-scale, integrated observations of the early 21st century multiyear drought in southeast Australia. *Water Resources Research*, **45**, W04408, doi:10.1029/2008WR007333.

Leemis, L.M. (1986) Relationships among common univariate distributions. *The American Statistician*, **40**, 143–146.

Lehmann, B.E. and Purtschert, R. (1997) Radioisotope dynamics – the origin and fate of nuclides in ground water. *Applied Geochemistry*, **12**, 727–738.

Lennox, D.H. and Vanden Berg, A. (1962) Drawdown due to cyclic pumping. *Journal Hydraulics Division, American Society of Civil Engineers*, **93**, 35–51.

Leopold, A. (1997) A Sand County Almanac, in *Ethics in Practice* (ed H. La Follette), Blackwell, Oxford, pp. 634–643.

Leopold, L.B. and Wolman, M.G. (1957) River channel patterns – braided, meandering, and straight. *US Geological Survey Prof. Paper*, **282-B**, 39–85.

Leopold, L.B., Wolman, M.G. and Miller, J.P. (1964) *Fluvial Processes in Geomorphology*, W.H. Freeman, New York, p. 522.

Lerner, D.N., Issar, A.S. and Simmers, I. (1990) Groundwater Recharge, a Guide to Understanding and Estimating Natural Recharge. Kenilworth, International Association of Hydrogeologists, *Report No. 8*, p. 345.

Liggett, J.A. and Liu, P.L.-F. (1983) *The Boundary Integral Equation Method for Porous Media Flow*, George Allen & Unwin, London, p. 255.

Lindberg, R.D. and Runnels, D.D. (1984) Groundwater redox reactions: an analysis of equilibrium state applied to Eh measurements and geochemical modeling. *Science*, **225**, 925–927.

Liu, C. (1997) Potential impact of climate change on hydrology and water resources in China. *Advance of Water Science*, **8**, 220–225.

Liu, C. (1998) Environmental issues and the South-North Water Transfer Scheme. *The China Quarterly*, **156**, 899–910.

Liu, C. and Zheng, H. (2004) Changes in components of the hydrological cycle in the Yellow River basin during the second half of the 20th century. *Hydrological Processes*, **18**, 2337–2345.

Liu, D.H.F. and Lipták, B.G. (2000) *Wastewater Treatment*, Lewis, Boca Raton, FL, ISBN 1-56670-515-0.

Logan, B.E., Hamelers, B. *et al.* (2006) Microbial fuel cells: methodology and technology. *Environmental Science and Technology*, **40**, 5181–5192.

Lovelock, J.E. and Margulis, L. (1974) Atmospheric homeostasis by and for the biosphere: the Gaia hypothesis. *Tellus*, **26**, 2–10.

Lucas, L.L. and Unterweger, M.P. (2000) Comprehensive review and critical evaluation of the half-life of tritium. *Journal of Research of the National Institute of Standards and Technology*, **105**, 541–549.

Ludin, A.I., Wepperning, R., Bönisch, G. and Schlosser, P. (1997) *Mass Spectrometric Measurement of Helium Isotopes and Tritium in Water Samples*, Lamont-Doherty Earth Observatory, Palisades, Report 98.6.

Luke, J. (1997) *The Effect of Fluoride on the Physiology of the Pineal Gland*, Ph.D. Thesis, University of Surrey, Guildford, p. 177.

Lupton, J.E., Weiss, R.F. and Craig, H. (1977) Mantle helium in the Red Sea. *Nature*, **266**, 244–246.

Mackay, D.M. and Cherry, J.A. (1989) Groundwater contamination: pump-and-treat remediation. *Environmental Science and Technology*, **23**, 630–636.

Mackinnon, P.A., Elliot, T., Zhao, Y.Q., Murphy, J.L. and Kalin, R.M. (2002) Evaluation of a novel technique for measuring re-aeration in rivers, in *Development and Application of Computer Techniques to Environmental Studies IX* (eds C.A. Brebbia and P. Zanetti), WIT Press, Southampton, UK, pp. 531–540.

Macler, B.A. and Merkle, J.C. (2000) Current knowledge on groundwater microbial pathogens and their control. *Hydrogeology Journal*, **8**, 29–40.

Malard, F., Reygrobellet, J.-L. and Soulie, M. (1994) Transport and retention of faecal bacteria at sewage-polluted fractured rock sites. *Journal of Environmental Quality*, **23**, 1352–1363.

Manning, A.H., Solomon, D.K. and Thiros, S.A. (2005) H-3/He-3 age data in assessing the susceptibility of wells to contamination. *Ground Water*, **43**, 353–367.

Manning, R. (1891) On the flow of water in open channels and pipes. *Transactions Institute of Civil Engineering, Ireland*, **20**, 161–207.

Manov, G.G., Bates, R.G., Hamer, W.J. and Acree, S.F. (1943) Values of constants in the Debye-Hückle equation for activity coefficients. *Journal of the American Chemical Society*, **65**, 1765–1767.

Mantua, N.J., Hare, S.R., Zhang, Y., Wallace, J.M. and Francis, R.C. (1997) A Pacific interdecadal climate oscillation with impacts on salmon production. *Bulletin of the American Meteorological Society*, **78**, 1069–1079.

Martinec, J., Rango, A. and Major, E. (1983) *The Snowmelt-runoff (SRM) User's Manual*, NASA, Washington DC, p. 118.

Marty, B., Torgersen, T., Meynier, V., Onions, R.K. and Demarsily, G. (1993) Helium isotope fluxes and groundwater ages in the Dogger Aquifer, Paris Basin. *Water Resources Research*, **29**, 1025–1035.

Mather, P.M. (1999) Land cover classification revisited, in *Advances in Remote Sensing and GIS* (eds P.M. Atkinson and N.J. Tate), John Wiley & Sons, New York, pp. 7–16.

Mauser, W. (1991) *The Use of Multitemporal TM- and Spot-Data in Geographical Information System to Model the Spatial Variability of Evapotranspiration, Geoscience and Remote Sensing, IGARSS '9, Remote Sensing*. Global Monitoring for Earth Management, International Symposium, IEEE, pp. 59–62.

Mazor, E. (1972) Paleotemperatures and other hydrological parameters deduced from gases dissolved in groundwaters, Jordan Rift Valley, Israel. *Geochimica et Cosmochimica Acta*, **36**, 1321–1336.

Mazor, E. and Bosch, A. (1992) *Helium as Semi-Quantitative Tool for Groundwater Dating in the Range of 104-108 Years, Isotopes of Noble Gases as Tracers in Environmental Studies*, IAEA, Vienna, pp. 173–178.

McCabe, G.J. and Dettinger, M.D. (1999) Decadal variations in the strength of ENSO teleconnections with precipitation in the western United States. *International Journal of Climatology*, **19**, 1399–1410.

McCabe, G.J. and Palecki, M.A. (2006) Multidecadal climate variability of global lands and oceans. *International Journal of Climatology*, **26**, 849.

McCauley, J., Schaber, F. *et al.* (1982) Subsurface valleys and geoarcheology of the eastern Sahara revealed by Shuttle radar. *Science*, **318**, 1004–1020.

McDonald, M.G. and Harbaugh, A.W. (2003) The history of MODFLOW. *Ground Water*, **41**, 280–283.

McDonald, M.G., Harbaugh, A.W., Orr, B.R. and Ackreman, D.J. (1991) A Method of Converting No-Flow Cells to Variable Head Cells for the US Geological Survey Modular Finite-Difference Ground-Water Flow Model. *Open File Report 91-536*, US Geological Survey, p. 99.

McNeill, G.W., Yang, Y.S., Elliot, T. and Kalin, R.M. (2001) Krypton gas as a novel applied tracer of groundwater flow in a fissured sandstone aquifer, in *New Approaches to Characterizing Groundwater Flow*, Vol. **1** (eds K.-P. Seiler and S. Wohnlich), A.A.Balkema/Lisse, Rotterdam, pp. 143–148.

MDBC (1999) *Murray-Darling Basin Groundwater: A Resource for the Future*, Murray-Darling Basin Commission, Canberra, Australia, p. 31.

MDBC (2003) *Preliminary review of selected factors that may change future flow patterns in the River Murray System*, Murray-Darling Basin Commission, Canberra, Australia, p. 131.

MDBC (2004) Water Audit Monitoring Reports. Murray Darling Basin Commission, Canberra, available at http://www.mdbc.gov.au/nrm.

Meijerink, A.M.J. (1996) Remote sensing applications to hydrology: ground water. *Hydrological Sciences Journal*, **41**, 549–561.

Melesse, A.M., Weng, Q., Thenkabail, P.S. and Senay, G.B. (2007) Remote sensing sensors and applications in environmental resources mapping and modelling. *Sensors*, **7**, 3209–3241.

Menenti, M. (1984) Physical aspects and determination of evaporation in deserts, applying remote sensing techniques, Wageningen Agricultural University, Wageningen, The Netherlands.

Merot, P., Crave, A. and Gascuel-Odoux, C. (1994) Effect of saturated areas on backscattering coefficient of the ERS-1 synthetic aperture radar: First results. *Water Resources Research*, **30**, 175–179.

Metcalf and Eddy, Inc. (1991) *Wastewater Engineering: Treatment Disposal and Reuse*, McGraw-Hill, New York, p. 1334.

Meybeck, M., Chapman, D.V. and Helmer, R. (eds) (1989) *Global Freshwater Quality: A first assessment, GEMS Global Environment Monitoring System*, WHO/UNEP, Blackwell, Oxford, p. 306.

Michaud, J.P. (1991) *A Citizens' Guide to Understanding and Monitoring Lakes and Streams*, Washington State Department of Ecology, p. 66.

Miller, J.E. (1984) Basic Concepts of Kinematic-Wave Models. *US Geological Survey Professional Paper 1302*, p. 29.

Milliman, J.D. and Meade, R.H. (1983) Worldwide delivery of river sediment to the oceans. *Journal of Geology*, **91**, 1–21.

Minissale, A., Vaselli, O., Chandrasekharam, D., Magro, G., Tass, F. and Casiglia, A. (2000) Origin and evolution of 'intracratonic' thermal fluids from central-western peninsular India. *Earth and Planetary Sciences Letters*, **181**, 377–397.

Minshall, W.G., Cummins, K.W. *et al.* (1985) Developments in stream ecosystem theory. *Canadian Journal of Fisheries and Aquatic Sciences*, **42**, 1045–1055.

Mockus, V. (1972) Estimation of direct runoff from storm rainfall, *National Engineering Handbook*. NEH Notice 4-102, p. 10.1–10.22.

Moench, A. (1971) Groundwater fluctuations in response to arbitrary pumpage. *Ground Water*, **9**, 4–8.

Moench, A. and Prikett, T.A. (1972) Radial flow in an infinite aquifer undergoing conversion from artesian to water table conditions. *Water Resources Research*, **8**, 494–499.

Mohrlok, U., McNeill, G., Elliot, T. and Kalin, R. (2002) Modelling tracer injection for the interpretation of a tracer test in layered fluvial sediments. *Acta Universitatis Carolinae – Geologica*, **46**, 395–399.

Monteith, J.L. (1965) *Evaporation and the Environment. The State and Movement of Water in Living Organisms*. Proceedings of the XIXth Symposium of the Society for Experimental Biology, Cambridge University Press, Cambridge, pp. 205–234.

Montgomery, D.R. (1999) Process domains and the river continuum. *Journal of the American Water Resources Association*, **35**, 397–410.

Mook, W.G. (1976) *The Dissolution-Exchange Model for Dating Groundwater with ^{14}C, Interpretation of Environmental Isotopes and Hydrochemical Data in Groundwater Hydrology*. IAEA, Vienna, pp. 213–225.

Mook, W.G., (2000) Introduction: Theory, Methods, Review, *in* W.G. Mook, ed, Environmental isotopes in the hydrological cycle: principles and applications, v. 1: Paris, UNESCO/IAEA, IHP-V, Technical Documents in Hydrology, No. 39, p. 280.

Mook, W.G. (ed) (2000–2001) Environmental isotopes in the hydrological cycle: principles and applications, vol. IHP-V, Technical Documents in Hydrology, No. 39: UNESCO/IAEA, Paris.

Morel, F.M.M. and Hering, J.G. (1993) *Principles and Applications of Aquatic Chemistry*, John Wiley & Sons, New York, p. 588.

Morgenstern, U., Fifield, L.K. and Zondervan, A. (2000) New frontiers in glacial ice dating: Measurement of natural ^{32}Si by AMS. *Nuclear Instruments and Methods in Physics Research*, **B172**, 605–609.

Morris, D.A. and Johnson, A.I. (eds) (1967) Summary of hydrologic and physical properties of rock and soil materials as analysed by the Hydrologic Laboratory of the US Geological Survey 1948-60, US Geological Survey, p. 42.

Münnich, K.O. (1957) Messung des ^{14}C – Gehaltes von hartem Grundwasser. *Naturwissenschaften*, **34**, 32–33.

Münnich, K.O. (1968) Isotopen-datierung von grundwasser. *Naturwissenschaften*, **55**, 158–163.

Murphy, J.L., Mackinnon, P.A., Zhso, Y.Q., Kalin, R.M. and Elliot, T. (2001) Measurement of the surface water re-aeration coefficient using Krypton as a gas tracer, in *Water Pollution VI* (ed C.A. Brebbia), WIT Press, Southampton, UK, pp. 505-514.

Murray Rust, H. and Vander Velde, E. (1992) Conjunctive use of canal and ground water in Punjab, Pakistan: management and policy options, *Advancements in IIMI's Research 1992*. A Selection of papers presented at the Internal Program Review, Colombo, Sri Lanka, International Irrigation Management Institute, pp. 161-188.

Navada, S.V. and Rao, S.M. (1991) Study of Ganga River – groundwater interaction using environmental ^{18}O. *Isotopenpraxis*, **27**, 380-384.

Navada, S.V., Nair, A.R., Rao, S.M., Paliwall, L. and Dashy, C.S. (1993) Groundwater recharge studies in arid region of Jalore, Rajasthan using isotope techniques. *Journal of Arid Environments*, **24**, 125-133.

Navalgund, R.R., Jayaraman, V. and Roy, P.S. (2007) Remote sensing applications: an overview. *Current Science*, **93**, 1747-1766.

Nazaroff, W.W., Moed, B.A. and Sextro, R.G. (1988) Soil as a source of indoor radon: generation, migration and entry, in *Radon and its Decay Products in Indoor Air* (eds W.W. Nazaroff and A.V.J. Nero), John Wiley & Sons, New York, pp. 57-112.

NDWM (1990) *National Drinking Water Mission*, Goverment of India, New Delhi.

Negi, J.G., Agrawal, P.K., Singh, A.P. and Pandey, O.P. (1992) Bombay gravity high and eruption of Deccan flood basalts (India) from a shallow secondary plume. *Tectonophysics*, **206**, 341-350.

Neuman, S.P. (1972) Theory of flow in unconfined aquifers considering delayed response of water table. *Water Resources Research*, **8**, 1031-1045.

Neuman, S.P. (1975) Analysis of pumping test data from anisotropic unconfined aquifers considering delayed gravity response. *Water Resources Research*, **11**, 329-342.

Neuman, S.P. and Witherspoon, P.A. (1972) Field determination of hydraulic properties of leaky multiple aquifer systems. *Water Resources Research*, **6**, 1284-1298.

Neumann, P., Fett, W. and Schultz, G.A. (1990) A Geographic Information System as Data Base For Distributed Hydrological Models. *International Symposium, Remote Sensing and Water Resources, Enschede, The Netherlands*, pp. 781-791.

Nijampurkar, V.N. (1974) Application of Cosmic Ray Produced Isotope Silicon-32 with Special Reference to Dating of Ground water. Ph.D. Thesis, University of Bombay, Mumbai, India.

Nijampurkar, V.N., Amin, B.S., Kharkar, D.P. and Lal, D. (1966) 'Dating' ground waters of ages younger than 1000-1500 years using natural Silicon-32. *Nature*, **210**, 478-480.

Nordstrom, D.K., Plummer, L.N. *et al.* (1979) Comparison of computerized chemical models for equilibrium calculations in aqueous systems; chemical modeling in aqueous systems; speciation, sorption, solubility, and kinetics, in *Chemical Modelling in Aqueous Systems, American Chemical Society Symposium Series 93* (ed E.A. Jenne), pp. 857-892.

Nordstrom, D.K., Plummer, L.N. Langmuir, D., Busenberg, E., May, H.M., Jones, B.F., Parkhurst, D.L. (1990) Revised chemical equilibrium data from major mineral reactions and their limitations, in *Chemical Modelling of Aqueous Systems II*, Vol. ACS Ser. **416** (eds D.C. Melchior and R.L. Bassett), American Chemical Society, Washington DC, pp. 398-413.

NPS (2005) Fight the Invasion: Controlling invasive plant species at Grand Canyon National Park. Grand Canyon, AZ, National Park Service, Grand Canyon National Park Foundation, and the Arizona Water Protection Fund Commission.

NRC (1992) *Water Transfers in the West: Efficiency, Equity, and the Environment*, National Academy Press, Washington DC.

NRC (2007) *Colorado River Basin Water Management: Evaluating and Adjusting to Hydroclimatic Variability: Washington, DC, National Research Council*, The National Academies Press, Washington DC, p. 222.

NRSA (2003) Rajiv Gandhi National Drinking Water Mission (RGNDWM), *Report No. NRSA/HD/RGNDWM/TR: 02:2003.*

Oke T.R. (1982) The energetic basis of the urban heat island. *Quarterly Journal of the Royal Meteorological Society*, **108**, 1-24.

Ongley E.D. (2000) The Yellow River: managing the unmanageable. *Water International*, **25**, 227-331.

Orth J.P., Netter, R. and Merkl, G. (1997) Bacterial and chemical contaminant transport tests in a confined karst aquifer (Danube Valley, Swabian Jura, Germany), in *Karst Water and Environmental Impacts* (eds G. Gunay and A.I. Johnson), Balkema, Rotterdam, pp. 173-180.

Ott, M., Su, Z., Schumann, A.H. and Schultz, G.A. (1991) *Development of a Distributed Hydrological Model for Flood Forecasting and Impact Assessment of Land Use Change in the International Mosel River Basin, Hydrology for the Water Management of Large River Basins*. Proceedings of the Vienna Symposium, vol. IAHS Publ. no. 201: International Association of Hydrological Sciences, Wallingford, UK, pp. 183-184.

Owe, M., Chang, A.T.C. and Golus, R.E. (1988) Estimating surface soil moisture from satellite microwave measurements and a satellite derived vegetation index. *Remote Sensing of Environment*, **24**, 331-345.

Panda, P.K. (1985) Geothermal maps of India and their significance in resources assessments. *Petroleum Asia Journal*, **VIII**, 202-210.

Pandarinath, K., Prasad, S., Deshpande, R.D. and Gupta, S.K. (1999) Late Quaternary sediments from Nal Sarovar, Gujarat, India: Distribution and Provenance. *Proceedings Indian Academy Sciences (Earth & Planetary Science)*, **108**, 107-116.

Pankow, J.F. and Cherry, J.A. (1996) *Dense Chlorinated Solvents and Other DNAPLs in Groundwater*, Waterloo Press, Portland, OR, p. 525.

Pankratz, A. (1983) *Forecasting with Univariate Box-Jenkins Models: Concepts and Cases*, John Wiley & Sons, New York, p. 576.

Papadopulos, I.S. and Cooper, H.H. (1967) Drawdown in a well of large diameter. *Water Resources Research*, **3**, 241-244.

Patel, P.P. (1986) A Note on Drinking Water Problem in The Patan-Chanasma Area of Mehsana District, Vadodara, India, Department of Geology, MS, University of Baroda.

Peck, A.J., Johnston, C.D. and Williamson, D.R. (1981) Analyses of solute distributions in deeply weathered soils. *Agricultural Water Management*, **4**, 83-102.

Pedley, S., Yates, M., Schijven, J.F., West, J., Howard, G. and Barrett, M. (2006) Pathogens: health relevance, transport and attenuation, in *Protecting Groundwater for Health: Managing the Quality of Drinking-water Sources* (eds O. Schmoll, G. Howard, J. Chilton and I. Chorus), IWA Publishing for World Health Organization, London, pp. 49-80.

Penman, H.L. (1948) Natural evaporation from open water, bare soil and grass. *Proceedings of the Royal Society of London*, **A193**, 129-145.

Perel'man, A.I. (1977) *Geochemistry of Elements in the Supergene Zone*, Keter Publishing House Ltd, Jerusalem, p. 266.

Perry, J.A. and Schaeffer, D.J. (1987) The longitudinal distribution of riverine benthos: a river discontinuum? *Hydrobiologia*, **148**, 257-268.

Peters, N.E. (1991) Chloride cycling in two forested lake watersheds in the west-central Adirondack Mountains, New York, USA. *Water Air Soil Pollution*, **59**, 201-215.

Philip, J.R. (1957) The theory of infiltration: 1. The infiltration equation and its solution. *Soil Science*, **83**, 345-357.

Piechota, T.C., Timilsena, J., Tootle, G. and Hidlago, H. (2004) The western US drought: How bad is it? *EOS Transactions AGU*, **85**, 301-308.

Piexoto, J.P. and Oort, A.H. (1992) *Physics of Climate*, American Institute of Physics, New York, p. 520.

Pinder, G.F. and Jones, J.F. (1969) Determination of the groundwater component of peak discharge from the chemistry of total runoff. *Water Resources Research*, **5**, 438-445.

Piper, A.M. (1944) A graphic procedure in the geochemical interpretation of water analyses. *American Geophysical Union Transactions*, **25**, 914-923.

Piper, A.M., Garrett, A.A. *et al.* (1953) Native and Contaminated Ground Waters in the Long Beach-Santa Ana Area, California. *US Geological Survey Water-Supply Paper 1136*, p. 320.

Plummer, L.N. and Busenberg, E. (2000) Chlorofluorocarbons, in *Environmental Tracers in*

Subsurface Hydrology (eds P. Cook and A.L. Herczeg), Kluwer Academic Publishers, Boston, pp. 441–478.

Plummer, L.N., Michel, R.L., Thurman, E.M. and Glynn, P.D. (1993) Environmental tracers for age dating young ground water, in *Regional Ground-Water Quality* (ed W.M. Alley), Van Nostrand Reinhold, New York, pp. 255–294.

Pollock, D.W. (1989) Documentation of Computer Programs to Complete and Display Pathlines using Results from the US Geological Survey Modular Three-dimentional Finite-Difference Groundwater Model. Open File Report 89-381, US Geological Survey, p. 188.

Prasad, S. and Gupta, S.K. (1999) Role of eustasy, climate and tectonics in late Quaternary evolution of Nal-Cambay region, NW India. *Zeitschrift fur Geomorphologie NF*, **43**, 483–504.

Prickett, T.A. (1975) Modelling techniques for groundwater evaluation, in *Advances in Hydrosciences* Vol. 10 (ed V.T. Chow), Academic Press, New York, pp. 1–143.

Priestly, C.H.B. and Taylor, R.J. (1972) On the assessment of surface heat flux and evaporation using large-scale parameters. *Monthly Weather Review*, **100**, 82–92.

Prinz, D., Wolfer, S. and Siegert, K. (2000) Water harvesting for crop production –FAO Training Course, Food and Agriculture Organization, United Nations, http://www.fao.org/ag/AGL/AGLW/wharv/whtoc/sld001.htm (accessed 19 July 2007).

Priscoli, J.D., Dooge, J. and Llamas, R. (2004) *Water and Ethics: Overview*, UNESCO, Paris, p. 32.

Qasim, S.R. (1999) *Waste-water Treatment Plants: Planning, Design, and Operation*, Technomic Publishing Company, Lancaster, PA, p. 1107.

Qureshi, M.E., Kirby, M. and Mainuddin, M. (2004) Integrated Water Resources Management In The Murray Darling Basin, Australia, International Conf. on Water Resources & Arid Environment (2004), King Saud University, Riyadh, Kingdom of Saudi Arabia, 5–8 December 2004.

Raja, M. (2006) India grows a grain crisis, *Asia Times Online*, Hong Kong.

Rama and Moore, W.S. (1984) Mechanism of transport of U-Th series radioisotopes from solids into groundwater. *Geochimica et Cosmochimica Acta*, **48**, 395–399.

Randel, D.L., Vonder Haar, T.H., Ringerud, M.A., Stephens, G.L., Greenwald, T.J. and Combs, C.L. (1996) A new global water vapor dataset. *Bulletin of American Meteorological Society*, **77**, 1233–1246.

Rango, A. (1992) Worldwide testing of the snowmelt runoff model with applications for predicting the effects of climate change. *Nordic Hydrology*, **23**, 155–172.

Rango, A. (1993) Snow hydrology processes and remote sensing. *Hydrological Processes*, **7**, 121–138.

Rango, A., Martinec, J., Chang, A.T.C., Foster, J.L. and van Katwijk, V.F. (1989) Average areal water equivalent of snow in a mountain basin using microwave and visible satellite data. *IEEE Transactions on Geoscience and Remote Sensing*, **27**, 740–745.

Rank, D., Volkl, G., Maloszewski, P. and Stichler, W. (1992) *Flow Dynamics in an Alpine Karst Massif Studied by Means of Environmental Isotopes, Isotope techniques in Water Resources Development*. Proceedings of the IAEA Symposium 319: IAEA, Vienna, p. 327–343.

Rantz, S.E. *et al.* (1982) Measurement and Computation of Streamflow; vol. 1, Measurement of Stage and Discharge; vol. 2, Computation of Discharge, *US Geological Survey Water-Supply Paper 2175*, p. 631.

Rao, A.R. and Hamed, K.H. (2000) *Flood Frequency Analysis*, CRC Press, Boca Raton, FL, p. 35.

Rao, G.V., Reddy, G.K., Rao, R.U.M. and Gopalan, K. (1994) Extraordinary helium anomaly over surface rupture of September 1993 Killari earthquake, India. *Current Science*, **66**, 933–935.

RaviShankar (1988) Heat flow map of India and discussions on its geological and economic significance. *Indian Minerals*, **42**, 89–110.

Reddy, D.V., Sukhija, B.S., Nagabhushanam, P., Reddy, G.K., Kumar, D. and Lachassagne, P. (2006) Soil gas radon enamometry: a tool for delineation of fractures for groundwater in granitic terrains. *Journal of Hydrology*, **329**, 186–195.

Redmond, K.T. and Koch, R.W. (1991) Surface climate and stream-flow variability in the western United States and their relationship to large-scale circulation indexes. *Water Resources Research*, **27**, 2381-2399.

Reed, S.C., Middlebrooks, E.J. and Crites, R.W. (1988) *Natural Systems for Waste Management and Treatment*, McGraw-Hill, New York, p. 433.

Reilly, T.E., Franke, O.L., Buxto, H.T. and Bennett, G.D. (1987) A Conceptual Framework for Ground-Water Solute-Transport Studies with Emphasis on Physical Mechanisms of Solute Movement. *US Geol. Survey Water-Res.* Inv. Rept. 87-4191, US Geol. Survey, p. 44.

Reimer, G.M. (1976) Design and Assembly of a Portable Helium Leak Detector for Evaluation as a Uranium Exploration Instrument. *US Geological Survey, Open-File Report 18*, pp. 76-398.

Reimer, G.M. (1984) Prediction of central California earthquakes from soil-gas helium. *PAGEOPH*, **122**, 369-375.

Ren, M. and Walker, H.J. (1998) Environmental consequences of human activity on the Yellow river and its delta, China. *Physical Geography*, **19**, 421-432.

Renault, D. (2003) *Value of Virtual Water in Food: Principles and Virtues.* in A.Y. Hoekstra (ed), Virtual Water Trade, Proceedings Expert Meeting on Virtual Water, Dec 2002, FAO, Delft, pp. 77-91.

Rice, S.P., Greenwood, M.T. and Joyce, C.B. (2001) Tributaries, sediment sources, and the longitudinal organization of macro-invertebrate fauna along river systems. *Canadian Journal of Fisheries and Aquatic Sciences*, **58**, 828-840.

Richards, R.A. (1931) Capillary conduction of liquid through porous media. *Physics*, **1**, 318-333.

Riley, J.P. (1965) The occurrence of anomalously high fluoride concentrations in the North Atlantic. *Deep Sea Research*, **12**, 219-220.

Rivett, M., Drewes, J. Barrett, M., Chilton, J., Appleyard, S., Dieter, H.H., Wauchope, D., Fastner, J. (2006) Chemicals: Health relevance, transport and attenuation, in *Protecting Groundwater for Health: Managing the Quality of Drinking-water Sources* (eds O. Schmoll, G. Howard, J. Chilton and I. Chorus), IWA Publishing for World Health Organization, London, pp. 81-137.

Robson, A. and Neal, C. (1990) Hydrograph separation using chemical techniques: an application to catchments in Mid-Wales. *Journal of Hydrology*, **116**, 345-363.

Rodda, A. (1991) *Women and the Environment*, London, Zed Books Ltd, p. 180.

Ronan, C.A. (1995) *The Shorter Science and Civilization in China: an Abridgement of Joseph Needham's Original Text*, Cambridge University Press, Cambridge.

Rosegrant, M. and Ringler, C. (1999) *Impact on Food Security and Rural Development of Reallocating Water from Agriculture*, EPTD Discussion Paper No. 47. Environment and Production Technology Division, International Food Policy Research Institute (IFPRI), Washington DC, p. 48.

Rozanski, K., Araguas-Araguas, L. and Gonfiantini, R. (1993) Isotopic patterns in modern global precipitation, climate change in continental isotopic records, vol. 78, *American Geophysical Union Monograph*, pp. 1-36.

Rudolph, J. (1981) Edelgastemperaturen und Heliumalter ^{14}C-datierter Paläowässer. Ph.D. thesis, Universität Heidelberg, Heidelberg.

Rushton, K.R. and Chan, Y.K. (1976) Pumping test analysis when parameters vary with depth. *Ground Water*, **14**, 82-87.

Salas, J.D., Delleur, J.W., Yevjevich, V.M. and Lane, W.L. (1980) *Applied Modeling of Hydrologic Time Series*, Water Resources Publications, Littleton, CO, p. 484.

Sanford, W.E. and Konikow, L.F. (1985) A two-constituent solute transport model for ground water having variable density. *Water Resources Investigation Report 85-4279*, US Geological Survey, p. 88.

Sanks, R.L. and Asano, T. (1976) *Land Treatment and Disposal of Municipal and Industrial Wastewater*, Ann Arbor Science, Michigan, p. ——????.

Scanlon, B R., Keese, K.E. *et al.* (2006) Global synthesis of groundwater recharge in semi-arid and arid regions. *Hydrological Process*, **20**, 3335-3370.

Scheidegger, A.E. (1961) General theory of dispersion in porous media. *Journal of Geophysical Research*, **66**, 3273-3278.

Schoeller, H. (1935) Utilité de la notion des exchanges de bases pour la comparison des eaux souterraines: France. *Société Géologie Comptes rendus Sommaire et Bulletin*. Série 5 (5), 651–657.

Schultz, G.A. (1993) Hydrological modeling based on remote sensing information. *Advances in Space Research*, **13**, (5)149–(5)166.

Schumm, S.A. (1956) Evolution of drainage systems and slopes in badlands at Perth Amboy, NJ. *Geological Society of America Bulletin*, **67**, 597–646.

Schwarzenbach, R.P., Gschwend, P.M. and Imboden, D.M. (2003) *Environmental Organic Chemistry*, John Wiley & Sons, Hoboken, NJ, p. 1313.

SCS (1983) TR-20 Project Formulation-hydrology (1982 version), Technical Release No. 20, Soil Conservation Service, p. 296.

SCS (1986) *Urban Hydrology for Small Watersheds, 210-VI-TR-55*, 2nd edn, United States Department of Agriculture, USDA, Washington, DC, p. 164.

Selim, S.M. and Kirkham, D. (1974) Screen theory of wells and soil drainpipes. *Ground Water*, **4**, 33–34.

Sellers W.D. (1965) *Physical Climatology*, University of Chicago, Chicago, p. 272.

Sen, A. (1970) *Collective Choice and Social Welfare*, Halden-Day, San Francisco, CA, p. 225.

Sen, A. (1973) *On Economic Inequality*, Oxford University Press, Oxford, p. 280.

Sen, A. (1981) *Poverty and Famines*, Oxford University Press, Oxford, p. 257.

Sen, A. (1982) *Choice, Welfare and Measurements*, Oxford University Press, Oxford, p. 480.

Sen, A. (1987) *On Ethics and Economics*, Basil Blackwell, Oxford, p. 131.

Shabri, A. (2002) Nonparametric kernel estimation of annual maximum stream flow quantiles. *Matematika*, **18**, 99–107.

Shapiro, M.H. (1981) Relationship of the 1979 southern California radon anomaly to a possible regional strain event. *Journal of Geophysical Research*, **86**, 1725–1730.

Sharma, M.L. and Hughes, M.W. (1985) Groundwater recharge estimating using chloride, deuterium and oxygen-18 profiles in the deep coastal sands of Western Australia. *Journal of Hydrology*, **81**, 93–109.

Shaw, E.M. (1988) *Hydrology in Practice*, Van Nostrand Reinhold (International) Co. Ltd, London, p. 592.

Sheahan, N.T. (1971) Type-curve solution of step-drawdown test. *Ground Water*, **9**, 25–29.

Shen, H.W. and Bryson, M. (1979) Impact of extremal distributions on analysis of maximum loading, in *Reliability in Water Resources Management* (eds E.A. McBean, K.W. Hipel and T.E. Unny), Water Resources Publications, Littleton, CO.

Sherman, L.K. (1932) Stream flow from Rainfall by the Unit-Graph Method. *Engineering News-Rec.*, **10**, 501–505.

Shi, J.C. and Dozier, J. (1992) Radar response to snow wetness. *IGARSS 92, IEEE 92 CH3041-1*, p. 927–929.

Shi, J.C., Dozier, J. and Rott, H. (1994) Snow mapping in alpine regions with synthetic aperture radar. *IEEE Transactions on Geoscience and Remote Sensing*, **32**, 152–158.

Shiklomanov, I.A. (1999) *World Water Resources and their Use*, Database on CD Rom, UNESCO, Paris.

Shivanna, K., Kulkarni, U.P., Joseph, T.B. and Navada, S.V. (2004) Contribution of storm to groundwater in semi-arid regions of Karnataka, India. *Hydrological Processes*, **18**, 473–485.

Shreve, R.L. (1966) Statistical law of stream numbers. *Journal of Geology*, **74**, 17–37.

Shreve, R.L. (1967) Infinite topologically random channel networks. *Journal of Geology*, **75**, 178–186.

Simpson, H.J. and Herczeg, A.L. (1991) Salinity and evaporation in the River Murray basin. *Australia: Journal of Hydrology*, **124**, 27.

Singer, P. (1997) Famine, affluence and morality, in *Ethics in Practice* (ed H. la Follette), Blackwell, Oxford, pp. 585–595.

Smith, D.B., Wearn, P.L., Richards, H.J. and Rowe, P.C. (1970) *Water Movement in the Unsaturated Zone of High and Low Permeability Strata by Measuring Natural Tritium*. Proceedings Symposium on Isotope Hydrology, pp. 73–86.

Smith, L. and Schwartz, F.W. (1980) Mass transport: 1. a stochastic analysis of macroscopic dispersion. *Water Resources Research*, **16**, 303–313.

Smith, L. and Wheatcraft, S.W. (1992) *Groundwater Flow, Handbook of Hydrology*, Chapter 6, (ed D.R. Maidment) McGraw Hill, New York, p. 6.1–6.58.

Soares, J.V. Bernard, R., Taconet, O., Vidal-Madjar, D. and Weill, A. (1988) Estimation of bare soil evaporation from airborne measurements. *Journal of Hydrology*, **99**, 281–296.

Solomon, D.K. and Cook. P.G. (2000) ^{3}H and ^{3}He, in *Environmental Tracers in Subsurface Hydrology* (eds P.G. Cook and A.L. Herczeg), Kluwer Academic Publishers, Boston, pp. 397–424.

Solomon, D.K., Hunt, A. and Poreda, R.J. (1996) Source of radiogenic helium-4 in shallow aquifers: implications for dating young ground water. *Water Resources Research*, **32**, 1805–1813.

Spencer, R.W., Goodman, H.M. and Wood, R.E. (1988) Precipitation retrieval over land and ocean with SSM/I, Part 1: Identification and characteristics of the scattering signal. *Journal of Atmospheric and Oceanic Technology*, **2**, 254–263.

Srivastava, B.N., Deshmukh, A.N. and Kodate, J.K. (1995) The impact of groundwater composition of fluorite solubility product: a case study of selected Indian aquifers. *Gondwana Geological Magazine*, **9**, 31–44.

Stedinger, J.R., Vogel, R.M. and Foufoula-Georgoiu, E. (1993) Frequency analysis of extreme events, in *Handbook of Hydrology* (ed D.R. Maidment), McGraw-Hill, New York, pp. 18.1–18.66.

Steffen, W., Sanderson, A. *et al.* (2004) *Global Change and the Earth System: A Planet Under Pressure: Global Change*, The IGBP Series, Springer-Verlag, Berlin, Heidelberg, New York, p. 336.

Sternberg, Y.M. (1967) Transmissibility determination from variable discharge pumping tests. *Ground Water*, **5**, 27–29.

Sternberg, Y.M. (1968) Simplified solution for variable rate pumping test. *Journal Hydraulics Division, American Society of Civil Engineers*, **94**, 177–180.

Sternberg, Y.M. (1973) Efficiency of partially penetrating wells. *Ground Water*, **11**, 5–8.

Stevenson, F.J. (1994) *Humus Chemistry: Genesis, Composition, Reactions*, John Wiley & Sons, New York, p. 516.

Stiff, H.A. (1951) The interpretation of chemical water analysis by means of patterns. *Journal of Petroleum Technology*, **3**, 15–17.

Stockton, C.W. and Jacoby, G.C. (1976) Long-term surface-water supply and streamflow trends in the Upper Colorado River basin based on tree-ring analyses. *Lake Powell Research Project Bulletin*, **18**, 1–70.

Strack, O.D.L. (1989) *Groundwater Mechanics*, Prentice Hall, Englewood Cliffs, NJ, p. 732.

Strack, O.D.L. (1999) Principles of the analytic element method. *Journal of Hydrology*, **226**, 128–138.

Strahler, A.N. (1957) Quantitative analysis of watershed geomorphology. *EOS Trans. AGU*, **38**, 913–920.

Strunz, H. (1970) *Mineralogische Tabellen*, Vol. **5**, Geest und Portig, Leipzig, p. 487.

Stumm, W. and Morgan, J.J. (1996) *Aquatic Chemistry*, John Wiley & Sons, New York, p. 1022.

Stute, M. (1989) Edelgase im Grundwasser: Bestimmung von Paläotemperaturen und Untersuchung der Dynamik von Grundwasserfliesssystemen: Ph.D. thesis, Universität Heidelberg.

Stute, M., Clark, J.F., Schlosser, P. and Broecker, W.S. (1995a) A 30000 yr continental paleo-temperature record derived from noble gases dissolved in ground water from the San Juan Basin, New Mexico. *Quaternary Research*, **43**, 209–220.

Stute, M., Forster, M. *et al.* (1995b) Cooling of tropical Brazil (5°C) during the Last Glacial Maximum. *Science*, **269**, 379–383.

Stute, M. and Schlosser, P. (2000) Atmospheric noble gases, in *Environmental Tracers in Subsurface Hydrology* (eds P. Cook and A.L. Herczeg), Kluwer Academic Publishers, Boston, pp. 349–377.

Stute, M., Sonntag, C., Déak, J. and Schlosser, P. (1992) Helium in deep circulating ground water in the Great Hungarian Plain: flow dynamics and crustal and mantle helium fluxes. *Geochimica et Cosmochimica Acta*, **56**, 2051–2067.

Su, X., Lin, X., Liao, Z. and Wang, J. (2004) The main factors affecting isotopes of Yellow River Water in China, *Water International*, **29**, 475–482.

Sugawara, K. (1967) Migration of elements through phases of the hydrosphere and atmosphere, in *Chemistry of the Earth's Crust*, vol. **2** (ed A.P. Vinogradov), Israel Program for Scientific Translations, Jerusalem, p. 501.

Sukhija, B.S. and Rama (1973) Evaluation of groundwater recharge in the semi-arid region of India using environmental tritium. *Proceedings Indian Academy of Sciences*, 77, 279–292.[A1.22]

Sukhija, B.S., Reddy, D.V. and Nagabhushanam, P. (1998) Isotopic fingerprint of palaeoclimates during the last 30,000 years in deep confined groundwaters of Southern India. *Quaternary Research*, **50**, 252–260.

Sukhija, B.S., Reddy, D.V., Nagabhushanam, P. and Hussain, S. (2003) Recharge processes: piston flow vs. preferential flow in semi-arid aquifers of India. *Hydrogeology Journal*, **11**, 387–395.

Sun, R., Gao, X., Liu, C.-M. and Li, X.-W. (2004) Evapotranspiration estimation in the Yellow River Basin, China using integrated NDVI data. *International Journal of Remote Sensing*, **25**, 2523–2534.

Svoma, J. (1990) Remote Sensing Methods for the Detection of Rock Medium and Groundwater Contamination. *International Symposium on Remote Sensing and Water Resources, IAH/Netherlands Society of Remote Sensing, Enschede, The Netherlands*, pp. 465–472.

Taconet, O., Bernard, R. and Vidal-Madjar, D. (1986) Evapotranspiration over an agricultural region using a surface flux/temperature model based on NOAA-AVHRR data. *Journal of Climate and Applied Meteorology*, **25**, 284–307.

Taconet, O. and Vidal-Madjar, D. (1988) Applications of a flux algorithm to a field-satellite campaign over vegetated area. *Remote Sensing of Environment*, **26**, 227–239.

Tate, C.H. and Arnold, K.F. (1990) Health and aestheic aspects of water quality, in *Water Quality and Treatment: A Handbook of Community Water Supplies* (ed F.W. Pontius), American Water Works Association, New York; McGraw Hill, Washington DC, p. 617.

Theis, C.V. (1935) The relation between the lowering of the piezometric surface and the rate and duration of discharge of a well using groundwater storage. *Transactions of American Geophysical Union*, **16**, 519–524.

Thornthwaite, C.W. (1948) An approach towards a rational classification of climates. *Geographical Review*, **38**, 55–94.

Thurman, E.M. (1985) *Organic Geochemistry of Natural Waters*, Martinus Nijoff/Dr. W. Junk Publishers, Dordrecht, Netherlands, p. 497.

Thyne, G. (2005) Geochemical Modeling – Computer Codes, GW-246, in *The Encyclopedia of Water* (ed J.H. Lehr), John Wiley & Sons, New York.

Tietjen, G. (1994) Recursive schemes for calculating cumulative binomial and Poisson probabilities. *The American Statistician*, **48**, 136–137.

Todd, D.K. (2006) *Groundwater Hydrology*, John Wiley & Sons, New Delhi, p. 537.

Tolstikhin, I.N. and Kamenskiy, I.L. (1969) Determination of groundwater ages by the T- ^{3}He Method. *Geochemistry International*, **6**, 810–811.

Torgersen, T. and Ivey, G.N. (1985) Helium accumulation in groundwaters, II. A model for the accumulation of the crustal ^{4}He degassing flux. *Geochimica et Cosmochimica Acta*, **49**, 2445–2452.

Torgersen, T. (1980) Controls on pore-fluid concentration of ^{4}He and ^{222}Rn and the calculation of ^{4}He/^{222}Rn ages. *Journal Geochemical Exploration*, **13**, 57–75.

Torgersen, T. and Clarke, W.B. (1985) Helium accumulation in groundwater, I : an evaluation of sources and the continental flux of crustal ^{4}He in the Great Artesian Basin, Australia. *Geochimica et Cosmochimica Acta*, **49**, 1211–1218.

Trenberth, K.E. (1988) Atmospheric moisture residence times and cycling: implications for rainfall rates with climate change. *Climate Change*, **39**, 667–694.

Trenberth, K.E. (1999) Conceptual framework for changes of extremes of the hydrological cycle with climate change. *Climate Change*, **42**, 327–339.

Trenberth, K.E. and Hurrel, J.W. (1994) Decadal atmospheric-ocean variations in the Pacific. *Climate Dynamics*, **9**, 303–319.

Tricker, A.S. (1978) The infiltration cylinder: some comments on its use. *Journal of Hydrology*, **36**, 383–391.

Troldborg, L. (2004) The influence of conceptual geological models on the simulation of flow and transport in Quaternary aquifer systems, Technical University of Denmark, p. 107.

Tucker, C.J., Dregne, H.E. and Newcom, W.W. (1991) Expansion and contraction of the Sahara Desert from 1980 to 1990. *Science*, **253**, 299–302.

Tung, Y.K. and Yen, B.C. (2005) *Hydrosystems Engineering Uncertainty Analysis*, McGraw-Hill, New York, p. 285.

Tung, Y.K., Yen, B.C. and Melching, C.S. (2006) *Hydrosystems Engineering Reliability Assessment and Risk Analysis*, New York, McGraw-Hill, 512 p.

Udall, B. and Hoerling, M. (2005) *Colorado River Basin Climate/Paleo, Present and Future*, California Department of Water Resources, pp. 23–29.

UDFCD (1984) Rainfall and Runoff, *Urban Storm Drainage Criteria Manual*, vol. 1, 1984 revision, Urban Drainage and Flood Control District, Denver, CO, p. 120.

UN (1955) Economic Commission for Asia and the Far East. Multipurpose River Basin Development, Part 1, Manual of River Basin Planning Flood Control. Flood Control Series No. 7. *United Nations Publication ST/ECAFE/SERF/7*, New York.

UNESCWA (2003) Waste-Water Treatment Technologies: A General Review, United Nations Economic and Social Commission for West Asia, p. 132.

UNESCO–WWAP (2003) Water for People, Water for Life. *The United Nations World Water Development Report – Executive Summary*, United Nations, p. 173.

US-EPA (1997) Guidelines for Wellhead and Spring Protection in Carbonate Rocks. *EPA 904-B-97-0003*, Washington DC.

USEPA (2002) EPA 625/R-00/008 Onsite Wastewater Treatment Systems Technology Fact Sheet 12 – Land Treatment Systems, US Environmental Protection Agency, p. TFS-71-TFS-76.

US-SL (1954) Diagnosis and improvement of saline and alkali soils, *Handbook 60, US Salinity* Laboratory, US Department of Agriculture, p. 160.

USDA (1972) *National Engineering Handbook, Section 4, Hydrology*. US Department of Agriculture, Soil Conservation Service, US Government Printing Office, Washington, DC, p. 544.

USGS (2005) The State of the Colorado River Ecosystem in Grand Canyon. *A Report of the Grand Canyon Monitoring and Research*, United States Geological Survey, Reston, VA, p. 220.

USWRC (1967) *A Uniform Technique for Determining Flood Flow Frequencies*, US Water Resources Council, Bulletin No. 15, US Geological Survey, Reston, VA, p. 16

USWRC (1982) *Guideline in Determining Flood Flow Frequency*, US Water Resources Council, Bulletin 17B, US Geological Survey, Reston, VA, p. 48.

Van der Meer, J.R. and Kohler, H.P.E. (2000) Groundwater pollution – the limits of biodegradability. *EAWAG News*, **49e**, 23–25.

Vannote, R.L., Minshall, G.W., Cummins, K.W., Sedell, J.R. and Cushing, C.E. (1980) The river continuum concept. *Canadian Journal of Fisheries and Aquatic Sciences*, **37**, 130–137.

Vasavada, B J. (1998) Excessive fluoride in Gujarat. A perspective plan of solution. *Journal Indian Water Works Association*, **30**, 191–198.

Vasudevan, L. (2002) *A Study of Biological Contaminants in Rainwater Collected from Rooftops in Bryan and College Station, Texas*. Texas A&M University, College Station (TX), p. 180.

Verruijt, A. (1969) Elastic storage of aquifers, in *Flow Through Porous Media* (ed R.J.M. DeWiest), Academic Press, New York, pp. 331–376.

VICAIRE (2003) Virtual campus in hydrology and water resources, Module 1b, Chapter 10, Urban Hydrology.

Viessman, W., Knapp, J.W., Lewis, G.K. and Harbaugh, T.E. (1977) *Introduction to Hydrology*, Harper & Row, New York, p. 704.

Villasenor, J.D., Fatland, D.R. and Hinzman, L.D. (1993) Change detection on Alaska's North Slope using repeat-pass ERS-1 SAR images. *IEEE Transactions on Geoscience and Remote Sensing*, **31**, 227–236.

Virk, H.S., Walia, V. and Kumar, N. (2001) Helium/radon precursory anomalies of Chamoli earthquake, Garhwal Himalaya, India. *Journal of Geodynamics*, **31**, 201–210.

Vivoni, E.R. (2005a) Lecture 7: Introduction to Hydrology – Infiltration and Soil Water (Unpublished), Department of Earth and Environmental Science, New Mexico Institute of Mining and Technology.

Vivoni, E.R. (2005b) Lecture 13: Introduction to Hydrology – Runoff Processes and Conceptual Models (Unpublished), Department of Earth and Environmental Science, New Mexico Institute of Mining and Technology.

Vivoni, E.R. (2005c) Lecture 14: Introduction to Hydrology – Estimating the hydrograph and its properties (Unpublished), Department of Earth and Environmental Science, New Mexico Institute of Mining and Technology.

Vivoni, E.R. (2005d) Lecture 18: Introduction to Hydrology – Concepts in stream processes (Unpublished), Department of Earth and Environmental Science, New Mexico Institute of Mining and Technology.

Vogel, J.C. (1967) *Investigation of Groundwater Flow with Radiocarbon, Isotopes in Hydrology*, IAEA, Vienna, pp. 355–368.

Vogel, J.C. (1970) *Carbon-14 Dating of Groundwater, Isotope Hydrology*, IAEA, Vienna, pp. 225–237.

von Gunten, U. and Zobrist, J. (1993) Biogeochemical changes in groundwater infiltration systems: column studies. *Geochemica et Cosmochemica Acta*, **57**, 3895–3906.

Walker, G.T. (1925) On periodicity (with discussion). *Quarterly Journal of the Royal Meteorological Society*, **51**, 337–346.

Walton, W.C. (1960) Leaky artesian aquifer condition in Illinois, Illinois State Water. *Survey Report Invest.*, **39**, 27.

Walton, W.C. (1970) *Groundwater Resource Evaluation*, McGraw Hill, New York, p. 576.

Wang, G. and Shi, Z. (2000) Impacts of climate change on hydrology in upper reaches of the Yellow River. *Meteorology Journal of Henan*, **4**, 20–22.

Wang, G., Wang, Y. and Kang, L. (2002) Analysis on the sensitivity of runoff in Yellow River to climate change. *Journal of Applied Meteorology*, **13**, 117–121.

Wang, S., Song, X., Wang, Q., Xiao, G., Liu. C. and Liu, J. (2009) Shallow groundwater dynamics in North China Plain. *Journal of Geographical Sciences*, **19**, 175–188.

Ward, A.D. and Trimble, S.W. (2003) *Environmental Hydrology*, CRC Press, Boca Raton, FL, p. 465.

Warner, M.J. and Weiss, R.F. (1985) Solubilities of Chlorofluorocarbons 11 and 12 in water and seawater. *Deep-Sea Research*, **32**, 1485–1497.

Wasson, R.J., Rajguru, S.N., Misra, V.N., Agrawal, D.P., Dhir, R.P. and Singhvi, A.K. (1983) Geomorphology, Late Quaternary stratigraphy and palaeoclimatology of the Thar dune field. *Zeitschrift fur Geomorphologie NF*, **45**, 117–151.

WB (1993) China: Yellow River Basin Investment Planning Study, Report No. 11146-CHA.

Wedepohl, K.H. (1974) *Handbook of Geochemistry*, vol. **II-4**: Springer-Verlag, Berlin, Heidelberg, New York, p. 9K-1.

Weeks E.P. (1969) Determining ratio of horizontal to vertical permeability by aquifer-test analysis. *Water Resources Research*, **5**, 196–214.

Weismann, T.J. (1980) Developments in Geochemistry and Their Contribution to Hydrocarbon Exploration. 10th World Petroleum Congress, pp. 369–386.

Weiss, R.F. (1970) The solubility of nitrogen, oxygen and argon in water and seawater. *Deep-Sea Research*, **17**, 721–735.

Weiss, R.F. (1971) Solubility of helium and neon in water and seawater. *Journal of Chemical. Engineering Data*, **16**, 235–241.

Weiss, R.F. and Kyser, T.K. (1978) Solubility of krypton in water and seawater. *Journal of Chemical and Engineering Data*, **23**, 69–72.

Weissmann, G.S., Zhang, Y., LaBolle, E.M. and Fogg, G.E. (2002) Dispersion of groundwater age in an alluvial aquifer system. *Water Resources Research*, **38** (10), 1198–1211.

Weng, Q. (2007) *Remote Sensing of Impervious Surfaces*, CRC Press/Taylor & Francis, p. 454.

Weng, Q. and Quattrochi, D.A. (2006) *Urban Remote Sensing*, CRC Press/Taylor & Francis, p. 448.

Wenzel, L.K. (1942) Methods for Determining Permeability of Water-bearing Materials with Special Reference to Discharging-Well Methods, US Geological Survey, Water-Supply Paper 887, pp. 1–192.

West, J.M., Pedley, S., Baker, S.J., Barrott, L., Morris, B. and Storey, A. (1998) A Review of the Impact of Microbiological Contaminants in Ground water. *UK Environment Agency, Bristol, R and D Technical Report*, p. 139.

Weyhenmeyer, C.E., Burns, S.J. *et al.* (2000) Cool glacial temperatures and changes in moisture source recorded in Oman ground waters. *Science*, **287**, 842–845.

WHO (1970) *Fluorides and Human Health*, Monograph 59, World Health Organization, Geneva, p. 273.

WHO (1984) *Fluorine and Fluorides: Environmental Health Criteria 36*, World Health Organization, Geneva, p. 136.

WHO (2009) Hardness in Drinking-water. Background document for development of WHO Guidelines for Drinking-water Quality, WHO, p. 9.

Wichelns, D. (2001) The role of 'virtual water' in efforts to achieve food security and other national goals, with an example from Egypt. *Agricultural Water Management*, **49**, 131–151.

Wigley, T.M.L. (1968) Flow into a finite well with arbitrary discharge. *Journal of Hydrology*, **6**, 209–213.

Wild, C.J. and Seber, G.A.F. (2000) *Chance Encounters: A First Course In Data Analysis And Inference synopsis*, New York, John Wiley and Sons Ltd, 632 p.

Wilheit, T.T., Chang, A.T.C., Rao, M.S.V., Rodgers, E.B. and Theon, J.S. (1977) A satellite technique for quantitatively mapping rainfall rates over the ocean. *Journal of Applied Meteorology*, **16**, 551–560.

Wilsenach, J.A., Maurer, M., Larsen, T.A. and van Loosdrecht, M.C. (2003) From waste treatment to integrated resource management. *Water Science and Technology*, **48**, 1–9.

Winter, T.C., Harvey, J.W., Franke, O.L. and Alley, W.M. (1998) Ground Water and Surface Water A Single Resource-USGS Circular 1139, U.S. Geological Survey, p. 87.

WIRG (2005) Outline of the Yellow River Groundwater Project, Water Environment Research Group, Institute for Geo-Resources and Environment (GREEN), AIST, http://unit.aist.go.jp/georesenv/gwrg/outline%20of%20Yellow%20River.html (accessed 4 March 2010).

WMO (1966) *Technical Note No. 79: Climatic Change*, World Meterorological Organization, Geneva, p. 80.

WMO (1992) *Simulated Real-Time Intercomparison of Hydrological Models*, World Meteorological Organization, Geneva, p. 241.

WMO (1994) *Guide to Hydrological Practices*: data acquisition and processing, analysis, forecasting and other applications, WMO No. 168, World Meteorological Organization, Geneva, p. 729.

Woodhouse, C.A. and Lukas, J.J. (2006) Drought, tree rings, and water resource management in Colorado. *Canadian Water Resources Journal*, **31**, 1–14.

Woodhouse, C.A. and Meko, D.M. (2007) Dendroclimatology and water resources management, in *Dendroclimatology: Progress and Prospects* (eds M.K. Hughes, T.W. Swetman and H.F. Diaz), Springer, New York, p. —-????.

Woodhouse, C.A., Gray, S.T. and Meko, D.M. (2006) Updated streamflow reconstructions for the Upper Colorado River basin. *Water Resources Research*, **42**, p. W05415, doi: 10.1029/2005WR 004455 (accessed 4 March 2010).

WRI (1996) *World Resources 1996–1997*, World Resources Institute–Oxford University Press, New York, p. 190.

Wu, J.G., Jelinski, D.E., Luck, M. and Tueller, P.T. (2000) Multiscale analysis of landscape heterogeneity: scale variance and pattern metrics. *Geographic Information Sciences*, **6**, 6–19.

Xia, J., Ye, A. and Wang, G. (2005) A distributed time-variant gain model applied to Yellow River

(I): Model theories and structures. *Engineering Journal of Wuhan University*, **38**, 10-15.

Yadav, D.N. (1997) Oxygen isotope study of evaporating brines in Sambar Lake, Rajasthan (India). *Chemical Geology*, **138**, 109-118.

Yevjevich, V. (1972) Stochastics processes in hydrology: Fort Collins, CO, Water Resources Publications, 276 p.

Ye Aizhong, X. and Wang, G. (2006) A distributed time-varying gain model applied to Yellow River Basin. *Engineering Journal of Wuhan University*, **39**, 29-32.

Yong, L. and Hua, Z.W. (1991) Environmental characteristics of regional ground water in relation to fluoride poisoning in North China. *Environment Geology and Water Science*, **18**, 3-10.

Young, G.J. (1976) A Portable Profiling Snow-Gauge: Results of Field Tests on Glaciers. *Forty-fourth Annual Western Snow Conference, Atmospheric Environment Service, Canada*, pp. 7-11.

YRCC (2002) Yellow River Basin Planning (in Chinese), Yellow River Conservancy Commission http:// www.yrcc.gov.cn/ (accessed 4 March 2010).

Zarriello P.J. (1998) Comparison of Nine Uncalibrated Runoff Models to Observed Flows in Two Small Urban Watersheds. *Proceedings of the First Federal Interagency Hydrologic Modeling Conference*, pp. 7-163-7-170.

Zebker, H.A., Madsen, S.N. *et al.* (1992) The TOPSAR interferometric radar topographic mapping instrument. *IEEE Transactions on Geoscience and Remote Sensing*, **30**, 933-940.

Zero, D.T., Raubertas, R.F., Pedersen, A.M., Fu, J. and Hayes, A.L. (1992) Studies of fluoride retention by oral soft tissues after the application of home-use topical fluorides. *Journal of Dental Research*, **71**, 1546-1552.

Zhu, C. and Anderson, G. (2002) *Environmental Applications of Geochemical Modeling*, Cambridge University Press, London, p. 304.

Zimmer, D. and Renault, D. (2003) *Virtual Water in Food Production and Trade at Global Scale: Review of Methodological Issues and Preliminary Results*. Expert Meeting on Virtual Water, December 12-13, 2002.

Zimmermann, U., Münnich, K.O. and Roether, W. (1967) Downward Movement of the Soil Moisture Traced by Means of Hydrogen Isotopes, Isotope Techniques in the Hydrologic Cycle. Geophysical Monograph Series 11, American Geophysical Union, Washington DC, p. 28-36.

Zisheng, L., Xueyu, L., Qinzhou, S., Raoul, G. and Xinqiang, D. (2004) Study on the groundwater exploitation test in the Yellow River lower reaches. *Science in China Series E Engineering and Materials Science*, **47**, 14-24.

Zobrist, J., Mengis, M. and Hug, S. (2000) Ground Water Quality - the Result of Biogeochemical Processes, *EAWAG News*, **49e**, 15-17.

Zongyu, C., Zhenlong, N., Zhaoji, Z., Jixiang, Q. and Yunju, N. (2005) Isotopes and sustainability of ground water resources, North China Plain. *Ground Water*, **43**, 485-493.

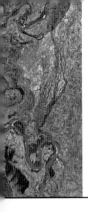

Index